Fourier Analysis and Boundary Value Problems

WITHDRAWI

Fourier Analysis and Boundary Value Problems

Enrique A. González-Velasco
University of Massachusetts
Lowell, Massachusetts

ACADEMIC PRESS
San Diego New York Boston London Sydney Tokyo Toronto

Academic Press, Inc.
A Division of Harcourt Brace & Company
525 B Street, Suite 1900, San Diego, California 92101-4495

United Kingdom Edition published by
Academic Press Limited
24-28 Oval Road, London NW1 7DX

Library of Congress Cataloging-in-Publication Data

González-Velasco, Enrique.
    Fourier analysis and boundary value problems / by Enrique González -Velasco.
            p.   cm.
    Includes bibliographical references and index.
    ISBN 0-12-289640-8 (alk. paper)
    1. Fourier analysis.   2. Boundary value problems--Numerical solutions.   I. Title.
    QA403.5.066   1995
    515'.353--dc20   95-16532

Transferred to digital printing 2005

95 96 97 98 99 00 QW 9 8 7 6 5 4 3 2 1

TABLE OF CONTENTS

PREFACE

This is a book about the solution of boundary value problems involving the linear partial differential equations that appear in modeling many natural phenomena in
engineering and the physical sciences. In particular, it is about the development and
application of Fourier analysis and related techniques to the solution of such problems.
It does not, however, present results on Fourier analysis that are not directly applicable to
the solution of boundary value problems, nor does it include some additional techniques
to solve such problems, such as variational methods.

The necessary prerequisites to read this book have been kept to a minimum and
consist of the complete calculus sequence and some familiarity with the solution of
linear first- and second-order ordinary differential equations with constant coefficients.
There is no requirement of linear algebra or complex analysis, except that Cauchy's
residue theorem is used in the optional Section 8.5. However, the level of mathematical
maturity required of those students who expect to read the proofs of most theorems is
higher than that suggested by the prerequisites. In particular, the concept of uniform
convergence is central to many discussions, and for this reason Appendix A has been
included to develop the necessary facts at an elementary level. Thus, the book is primarily
aimed at advanced undergraduates and, in some instances, it is also appropriate for first
year graduate students.

To make the book accessible to as wide an audience as possible only Riemann
integration is used throughout, although its shortcomings are well known in comparison
with the Lebesgue theory.[1] For our purposes, the main difficulty created by the restriction
to Riemann integration is justifying the change of order of integration or differentiation
under the integral sign when dealing with certain improper integrals. It is not easy to
find proofs of such results in the literature that apply to discontinuous functions—which
appear frequently in many applications—and rely on Riemann integration only. For this
reason and to satisfy the very curious, Appendix B is devoted to such proofs, although
only a handful of students may want to read them and none need do it.

Many of the concepts of classical analysis had their origins in the study of physical
problems leading to the boundary value problems that are the subject of this book. In this
vein, and throughout the book, each mathematical topic is motivated by some physical
problem. Specifically, each topic starts by posing a physical problem, showing then how
the search for a solution leads to the discovery of new mathematical tools—tools that
are today of standard use in pure and applied mathematics—and returning later to apply

---

[1] Incidentally, this Lebesgue theory is easier to develop than is frequently assumed, and for those with
some basic knowledge of Lebesgue measure—as reviewed in the optional Section 2.9—it is assigned as
a set of exercises at the end of Chapter 2.

them to solve the stated problem. But also, and whenever possible, I have tried to go beyond such a closed circuit and attempted to whet the reader's appetite by indicating, mostly in optional sections, new problems and new techniques suggested by this type of discussion. Such are the developments of set theory and Lebesgue integration and the birth of functional analysis.

Painters sign their works, writers' names are printed on the covers of their books, and great composers are universally admired for their music. Why should the great creations in mathematics remain anonymous, as is frequently the case? To the extent of my knowledge and ability, I have integrated some of the history of the subject into the main text with the intention to present the various results as discoveries by their creators and to give notice of the times, the places and the circumstances, the trials and the errors, the arguments and the disputes that shaped the building of this body of knowledge. In order to do so, I have read a number of the original sources—which are quoted in the footnotes—to a larger or smaller extent; but in many cases I have had to rely on secondary sources, which are listed in Appendix D. I hope that these historical notes are valuable and inspiring to some future mathematicians and scientists. To aid the imagination I have included 22 photographs, which may be worth 22,000 words.

It may be argued that the historical approach is a straitjacket that restricts the presentation of physical applications to the well-known classic ones. This is as may be, but these classic applications suffice to develop the techniques that are also applicable to modern problems. The presentation of modern applications—modern today, old tomorrow—is not only unnecessary, but difficult because it would require specialized knowledge of these fields of application. The mature reader will realize that the techniques presented here are useful today. Otherwise, they would have been relegated to oblivion long ago, keeping company in the scientific firmament with Chaldean astronomy, the flogisto, and a host of other lesser-known entities.

There are different philosophies about the inclusion of proofs in a book of this type. My own is as follows: mathematics without proofs is not mathematics, and it is an elementary courtesy to the reader to include proofs whenever possible and to give references in the remaining cases. On the other hand, there are readers who need just the facts, have no interest in proofs, and would rather skip them. For them, a book such as this should be coherent and complete while skipping the proofs, and I have tried to write it in such a way. There is one exception that could not be avoided: it is necessary to examine the proofs of Theorems 3.1 and 3.2 as a motivation for the generalizations of Fourier series carried out in Chapter 4. Therefore, those readers who skip these proofs should also skip the starred portions of Sections 3.9 and 4.2 to 4.4 that refer to them. One of the other proofs is, hopefully, new at this level: that of the convergence of double Fourier series. Proofs of this fact are available in the literature, but mathematicians usually try to prove the best possible theorem with the weakest hypotheses. For this reason, the available proofs that I have seen are rather long and unpalatable for the intended reader of this book. While it is true that functions found in applications frequently have discontinuities, they are usually very smooth on any set on which they are continuous, and it is for such functions of two variables that I have included an elementary proof of the convergence of double Fourier series.

As for the physical features of this book perhaps the most obvious is its length, but, for the purposes of covering the material in class, it is shorter than it appears. One hundred fourteen pages are devoted to the exercises, which are both numerous and varied in difficulty. There are over 680 exercises, ranging from just drill and computational problems to those that extend and complement the theory and are meant for advanced students. I have taken care that none of the main text material depends on preceding exercises. The two exceptions are the optional Section 6.6, which depends on the Lebesgue theory developed in the exercises of Section 2.9, and the fact that the Green theorems of Exercise 1.6 of Chapter 9 are used in the last two chapters. But then, of course, a version of these theorems is frequently included in the usual calculus sequence.

The first eight chapters contain a number of optional sections or portions of sections. These are marked with symbol ⋆ in the margin. Any material in a given section is optional from the point marked with a ⋆ to the end of that section or until the symbol ◇ appears. In addition, some other sections throughout the book—especially those at the end of each chapter, such as Section 9.6—can also be considered optional. This is left to the judgement of the instructor.

Many pages of this book are devoted to historical material. Although many students, especially those in mathematics, would benefit from reading them, they are not meant for class coverage but for individual reading. Therefore, a significant portion of the book need not be covered in class, bringing its actual length to manageable proportions. By excluding a number of sections and some long proofs I have covered the entire book in one academic year. In a one-semester course I have covered the first five chapters, Appendix A, and some selected topics from Chapters 6 to 8. Different instructors will, of course, make their own choices. In this regard the following diagram of chapter interdependence may he helpful:

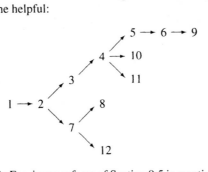

In addition, the multiple Fourier transform of Section 9.5 is mentioned in Section 12.10.

I have typeset this book in Times Roman and the TEXplorator's *Mathtime*™ using the LaTeX document preparation system. The particular implementation used is from Y&Y Inc. of Concord, Massachusetts.

I want to thank my colleague Yuly Makovoz who read portions of the manuscript and made numerous suggestions for improvement.

Dunstable, Massachusetts   *Enrique A. González-Velasco*
March, 1996

JOSEPH FOURIER

Portrait by Boilly.
Engraving by Geille.
Photograph by the author from
Œuvres de Fourier, Tome Second,
Gauthier-Villars, Paris, 1890.

1

A HEATED DISCUSSION

§1.1 Historical Prologue

Napoléon Bonaparte's expedition to Egypt took place in the summer of 1798, the expeditionary forces arriving on July 1 and capturing Alexandria the following day.[1] On the previous March 27—7 Germinal Year VI in the chronology of the French Republic—a young professor at the newly founded *École Polytechnique*, Jean Joseph Fourier (1768–1830), was summoned by the Minister of the Interior in no uncertain terms:

> Citizen, the Executive Directory having in the present circumstances a particular need of your talents and of your zeal has just disposed of you for the sake of public service. You should prepare yourself and be ready to depart at the first order.[2]

It was in this manner, perhaps not entirely reconcilable with the idea of *Liberté*, that Fourier joined the Commission of Arts and Sciences of Bonaparte's expedition and sailed for Egypt on May 19. While in temporary quarters in the town of Rosetta, near Alexandria, where he held an administrative position, the military forces marched on Cairo. They entered on July 24 after successfully defeating the Mameluks in the Battle of the Pyramids. By August 20 Bonaparte had decreed the foundation of the *Institut d'Egypte* in Cairo, modeled on the *Institut de France*,[3] of whose second *classe* (mechanical arts) he was a proud member, to serve as an advisory body to the administration, to engage on studies on Egypt, and, what is more important, to devote itself to the advancement of science in Egypt.

The first meeting of the *Institut d'Egypte*, with Fourier already appointed as its permanent secretary, was held on August 25. From this moment on until his departure

---

[1]This section, as well as portions of other sections in this book containing historical notes, is taken or adapted from the author's "Connections in Mathematical Analysis: The Case of Fourier Series," *Am. Math. Monthly*, **99** (1992) 427–441.

[2]Quoted from J. Herivel, *Joseph Fourier*, Oxford University Press, Oxford, 1975, page 64.

[3]The usual name for the *Institut National des Sciences et des Arts*.

in 1801 on the English brig *Good Design*, Fourier devoted his time not only to his administrative duties but also to scientific research, presenting numerous papers on several subjects. In the autumn of 1799 he was appointed leader of one of two scientific expeditions to study the monuments and inscriptions in Upper Egypt and was put in charge of cataloguing and describing all its discoveries.

After several military encounters the French surrendered to invading British forces on August 30, 1801. While forced to depart from Egypt, they were allowed to keep their scientific papers and collections of antiques with the exception of a precious find: the Rosetta stone.[1]

Upon his return to France in November of 1801, Fourier resumed his post at the *École Polytechnique* but only briefly. In February of 1802 Bonaparte himself appointed him *Préfet* of the Department of Isère in the French Alps. It was here, in the city of Grenoble, that Fourier returned to his physical and mathematical research, with which we shall presently occupy ourselves.[2]

But Fourier's stay in Egypt had left a permanent mark on his health which was to influence, perhaps, the direction of his research. He claimed to have contracted chronic rheumatic pains during the siege of Alexandria and that the sudden change of climate from that of Egypt to that of the Alps was too distressful for him. The facts are that he seemed to need large amounts of heat, that he lived in overheated rooms, that he covered himself with an excessive amount of clothing even in the heat of summer, and that his preoccupation with heat extended to the subject of heat propagation in solid bodies, heat loss by radiation, and heat conservation. It was then on the subject of heat that he concentrated his main research efforts, for which he had ample time after he settled down to the routine of his administrative duties. These efforts, of which there is documentary evidence as early as 1804, were first made public when Fourier read his work *Mémoire sur la propagation de la chaleur* before the first *classe* of the *Institut de France* on December 21, 1807.[3]

From our present point of view, approximately two centuries after the fact, this memoir stands as one of the most daring, original, complete, and influential works of the nineteenth century on mathematical physics. The methods that Fourier used to deal with heat problems were those of a true pioneer because he had to work with concepts that were not yet properly formulated. He worked with discontinuous functions when others dealt with continuous ones, used integral as an area when integral as a prederivative was

---

[1] On this slab of black basalt found at an excavation near Rosetta there is a triple inscription of the same text. The uppermost consists of Egyptian hieroglyphs, the central panel is in demotic script—a form of writing derived from the hieroglyphs which was used since 700 BC for secular documents—and the lower one is in Greek. Although the French took a copy of the inscription back to Paris, the original stone was transferred to the British Museum, where it can still be seen today.

[2] It was also at Grenoble that Fourier showed his collection of Egyptian curios and antiques to the boy Jean-François Champollion and awakened in him a desire to learn about the culture of old Egypt. As *Préfet* of Isère, Fourier was also able to free Champollion from military conscription, thus allowing him to pursue his studies on Egyptology. He was to spend fourteen years attempting to decipher the hieroglyphs on the Rosetta stone from the Greek translation.

[3] The royal academies were abolished by the revolution in 1793, but the *Académie Royale des Sciences* became the first *classe* of the *Institut*, devoted to the mathematical and physical sciences.

popular, and talked about the convergence of a series of functions before there was a definition of convergence. But the methods that Fourier used to deal with heat problems were to prove fruitful in many other physical disciplines such as electricity, acoustics and hydrodynamics. It was the success of Fourier's work in applications that made necessary a redefinition of the concept of function, the introduction of a definition of convergence, a reexamination of the concept of integral, and the ideas of uniform continuity and uniform convergence. It also provided motivation for the discovery of the theory of sets, was in the background of ideas leading to measure theory, and contained the germ of the theory of distributions.

However, back in 1807 his memoir was not well received. A committee consisting of Lacroix, Lagrange, Laplace, and Monge was to judge the memoir and publish a report on it, but never did so. Instead, criticisms were made personally to Fourier, either in 1808 or in 1809, on occasion of his visits to Paris to supervise the printing of his *Préface historique*—this title was personally chosen by the former First Consul who had since crowned himself Emperor Napoléon—to the *Description de l'Egypte*, a book on the Egyptian discoveries of the 1799 expedition. The criticisms came mainly from Lagrange and Laplace and referred to two major points: Fourier's derivation of the equations of heat propagation and his use of some series of trigonometric functions, known today as *Fourier series*. Fourier replied to their objections but, by this time, Jean-Baptiste Biot published some new criticisms in the *Mercure de France*, a fact that Fourier resented and moved him to write, in 1810, angry letters of protest and pointed attacks against Biot and Laplace, although Laplace had already become supportive of Fourier's work by 1809. In one of these letters—to unknown correspondents—he suggested that, as a means to settle the question, a public competition be set up and a prize be awarded by the *Institut* to the best work on the propagation of heat. If not because of this suggestion, it is at least possible that the *Institut* considered the question of a prize essay on the theory of heat in view of Fourier's vigorous defense of his own work. The fact is that in 1810 this was the subject chosen for a prize essay for the year 1811, and Laplace was probably instrumental in converting Fourier's suggestion into reality. A committee consisting of Haüy, Lagrange, Laplace, Legendre, and Malus was to judge on the only two entries. On January 6, 1812, the prize was awarded to Fourier's *Théorie du mouvement de la chaleur dans les corps solides*, an expanded version of his 1807 memoir. However, the committee's report expressed some reservations:

> This essay contains the correct differential equations of the transmission of heat, both in the interior of solid bodies or on their surface: and the novelty of the subject, added to its importance, has induced the Class to reward this Work, but observing meanwhile that the manner in which the Author arrives at his equations is not exempt from difficulties, and that his analysis, to integrate them, still leaves something to be desired in the realms of both generality and even rigor.[1]

Fourier protested but to no avail, and his new work, like his previous memoir, was not published by the *Institut* at this time. He was to ultimately prevail and enjoy a well

---

[1] Translated from I. Grattan-Guinness, *Joseph Fourier 1768–1830*, in collaboration with J. R. Ravetz, MIT Press, Cambridge, Massachusetts, 1972, page 452.

deserved fame, but the time has come when we should interrupt the telling of this story and present one of the problems, the earliest, considered by Fourier: that of a thin heated bar. This will show the originality of his methods and also the nature of those insidious *analytical difficulties*, as they were referred to by some of Fourier's opponents.

§1.2 The Heat Equation

Consider the problem of finding an equation describing the temperature distribution in a thin bar of some conducting material, which we suppose is located along the x-axis. We shall work under the following hypotheses:

1. the bar is insulated along its lateral surface so that there is no exchange of heat with the surrounding medium through this surface,

2. the bar has a uniform cross section, whose area is denoted by A, and constant density ρ,

3. at any given time t all the points of abscissa x have the same temperature, denoted by $u(x, t)$, and

4. the temperature varies so smoothly in time and along the bar that the function u has continuous first and second partial derivatives with respect to both variables.

In order to derive the desired equation we shall apply the law of conservation of energy to a small piece of the bar, the slice situated between the abscissas x and $x + h$ as shown in Figure 1.1. If c denotes the *specific heat capacity* of the material, which is defined as the

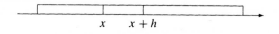

$$x \qquad x + h$$

Figure 1.1

amount of heat—the usual name for thermal energy—needed to raise the temperature of a unit mass by one degree, and if the temperature $u(t)$ were assumed to be the same at all points of the solid, then the amount of heat in the slice of volume Ah at time t would be $Q(t) = c\rho Ahu(t)$, where we have implicitly assumed that there is no heat at absolute zero temperature. But, of course, in our case the temperature is supposed to vary not only with time but also from point to point along the bar. Thus, the heat in the slice must be represented by the integral

$$Q(t) = \int_x^{x+h} c\rho Au(s, t)\, ds.$$

If u_t denotes the partial derivative of u with respect to time, the rate at which this heat changes with time is

$$Q'(t) = \int_x^{x+h} c\rho Au_t(s, t)\, ds,$$

a formula that can be obtained directly using the definition of derivative (Exercise 2.1).

Now, the only way for this heat to enter the slice is either through the sides at x and $x + h$ or by internal generation, which could happen, for instance, by resistance to an electrical current. If heat is generated internally at a constant rate q per unit volume then the heat is generated at a rate qAh in the slice. Finally, to consider the contribution through the sides of the slice, Fourier assumed that the rate at which heat flows through any cross section per unit area is proportional to the temperature gradient, which in the one-dimensional case is the partial derivative of u with respect to x. Accordingly, heat flows through the sides at the rates

$$-\kappa A u_x(x, t) \qquad \text{and} \qquad -\kappa A u_x(x + h, t).$$

The proportionality constant κ is called the *thermal conductivity* of the material and it is assumed to be positive. Then the negative sign expresses the fact that heat flows in the direction of decreasing temperature, that is, to the right where $u_x < 0$ and to the left where $u_x > 0$.

The law of conservation of energy, which is of course valid for heat rates, is then expressed by the equation

$$\int_x^{x+h} c\rho A u_t(s, t)\, ds = qAh - \kappa A u_x(x, t) + \kappa A u_x(x + h, t).$$

Dividing by Ah and finding the limit as $h \to 0$ we obtain

$$\lim_{h \to 0} \frac{1}{h} \int_x^{x+h} c\rho u_t(s, t)\, ds = q + \kappa \lim_{h \to 0} \frac{u_x(x + h, t) - u_x(x, t)}{h} = q + \kappa u_{xx}(x, t).$$

But the left-hand side is recognized to be the derivative of

$$\int_0^x c\rho u_t(s, t)\, ds$$

with respect to x, which is known to equal $c\rho u_t(x, t)$ by the fundamental theorem of calculus. Hence, the temperature distribution in the bar is described by the equation

$$c\rho u_t(x, t) = q + \kappa u_{xx}(x, t).$$

If we define a new constant

$$k = \frac{\kappa}{c\rho} > 0,$$

called the *thermal diffusivity* of the material, the simpler form of this equation when there is no internal generation of heat

$$u_t(x, t) = k u_{xx}(x, t)$$

is called the *heat equation* or the *diffusion equation*.

This derivation is somewhat different from the original one by Fourier. However, although it is true that a small imperfection still existed in his treatment in the 1807 memoir, it is well to remark at this point that there was already nothing wrong in the derivation contained in his prize essay.

Before presenting Fourier's method of solution we shall make some general remarks concerning this type of problem.

§1.3 Boundary Value Problems

An equation involving a function of several variables and some of its partial derivatives, such as the two derived at the end of the previous section, is called a *partial differential equation* and a function that satisfies the equation is called a *solution*. We shall take the heat equation as our starting point. Besides the trivial solution, $u \equiv 0$, it is easy to see that

$$u_1(x, t) = x,$$
$$u_2(x, t) = x(x^2 + 6kt),$$
$$u_3(x, t) = e^{-kt} \cos x,$$
$$u_4(x, t) = \frac{1}{\sqrt{t}} e^{-x^2/4kt}$$

are also solutions. Moreover, it is easily verified that the sum of two solutions and the product of one of them by a constant are also solutions. Which of these represents the actual temperature distribution in the bar? The unavoidable answer to this is that the question itself is not well posed. The temperature on the bar will also depend on several additional conditions. For instance, from a certain time on these temperatures will depend on the initial temperature distribution at the given instant and, although the bar is insulated along its lateral surface, on any amount of heat that may enter or leave at the endpoints. Therefore, we see that both initial conditions and boundary conditions will affect the actual temperature distribution. These conditions can be specified in a variety of ways. For example, the temperature may be given at $t = 0$ as a function of x

$$u(x, 0) = f(x),$$

and it may be further specified that the temperatures at the endpoints $u(0, t)$ and $u(a, t)$ will remain fixed for $t > 0$. This can be accomplished, for instance, by the application of electric heaters with well-regulated thermostats at the ends of the bar.

A problem consisting of finding solutions of a partial differential equation subject to some initial and boundary conditions is normally referred to as a *boundary value problem*. In order to pose some boundary value problems for the heated bar we shall find the following notation and terminology to be convenient. Suppose that the bar is located between $x = 0$ and $x = a$ and, if $\mathbb{R}^2$ is the Euclidean plane, define the sets

and
$$D = \{ (x, t) \in \mathbb{R}^2 : 0 < x < a, \ t > 0 \}$$
$$\overline{D} = \{ (x, t) \in \mathbb{R}^2 : 0 \leq x \leq a, \ t \geq 0 \}.$$

Then a real-valued function that is defined and has continuous first- and second-order partial derivatives in D is said to be of *class C^2* in D.

Example 1.1 The problem of finding a continuous function $u : \overline{D} \to \mathbb{R}$ that is of class C^2 in D and such that

$$\begin{array}{ll} u_t = k u_{xx} & \text{in } D \\ u(0, t) = 0 & t \geq 0 \\ u(a, t) = 0 & t \geq 0 \end{array}$$

has the solution

$$u(x, t) = e^{-\pi^2 kt/a^2} \sin \frac{\pi}{a} x,$$

as is easily verified by differentiation.

Sometimes the boundary and initial conditions are given as limits as the boundary of the domain of definition of the equation is approached in a perpendicular direction.

Example 1.2 The problem of finding a function $u : D \to \mathbb{R}$ that is of class C^2 in D and such that

$$
\begin{aligned}
u_t &= ku_{xx} & &\text{in } D \\
u(x, t) &\to 0 & &\text{as } x \to 0, \quad t > 0 \\
u(x, t) &\to 0 & &\text{as } t \to 0, \quad 0 < x < a
\end{aligned}
$$

has the solution

$$u(x, t) = \frac{x}{t\sqrt{t}} e^{-x^2/4kt},$$

as is verified by differentiation (Exercise 3.2). Note that u is not defined for $t = 0$ but has the required limit. This is one reason why the boundary conditions may be given as limits. It allows us to exhibit solutions that would otherwise not exist.

A close look at these examples will show that care must be exercised in specifying the boundary conditions. Each of these boundary value problems also admits the trivial solution $u \equiv 0$. Besides, for any positive integer n, the function

$$u(x, t) = e^{-n^2\pi^2 kt/a^2} \sin \frac{n\pi}{a} x$$

is also a solution of the problem in Example 1.1, as will be carefully shown in §1.4. Which, then, is the solution that, in each case, represents the actual temperature distribution on the bar? The fact is that one expects a well-posed physical problem to have a unique solution. If more than one solution of its mathematical counterpart exists, we infer that the given conditions are insufficient to determine which one represents the answer to the problem.

In the case of Example 1.1 the insufficiency consists of not having specified the initial temperature distribution at $t = 0$. If we require that

$$u(x, 0) = \sin \frac{\pi}{a} x,$$

then only the solution on page 6 satisfies the complete boundary value problem. In fact, it will be shown later that no other solution is possible. In Example 1.2 the temperature at the right end of the bar has not been specified for $t > 0$. If we require that

$$u(x, t) \to 0 \quad \text{as} \quad x \to a$$

then the first proposed solution must be discarded because

$$u(x, t) \to \frac{a}{t\sqrt{t}} e^{-a^2/4kt} \neq 0 \quad \text{as} \quad x \to a$$

for $t > 0$. Under the given conditions it is to be expected that the trivial solution $u \equiv 0$ will be the answer to the complete boundary value problem. Of course, in any given situation it must be proved mathematically that the solution of the problem is unique.

There is still another requirement that physical considerations impose on the solution of a boundary value problem. Physical measurements or observations of the initial temperature of the bar would result in only approximate values. In a similar manner, the temperatures at the endpoints cannot be maintained with perfect accuracy. Thus, the mathematical formulation of a problem will contain small errors in the initial and boundary values, and the corresponding solution can only approximate the true one. What we must require is that it be a good approximation. That is, if the initial or boundary values change by a small amount the solution of a well posed problem should change only by a small amount.

Jacques Hadamard (1865–1963), of the *École Polytechnique*, is credited with first realizing that this condition should be imposed and with exhibiting the first example of a boundary value problem whose solution does not satisfy this requirement (Exercise 3.10). For the case of the heat equation the following example, originally given by E. Rothe in 1928 for an ideal bar of infinite length, shows a violation of this condition.

Example 1.3 Consider the boundary value problem

$$u_t = k u_{xx} \qquad\qquad -\infty < x < \infty, \quad t > 0$$
$$u(x, 1) = f(x) \qquad\quad -\infty < x < \infty.$$

If $f \equiv 0$ then $u \equiv 0$ is the solution. But if c is a small positive number and if we define

$$f(x) = c \sin \frac{x}{c\sqrt{k}},$$

then the problem has the solution

$$u(x, t) = c\, e^{(1-t)/c^2} \sin \frac{x}{c\sqrt{k}}$$

(Exercise 3.4). By choosing c small enough f can be made as close to the zero function as desired. However, if $0 < t < 1$ this solution does not tend to the trivial solution as $c \to 0$. Instead it becomes unbounded. Thus, a small change in the initial condition from the zero function to f results in a large change in the solution.

In conclusion, when solving a boundary value problem it must be shown that the following three requirements are satisfied:

1. there exists a solution,

2. the solution is unique, and

3. the solution is stable; that is, it depends continuously on the boundary conditions in the sense that a small change in the initial and boundary values results in only a small change in the solution.

A boundary value problem which satisfies these three requirements is said to be *well posed in the sense of Hadamard*, who introduced this concept in his 1920 lectures at Yale University.

§1.4 The Method of Separation of Variables

We shall now seek a solution for a specific boundary value problem involving the heated bar. If D and $\bar{D}$ are as in Example 1.1, find a continuous function $u: \bar{D} \to \mathbb{R}$ that is of class C^2 in D and such that

$$
\begin{aligned}
u_t &= k u_{xx} & &\text{in } D \\
u(0, t) &= 0 & &t \geq 0 \\
u(a, t) &= 0 & &t \geq 0 \\
u(x, 0) &= f(x) & &0 \leq x \leq a,
\end{aligned}
$$

where f is a given continuous function.

We shall look for a solution of the form

$$u(x, t) = X(x)T(t),$$

where X and T are unknown functions to be determined. There is no reason to assume *a priori* that the solution will be the product of a function of x alone by a function of t alone. In searching for a solution of this form Fourier was inspired by a similar method of attack employed by other mathematicians before him, notably d'Alembert, whose work we shall present in Chapter 5. The ultimate justification of such a method—which, as might be expected, is not applicable to all boundary value problems—is provided by the fact that it does indeed work in many cases of interest in mathematical physics.

If such a solution exists the heat equation implies that

$$XT' = u_t = k u_{xx} = kX''T$$

for (x, t) in D. If we further assume that $X(x) \neq 0$ for $0 < x < a$ and that $T(t) \neq 0$ for $t > 0$ we obtain

$$\frac{T'}{T} = \frac{kX''}{X}.$$

Now, the only way for a function of t to equal a function of x is for both of them to be constant. Denoting this constant by $-\lambda$ leads to the pair of ordinary differential equations

$$T' + \lambda T = 0 \qquad \text{and} \qquad kX'' + \lambda X = 0.$$

Thus, the original equation in two variables has been replaced by two equations, each in one variable. That is, the variables have been separated, and this is what gives the method its name.

The first equation has the general solution

$$T(t) = Ce^{-\lambda t},$$

where C is any constant. Note that T is never zero as we assumed above.

The solution of the second equation depends on the sign of λ. If $\lambda < 0$ and if we put $\mu = \lambda/k$, the general solution is

$$X(x) = Ae^{\sqrt{-\mu}\,x} + Be^{-\sqrt{-\mu}\,x},$$

where A and B are arbitrary constants. But then the boundary conditions $u(0, t) = 0$ and $u(a, t) = 0$ imply that

$$A + B = 0$$
$$Ae^{\sqrt{-\mu}\,a} + Be^{-\sqrt{-\mu}\,a} = 0$$

which in turn imply that $A = B = 0$. This leads to the trivial solution $u = TX \equiv 0$, which is indeed a solution but only when $f \equiv 0$.

If $\lambda = 0$ the original equation reduces to $X'' = 0$, so that

$$X(x) = A + Bx.$$

As before, the boundary conditions lead to $A = B = 0$ and to the trivial solution.

Therefore, if $f \not\equiv 0$ it must be $\lambda > 0$. In this case and again with $\mu = \lambda/k$, the general solution is

$$X(x) = A \sin \sqrt{\mu}\, x + B \cos \sqrt{\mu}\, x.$$

We now deduce from the boundary conditions that

$$B = 0$$
$$A \sin \sqrt{\mu}\, a + B \cos \sqrt{\mu}\, a = 0.$$

Unless $A = 0$, in which case we obtain the trivial solution once more, it must be $\sqrt{\mu}\, a = n\pi$, where n can be any positive integer. Hence, we obtain infinitely many solutions, each of the form

$$X(x) = A_n \sin \frac{n\pi}{a} x.$$

The constant A_n can be chosen arbitrarily for each n. Note then that nontrivial solutions have been found only for values of λ of the form

$$\lambda = \mu k = \frac{n^2 \pi^2}{a^2} k.$$

For each of these, and if we recall the solution of the equation in t, we conclude that the heat equation has the nontrivial solution

$$u_n(x, t) = c_n e^{-n^2 \pi^2 kt/a^2} \sin \frac{n\pi}{a} x,$$

where $c_n = CA_n$ is just an arbitrary constant. Each of these solutions is of class C^2 and has the value zero for $x = 0$ and for $x = a$.

Perhaps it has been noticed that the solutions we found for X above have zeros, contrary to the starting assumption that $X(x) \neq 0$ for $0 < x < a$. But this is no obstacle because it is easily verified by differentiation that u_n is indeed a solution of the heat equation. What matters, then, is that we have found such a solution. What does not matter is that certain steps of our procedure were unjustified. In fact, this is not untypical of work in applied mathematics and will be a recurrent theme in this book. When looking for the solution of a particular problem we will not hesitate in making any number of reasonable assumptions, justified or unjustified, and if they lead to a solution that can be verified *a posteriori* the matter is settled, and that is that.

§1.5 Linearity and Superposition of Solutions

For each positive integer n, the solution of the heat equation

$$u_n(x, t) = c_n e^{-n^2\pi^2 kt/a^2} \sin \frac{n\pi}{a}x$$

satisfies the boundary conditions at $x = 0$ and $x = a$. It remains to see if any of these satisfies the initial condition. We have

$$u_n(x, 0) = c_n \sin \frac{n\pi}{a}x,$$

and it is clear that no u_n satisfies the initial condition unless this is of the form

$$f(x) = A \sin \frac{n\pi}{a}x$$

for some constant A and some n, a very unlikely occurrence. Does this mean that the method of separation of variables has failed and that we should look for a solution elsewhere? Not necessarily because, as we shall presently show, the heat equation is linear and this will enable us to try a linear combination of the preceding solutions in order to solve the complete boundary value problem.

A partial differential equation is defined to be *linear* if and only if the following conditions are satisfied:

1. if u and v are solutions, then $u + v$ is a solution, and

2. if u is a solution and r is a real number, then ru is a solution.

Example 1.4 The heat equation is linear because if $u_t = ku_{xx}$ and $v_t = kv_{xx}$, then

$$(u + v)_t = u_t + v_t = ku_{xx} + kv_{xx} = k(u + v)_{xx}$$

and, for any real number r,

$$(ru)_t = ru_t = rku_{xx} = kru_{xx} = k(ru)_{xx}.$$

Example 1.5 The equation $u_t + uu_x = u_{xx}$ is not linear, because if u is a non-constant solution and $r \neq 0, 1$, we have

$$(ru)_t + (ru)(ru)_x = ru_t + r^2 uu_x \neq ru_t + ruu_x = ru_{xx} = (ru)_{xx}.$$

This violates the second linearity condition.

It is now easy to see that the sum of three solutions u_1, u_2 and u_3 of a linear equation is also a solution, as it suffices to apply the first linearity property to the right-hand side of the equation

$$u_1 + u_2 + u_3 = (u_1 + u_2) + u_3.$$

In general, and by a similar reasoning, the sum of N solutions of a linear equation is a solution. Combining this with the second linearity condition we can formally state the

Principle of superposition of Solutions *If u_1, ..., u_N, are solutions of a linear partial differential equation, where N is a positive integer, and if c_1, ..., c_N are arbitrary real constants, then*

$$c_1 u_1 + \cdots + c_N u_N$$

is also a solution.

The relevance of linearity in our case is that the function

$$u(x, t) = \sum_{n=1}^{N} c_n e^{-n^2 \pi^2 kt/a^2} \sin \frac{n\pi}{a} x$$

is a solution of the heat equation for any positive integer N. Moreover, it satisfies the endpoint conditions because each term in the sum does. In fact, the endpoint conditions

$$u(0, t) = u(a, t) = 0$$

for $t \geq 0$ are also *linear*—the terms *linear homogeneous* or simply *homogeneous* are also frequently used to refer to this fact—in the sense that, if they are satisfied by functions $u_1, \ldots, u_N$, then they are satisfied by any linear combination of these. Thus, it remains to see if N can be chosen so that our proposed sum of solutions satisfies the initial condition

$$u(x, 0) = f(x).$$

Before we settle this matter it must be observed, because it is quite obvious, that the endpoint conditions are linear by design. If, instead, we had required that

$$u(0, t) = T_1 \neq 0$$
$$u(a, t) = T_2 \neq 0$$

for $t \geq 0$, then these more general conditions are no longer linear. In fact, if u and v satisfy these new conditions, $u + v$ does not because

$$(u + v)(0, t) = u(0, t) + v(0, t) = 2T_1 \neq T_1$$

and

$$(u + v)(a, t) = u(a, t) + v(a, t) = 2T_2 \neq T_2.$$

But this nonlinearity is only a minor obstacle, and the way around it is this. First define

$$U(x) = T_1 + (T_2 - T_1)\frac{x}{a},$$

which is easily seen to be a solution of the heat equation such that $U(0) = T_1$ and $U(a) = T_2$. If we now find another solution v of the heat equation such that $v(0) = v(a) = 0$ and $v(x) = f(x) - U(x)$, which is precisely the type of problem we are in the process of solving, then the function $u = U + v$ is clearly a solution of the boundary value problem corresponding to the new nonlinear endpoint conditions. Thus, no generality is lost in assuming linear endpoint conditions.

Returning then to the problem as originally stated, the question that arises regarding the initial condition is whether a positive integer N and real constants $c_1, \ldots, c_N$ can be chosen so that

$$f(x) = \sum_{n=1}^{N} c_n \sin \frac{n\pi}{a} x.$$

Again the answer is no in general, since the sum of trigonometric functions is a very particular kind of function. The answer is definitely no if f is not differentiable, even if it is continuous, because the sum of the trigonometric functions on the right is differentiable.

What Fourier proposed is an infinite sum, and claimed that for any arbitrary function f it is possible to find constants c_n such that

$$f(x) = \sum_{n=1}^{\infty} c_n \sin \frac{n\pi}{a} x,$$

even if f is not differentiable. The same assertion had been made much earlier by Daniel Bernoulli in connection with a different problem. But, unlike Fourier later on, Bernoulli did not attempt any mathematical justification of this idea, which was overwhelmingly rejected by the mathematical authorities of the times. As we have already seen, Fourier experienced a similar kind of skeptical opposition.

The main obstacle in accepting that an infinite series could add up to an arbitrary function was precisely the concept of function then in vogue at the turn of the century. Because of the success of the calculus of Newton and Leibniz, mathematicians grew accustomed to functions being given by analytical expressions such as powers, roots, logarithms, and so on. How, they demanded, can a function such as $f(x) = e^x$ be represented by the sum of an infinite series of sines? Why, this function is not even periodic while the sine functions are and, consequently, so is the sum of a series of sines. Since a function such as the exponential cannot coincide with a periodic function over the whole real line, they failed to realize that this could happen over a bounded interval.

Of course, things change if a different, geometric as opposed to analytic, concept of function is adopted. For instance, a function may simply be given by its graph in the plane, whether or not there is an analytical expression for it. Then two different functions, such as those in Figure 1.2, can be identical over a bounded interval but not outside.

Fourier was confident of the basic truth of his assertion, that the sum of a trigonometric series can equal an arbitrary function over a bounded interval, and gave numerous examples in which this is geometrically evident.

Example 1.6 Figure 1.3 represents several stages in the addition of terms of the series

$$\sum_{n=1}^{\infty} 2 \frac{(-1)^{n+1}}{n} \sin nx$$

over the interval $[0, \pi)$. The larger the number of terms incorporated into the sum the closer the addition seems to get to the function $f(x) = x$ on $[0, \pi)$. The same is true,

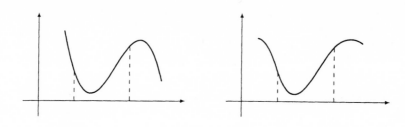

Figure 1.2

since every term of the sum is odd, on $(-\pi, 0]$. Of course the sum of the series is periodic of period 2π and its graph, discontinuous at each point of the form $\pi + 2k\pi$ with k any integer, is the straight line $y = x$ only on $(-\pi, \pi)$.

   While Fourier's arguments and examples were overwhelming and sufficiently convincing to merit the *Institut*'s Prize for 1811, the fact remains that he did not provide a mathematical proof of his assertions. We shall devote the next chapter to study the convergence of trigonometric series and the conditions under which their sum can equal an 'arbitrary' function. It is only after these facts are established on solid mathematical ground that we shall return to the problem of the heated bar and to other problems based on the use of identical or similar series.

§1.6 Historical Epilogue

In 1812, dissapointed by the committee's reaction to his memoir, Fourier reluctantly had to return to Grenoble, and, being far from Paris, he lacked the power and influence to have his prize essay published by the *Institut de France*. But new political events would soon change his fortune. A European alliance against Napoléon forced the unconditional abdication of the Emperor on April 11, 1814, restoring the monarchy in the person of Louis XVIII. Fourier remained as *Préfet* of Isère under the new regime, a tribute to his diplomatic abilities, but early the following March he learned that Napoléon had returned from his exile at Elba and landed at Cannes at the head of 1700 men. Fearful of the consequences of his temporary allegiance to the Crown he departed from Grenoble on March 7, but not without leaving a room ready for an approaching Napoléon and a letter of apology. By the 12th he reached Lyons, and on that very same day he was appointed *Préfet* of the Rhône by the Emperor, who had forgiven his ungrateful behavior. This position was short-lived, as he was relieved from it by a decree dated May 17, supposedly for failing to comply with orders of a certain political purge. Upon his dismissal, and having been granted a pension of 6000 francs by Napoléon, Fourier finally returned to Paris.
   A new allied army was victorious over Napoléon on June 18, 1815, at the Battle of Waterloo. Thus ended his second period in power, known as the Hundred Days, as

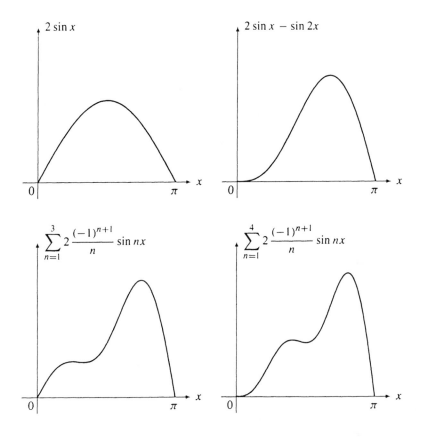

Figure 1.3

he was forced to abdicate again on June 22 and was forever banished to the island of St. Helena. Fourier's pension, which was to start on July 1, never materialized under the King's restored government and he found himself penniless. With the influence of a friend and former student at the *École Polytechnique*, the Count of Chabrol de Volvic, he secured the position of Director of the Bureau of Statistics of the Department of the Seine, and this allowed him to remain in Paris permanently.

Now he set down to business. First, there was the publication of the prize essay, a matter in which he succeeded after a considerable amount of persistence and determination, a personality trait that served him well throughout his life. It finally appeared in 1824 and 1826, officially dated 1819–1820 and 1821–1822, in volumes 4 and 5 of the *Mémoires de l'Académie Royale des Sciences de l'Institut de France* (with the restoration of the monarchy, the academies were reopened as sections of the *Institut*). But before this, in May of 1816, two new members of the Academy of Sciences were to be elected.

Fourier lobbied vigorously on his own behalf, and, after several rounds of voting, he was elected to the second position. The King, resentful of Fourier's activities during the Hundred Days, refused to give his approval. But a vacancy for physics became available in 1817, and on the election of May 12 Fourier obtained 47 of the 50 votes. The King was then compelled to grant his approval, but he still waited until May 21 to do so.

Fourier's scientific standing was no longer in doubt. From now on he devoted his full attention to research and presented numerous papers on the theory of heat and other mathematical subjects. In 1822 his monumental work *Théorie analytique de la chaleur*, a compendium of the larger part of his researches on heat, was printed in Paris, and on November 18 of the same year he became permanent secretary of the mathematics section of the Academy of Sciences.[1]

His last years were marked by honors and poor health. He was elected to the Royal Society of London in 1823 and to the *Académie Française* in 1826. Then, the next year, on the death of Laplace, he became president of the *Conseil de Perfectionnement de l'École Polytechnique*. But already in 1826, in a letter to Auger, permanent secretary of the French Academy, he claimed to *see the other bank where one is healed of life*.[2] In addition to his rheumatism, which never left him, he developed a shortness of breath that was particularly acute if not standing up. Resourceful to the very end, he invented a contraption in the form of a box with holes for his arms and head to protrude and carried on in this fashion. The end came at about four o'clock in the afternoon of May 16, 1830, in the form of a heart attack, and shortly afterward he died.

EXERCISES

2.1   Derive the formula for $Q'(t)$ on page 4 by showing that

$$\int_x^{x+h} \frac{u(s, t+l) - u(s, t)}{l} \, ds - \int_x^{x+h} u_t(s, t) \, ds$$

approaches zero as $l \to 0$ (**Hint:** use Taylor's theorem with remainder to estimate the difference $u(s, t+l) - u(s, t)$ and then use the continuity of u_{tt} on the interval $[x, x+h]$).

---

[1] Two months before, on September 14, a breakthrough took place in another corner of science to which Fourier was no stranger. Champollion, who had continued to work without success on the hyeroglyphs of the Rosetta stone, had received in January a copy of the inscriptions on an obelisk found at Philae. He managed to decipher the names of Cleopatra and Ptolemy, later that of Alexander. On this basis he begun the construction of a list of symbols whose meaning he knew, and to put this knowledge to the test he turned his attention to a new inscription on a *bas relief* from an Egyptian temple. On September 14, 1822, he deciphered his first all-Egyptian names: those of Ramses and Tutmosis. The Egyptian hyeroglyphs had surrendered their secret.

[2] Quoted from J. Herivel, *Joseph Fourier*, Oxford University Press, Oxford, 1975, page 137.

2.2 Modify the derivation of the equation of heat propagation when the internal generation of heat is not constant but a continuous function of x. Indicate at which point in your derivation you use the continuity of this function.

2.3 Show that the value of the thermal diffusivity in the heat equation can always be chosen to be 1 by introducing a new time variable $\tau = kt$, defining a new function $v(x, \tau) = u(x, t) = u(x, \tau/k)$ and finding the equation that v and v_{xx} satisfy.

2.4 Assume that there are a positive constant C and a function v of x only such that $e^{-Ckt}v$ satisfies the heat equation. Show that v satisfies the *Helmholtz* equation $v_{xx} + Cv = 0$.

2.5 Assume that the heated bar is not insulated along its lateral surface—but it is still so thin that the temperature is a function of x and t only—and that the rate of heat loss per unit length along this surface is given by $C[u(x, t) - T]$, where T is the temperature of the medium surrounding the bar and C is a positive constant. This type of heat loss is called *convection*. With c, q, κ, ρ, and A as before, show that the equation of heat propagation is now

$$c\rho u_t = q + \kappa u_{xx} - \frac{C}{A}(u - T).$$

2.6 If u satisfies the equation in Exercise 2.5 and if v is a function of x and t such that

$$u = T + e^{-Ct/c\rho A}v,$$

show that v satisfies the heat propagation equation $c\rho u_t = \kappa u_{xx} + q$.

2.7 Assume that the thermal coefficients c and κ of the bar, its density ρ, and the rate of internal heat generation q are all functions of x. Derive the new equation of heat propagation.

2.8 Assume that, from a local point of view, the Earth is flat and extends indefinitely in any horizontal direction and in depth, and also that its heat capacity and thermal conductivity are constant. In the absence of internal heat generation, derive the equation of heat propagation in terms of time and depth from the surface.

3.1 Find all solutions of the heat equation that are independent of time. They are called *steady-state solutions*. Does any of these satisfy the boundary conditions
  (i) $u(0, t) = 0$ and $u(a, t) = 5$?
  (ii) $u(0, t) = 3$ and $u(a, t) = 0$?
What boundary conditions does the sum of the solutions in (i) and (ii) satisfy?

3.2 Verify that

$$u(x, t) = \frac{Cx}{t\sqrt{t}}e^{-x^2/4kt},$$

C being any constant, is a solution of the boundary value problem

$$\begin{aligned}
u_t &= ku_{xx} & 0 < x < a, \quad t > 0 \\
u(x, t) &\to 0 \quad \text{as} \quad t \to 0 & 0 < x < a \\
u(x, t) &\to 0 \quad \text{as} \quad x \to 0 & t > 0.
\end{aligned}$$

3.3 Show that the solution in Exercise 3.2 cannot represent the physical temperature of the bar because it is unbounded in a neighborhood of $(0, 0)$ (**Hint:** examine its values on a suitably chosen parabola through the origin).

3.4 Verify that

$$u(x, t) = ce^{(1-t)/c^2}\sin\frac{x}{c\sqrt{k}},$$

where c is a positive constant, is a solution of the boundary value problem

$$u_t = k u_{xx} \qquad\qquad -\infty < x < \infty, \quad t > 0$$
$$u(x, 1) = c \sin \frac{x}{c\sqrt{k}} \qquad -\infty < x < \infty.$$

3.5 State the boundary value problem corresponding to a heated bar with insulated endpoints at $x = 0$ and $x = a$—that is, there is no heat flow through these sides—and a given initial temperature distribution $u(x, 0) = f(x)$.

3.6 Modify the boundary value problem of Exercise 3.5 if there is heat loss by convection at $x = a$, that is, if the rate of heat flow at $x = a$ is proportional to the difference between the temperature of the bar and the temperature $T(t)$ of the air.

3.7 Modify the boundary value problem of Exercise 3.6 if the heat loss at $x = a$ is by radiation, in which case and according to the *Stefan-Boltzmann law* of radiation, the rate of heat flow at $x = a$ is proportional to the difference of the fourth powers of the temperatures of the bar and the air.

3.8 Is there more than one nontrivial solution of the boundary value problem

$$u_{tt} = u_{xx} \qquad\qquad 0 < x < \pi, \quad t > 0$$
$$u(x, 0) = 0 \qquad\qquad 0 \le x \le \pi$$
$$u(0, t) = u(\pi, t) = 0 \qquad t \ge 0?$$

3.9 Show that the boundary value problem

$$u_{tt} + u_{xx} = 0 \qquad\qquad -\infty < x < \infty, \quad t > 0$$
$$u(x, 0) = 0 \qquad\qquad -\infty < x < \infty$$
$$u_t(x, 0) = c \sin \frac{x}{c} \qquad -\infty < x < \infty,$$

where c is a positive constant, has the solution $u(x, t) = c^2 \sinh(t/c) \sin(x/c)$

3.10 Use the result of Exercise 3.9 to show that the boundary value problem

$$u_{tt} + u_{xx} = 0 \qquad\qquad -\infty < x < \infty, \quad t > 0$$
$$u(x, 0) = 0 \qquad\qquad -\infty < x < \infty$$
$$u_t(x, 0) = 0 \qquad\qquad -\infty < x < \infty$$

is not well posed in the sense of Hadamard.

3.11 Show that the boundary value problem

$$t u_t - x u_x = 0 \qquad x > 0, \quad t > 0$$
$$u(x, 0) = 1 \qquad\qquad x \ge 0$$
$$u(0, t) = 1 \qquad\qquad t \ge 0$$

is not well posed in the sense of Hadamard.

3.12 Is the boundary value problem

$$u_t = k u_{xx} \qquad\qquad 0 < x < a, \quad t > 0$$
$$u_x(0, t) = 0 \qquad\qquad t \ge 0$$
$$u_x(a, t) = 0 \qquad\qquad t \ge 0$$

well posed in the sense of Hadamard?

4.1   Use separation of variables to find a solution of the boundary value problem

$$u_t + u_x = u \qquad\qquad x > 0, \quad t > 0$$
$$u(0, t) = e^{t/2} \qquad\qquad t \geq 0.$$

4.2   Use separation of variables to find a solution of the boundary value problem

$$2x(1 + t)u_t - u_x = 0 \qquad\qquad x > 0, \quad t > 0$$
$$u(x, 0) = e^{x^2} \qquad\qquad x \geq 0$$
$$u(0, t) = 1 + t \qquad\qquad t \geq 0.$$

4.3   The equations $T' + \lambda kT = 0$ and $X'' + \lambda X = 0$ have been solved in the text under the assumption that λ was a real number. Assume now that λ is a purely imaginary number and find the corresponding real-valued product solutions of the heat equation (**Hint:** if a complex valued solution is found, then its real and imaginary parts are also solutions).

4.4   If A and ω are positive constants, find a real-valued solution of the boundary value problem

$$u_t = ku_{xx} \qquad\qquad x > 0, \quad t > 0$$
$$u(x, t) \to 0 \quad \text{as} \quad x \to \infty \qquad\qquad t \geq 0$$
$$u(0, t) = A \cos \omega t \qquad\qquad t \geq 0$$

(**Hint:** use the solution of Exercise 4.3).

4.5   Consider the preceding boundary value problem when $2\pi/\omega = 31.536 \times 10^6$ sec and when $2\pi/\omega = 86{,}400$ sec. Refering now to the situation described in Exercise 2.8, use a value of $k = 2 \times 10^{-3}$ cm^2/sec for the thermal diffusivity, which is typical for an average soil, and give a reasonable physical interpretation of this problem and of its solution, estimating in each case the smallest value of x for which the temperature fluctuations with time are of the form $u(x, t) = B(x) \cos \omega t$ with $B(x)/A < 0$—this represents a *phase lag* of π radians with respect to $u(0, t)$. In each case, find the ratio $|B(x)/A|$.

4.6   If ω is a positive constant, find an unbounded solution of the boundary value problem

$$u_t = ku_{xx} \qquad\qquad x > 0, \quad t > 0$$
$$u(0, t) = \sin \omega t \qquad\qquad t \geq 0$$

(**Hint:** use the solution of Exercise 4.3).

4.7   Solve the equations $T' + \lambda kT = 0$ and $X'' + \lambda X = 0$ when λ is a complex number and find the corresponding product solutions of the heat equation.

4.8   Use the result of Exercise 4.7 to find a solution of the boundary value problem

$$u_t = ku_{xx} \qquad\qquad x > 0, \quad t > 0$$
$$u(x, t) \to 0 \quad \text{as} \quad x \to \infty \qquad\qquad t \geq 0$$
$$u(0, t) = e^{-4t} \cos 3t \qquad\qquad t \geq 0.$$

5.1   Which of the equations in Exercises 2.4 to 2.6 are linear?

5.2   Show that the second-order partial differential equation in two variables x and y

$$Au_{xx} + Bu_{xy} + Cu_{yy} + Du_x + Eu_y + Fu = 0,$$

where the coefficients are functions of x and y, is linear.

5.3   Are there any linear second-order partial differential equations in two variables x and y which are not of the form given in Exercise 5.2? Justify your answer.

5.4   Show that an equation of the form given in Exercise 5.2 cannot be solved by separation of variables if $B \neq 0$. This motivates the next exercise.

5.5   Consider the equation

$$Au_{xx} + Bu_{xy} + Cu_{yy} + Du_x + Eu_y + Fu = 0,$$

where the coefficients are constants. Show that if $B \neq 0$ the change variables

$$\begin{aligned} z &= \quad x \cos\theta + y \sin\theta \\ w &= -x \sin\theta + y \cos\theta \end{aligned}$$

(this is a counterclockwise rotation of the plane through an angle θ), where

$$\cot 2\theta = \frac{A - C}{B},$$

reduces the given equation to another of the form

$$au_{zz} + cu_{ww} + du_z + eu_w + fu = 0.$$

Show also that

$$a = \tfrac{1}{2}\left(A + C + \sqrt{(A + C)^2 + B^2 - 4AC}\right)$$

and

$$c = \tfrac{1}{2}\left(A + C - \sqrt{(A + C)^2 + B^2 - 4AC}\right).$$

5.6   Using the method of Exercise 5.5, transform the partial differential equations
     (*i*) $5u_{xx} + 4u_{xy} + 2u_{yy} = 0$,
     (*ii*) $3u_{xx} + 8u_{xy} - 3u_{yy} = 0$, and
     (*iii*) $-u_{xx} + 2\sqrt{3}\,u_{xy} - 3u_{yy} = 0$
to the form $au_{zz} + cu_{ww} = 0$ by a suitable rotation of the plane.

5.7   The methods of solution and the types of boundary conditions that can be imposed on an equation of the form

$$Au_{xx} + Bu_{xy} + Cu_{yy} + Du_x + Eu_y + Fu = 0$$

depend on the sign of the product ac after it has been reduced to the form described in Exercise 5.5; that is, they depend on the sign of $B^2 - 4AC$. Thus, there are three classes of equations of the form given above, which are called *elliptic* if $B^2 - 4AC < 0$, *parabolic* if $B^2 - 4AC = 0$, and *hyperbolic* if $B^2 - 4AC > 0$. Determine whether each of the following equations is elliptic, parabolic, or hyperbolic.
     (*i*) The *heat* equation $u_t = ku_{xx}$.
     (*ii*) The *potential* equation $u_{xx} + u_{yy} = 0$.
     (*iii*) The *wave* equation $u_{tt} - u_{xx} = 0$.
     (*iv*) The *telegraph* equation $u_{xx} = clu_{tt} + (cr + gl)u_t + gru$, where the constants c, l, r, and g are nonnegative.

5.8 If the coefficients of the equation in Exercise 5.7 are functions of x and y, the equation is called *elliptic, parabolic,* or *hyperbolic* at a point (x, y) if the value of $B^2 - 4AC$ at that point is negative, zero, or positive, respectively.

(*i*) Determine the regions of the plane in which the *Tricomi* equation $u_{xx} + xu_{tt} = 0$ is elliptic, parabolic, and hyperbolic.

(*ii*) Choose nonconstant functions A and C of x and y such that the equation $Au_{xx} + Cu_{yy} + u = 0$ is elliptic for $x > y$, parabolic for $x = y$, and hyperbolic for $x < y$.

5.9 Find a solution of the boundary value problem

$$
\begin{aligned}
u_t &= ku_{xx} & 0 < x < 1, \quad t > 0 \\
u(0, t) &= u(1, t) = 1 & t \geq 0 \\
u(x, 0) &= 1 + \sin \pi x \cos \pi x & 0 \leq x \leq 1
\end{aligned}
$$

(**Hint:** pose a boundary value problem for the function $v = u - 1$).

5.10 Find a solution of the boundary value problem

$$
\begin{aligned}
u_t &= ku_{xx} & 0 < x < 2, \quad t > 0 \\
u(0, t) &= 0 & t \geq 0 \\
u(2, t) &= 2 & t \geq 0 \\
u(x, 0) &= x + \sin \frac{\pi x}{2} & 0 \leq x \leq 2.
\end{aligned}
$$

5.11 Propose a method of solution of the boundary value problem

$$
\begin{aligned}
u_t &= ku_{xx} + q & 0 < x < a, \quad t > 0 \\
u(0, t) &= T_1 & t \geq 0 \\
u(a, t) &= T_2 & t \geq 0 \\
u(x, 0) &= f(x) & 0 \leq x \leq a
\end{aligned}
$$

when the rate of internal heat generation q is
(*i*) a constant and
(*ii*) a function of x
(**Hint:** Find first a solution U of the first three equations that is independent of time).

5.12 Give some thought to the boundary value problem

$$
\begin{aligned}
u_t &= ku_{xx} + F(x, t) & 0 < x < a, \quad t > 0 \\
u(0, t) &= g_1(t) & t \geq 0 \\
u(a, t) &= g_2(t) & t \geq 0 \\
u(x, 0) &= f(x) & 0 \leq x \leq a,
\end{aligned}
$$

where F is a function of x and t and g_1 and g_2 are differentiable functions of t.
(*i*) What suggestions can you make to reduce this problem to a simpler one?
(*ii*) Can you deal with the particular case

$$
\begin{aligned}
u_t &= ku_{xx} + \cos t & 0 < x < a, \quad t > 0 \\
u(0, t) &= \sin t & t \geq 0 \\
u(\pi, t) &= \sin t & t \geq 0 \\
u(x, 0) &= \sin x & 0 \leq x \leq \pi?
\end{aligned}
$$

5.13   Make a graphical representation of the sum

$$\frac{3}{2} + \frac{2}{\pi} \sum_{n=1}^{N} \frac{1}{2n-1} \sin(2n-1)\pi(x-1)$$

on the interval $[0, 2]$ for $N = 1, 2, 3$. On this basis draw a graph of the possible sum of the infinite series obtained when N is replaced by ∞.

5.14   Repeat Exercise 5.13 for the sum

$$\frac{4}{\pi} \sum_{n=1}^{N} \frac{(-1)^n}{2n-1} \cos(2n-1)\pi(x-1).$$

2

FOURIER SERIES

§2.1 Introduction

Fourier died without having been able to prove that an arbitrary function can be represented by a convergent series of trigonometric functions. Actually, when his prize essay of 1811 was found to be faulty on these grounds there was no workable definition of convergence. Surely the concept did exist. We have only to recall that about 1360 Nicole Oresme, Bishop of Lisieux (c.1323–1382), had shown the divergence of the harmonic series. Even the words *convergent* and *divergent* were available, introduced by James Gregory (1638–1675) in 1668. But vague ideas and words can get one only so far in mathematical proofs. Mathematics deals with quantities and comparisons between quantities, with equalities and inequalities. What was needed was a definition of convergence involving comparisons between the partial sums of a series and its proposed sum, such comparisons to be established by means of inequalities.

As it happened, one of the first workable definitions of convergence was given by Fourier himself in his prize essay of 1811, later incorporated into his book of 1822. He stated that to establish the convergence of a series

> it is necessary that the values at which we arrive on increasing continually the number of terms, should approach more and more a fixed limit, and should differ from it only by a quantity which becomes less than any given magnitude: this limit is the value of the series.[1]

The use of inequalities is already implicit in his *less than any given magnitude*.

More precise, and definitely more influential, was the definition of convergence given by Augustin Louis Cauchy (1789–1857), often referred to as the father of modern analysis. And indeed he was, for he was the first to understand the importance of rigor

---

[1] See the English translation of his book by A. Freeman, *The Analytical Theory of Heat*, Dover, New York, 1955, pages 196–197.

in analysis and the first to consistently use the language of inequalities, starting with his definitions of limit and continuity—to him we owe the use of the letter ϵ to denote the *error* of the approximation between a variable quantity and its limit. Cauchy was well aware of the work of his predecessors and contemporaries, and in particular of Fourier's work, but he was not given to quote any of his sources of inspiration. We do not know whether or not Fourier's earlier definition of convergence was instrumental in shaping Cauchy's own ideas. What we know is that, once in possession of a rigorous definition of limit, Cauchy published what is essentially the modern definition of convergence in his 1821 textbook *Cours d'analyse de l'École Royale Polytechnique.* It was later

AUGUSTIN LOUIS CAUCHY IN 1821
Portrait by Boilly.
From D. E. Smith, *Portraits of Eminent Mathematicians,*
Portfolio II, Scripta Mathematica, New York, 1938.

written in fully modern terminology by Eduard Heine in 1872, and this is the version reproduced in Appendix A. But what is more remarkable about Cauchy's contribution is that he did not limit himself to stating the definition of convergence, but immediately, one page later, he started giving theorems containing tests for convergence: the so-called Cauchy criterion—already anticipated by Bolzano in 1817 along with the definition of continuity—and the root and ratio tests. Further tests were given later by several other mathematicians, among them Raabe, Abel, Dirichlet, and Weierstrass, thereby building a solid theory of the convergence of series that could be applied in many practical situations. A proof of the convergence of Fourier's trigonometric series was attempted by Poisson in 1820, by Cauchy in 1823, and, of course, by Fourier himself throughout his life. He never succeeded, but one of his sketches for a proof[1] would be of value to the man who finally did.

In the summer of 1822 a Prussian teenager, Johann Peter Gustav Lejeune Dirichlet (1805–1859), from Düren in the *Rheinprovinz*, came to Paris, attracted by the presence there of so many luminaries in the field, to study mathematics. He became acquainted with Fourier, who encouraged him to work on trigonometric series and showed him his sketch of the convergence proof. It would be some time, however, before Dirichlet could complete a valid proof, his main interest being outside the field of applied mathematics. He left Paris in 1827 and the following year started teaching mathematics at the General Military School in Berlin. Then, in 1829 he published a paper entitled *Sur la convergence des séries trigonométriques qui servent a représenter une fonction arbitraire entre des limites données*. Here, after replacing a certain trigonometric identity in Fourier's sketch of proof with one of his own—see Lemma 2.1—he suceeded in giving sufficient conditions for the convergence of Fourier's series, namely, that f have a finite number of jump discontinuities and a finite number of maxima and minima.

In this chapter we shall give sufficient conditions for the convergence of these trigonometric series and present some related results.

§2.2 Fourier Series

The main question is whether or not a given function f defined on an arbitrary interval $(\alpha - a, \alpha + a), a > 0$, can be represented by the sum of a convergent series of trigonometric functions. Actually, we can reduce this to a somewhat simpler problem: given a function $f : (-\pi, \pi) \to \mathbb{R}$, is it possible to determine coefficients a_n and b_n such that

$$(1) \qquad f(x) = \tfrac{1}{2}a_0 + \sum_{n=1}^{\infty} (a_n \cos nx + b_n \sin nx)$$

for every x in $(-\pi, \pi)$? The reason why a_0 is multiplied by $1/2$ is purely technical and will be seen below. As for the restriction to the interval $(-\pi, \pi)$, motivated by the fact that the trigonometric functions are periodic of period 2π, it is actually not significant. It will be dealt with in §2.5 by introducing a new variable

$$y = \frac{\pi}{a}(x - \alpha)$$

that varies from $-\pi$ to π as x varies from $\alpha - a$ to $\alpha + a$.

The determination of the coefficients a_n and b_n is a simple matter if certain assumptions are made. Assume that the series in (1) converges and that equality holds—all this, of course, remains to be settled—and assume further that the series can be integrated term by term from $-\pi$ to π and that the same is true when both sides of (1) are multiplied

---

[1] *The Analytical Theory of Heat*, pages 438–440.

by either $\cos mx$ or by $\sin mx$, where m a nonnegative integer. Then, if we multiply (1) by $\cos mx$, we get

$$\int_{-\pi}^{\pi} f(x)\cos mx \, dx = \tfrac{1}{2}a_0 \int_{-\pi}^{\pi} \cos mx \, dx$$

$$+ \sum_{n=1}^{\infty}\left[a_n \int_{-\pi}^{\pi} \cos nx \cos mx \, dx + b_n \int_{-\pi}^{\pi} \sin nx \cos mx \, dx \right].$$

If we recall the well-known formulas

$$\int_{-\pi}^{\pi} \cos mx \, dx = \begin{cases} 2\pi & \text{if } m = 0 \\ 0 & \text{if } m \neq 0 \end{cases}$$

$$\int_{-\pi}^{\pi} \cos nx \cos mx \, dx = \begin{cases} \pi & \text{if } m = n \\ 0 & \text{if } m \neq n \end{cases}$$

$$\int_{-\pi}^{\pi} \sin nx \cos mx \, dx = 0,$$

we obtain

$$a_m = \frac{1}{\pi}\int_{-\pi}^{\pi} f(x)\cos mx \, dx, \qquad m = 0, 1, 2, \dots.$$

The reason for the coefficient $1/2$ in (1) is now apparent: without it the previous formula would not be valid for $m = 0$ and a separate formula would be needed. A similar procedure, but multiplying (1) by $\sin mx$, yields

$$b_m = \frac{1}{\pi}\int_{-\pi}^{\pi} f(x)\sin mx \, dx, \qquad m = 1, 2, 3, \dots.$$

This simple method was devised by Leonhard Euler (1707–1783) in 1777, but only after having first obtained these formulas in a more complicated manner. Fourier arrived at the same result independently, also discovering the simple method after a much more complicated one.

There is no reason to believe at this moment that the assumptions made above on the convergence of the series (1) and on the term by term integration are valid. But, whether or not they are—this will be decided later on—we take the resulting formulas as our starting point.

Definition 2.1 *If the following integrals exist, the numbers*

(2) $$a_n = \frac{1}{\pi}\int_{-\pi}^{\pi} f(x)\cos nx \, dx, \qquad n = 0, 1, 2, \dots$$

and

(3) $$b_n = \frac{1}{\pi}\int_{-\pi}^{\pi} f(x)\sin nx \, dx, \qquad n = 1, 2, 3, \dots$$

are called the **Fourier coefficients** *of* f *on* $(-\pi, \pi)$. *The series*

$$\tfrac{1}{2}a_0 + \sum_{n=1}^{\infty} (a_n \cos nx + b_n \sin nx),$$

where a_n *and* b_n *are given by* (2) *and* (3), *is called the* **Fourier series** *of* f *on* $(-\pi, \pi)$.

Since we do not know yet if and when the equality in (1) is valid, we shall use the symbol $\sim$ with the meaning *has Fourier series* and write

$$f(x) \sim \tfrac{1}{2}a_0 + \sum_{n=1}^{\infty} (a_n \cos nx + b_n \sin nx).$$

Example 2.1 We shall find the Fourier series on $(-\pi, \pi)$ of the function defined by

$$f(x) = \begin{cases} 0 & \text{if} \quad -\pi < x < -\pi/2 \\ 1 & \text{if} \quad -\pi/2 < x < \pi, \end{cases}$$

whose graph is represented in Figure 2.1.

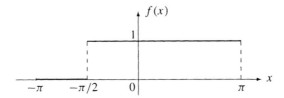

Figure 2.1

According to (2) and (3) we have

$$a_0 = \frac{1}{\pi} \int_{-\pi/2}^{\pi} dx = \frac{3}{2}$$

and, for $n > 0$,

$$a_n = \frac{1}{\pi} \int_{-\pi/2}^{\pi} \cos nx\, dx = \frac{\sin nx}{\pi n}\Big|_{-\pi/2}^{\pi} = \frac{\sin \dfrac{n\pi}{2}}{\pi n} = \begin{cases} \dfrac{(-1)^{\frac{n-1}{2}}}{\pi n} & \text{if } n \text{ is odd} \\ 0 & \text{if } n \text{ is even} \end{cases}$$

and

$$b_n = \frac{1}{\pi} \int_{-\pi/2}^{\pi} \sin nx\, dx = \frac{-\cos nx}{\pi n}\Big|_{-\pi/2}^{\pi} = \frac{-\cos n\pi}{\pi n} + \frac{\cos \dfrac{n\pi}{2}}{\pi n}$$

$$= \begin{cases} \dfrac{1}{\pi n} & \text{if } n \text{ is odd} \\ -\dfrac{1}{\pi n}\left[1 - (-1)^{n/2}\right] & \text{if } n \text{ is even}. \end{cases}$$

Hence,

$$f(x) \sim \frac{3}{4} + \frac{1}{\pi}\left(\cos x + \sin x - \sin 2x - \frac{1}{3}\cos 3x + \frac{1}{3}\sin 3x + \frac{1}{5}\cos 5x \right.$$

$$\left. + \frac{1}{5}\sin 5x - \frac{1}{3}\sin 6x - \cdots \right)$$

$$= \frac{3}{4} + \frac{1}{\pi}\sum_{n=1}^{\infty}\frac{1}{2n-1}[(-1)^{n-1}\cos(2n-1)x + \sin(2n-1)x - \sin(4n-2)x].$$

Sometimes a careful observation of the function under consideration can greatly simplify the computation of its Fourier coefficients. A function f such that $f(-x) = f(x)$ for every x in $(-\pi, \pi)$ is called *even*, and its graph is symmetric about the y-axis as shown on the left in Figure 2.2. If $f(-x) = -f(x)$, the function is called *odd*, and

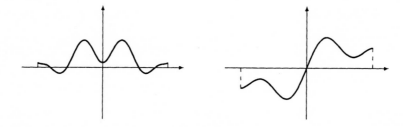

Figure 2.2

its graph is symmetric with respect to the origin as shown on the right in Figure 2.2. It should be clear from these graphs that the integral of an even function from $-\pi$ to π is equal to twice the value of its integral from 0 to π while the integral of an odd function from $-\pi$ to π is zero. Since the cosine is even and the sine is odd, and since these facts remain true upon multiplication of these by an even function, but they are reversed upon multiplication by an odd function, the following result has been established.

Proposition *If $f: (-\pi, \pi) \to \mathbb{R}$ is such that its Fourier coefficients are defined, that is, if the integrals in (2) and (3) exist, then*
(i) *if f is even,*

$$f(x) \sim \tfrac{1}{2}a_0 + \sum_{n=1}^{\infty} a_n \cos nx,$$

where

$$a_n = \frac{2}{\pi}\int_0^{\pi} f(x)\cos nx\,dx. \qquad n = 0, 1, 2, \ldots,$$

and

(ii) *if f is odd,*

$$f(x) \sim \sum_{n=1}^{\infty} b_n \sin nx,$$

where

$$b_n = \frac{2}{\pi} \int_0^{\pi} f(x) \sin nx \, dx. \qquad n = 1, 2, 3, \ldots .$$

Example 2.2 The function defined by $f(x) = x$ on $(-\pi, \pi)$ is odd since $f(-x) = -x = -f(x)$, and its Fourier series contains only sine terms. The coefficients are

$$b_n = \frac{2}{\pi} \int_0^{\pi} x \sin nx \, dx$$

$$= \frac{2}{\pi} \left[-x \frac{\cos nx}{n} \Big|_0^{\pi} + \int_0^{\pi} \frac{\cos nx}{n} \, dx \right]$$

$$= -2 \frac{\cos n\pi}{n}$$

$$= 2 \frac{(-1)^{n+1}}{n}.$$

Then

$$f(x) \sim \sum_{n=1}^{\infty} 2 \frac{(-1)^{n+1}}{n} \sin nx.$$

Compare this result with the series in Example 1.6 and observe how the sum of the series approaches the given function as the number of terms increases.

Example 2.3 The function defined by $f(x) = e^{|x|}$ on $(-\pi, \pi)$ is even, and we have

$$a_0 = \frac{2}{\pi} \int_0^{\pi} e^x \, dx = \frac{2}{\pi} (e^{\pi} - 1),$$

and

$$a_n = \frac{2}{\pi} \int_0^{\pi} e^x \cos nx \, dx = \frac{2}{\pi} \frac{e^x}{1 + n^2} (n \sin nx + \cos nx) \Big|_0^{\pi} = \frac{2}{\pi} \frac{e^{\pi} \cos n\pi - 1}{1 + n^2}.$$

Then

$$f(x) \sim \frac{1}{\pi} (e^{\pi} - 1) + \frac{2}{\pi} \sum_{n=1}^{\infty} \frac{(-1)^n e^{\pi} - 1}{1 + n^2} \cos nx.$$

§2.3   The Riemann-Lebesgue Theorem

Since the Fourier coefficients are defined by integrals, a brief examination of the concept of integral is in order before we discuss the convergence of Fourier series in §2.4. Moreover, once the concept of integral is established on a firm foundation, we shall be able to present a result—obtained later by a former student of Dirichlet—that will allow us to give a shorter proof of a sharper convergence theorem than the original one by Dirichlet. This result, the so-called Riemann-Lebesgue theorem, is a very powerful tool that we shall use frequently.

The popular concept of integral in the eighteenth century was that of prederivative or indefinite integral. It was sufficient for all their applications and easy to explain. Of course, Leibniz had defined the integral much earlier as a sum and even introduced the integral sign $\int$ as an elongated S for sum. However, his idea did not quite catch at that time. How could it, involving, as it does, the sum of infinitely many infinitely small quantities? Thus we come to the nineteenth century with the notion of integral as prederivative reigning supreme. Fourier changed that. As we have seen, he was used to handling functions not given by analytic expressions, but by curves and pieces of curves, and found the use of prederivatives to be impractical. Instead he remarked that, for instance, the constant b_n can be viewed as the area under the graph of

$$\frac{1}{\pi} f(x) \sin nx$$

from $-\pi$ to π, and this whether or not f is continuous.[1] It may have been responding to this interpretation of the integral as an area that Cauchy gave the following definition of the definite integral in his text *Resumé des leçons donnés à l'École Royale Polytechnique sur le calcul infinitésimal* of 1823, which we reproduce in the current, more usual notation. If f is a continuous real-valued function on an interval $[a, b]$ and if $x_0, x_1, \ldots, x_n$ are points such that $a = x_0 < x_1 < \cdots < x_n = b$, then

$$\int_a^b f = \lim_{n \to \infty} \sum_{i=1}^n f(x_{i-1})(x_i - x_{i-1}),$$

where it is assumed that, for each i, the distance between x_{i-1} and x_i approaches zero as $n \to \infty$. Cauchy was then able to prove—not quite rigorously because he lacked the concept of uniform continuity—the existence of this limit and its independence of the mode of division of $[a, b]$ into subintervals. Incidentally, the notation $\int_a^b f$ for the definite integral, which Cauchy adopted, is due to Fourier.[2] Cauchy's definition is quite adequate to prove Dirichlet's convergence theorem for Fourier series. It is true that the functions in this theorem may have a finite number of discontinuities—as we shall see in §2.4—but this is no obstacle because the given interval can be subdivided into a finite number of subintervals where the function is continuous, and then the integrals over these

---

[1] *The Analytical Theory of Heat*, page 186.
[2] *The Analytical Theory of Heat*, page 463.

subintervals added together. But Dirichlet knew that he had introduced the requirement of continuity—except a finite number of points—precisely so that the definite integrals defining the Fourier coefficients exist. Consequently, in order to generalize his theorem to functions with an infinite number of discontinuities—which he chose as his next task—he needed only to make sure that they could be integrated. That is, what Dirichlet needed is what Cauchy's definition did not provide, namely, a condition for integrability. Cauchy had defined the integral for continuous functions, but maybe other functions can be integrated as well. Dirichlet never achieved his goal of integrating functions with an infinite number of discontinuities nor was able to give a condition for integrability. But during the years 1847–1849, he had the good fortune of counting a very gifted young man among his students at the University of Berlin. Georg Friedrich Bernhard Riemann (1826–1866) had transferred from the Univeristy of Göttingen to Berlin, switching from

BERNHARD RIEMANN IN 1863
Photograph by the author from Riemann's
Gesammelte Mathematische Werke of 1892.

the study of theology to mathematics, and the University of Berlin had a better reputation in this field at that time. Here, Dirichlet was his favorite teacher and was instrumental in shaping some of Riemann's research interests. In particular, those on Fourier series and the definite integral. Riemann returned to Göttingen and received his doctorate in 1851. In December of 1853, wishing to qualify for a position without pay at the university—as *Privatdozent* he would receive a fee from any students his lectures might attract, but no university stipend—he wrote an *Habilitationsschrift*, which at Dirichlet's suggestion was *Über die Darstellbarkeit einer Function durch eine trigonometrische Reihe*. One of Riemann's goals in this work—printed posthumously in 1867—was to give conditions for integrability that would accomodate discontinuous functions, even if this happens at infinitely many points. To this end he modified Cauchy's definition as follows. First he replaced the factor $f(x_{i-1})$ by $f(t_i)$, where t_i is any point in the subinterval from x_{i-1} to x_i. Then—and this is more important—he removed the continuity requirement on f. Instead, he turned things around and *defined* f to be integrable if the limit

$$\lim_{n \to \infty} \sum_{i=1}^{n} f(t_i)(x_i - x_{i-1})$$

exists, provided that the distance between x_{i-1} and x_i approaches zero as $n \to \infty$. Next he stated a theorem giving necessary and sufficient conditions for the integral to exist and, to show the wide applicability of his definition, he gave an example of an integrable function that has infinitely many discontinuities on every arbitrarily small interval.[1]

Riemann's definition was perfected in 1875 by Jean Gaston Darboux (1842–1917), the man responsible for the dissemination of Riemann's ideas in France, who also proved the corresponding version of the fundamental theorem of calculus. In modern terminology, this definition, which we shall need later, can be introduced as follows. Let x_0, $x_1, \ldots, x_n$ be points in $[a, b]$ such that $a = x_0 < x_1 < \cdots < x_n = b$. Such a collection of points is called a *partition* of $[a, b]$. Then, for $1 \le i \le n$, define M_i to be the smallest real number that is not smaller than any value of f on (x_{i-1}, x_i). Of course, if f has a maximum on (x_{i-1}, x_i), then M_i is this maximum value. Similarly, let m_i be the largest real number that is not larger than any value of f on (x_{i-1}, x_i).

Definition 2.2 *If $f : (a, b) \to \mathbb{R}$ is a bounded function, then it is said to be **Riemann integrable**, or simply **integrable**, on (a, b) if and only if there is a unique number I such that*

$$\sum_{i=1}^{n} m_i (x_i - x_{i-1}) \le I \le \sum_{i=1}^{n} M_i (x_i - x_{i-1})$$

for every partition $a = x_0 < x_1 < \cdots < x_n = b$ of $[a, b]$. This number I is called the **Riemann integral** *of f from a to b and is denoted by $\int_a^b f$.*

In 1875 Darboux proved that a function is integrable in the sense of Definition 2.2 if and only if it is integrable in Riemann's sense, as explained above. Thus, the properties of this integral are those already known from calculus, and we shall freely use them without further comment.

What is not easy to determine through direct application of Definition 2.2 is whether or not a given bounded function is Riemann integrable. A simple criterion for integrability was given by Lebesgue in 1904 on the basis of the following concept: a subset

---

[1] See, for instance, I. N. Pesin, *Classical and Modern Theories of Integration*, Academic Press, New York, 1970, page 8.

of $\mathbb{R}$ is said to have *measure zero* if and only if it can be enclosed in a finite or infinite sequence of open intervals whose combined total length—the sum of a finite or infinite series whose terms are the lengths of the individual intervals—is arbitrarily small, that is, smaller than any preassigned positive number. Then Lebesgue showed that f is Riemann integrable on (a, b) if and only if the set of points where f is discontinuous has measure zero.[1]

It is clear, in particular, that a bounded continuous function is integrable on (a, b). A type of integrable function with discontinuities that is frequently encountered in the applications is defined as follows.

Definition 2.3 *A function* f *is* **piecewise continuous** *on* (a, b) *if and only if there is a partition* $a = x_0 < x_1 < \cdots < x_n = b$ *of* $[a, b]$ *such that* f *is continuous on each interval* (x_{i-1}, x_i) *for* $1 \le i \le n$, *and, for each such* i, *the limits*

$$\lim_{\substack{x \to x_{i-1} \\ x > x_{i-1}}} f(x) \qquad \text{and} \qquad \lim_{\substack{x \to x_i \\ x < x_i}} f(x)$$

exist and are finite. These limits are denoted by $f(x_{i-1}+)$ *and* $f(x_i-)$, *respectively.*

Note that the existence of the one-sided limits $f(x_{i-1}+)$ and $f(x_i-)$ implies that f can be extended as a continuous function to each closed interval $[x_{i-1}, x_i]$. Thus, f is bounded, and, since it has only a finite number of discontinuities, it is integrable by Lebesgue's theorem—in this simple particular case integrability can also be proved directly using Definition 2.2.

Another type of integrable function frequently used in applications is described in Exercise 3.1.

Example 2.4 The function $f : (0, 1) \to \mathbb{R}$ defined by

$$f(x) = \sin \frac{1}{x}$$

is bounded and continuous, and therefore integrable, on $(0, 1)$. But it is not piecewise continuous because $f(0+)$ does not exist.

Example 2.5 The function $f : (0, 1) \to \mathbb{R}$ defined by

$$f(x) = \begin{cases} 0 & \text{if} \quad 0 < x \le \frac{1}{4} \\ \frac{1}{2} & \text{if} \quad \frac{1}{4} < x \le \frac{1}{2} \\ 1 & \text{if} \quad \frac{1}{2} < x < 1 \end{cases}$$

is not continuous on $(0, 1)$ but it is piecewise continuous there.

---

[1] For a more complete treatment of the Riemann integral and its properties see, for instance, W. R. Parzynski and P. W. Zipse, *Introduction to Mathematical Analysis*, McGraw-Hill, New York, 1982. The theorems of Darboux and Lebesgue quoted here are on pages 163 and 182, respectively.

Example 2.6 The function $f : (0, 1) \to \mathbb{R}$ defined by

$$f(x) = \begin{cases} \dfrac{1}{2^n} & \text{if} \quad \dfrac{1}{2^{n+1}} < x \le \dfrac{1}{2^n} \quad \text{for} \quad n = 1, 2, \ldots \\[3mm] 1 & \text{if} \quad \frac{1}{2} < x < 1 \end{cases}$$

is neither continuous nor piecewise continuous on $(0, 1)$ because it is discontinuous at infinitely many points. However, f is integrable on $(0, 1)$. To see this, let ϵ be an arbitrary positive number smaller than $1/2$ and for every positive integer n let I_n be the open interval of length $\epsilon/2^n$ centered at $1/2^n$. Then all the points at which f is discontinuous are contained in the sequence of intervals $I_1, I_2, I_3, \ldots$ whose combined total length does not exceed

$$\sum_{n=1}^{\infty} \frac{\epsilon}{2^n} = \epsilon \sum_{n=1}^{\infty} \left(\frac{1}{2}\right)^n.$$

Since the sum of the geometric series on the right is known to be 1, this total length is ϵ, an arbitrarily small positive number.

Riemann's 1854 paper also contains, as an application of his new definition of integral, a very powerful result that will serve us well later on. In its simplest form it says that the Fourier coefficients a_n and b_n approach zero as $n \to \infty$. This should be clear from the graphical representations in Figure 2.3 and the interpretation of integral as area. But it should be equally clear that the interval $(-\pi, \pi)$ and the positive integer n can be replaced by an arbitrary interval (a, b) and an arbitrary positive real number r. Moreover, in 1903 Lebesgue showed that the result is still valid using a broader and more powerful definition of integral that he had introduced himself the previous year (see §2.9). This gives the theorem its present name.

Theorem 2.1 (**The Riemann-Lebesgue theorem**) *If f is integrable on (a, b), then*

$$\lim_{r \to \infty} \int_a^b f(x) \sin rx \, dx = \lim_{r \to \infty} \int_a^b f(x) \cos rx \, dx = 0.$$

Proof. Let $a = x_0 < x_1 < \cdots < x_n = b$ be a partition of $[a, b]$ and, if m_i is the largest real number that is no larger than any value of f on (x_{i-1}, x_i)—such a number is called the *infimum* of f on (x_{i-1}, x_i)—define $g : (a, b) \to \mathbb{R}$ by $g(x) = m_i$ for x in (x_{i-1}, x_i) and $g(x_i) = f(x_i)$, $i = 1, \ldots, n$. Clearly, $g \le f$ on (a, b). We have

$$\left| \int_a^b f(x) \sin rx \, dx - \int_a^b g(x) \sin rx \, dx \right| \le \int_a^b |f(x) \sin rx - g(x) \sin rx| \, dx$$

$$\le \int_a^b [f(x) - g(x)] \, dx$$

$$= \int_a^b f(x) \, dx - \sum_{i=1}^{n} m_i (x_i - x_{i-1}).$$

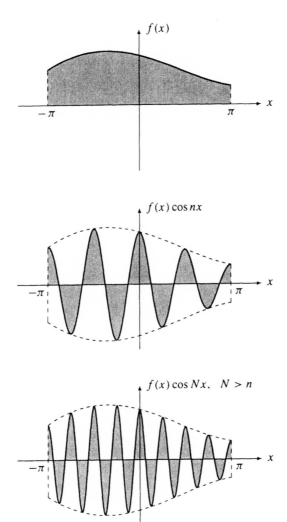

Figure 2.3

By Definition 2.2, given $\epsilon > 0$ the partition can be chosen so fine that the right-hand side is smaller than $\epsilon/2$ and then, using the inequality $|x| \leq |y| + |x - y|$, valid for any real numbers x and y,

$$\left| \int_a^b f(x) \sin rx \, dx \right| < \left| \int_a^b g(x) \sin rx \, dx \right| + \frac{\epsilon}{2}.$$

But, for $r > 0$,

$$\left| \int_a^b g(x) \sin rx \, dx \right| = \left| \sum_{i=1}^n \int_{x_{i-1}}^{x_i} m_i \sin rx \, dx \right|$$

$$= \left| \sum_{i=1}^n \frac{m_i}{r} (\cos rx_{i-1} - \cos rx_i) \right|$$

$$\leq \frac{2}{r} \sum_{i=1}^n |m_i|.$$

For r large enough the right-hand side is smaller than $\epsilon/2$ and then

$$\left| \int_a^b f(x) \sin rx \, dx \right| < \frac{\epsilon}{2} + \frac{\epsilon}{2} = \epsilon$$

for such large r. Since ϵ is arbitrarily small the first limit follows, and the second is obtained in an analogous manner, Q.E.D.

§2.4 The Convergence of Fourier Series

Back in 1829 Dirichlet had proved that the Fourier series of a bounded function f that has a finite number of jump discontinuities and a finite number of maxima and minima on $(-\pi, \pi)$ converges to $f(x)$ at every point x at which f is continuous. In 1854 Riemann discovered that the convergence of such a Fourier series depends exclusively on the values of f on an arbitrarily small interval centered at x, a fact usually referred to as *Riemann's localization principle*. We shall combine these two results into a single convergence theorem, and the use of the Riemann-Lebesgue theorem will allow us to give a shorter and far more palatable proof than the original one by Dirichlet.

The first step is to rewrite an arbitrary partial sum of its Fourier series, that is,

$$(4) \qquad s_N(x) = \tfrac{1}{2} a_0 + \sum_{n=1}^N (a_n \cos nx + b_n \sin nx),$$

where $N \geq 1$ is arbitrary, as an integral, whose limit is later found as $N \to \infty$. Using the definition of Fourier coefficients, (4) becomes

$$s_N(x) = \tfrac{1}{2} a_0 + \sum_{n=1}^N (a_n \cos nx + b_n \sin nx)$$

$$= \frac{1}{2\pi} \int_{-\pi}^{\pi} f(t) \, dt + \frac{1}{\pi} \sum_{n=1}^N \left[\cos nx \int_{-\pi}^{\pi} f(t) \cos nt \, dt + \sin nx \int_{-\pi}^{\pi} f(t) \sin nt \, dt \right]$$

$$= \frac{1}{2\pi} \int_{-\pi}^{\pi} f(t) \left[1 + 2 \sum_{n=1}^N \cos n(t - x) \right] dt.$$

Both Fourier and Dirichlet proceeded in this manner, and then each chose a certain trigonometric identity to compute the sum in the square brackets on the right. But Dirichlet's choice of trigonometric identity, namely,

$$2 \cos nt \sin \tfrac{1}{2}t = \sin \left(n + \tfrac{1}{2}\right)t - \sin \left(n - \tfrac{1}{2}\right)t,$$

resulted in a better simplification of the integrand above, making it easier—in fact

J. P. G. L. DIRICHLET IN 1853
Portrait by Schrader.
Photograph by the author from Dirichlet's
Mathematische Werke of 1889.

possible—to find its limit as $N \to \infty$. Adding the equations that result from substituting $n = 1, \dots, N$ in the previous identity we obtain

$$\left(2 \sin \tfrac{1}{2}t\right) \sum_{n=1}^{N} \cos nt = - \sin \tfrac{1}{2}t + \sin \left(N + \tfrac{1}{2}\right)t.$$

Hence, if $-\pi \le t \le \pi$ and $t \ne 0$, we have

(5)
$$1 + 2 \sum_{n=1}^{N} \cos nt = \frac{\sin \left(N + \tfrac{1}{2}\right)t}{\sin \tfrac{1}{2}t},$$

and as $t \to 0$ the right-hand side has the limit $2N + 1$, which is precisely the value of the left-hand side for $t = 0$. Then the right-hand side is piecewise continuous and, since the previous identity is equally valid with $t - x$ in place of t, we obtain

(6)
$$s_N(x) = \frac{1}{2\pi} \int_{-\pi}^{\pi} f(t) \frac{\sin\left(N + \frac{1}{2}\right)(t - x)}{\sin\frac{1}{2}(t - x)} \, dt.$$

This is the first result obtained by Dirichlet. We can simplify the notation by defining

$$D_N(t) = \frac{\sin\left(N + \frac{1}{2}\right)t}{\sin\frac{1}{2}t},$$

which is called the *Dirichlet kernel*. If we use it to rewrite (5) and (6), and if we observe that integrating the left-hand side of (5) from zero to π yields the value π, then we can sum up the results obtained so far as follows.

Lemma 2.1 *Let f be an integrable function on $(-\pi, \pi)$; for $N \geq 1$, let $s_N(x)$ be the Nth partial sum of its Fourier series on $(-\pi, \pi)$, as in (4); and let D_N be the Dirichlet kernel as defined above. Then*

(7)
$$\int_0^\pi D_N(t) \, dt = \pi$$

and

(8)
$$s_N(x) = \frac{1}{2\pi} \int_{-\pi}^{\pi} f(t) D_N(t - x) \, dt.$$

Since the statement of this lemma may look a bit technical and unexciting, we shall illustrate its importance and that of the Dirichlet kernel by some intuitive remarks, which actually form the basis of the convergence proof. For N large enough the graph of $D_N(t - x)$ versus t is represented in Figure 2.4. Its zeros are at the points $t = x + k\delta$, where k is a nonzero integer and

$$\delta = \frac{2\pi}{2N + 1}.$$

For very large N we can think of the main arch in the graph of $D_N(t - x)$ as a quasi-triangular window, resting on the interval $(x - \delta, x + \delta)$, through which we can see a very narrow portion of the graph of f, also represented in Figure 2.4 for a continuous function. As x varies the window moves, revealing only values of f very close to $f(x)$. When the product $f(t)D_N(t - x)$ is integrated from $-\pi$ to π, as required by Equation (8), the representation in the figure suggests that we can neglect the contribution to the integral from outside the interval $(x - \delta, x + \delta)$. Inside this interval $f(t) \approx f(x)$, while the area under the main arch in the graph of $D_N(t - x)$ is approximately $\delta(2N + 1) = 2\pi$. Then (8) implies that

$$s_N(x) \approx \frac{1}{2\pi} \int_{x-\delta}^{x+\delta} f(t) D_N(t - x) \, dt \approx \frac{f(x)}{2\pi} \int_{x-\delta}^{x+\delta} D_N(t - x) \, dt \approx f(x)$$

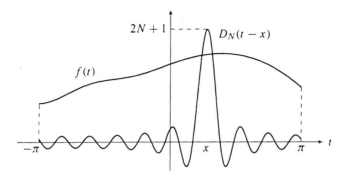

Figure 2.4

for N very large. Hence, it is reasonable to expect that $s_N(x) \to f(x)$ as $N \to \infty$, as is our wish to prove. Moreover, this approximation suggests that, if the right-hand side has a limit as $N \to \infty$, then both this limit and the convergence of $s_N(x)$ at a given point x depend exclusively on the values of f on the interval $(x - \delta, x + \delta)$, which becomes arbitrarily small as $N \to \infty$.

It will be easier to state and prove the convergence theorem if f is extended beyond the interval $(-\pi, \pi)$ as a periodic function of period 2π. This will simplify the computations and, of course, it does not alter its Fourier series on $(-\pi, \pi)$ since the coefficients are still evaluated as integrals from $-\pi$ to π. In fact, if f is extended as stated above, making the change $u = x - t$ in (8) and noticing that D_N is an even function, we obtain

$$s_N(x) = -\frac{1}{2\pi} \int_{x+\pi}^{x-\pi} f(x-u) D_N(-u)\, du = \frac{1}{2\pi} \int_{x-\pi}^{x+\pi} f(x-u) D_N(u)\, du.$$

Since the integrand is now periodic of period 2π, the integral has the same value over any other interval of length 2π, and we have

$$s_N(x) = \frac{1}{2\pi} \int_{-\pi}^{\pi} f(x-u) D_N(u)\, du.$$

In a similar way, but making the change $u = t - x$, we would obtain

$$s_N(x) = \frac{1}{2\pi} \int_{-\pi}^{\pi} f(x+u) D_N(u)\, du.$$

Adding the last two equations, observing that the resulting integrand $[f(x + u) + f(x - u)]D_N(u)$ is an even function of u and replacing u by t, we arrive at

$$(9) \qquad s_N(x) = \frac{1}{2\pi} \int_{0}^{\pi} [f(x+t) + f(x-t)] D_N(t)\, dt.$$

Before we state the convergence theorem we need to impose a differentiability requirement on the function f. To this end we recall the following concept.

Definition 2.4 *Let* $f: (a, b) \to \mathbb{R}$ *be a piecewise continuous function and let* x *be in* (a, b). *Then the* **right-hand derivative** *of* f *at* x *is*

$$\lim_{\substack{h \to 0 \\ h > 0}} \frac{f(x + h) - f(x+)}{h}$$

if this limit exists. Similarly, the **left-hand derivative** *of* f *at* x *is*

$$\lim_{\substack{h \to 0 \\ h < 0}} \frac{f(x + h) - f(x-)}{h}$$

if this limit exists.

Note that the second limit in this definition can also be written as

$$-\lim_{\substack{h \to 0 \\ h > 0}} \frac{f(x - h) - f(x-)}{h},$$

a fact that we shall use in the proof of Theorem 2.2, and also that f is differentiable at x if and only if it is continuous at x and both its right-hand and left-hand derivatives at x exist and are equal.[1]

Example 2.7 Let $f: (-1, 1) \to \mathbb{R}$ be defined by

$$f(x) = \begin{cases} \sqrt{x + 1} & \text{if} \quad -1 < x \leq 0 \\ x^3 & \text{if} \quad 0 < x < 1. \end{cases}$$

This function is piecewise continuous, and its right-hand and left-hand derivatives at $x = 0$ are

$$\lim_{\substack{h \to 0 \\ h > 0}} \frac{h^3 - 0}{h} = 0$$

and

$$\lim_{\substack{h \to 0 \\ h < 0}} \frac{\sqrt{h + 1} - \sqrt{1}}{h} = \lim_{\substack{h \to 0 \\ h < 0}} \frac{h}{h(\sqrt{h + 1} + \sqrt{1})} = \frac{1}{2},$$

respectively. Note that f is neither continuous nor differentiable at $x = 0$.

It is not always easy to use Definition 2.4 to determine whether or not a given function has one-sided derivatives at every point of an interval. The following result provides an easy to use sufficient condition.

Lemma 2.2 *If* f *and* f' *are both piecewise continuous on a closed interval* $[a, b]$, *then* f *has one-sided derivatives at every point* x *in* $[a, b]$—*only one, of course, at each endpoint*—*and they are equal to the one-sided limits of* f' *at* x.

---

[1] See, for instance, Parzynski and Zipse, *op. cit.*, pages 146–147.

Proof. If $x \neq b$ and if f is not continuous at x, define—or redefine—$f(x) = f(x+)$. Then there is a number $h > 0$ so small that f is continuous on $[x, x + h]$ and f' is continuous on $(x, x + h)$. By the mean value theorem of the differential calculus, there is a point c between x and $x + h$ such that

$$\frac{f(x + h) - f(x+)}{h} = f'(c).$$

As $h \to 0$ we see that $c \to x$ and $f'(c) \to f'(x+)$. Hence, the previous equation shows that f has a right-hand derivative at x and that its value is $f'(x+)$. Similarly, if $x \neq a$, then f has a left-hand derivative at x and it is equal to $f'(x-)$, Q.E.D.

It should be remarked that the sufficient condition in Lemma 2.2 is not a necessary condition.

Example 2.8 The function defined by

$$f(x) = \begin{cases} x^2 \sin \dfrac{1}{x} & \text{if} \quad x \neq 0 \\ 0 & \text{if} \quad x = 0 \end{cases}$$

is differentiable, and therefore has one-sided derivatives, at every x. But, if $x \neq 0$,

$$f'(x) = 2x \sin \frac{1}{x} - \cos \frac{1}{x}$$

has no limit as $x \to 0$; that is, f' is not piecewise continuous.

We are now ready to state and prove the convergence theorem of Dirichlet and Riemann.

Theorem 2.2 *Let $f : \mathbb{R} \to \mathbb{R}$ be a periodic function of period 2π that is integrable on $(-\pi, \pi)$. Then its Fourier series on $(-\pi, \pi)$ converges to the value*

$$\tfrac{1}{2}[f(x+) + f(x-)]$$

at every point x in $\mathbb{R}$ such that f has right-hand and left-hand derivatives at x and such that f is piecewise continuous on an arbitrarily small interval centered at x. If, in addition, f is continuous at x, then its Fourier series converges to $f(x)$.

Proof. The last statement is obvious. Now, if $\delta < \pi$ is a positive number so small that f is piecewise continuous on $(x - \delta, x + \delta)$, then using (9) and (7) we obtain

$$s_N(x) - \tfrac{1}{2}[f(x+) + f(x-)]$$

$$= \frac{1}{2\pi} \int_0^\pi [f(x + t) + f(x - t) - f(x+) - f(x-)] D_N(t) \, dt$$

$$= \frac{1}{\pi} \int_0^\delta \left[\frac{f(x + t) - f(x+)}{t} + \frac{f(x - t) - f(x-)}{t} \right] \frac{\frac{1}{2}t}{\sin \frac{1}{2}t} \sin \left(N + \tfrac{1}{2} \right) t \, dt$$

$$+ \frac{1}{2\pi} \int_\delta^\pi \frac{f(x + t) + f(x - t) - f(x+) - f(x-)}{\sin \frac{1}{2}t} \sin \left(N + \tfrac{1}{2} \right) t \, dt.$$

If f has right-hand and left-hand derivatives at x, each quotient in the brackets in the first of the last two integrals on the right has a limit as $t \to 0$ from the right. Then the expression in the brackets is a piecewise continuous function of t on $(0, \delta)$ because f is. The next factor is continuous on $(0, \delta)$, and it is well known that

$$\lim_{t \to 0} \frac{\frac{1}{2}t}{\sin \frac{1}{2}t} = 1.$$

Thus, this integrand is the product of a piecewise continuous, and therefore integrable, function by $\sin\left(N + \frac{1}{2}\right)t$. The quotient in the second integral is integrable on (δ, π) because both the numerator and the denominator are. Therefore, each integrand on the right-hand side is the product of an integrable function by $\sin\left(N + \frac{1}{2}\right)t$, and we conclude from the Riemann-Lebesgue theorem that each integral tends to zero as $N \to \infty$, Q.E.D.

The remarkable thing is that, while the Fourier coefficients that make up the series depend on the values of f over the whole interval $(-\pi, \pi)$, the convergence of the resulting Fourier series at a given point does not. It is a purely local property due to the fact that, for large N, the main arch in the graph of D_N is concentrated over a very small interval while the others are negligible.

Example 2.9 It was seen in Example 2.2 that the Fourier series of $f(x) = x$ on $(-\pi, \pi)$ is

$$\sum_{n=1}^{\infty} 2 \frac{(-1)^{n+1}}{n} \sin nx.$$

Since f satisfies the conditions of Theorem 2.2 at any point, this series has sum x on $(-\pi, \pi)$. This confirms the result suggested by Example 1.6.

Example 2.10 The periodic extension of period 2π of the function defined by $f(x) = e^{|x|}$ on $(-\pi, \pi)$ satisfies the conditions of Theorem 2.2 because both e^x and e^{-x} are differentiable everywhere, and then $e^{|x|}$ has a right-hand derivative at $-\pi$ and a left-hand derivative at π. Moreover, f is continuous on $\mathbb{R}$ since $f(-\pi+) = f(\pi-) = e^\pi$. Hence, the Fourier series found in Example 2.3 converges for any x in $\mathbb{R}$, and we have

$$e^{|x|} = \frac{1}{\pi}(e^\pi - 1) + \frac{2}{\pi} \sum_{n=1}^{\infty} \frac{(-1)^n e^\pi - 1}{1 + n^2} \cos nx$$

for $-\pi \leq x \leq \pi$.

Of course, if f is piecewise continuous on $(-\pi, \pi)$, it is integrable there, and we can state a simpler looking version of Theorem 2.2—one that does not highlight the local aspect of convergence but that is quite sufficient in most applications—as follows.

Theorem 2.3 *Let* $f : \mathbb{R} \to \mathbb{R}$ *be a periodic function of period* 2π *that is piecewise continuous on* $(-\pi, \pi)$. *Then its Fourier series on* $(-\pi, \pi)$ *converges to the value*

$$\tfrac{1}{2}[f(x+) + f(x-)]$$

at every point x in $\mathbb{R}$ *at which* f *has right-hand and left-hand derivatives. If, in addition,* f *is continuous at* x, *then its Fourier series converges to* $f(x)$.

★   There are several other theorems giving sufficient conditions for the convergence of a Fourier series at a point, and some are applicable when others fail. Most of them are beyond the scope of this book and will not be presented here.[1] We shall prove only one such theorem discovered in 1867 by Rudolph Otto Sigismund Lipschitz (1832–1903) of Bonn. It is actually a particular case of a result obtained in 1880 by Ulisse Dini (1845–1918) of Pisa (Exercise 4.9).

Theorem 2.4 (**The Lipschitz test**) *Let* $f : \mathbb{R} \to \mathbb{R}$ *be a periodic function of period* 2π *that is integrable on* $(-\pi, \pi)$. *Then its Fourier series on* $(-\pi, \pi)$ *converges to* $f(x)$ *at a given point* x *in* $\mathbb{R}$ *if* f *is continuous on an arbitrarily small interval centered at* x *and if there are positive numbers* M *and* δ *such that*

$$|f(x+t) - f(x)| \le M|t|$$

for t *in* $(-\delta, \delta)$.

Proof. Using (9) and (7) we obtain

$$s_N(x) - f(x) = \frac{1}{2\pi} \int_0^\pi [f(x+t) + f(x-t) - 2f(x)]D_N(t)\,dt.$$

If M and δ are as above and, in addition, δ is so small that f is continuous on $(x-\delta, x+\delta)$, then the function of t defined by

$$\frac{f(x+t) + f(x-t) - 2f(x)}{t}$$

is integrable on $(0, \delta)$ because it is continuous there and bounded by $\pm 2M$. Indeed,

$$|f(x+t) + f(x-t) - 2f(x)| \le |f(x+t) - f(x)| + |f(x-t) - f(x)| \le 2M|t|.$$

The quotient above is also integrable on (δ, π) because f is integrable and $t > \delta > 0$ on this interval, and therefore it is integrable on $(0, \pi)$. Since $t/\sin \frac{1}{2}t$ is also bounded and continuous, on $(0, \pi)$, the Riemann-Lebesgue theorem gives

$$\lim_{N\to\infty} \int_0^\pi \frac{f(x+t) + f(x-t) - 2f(x)}{t} \frac{t}{\sin \frac{1}{2}t} \sin\left(N + \tfrac{1}{2}\right)t\,dt = 0,$$

Q.E.D.

---

[1] See, for instance, A. Zygmund, *Trigonometric Series*, 2nd ed., Vols. I and II, Cambridge University Press, Cambridge, Massachusetts, 1988.

Example 2.11 The function $f : (-\pi, \pi) \to \mathbb{R}$ defined by

$$f(x) = \begin{cases} x \sin \dfrac{1}{x} & \text{if } x \neq 0 \\ 0 & \text{if } x = 0 \end{cases}$$

is bounded and continuous on $(-\pi, \pi)$, but does not have one-sided derivatives at the origin since

$$\frac{h \sin \dfrac{1}{h}}{h} = \sin \frac{1}{h}$$

has no limit as $h \to 0$ either for $h > 0$ or for $h < 0$. Thus, Theorem 2.2 is not applicable in this case at $x = 0$. But f satisfies the inequality in Theorem 2.4 at the origin with $M = 1$ because

$$|f(0 + t) - f(0)| = \left| t \sin \frac{1}{t} \right| \leq |t|.$$

Hence, the Fourier series of f on $(-\pi, \pi)$ converges to $f(0) = 0$ at $x = 0$.

We conclude this section by outlining some of the most relevant results on the pointwise convergence of Fourier series since the times of Dirichlet and Riemann. Even in the absence of proof, it was thought for some time that the continuity of f was a sufficient condition for the convergence of its Fourier series at all points. This conjecture was disproved in 1873, when Paul Du Bois-Reymond gave an example of a function bounded and continuous on $(-\pi, \pi)$ but whose Fourier series diverges at the origin. Then, three years later, he exhibited another continuous function such that in any neighborhood of an arbitrary point of its domain there is at least one point where its Fourier series diverges.

For a long time it was not known whether or not the Fourier series of a continuous function must converge even at a single point. This question was settled by Lennart Carleson in 1966, when he showed that if f^2 is integrable on $(-\pi, \pi)$, even in the wider sense of Lebesgue to be explained in §2.9, its Fourier series converges except, possibly, on a set of measure zero—refer to the remarks that follow Definition 2.2 for an explanation of this term. Of course, a bounded continuous function is square integrable, and Carleson's theorem applies in this case. However, Carleson's result is the best possible for continuous functions in the sense that, given a subset S of $(-\pi, \pi)$ of measure zero, there is a function continuous on $(-\pi, \pi)$ and whose Fourier series on $(-\pi, \pi)$ diverges at every point of S. This was shown by Jean Pierre Kahane and Yitzhak Katznelson in 1966.[1] Outside the class of square-integrable functions the worse can happen; to wit, in 1926 Andrei Kolmogorov—then a young man of 23—gave an example of a function integrable in the sense of Lebesgue and whose Fourier series diverges at every point.[2]

---

[1] Advanced proofs of the results of Carleson, Kahane, and Katznelson, plus an extension by Richard Hunt of Carleson's theorem, can be seen in C. J. Mozzochi, *On the Pointwise Convergence of Fourier Series*, Springer-Verlag, Berlin/Heidelberg/New York, 1971, and also in O. G. Jørsboe and L. Mejlbro, *The Carleson-Hunt Theorem on Fourier Series*, Springer-Verlag, Berlin/Heidelberg/New York, 1982.

[2] Zygmund, *op. cit.*, pages 310–314.

§2.5 Fourier Series on Arbitrary Intervals

Up to this point we have considered a function defined on $(-\pi, \pi)$ and its periodic extension of period 2π. In general, we must consider a function defined on an arbitrary interval $f : (\alpha - a, \alpha + a) \to \mathbb{R}$, $a > 0$, and its periodic extension of period $2a$. If we introduce a new variable

$$y = \frac{\pi}{a}(x - \alpha)$$

and define a new function h by $h(y) = f(x)$, then h is defined on $(-\pi, \pi)$. If f is periodic of period $2a$, then h is periodic of period 2π. Also, h has one-sided derivatives at y if f has one-sided derivatives at x and is piecewise continuous on a small interval centered at y if f is piecewise continuous on a small interval centered at x. If y is such a point, Theorem 2.2 gives

$$\tfrac{1}{2}[h(y+) + h(y-)] = \tfrac{1}{2}a_0 + \sum_{n=1}^{\infty}(a_n \cos ny + b_n \sin ny),$$

where

$$a_n = \frac{1}{\pi}\int_{-\pi}^{\pi} h(y) \cos ny\, dy, \qquad n = 0, 1, 2, \ldots$$

and

$$b_n = \frac{1}{\pi}\int_{-\pi}^{\pi} h(y) \sin ny\, dy, \qquad n = 1, 2, 3, \ldots.$$

Undoing the change of variable, these formulas become

$$(10) \quad \tfrac{1}{2}[f(x+) + f(x-)] = \tfrac{1}{2}a_0 + \sum_{n=1}^{\infty}\left[a_n \cos \frac{n\pi}{a}(x - \alpha) + b_n \sin \frac{n\pi}{a}(x - \alpha)\right]$$

$$a_n = \frac{1}{a}\int_{\alpha-a}^{\alpha+a} f(x) \cos \frac{n\pi}{a}(x - \alpha)\, dx, \qquad n = 0, 1, 2, \ldots$$

and

$$b_n = \frac{1}{a}\int_{\alpha-a}^{\alpha+a} f(x) \sin \frac{n\pi}{a}(x - \alpha)\, dx, \qquad n = 1, 2, 3, \ldots.$$

Definition 2.5 *The constants a_n and b_n defined above are called the* **Fourier coefficients** *of f on $(\alpha - a, \alpha + a)$, and the right-hand side of (10) is called the* **Fourier series** *of f on $(\alpha - a, \alpha + a)$.*

The result we have just obtained can be stated as follows.

Theorem 2.5 *Let $f : \mathbb{R} \to \mathbb{R}$ be a periodic function of period $2a$, $a > 0$, that is integrable on $(\alpha - a, \alpha + a)$. Then its Fourier series on $(\alpha - a, \alpha + a)$ converges to the value*

$$\tfrac{1}{2}[f(x+) + f(x-)]$$

at every point x in $\mathbb{R}$ *such that f has right-hand and left-hand derivatives at x and f is piecewise continuous on an arbitrarily small interval centered at x. If, in addition, f is continuous at x its Fourier series converges to* $f(x)$.

★   In a similar manner we can prove the analog of the Lipschitz test.

Theorem 2.6 *Let* $f: \mathbb{R} \to \mathbb{R}$ *be a periodic function of period 2a, a > 0, that is integrable on* $(\alpha - a, \alpha + a)$. *Then its Fourier series on* $(\alpha - a, \alpha + a)$ *converges to* $f(x)$ *at a given point x in* $\mathbb{R}$ *if f is continuous on an arbitrarily small interval centered at x and if there are positive numbers M and* δ *such that*

$$|f(x + t) - f(x)| \le M|t|$$

for t in $(-\delta, \delta)$.

Proof. If h is as above and if $y = \pi(x - \alpha)/a$, then h is continuous about y if f is continuous about x. Now, if u is in the interval $(-\pi\delta/a, \pi\delta/a)$, we have

$$|h(y + u) - h(y)| = \left| f\left(x + \frac{a}{\pi}u\right) - f(x)\right| \le M\left|\frac{a}{\pi}u\right| \le \frac{M}{\pi}|a||u|.$$

This shows that h satisfies the hypotheses of Theorem 2.4, and then its Fourier series on $(-\pi, \pi)$ converges to $h(y)$ at y. Undoing the change of variable, (10) holds with $f(x)$
◇   in place of $\frac{1}{2}[f(x+) + f(x-)]$, Q.E.D.

Example 2.12 Find the Fourier series on $(4, 6)$ of the function defined by

$$f(x) = \begin{cases} 1 & \text{if} \quad 4 < x \le 5 \\ 2 & \text{if} \quad 5 < x < 6. \end{cases}$$

Here $\alpha = 5$ and $a = 1$ so that, according to Definition 2.5, the Fourier coefficients of f on $(4, 6)$ are

$$a_0 = \int_4^6 f(x)\, dx = \int_4^5 dx + \int_5^6 2\, dx = 3,$$

and, for $n > 0$,

$$a_n = \int_4^5 \cos n\pi(x - 5)\, dx + \int_5^6 2 \cos n\pi(x - 5)\, dx = 0,$$

while

$$b_n = \int_4^5 \sin n\pi(x - 5)\, dx + \int_5^6 2 \sin n\pi(x - 5)\, dx$$

$$= -\frac{\cos n\pi(x - 5)}{n\pi}\Big|_4^5 - \frac{2\cos n\pi(x - 5)}{n\pi}\Big|_5^6$$

$$= \begin{cases} 0 & \text{if} \quad n \quad \text{is even} \\ \dfrac{2}{n\pi} & \text{if} \quad n \quad \text{is odd.} \end{cases}$$

By Theorem 2.5,

$$f(x) = \frac{3}{2} + \frac{2}{\pi} \sum_{n=1}^{\infty} \frac{1}{2n-1} \sin(2n-1)\pi(x-5)$$

for $4 < x < 5$ and $5 < x < 6$. At $x = 5$ the sum of the series is 1.5.

Consider next a function f defined on an interval $(0, a)$. It can be extended to the interval $(-a, a)$ as an even function, as an odd function, or in any other way, and then each of these can be extended to the whole of $\mathbb{R}$ as a periodic function. Of course, the function must be defined to have the value $\frac{1}{2}[f(x+) + f(x-)]$ at any point x of discontinuity that may be created in this extension process. This is illustrated in Figure 2.5 for $f(x) = x$ on $(0, 1)$. Each of the functions (a) to (c) is periodic of period 2 and has a Fourier series

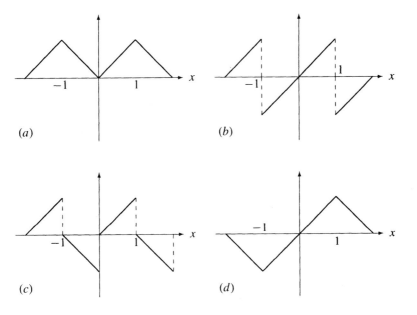

(a) (b) (c) (d)

Figure 2.5

on $(-1, 1)$ that converges to the value x at each x in $(0, 1)$. The three series are different, since their Fourier coefficients depend on all their values on $(-1, 1)$, and yet their sums coincide on the right half of this interval.

Note that (a) represents an even function while (b) represents an odd function. Thus, the Fourier series of the first will contain only cosine terms, and the Fourier series of the second only sine terms. The Fourier series of the function represented in (c), which is neither even nor odd, may contain both kinds of terms. The last function, represented in (d), is also odd but its periodic extension has period 4 instead of period 2 as in (a).

Its Fourier series on $(-2, 2)$ will contain only sine terms but is not identical to the Fourier series of the function in (a) on $(-1, 1)$. It is left to the reader to carry out the corresponding expansions in all four cases (Exercises 5.1 and 5.2).

In connection with the odd and even extensions mentioned above the following terminology is sometimes useful.

Definition 2.6 *If $a > 0$, let $f : (0, a) \to \mathbb{R}$ be an integrable function. Then,*

(i) *the* **Fourier sine series** *of f on $(0, a)$ is the Fourier series on $(-a, a)$ of its odd extension to $(-a, a)$; that is,*

$$\sum_{n=1}^{\infty} b_n \sin \frac{n\pi}{a} x,$$

where

$$b_n = \frac{2}{a} \int_0^a f(x) \sin \frac{n\pi}{a} x \, dx.$$

(ii) *the* **Fourier cosine series** *of f on $(0, a)$ is the Fourier series on $(-a, a)$ of its even extension to $(-a, a)$; that is,*

$$\tfrac{1}{2} a_0 + \sum_{n=1}^{\infty} a_n \cos \frac{n\pi}{a} x,$$

where

$$a_n = \frac{2}{a} \int_0^a f(x) \cos \frac{n\pi}{a} x \, dx.$$

Example 2.13 Find the Fourier cosine series on $(0, \pi)$ of $f(x) = \sin x$. We have

$$a_0 = \frac{2}{\pi} \int_0^\pi \sin x \, dx = \frac{4}{\pi}, \qquad a_1 = \frac{2}{\pi} \int_0^\pi \sin x \cos x \, dx = 0,$$

and, for $n > 1$,

$$\begin{aligned}
a_n &= \frac{2}{\pi} \int_0^\pi \sin x \cos nx \, dx \\
&= \frac{1}{\pi} \int_0^\pi [\sin(1 + n)x + \sin(1 - n)x] \, dx \\
&= -\frac{1}{\pi} \left[\frac{\cos(1 + n)x}{1 + n} + \frac{\cos(1 - n)x}{1 - n} \right]_0^\pi \\
&= \begin{cases} \dfrac{4}{\pi(1 - n^2)} & \text{if } n \text{ is even} \\ 0 & \text{if } n \text{ is odd}. \end{cases}
\end{aligned}$$

Therefore,

$$\sin x = \frac{2}{\pi} + \frac{4}{\pi} \sum_{n=1}^{\infty} \frac{1}{1 - 4n^2} \cos 2nx$$

on $(0, \pi)$. Actually, this Fourier series converges to $|\sin x|$ at every point x in $\mathbb{R}$. This is guaranteed by Theorem 2.2.

★   ## §2.6 The Gibbs Phenomenon

In a practical situation a Fourier series cannot be used in its entirety to approximate a given function since one must necessarily take only a finite number of terms. This raises the question of how well a partial sum of a Fourier series approximates the function to which it converges. In 1898 Albert A. Michaelson and S. W. Stratton, of the University of Chicago, expressed their dissatisfaction in a letter to *Nature* by pointing out that the approximation is not good enough near a jump discontinuity. Experiments with their harmonic analyzer showed this to be case for the function of Example 1.6, even after taking 160 terms of its Fourier series. An explanation by the Yale professor Josiah Willard Gibbs (1839–1903) appeared in the same magazine the following year. Gibbs considered the periodic extension of the function defined by $f(x) = x$ on the interval $(-\pi, \pi)$, which satisfies the conditions of Theorem 2.2, and pointed out that, in the neighborhood of any point of discontinuity at $x_0 = (2k + 1)\pi$, where k is an integer, and for N large enough, there are points $x_1 < x_0$ and $x_2 > x_0$ such that

$$s_N(x_1) \approx 1.18\pi \qquad \text{and} \qquad s_N(x_2) \approx -1.18\pi;$$

that is, close to x_0 and on its left the value of s_N exceeds that of $f(x_0-)$ by about 9% of the value 2π of the total jump at x_0. Similarly, close to x_0 and on its right, the value of s_N overswings the value $f(x_0+)$ by about 9% of the total fall at x_0. This situation is illustrated in Figure 2.6 for $x_0 = \pi$.

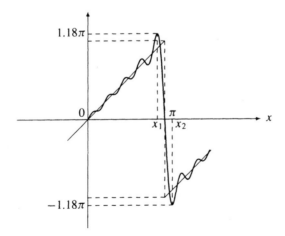

Figure 2.6

The remarkable thing is that, as we shall presently see, this overswing remains constant at about 9% independently of N, although the points x_1 and x_2 at which it occurs become closer to x_0 as N increases. Had Michaelson and Stratton added many more terms to their partial sum they would have been equally dissatisfied. This is an inherent imperfection of Fourier series and not just a matter of how many terms the approximating partial sum contains. The presence of this overswing is usually known as the *Gibbs phenomenon*, although it had been discovered earlier, in 1848, by the English mathematician Henry Wilbraham. Wilbraham's work remained largely unnoticed and Gibbs was unaware of it.

To demonstrate the appearance of the Gibbs phenomenon we shall find it convenient to consider the function defined by

$$f(x) = \begin{cases} 0 & \text{if} \quad -\pi < x < 0 \\ 2\pi & \text{if} \quad 0 \le x < \pi \end{cases}$$

rather than Gibbs' original example. By Theorem 2.2, its Fourier series converges to π at the origin and to $f(x)$ at any other x in $(-\pi, \pi)$. To study the behavior of $s_N(x)$ near the discontinuity at the origin we use Lemma 2.1. We obtain

$$(11) \qquad s_N(x) = \frac{1}{2\pi} \int_{-\pi}^{\pi} f(t) D_N(t - x)\, dt$$

$$= \int_0^{\pi} D_N(t - x)\, dt$$

$$= \int_{-x}^{\pi - x} D_N(u)\, du$$

$$= \int_{-x}^{0} D_N + \int_0^{\pi} D_N + \int_{\pi}^{\pi - x} D_N$$

$$= \int_0^{x} D_N + \pi - \int_{\pi - x}^{\pi} D_N,$$

where the value π for the second integral has been obtained from Equation (7) on page 38, and the reversal of the limits of integration in the first is possible because D_N is an even function. Since we are interested in values of x near zero, the interval between $\pi - x$ and π does not contain the origin and D_N is the product of $\csc \frac{1}{2} t$, which is integrable over this interval, by $\sin \left(N + \frac{1}{2} \right) t$. Thus the third integral tends to zero as $N \to \infty$ by the Riemann-Lebesgue theorem, and

$$s_N(x) \approx \int_0^{x} D_N + \pi$$

for x small and N large. The phenomenon is already apparent from this approximation and the geometric interpretation in Figure 2.7. When $x > 0$ the area under the graph of D_N from zero to x increases with x until $x = 2\pi/(2N + 1)$, where it reaches a maximum. Then it decreases as the area between the x-axis and the first ripple of D_N is incorporated into the integral with a negative sign. In this manner, the presence of the ripples of D_N, alternatively negative and positive, is translated into the maxima and minima of $\int_0^x D_N$ as shown in the lower part of Figure 2.7. The portion of the graph for

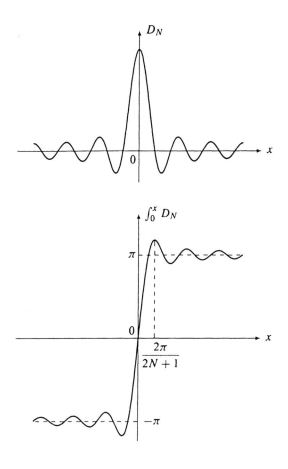

Figure 2.7

$x < 0$ is obtained from the fact that D_N is an even function. In order to estimate the value of the first maximum, which is clearly the largest, we must find the approximate value of $\int_0^x D_N$ for $x > 0$ by analytical means. We have

$$\int_0^x \frac{\sin\left(N + \frac{1}{2}\right)t}{\sin\frac{1}{2}t}\, dt = 2\int_0^x \frac{\sin\left(N + \frac{1}{2}\right)t}{t}\, dt + \int_0^x \frac{t - 2\sin\frac{1}{2}t}{t\sin\frac{1}{2}t}\sin\left(N + \frac{1}{2}\right)t\, dt.$$

Now, if x is small enough the factor

$$\frac{t - 2\sin \frac{1}{2}t}{t \sin \frac{1}{2}t}$$

is continuous on $(0, x)$ and, by l'Hôpital's rule, approaches zero as $t \to 0$. Hence, this factor is integrable on $(0, x)$, and the second integral on the right-hand side above approaches zero as $N \to \infty$ by the Riemann-Lebesgue theorem. We conclude that for x small and N large

$$s_N(x) \approx 2 \int_0^x \frac{\sin \left(N + \frac{1}{2}\right)t}{t} \, dt + \pi = 2 \int_0^{\left(N+\frac{1}{2}\right)x} \frac{\sin u}{u} \, du + \pi.$$

If we denote the right-hand side by $\varphi(x)$ we can construct its graph as follows. Clearly $\varphi(0) = \pi$. Then we compute

$$\varphi'(x) = 2\frac{\sin \left(N + \frac{1}{2}\right)x}{x}$$

for $x \neq 0$ and $\varphi'(0) = 2N + 1$ (Exercise 6.1). The last value is the slope of φ at the origin. Its relative maxima and minima occur where $\varphi'(x) = 0$; that is, at

$$x = \frac{k\pi}{N + \frac{1}{2}}$$

for every nonzero integer k. An examination of the second derivative shows that φ has a relative maximum for $k = \ldots -6, -4, -2, 1, 3, 5, \ldots$ and a relative minimum for $k = \ldots -3, -1, 2, 4, \ldots$. From the fact that, according to Theorem 2.2, $s_N(x)$ is close to zero for $x < 0$ and close to 2π for $x > 0$ we conclude that the same must be true of $\varphi(x)$, whose graph is then represented in Figure 2.8. The value of the maximum at

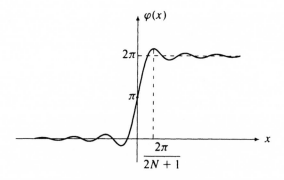

Figure 2.8

$x = \pi / \left(N + \frac{1}{2} \right)$ is

$$\varphi \left(\frac{\pi}{N + \frac{1}{2}} \right) = 2 \int_0^\pi \frac{\sin u}{u} \, du + \pi \approx 1.18\pi + \pi = 1.09(2\pi),$$

where the integral has been obtained from the tables. A similar calculation shows that the minimum at $-2\pi/(2N + 1)$ is approximately equal to $-0.09(2\pi)$.

To sum up, the abscissas of the maximum and minimum values of φ approach zero as $N \to \infty$, and, at the same time, the slope at the origin increases indefinitely. This allows $s_N(x)$ to approach $f(x)$ for each $x \neq 0$. However, for large N, we see that the amount of the overswing is always near 9% of the total jump at the discontinuity and this value is independent of N. This is the Gibbs phenomenon for f.

Of course, the presence of the phenomenon in a single example, or in a collection of examples, does not imply its existence in the general case. But in 1906 Maxime Bôcher (1867–1918), of Harvard University—who was not familiar with the work of either Wilbraham or Gibbs—showed that the partial sum of the Fourier series of an arbitrary function always has this overswing of about nine per cent at any jump discontinuity.[1] It is then an intrinsic defect of the convergence of Fourier series.

★   ## §2.7 Fejér Sums

So far the convergence of Fourier series has been found wanting on at least two counts. The Gibbs phenomenon is the most recent of these, but also remember that the Fourier series of a continuous function may diverge at infinitely many points. Both defects can be remedied using the following method due to Lipot Fejér (1880–1959). If we define $s_0(x) = \frac{1}{2}a_0$ and recall the well-known value of $s_n(x)$ for $n \geq 1$ given by (4) on page 36, then we shall use the arithmetic means

$$\sigma_N(x) = \frac{1}{N + 1} \sum_{n=0}^N s_n(x)$$

of these partial sums instead of the partial sums themselves—such arithmetic means had originally been introduced by Georg Frobenius in 1880 in his study of power series. That is, instead of the sequence $\{s_N\}$ we consider the sequence $\{\sigma_N\}$ to approximate the given function.

Recall that, by Lemma 2.1,

$$s_N(x) = \frac{1}{2\pi} \int_{-\pi}^\pi f(t) D_N(t - x) \, dt$$

for $n > 1$, where D_n is the Dirichlet kernel, and observe that this equation is also valid for $n = 0$ if we define $D_0 \equiv 1$ and recall the definition of a_0. Then we obtain the

---

[1] For a proof see, for instance, H. S. Carslaw, *An Introduction to the Theory of Fourier's Series and Integrals*, 3rd ed., Dover, New York, 1950, pages 305–307.

following expression for the Fejér sum,

$$\sigma_N(x) = \frac{1}{N+1} \sum_{n=0}^{N} \frac{1}{2\pi} \int_{-\pi}^{\pi} f(t) D_N(t-x)\, dt = \frac{1}{2\pi} \int_{-\pi}^{\pi} f(t) F_N(t-x)\, dt,$$

where

(12)
$$F_N = \frac{1}{N+1} \sum_{n=0}^{N} D_n$$

is called the *Fejér kernel*.

We collect in a technical lemma the most useful properties of F_N.

Lemma 2.3 *The Fejér kernel has the following properties*:

(13)
$$F_N(t) = \frac{1}{N+1} \frac{\sin^2 \frac{1}{2}(N+1)t}{\sin^2 \frac{1}{2}t} \geq 0$$

if $t \neq 0$,

(14)
$$\int_0^\pi F_N(t) = \pi,$$

and, for any number δ *such that* $0 < \delta < \pi$,

(15)
$$\int_\delta^\pi F_N \to 0$$

as $N \to \infty$.

Proof. Using the definition of D_N given on page 38,

$$F_N(t) = \frac{1}{(N+1)\sin \frac{1}{2}t} \sum_{n=0}^{N} \sin \left(n + \tfrac{1}{2}\right)t.$$

The sum on the right can now be evaluated using the trigonometric identity

$$2 \sin \tfrac{1}{2}nt \sin \tfrac{1}{2}t = \cos \tfrac{1}{2}(n-1)t - \cos \tfrac{1}{2}(n+1)t$$

and adding the equations that result from substituting $n = 1, 3, \ldots, 2N+1$ in it. This gives

$$\left(2 \sin \tfrac{1}{2}t\right)\left[\sin \tfrac{1}{2}t + \sin \tfrac{3}{2}t + \cdots + \sin \left(N + \tfrac{1}{2}\right)t \right] = 1 - \cos(N+1)t = 2 \sin^2 \tfrac{1}{2}(N+1)t;$$

that is, if $t \neq 0$,

$$\sum_{n=0}^{N} \sin \left(n + \tfrac{1}{2}\right)t = \frac{\sin^2 \frac{1}{2}(N+1)t}{\sin \frac{1}{2}t},$$

proving (13).

Equation (14) is a property that the Fejér kernel shares with the Dirichlet kernel. In fact, using Equation (7) on page 38, also valid for D_0, we get

$$\frac{1}{2\pi} \int_0^\pi F_N = \frac{1}{N+1} \sum_{n=0}^N \frac{1}{2\pi} \int_0^\pi D_n = \frac{1}{N+1} \sum_{n=0}^N \frac{1}{2} = \frac{1}{2}.$$

Finally, if $0 < \delta < \pi$, then the minimum value of $\sin^2 \frac{1}{2} t$ on $[\delta, \pi]$ is $\sin^2 \frac{1}{2}\delta$ and, using (13),

$$\left| \int_\delta^\pi F_N \right| < \int_\delta^\pi \frac{1}{(N+1)\sin^2 \frac{1}{2}\delta} \, dt = \frac{\pi - \delta}{(N+1)\sin^2 \frac{1}{2}\delta}.$$

It is clear that the expression on the right approaches zero as $N \to \infty$, Q.E.D.

We are now ready to prove that for any continuous function the sequence of its Fejér sums converges to the function. In fact, with no more effort, we can prove the following more general result.

Theorem 2.7 (**Fejér, 1904**) *Let* $f : \mathbb{R} \to \mathbb{R}$ *be a periodic function of period* 2π *that is integrable on* $(-\pi, \pi)$. *If the limits* $f(x+)$ *and* $f(x-)$ *exist at a point* x *in* $\mathbb{R}$, *then the sequence of Fejér sums* $\{\sigma_N(x)\}$ *converges to the value*

$$\tfrac{1}{2}[f(x+) + f(x-)]$$

at the given point x. *In particular, if* f *is continuous at* x *then* $\sigma_N(x) \to f(x)$. *Furthermore, if* f *is continuous on* $[-\pi, \pi]$ *then the sequence* $\{\sigma_N\}$ *converges uniformly on* $[-\pi, \pi]$.

Proof. Let x be such that the limits $f(x+)$ and $f(x-)$ exist. It easily follows from Equation (9) on page 39 that, for $N > 0$,

$$\sigma_N(x) = \frac{1}{N+1} \sum_{n=0}^N \frac{1}{2\pi} \int_0^\pi [f(x+t) + f(x-t)] D_N(t-x) \, dt$$

$$= \frac{1}{2\pi} \int_0^\pi [f(x+t) + f(x-t)] F_N(t-x) \, dt.$$

Then using (14) we obtain

$$\sigma_N(x) - \tfrac{1}{2}[f(x+) + f(x-)] = \frac{1}{2\pi} \int_0^\pi [f(x+t) + f(x-t) - f(x+) - f(x-)] F_N(t) \, dt.$$

To simplify the writing we shall find it convenient to define

$$g_x(t) = f(x+t) + f(x-t) - f(x+) - f(x-).$$

The existence of $f(x+)$ and $f(x-)$ means that given any number $\epsilon > 0$ we can find another number δ_x such that $0 < \delta_x < \pi$ and such that

$$|f(x+t) - f(x+)| < \frac{\epsilon}{2} \quad \text{and} \quad |f(x-t) - f(x-)| < \frac{\epsilon}{2}$$

for $0 < t < \delta_x$. Then $|g_x(t)| < \epsilon$ for $0 < t < \delta_x$. Also g_x is bounded because f is, and then there is a positive constant M such that $|g_x(t)| < M$ for $\delta_x \le t \le \pi$. Hence,

$$\left| \sigma_N(x) - \tfrac{1}{2}[f(x+) + f(x-)] \right| = \left| \frac{1}{2\pi} \int_0^\pi g_x F_N \right|$$

$$\le \frac{1}{2\pi} \int_0^\pi |g_x| F_N$$

$$< \frac{\epsilon}{2\pi} \int_0^{\delta_x} F_N + \frac{M}{2\pi} \int_{\delta_x}^\pi F_N$$

$$< \frac{\epsilon}{2\pi} \int_0^\pi F_N + \frac{M}{2\pi} \int_{\delta_x}^\pi F_N.$$

Using (14) and (15) we see that the right-hand side is smaller than ϵ for N large enough. This proves the first assertion, and, if f is continuous at x, the second assertion follows from the first.

Finally, if f is continuous on $[-\pi, \pi]$, then it is uniformly continuous there and the same δ_x can be chosen for every x. Therefore, the number N at the end of the preceding proof can be chosen independent of x, showing that the convergence is uniform, Q.E.D.

It is possible to obtain a number of results on Fourier series from Fejér's theorem, and we prove some of these below as corollaries. In fact, even the original Dirichlet theorem follows from Fejér's, as shown in 1910 by Godfrey Harold Hardy (1877–1947), but we shall not see this here.[1]

Corollary 1 *If* $f:(-\pi, \pi) \to \mathbb{R}$ *is an integrable function and if its Fourier series on* $(-\pi, \pi)$ *converges at a point* x *at which* f *is continuous, then it converges to* $f(x)$.

Proof. If s_N and σ_N denote the Nth Fourier and Fejér sums, respectively, it is enough to show that, if $s_N(x)$ converges to a limit s and f is continuous at x, then $\sigma_N(x)$ also converges to s. For, according to the theorem, $\sigma_N(x) \to f(x)$, and then $s = f(x)$.

Now, if s is as above, M is a positive integer, and $N > M$, we have

$$|\sigma_N(x) - s| = \left| \frac{s_0(x) + \cdots + s_N(x)}{N+1} - s \right|$$

$$= \left| \frac{[s_0(x) - s] + \cdots + [s_N(x) - s]}{N+1} \right|$$

$$\le \frac{|s_0(x) - s| + \cdots + |s_M(x) - s|}{N+1} + \frac{|s_{M+1}(x) - s| + \cdots + |s_N(x) - s|}{N+1}.$$

If $\epsilon > 0$ is arbitrary, M can be chosen so large that $|s_n(x) - s| < \epsilon/2$ for $n > M$, and then

$$|\sigma_N(x) - s| \le \frac{|s_0(x) - s| + \cdots + |s_M(x) - s|}{N+1} + \frac{N-M}{N+1} \frac{\epsilon}{2}.$$

---

[1] See, for instance, Carslaw, *op. cit.*, pages 262–264.

If N is now chosen so large that the first term on the right is smaller than $\epsilon/2$, it follows that $|\sigma_N(x) - s| < \epsilon$. This proves that $\sigma_N(x) \to s$ when $s_N(x) \to s$, Q.E.D.

Corollary 2 *If two functions f, $g: (-\pi, \pi) \to \mathbb{R}$ that are integrable and continuous have the same Fourier series on $(-\pi, \pi)$, then $f \equiv g$.*

Proof. By the theorem, $\sigma_N \to f$ and $\sigma_N \to g$ on $(-\pi, \pi)$ and then $f \equiv g$, Q.E.D.

This can be rephrased as follows.

Corollary 3 *If the Fourier coefficients of a function that is integrable and continuous on $(-\pi, \pi)$ are all zero, then the function is identically zero.*

We now turn our attention to the possibility of the Gibbs phenomenon for Fejér sums, which we shall study from the following relation between the Dirichlet and Fejér kernels

$$(2N + 1)F_{2N} = D_N^2.$$

This equation, which follows at once from the definition of D_N and (13), permits us to construct the graph of F_{2N}, represented in Figure 2.9, from that of D_N, represented

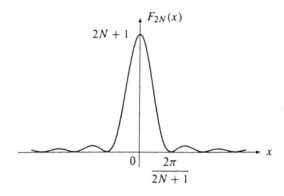

Figure 2.9

in Figure 2.7, by squaring and then dividing by $2N + 1$. Note that all the ripples are now above the x axis, and, if we remember that the Gibbs phenomenon was due to the presence of the ripples in the graph of D_N, alternatively above and below the x axis, it follows that it should now be completely eliminated. In fact, going back to the example in §2.6, an argument entirely analogous to the one leading to Equation (11) gives

$$\sigma_N(x) = \int_0^x F_N + \pi - \int_{\pi-x}^{\pi} F_N.$$

For x near zero the interval between $\pi - x$ and π does not contain the origin. Then, putting $\delta = \pi - x$ in the last integral above and using (15), we see that for N

sufficiently large and $|x|$ small

$$\sigma_N(x) \approx \int_0^x F_N + \pi.$$

If we denote the right-hand side by $\phi(x)$, the graph of ϕ, represented in Figure 2.10, can

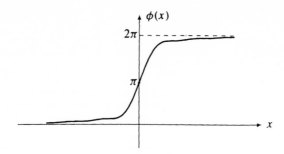

Figure 2.10

be constructed as follows. ϕ is nondecreasing because F_N is nonnegative, and, for N large enough, $\phi(x)$ is close to zero for $x < 0$ and close to 2π for $x > 0$ because $\sigma_N(x)$ is close to these values according to Theorem 2.7. We readily see that the overswing has been eliminated in this example. However, there is a price to pay for this. The slope of ϕ at the origin is

$$\phi'(0) = F_N(0) = N + 1,$$

while, at the same point, the slope of the function $\varphi \approx s_N$ of §2.6 was found to be $2N + 1$. Thus, twice as many terms are needed in the Fejér sum as in the corresponding Fourier sum to obtain the same slope at a point of discontinuity.

For a proof that there is no overswing for Fejér sums in the general case refer to Exercise 7.6.

§2.8   Integration of Fourier Series

In §2.2 the Fourier coefficients were obtained under the assumption that the equality in (1) held and that term by term integration of the series was possible. Once in possession of the Fourier coefficients these hypotheses were no longer necessary and were discarded. The convergence of the Fourier series was established under a different set of conditions. At this moment we have not yet shown whether or not a Fourier series can be integrated term by term.[1] In what follows we present a particular case of a theorem discovered in 1906 by Lebesgue, which asserts that the Fourier series of an integrable function can

---

[1] This follows from Theorem A.4 of Appendix A if a certain very strong condition—uniform convergence—is satisfied by the Fourier series. What we do in this section shows that this condition is not necessary at all.

always be integrated term by term, even if the Fourier series itself diverges. The particular version that we state has been selected because it is very easy to prove and sufficient for most applications. For a proof of the more general case refer to Exercise 4.11 of Chapter 6.

Theorem 2.8 *Let* $f : (-\pi, \pi) \rightarrow \mathbb{R}$ *be a piecewise continuous function that has Fourier series*

$$\tfrac{1}{2}a_0 + \sum_{n=1}^{\infty}(a_n \cos nx + b_n \sin nx)$$

on $(-\pi, \pi)$. *If a and x are points in* $[-\pi, \pi]$ *then*

$$\int_a^x f = \tfrac{1}{2}a_0(x - a) + \sum_{n=1}^{\infty}\int_a^x (a_n \cos nt + b_n \sin nt)dt,$$

whether or not the Fourier series of f converges.

Proof. It is enough to prove the theorem for $a = 0$. The general case then follows from the equation

$$\int_a^x f = \int_0^x f - \int_0^a f$$

and the fact that two convergent series of functions can be subtracted term by term. Then to establish the desired identity it suffices to prove that the Fourier series on $(-\pi, \pi)$ of the function $F : [-\pi, \pi] \rightarrow \mathbb{R}$ defined by

$$F(x) = \int_0^x f - \tfrac{1}{2}a_0 x$$

is

$$\sum_{n=1}^{\infty}\int_0^x (a_n \cos nt + b_n \sin t)dt$$

and that F equals its Fourier series.

First observe that the extension of F to $\mathbb{R}$ as a periodic function of period 2π is continuous on $\mathbb{R}$ because F is clearly continuous on $[-\pi, \pi]$ and because, recalling the definition of a_0,

$$F(\pi) = \int_0^\pi f - \tfrac{1}{2}a_0\pi = a_0\pi - \int_{-\pi}^0 f - \tfrac{1}{2}a_0\pi = \int_0^{-\pi} f + \tfrac{1}{2}a_0\pi = F(-\pi).$$

Also,

$$F'(x) = f(x) - \tfrac{1}{2}a_0$$

at every point x at which f is continuous. That is, F' is piecewise continuous on $(-\pi, \pi)$, and, by Lemma 2.2, F satisfies the conditions of Theorem 2.2. Hence, $F(x)$ is equal to the sum of its Fourier series on $(-\pi, \pi)$:

$$F(x) = \tfrac{1}{2}A_0 + \sum_{n=1}^{\infty}(A_n \cos nx + B_n \sin nx)$$

for any x in $\mathbb{R}$. For $n > 0$ the Fourier coefficients in this equation can be computed using integration by parts.

$$
\begin{aligned}
A_n &= \frac{1}{\pi} \int_{-\pi}^{\pi} F(x) \cos nx \, dx \\
&= \frac{1}{\pi} \left[\frac{F(x) \sin nx}{n} \bigg|_{-\pi}^{\pi} - \frac{1}{n} \int_{-\pi}^{\pi} F'(x) \sin nx \, dx \right] \\
&= -\frac{1}{n\pi} \int_{-\pi}^{\pi} \left[f(x) - \tfrac{1}{2} a_0 \right] \sin nx \, dx \\
&= -\frac{1}{n\pi} \int_{-\pi}^{\pi} f(x) \sin nx \, dx \\
&= -\frac{b_n}{n}.
\end{aligned}
$$

Analogously,

$$
B_n = \frac{1}{\pi} \int_{-\pi}^{\pi} F(x) \sin nx \, dx = \frac{a_n}{n}.
$$

Therefore,

(16) $$F(x) = \tfrac{1}{2} A_0 + \sum_{n=1}^{\infty} \frac{1}{n} (a_n \sin nx - b_n \cos nx).$$

To obtain the value of the remaining coefficient A_0 we set $x = 0$ and obtain

(17) $$0 = F(0) = \tfrac{1}{2} A_0 - \sum_{n=1}^{\infty} \frac{b_n}{n}.$$

Hence, taking this value of A_0 to (16),

$$
F(x) = \sum_{n=1}^{\infty} \frac{1}{n} [a_n \sin nx + b_n (1 - \cos nx)] = \sum_{n=1}^{\infty} \int_0^x (a_n \cos nt + b_n \sin nt) \, dt,
$$

Q.E.D.

Theorem 2.8 is not very useful for computation because term-by-term integration of $\sum_{n=1}^{\infty} b_n \sin nt$ from zero to x leads to the infinite series $\sum_{n=1}^{\infty} b_n/n$, whose evaluation may not be possible. For computational purposes this result can be restated in the following form, in which, for convenience and without loss of generality, we set $a = 0$.

Theorem 2.9 *Let f be as in Theorem 2.8. Then, for any x in $[-\pi, \pi]$,*

$$
\int_0^x f - \tfrac{1}{2} a_0 x = \frac{1}{2\pi} \int_{-\pi}^{\pi} \left(\int_0^x f \right) dx + \sum_{n=1}^{\infty} \frac{1}{n} (a_n \sin nx - b_n \cos nx),
$$

where the right-hand side is the Fourier series on $(-\pi, \pi)$ of the function on the left-hand side.

Proof. Use the definition of A_0 in (16), instead of the value furnished by (17), Q.E.D.

It is possible to state and prove the equivalent of Theorems 2.8 and 2.9 for Fourier series on arbitrary intervals. This is left to Exercise 8.4.

Example 2.14 In the Fourier series expansion of Example 2.9

$$x = \sum_{n=1}^{\infty} 2 \frac{(-1)^{n+1}}{n} \sin nx,$$

valid for any x in $(-\pi, \pi)$, we have $a_n = 0$ for all $n \geq 0$ and $b_n = 2(-1)^{n+1}/n$ for all $n \geq 1$. Since

$$\frac{1}{2\pi} \int_{-\pi}^{\pi} \left(\int_0^x t \, dt \right) dx = \frac{\pi^2}{6},$$

Theorem 2.9 gives

$$\frac{x^2}{2} = \frac{\pi^2}{6} + \sum_{n=1}^{\infty} 2 \frac{(-1)^n}{n^2} \cos nx.$$

Of course, the right-hand side is the Fourier series of $x^2/2$ on $(-\pi, \pi)$ and it can be integrated again. But now $a_0 = \pi^2/3$ while $a_n = 2(-1)^n/n^2$ and $b_n = 0$ for all $n \geq 1$. Using Theorem 2.9 again gives

$$\frac{x^3}{6} - \frac{\pi^2 x}{6} = \sum_{n=1}^{\infty} 2 \frac{(-1)^n}{n^3} \sin nx.$$

This procedure can be repeated any number of times and all the expansions so obtained are valid on $(-\pi, \pi)$.

Equation (17) clearly shows that, if f is as in Theorem 2.8, then the series

$$\sum_{n=1}^{\infty} \frac{b_n}{n}$$

converges. This result is sometimes useful to show that a certain trigonometric series cannot be a Fourier series (as in Definition 2.1 or 2.5). The following example was given by Pierre Fatou (1878–1929) in 1906.

Example 2.15 The series

$$\sum_{n=2}^{\infty} \frac{1}{\log n} \sin nx$$

is not a Fourier series because

$$\sum_{n=1}^{\infty} \frac{b_n}{n} = \sum_{n=2}^{\infty} \frac{1}{n \log n}$$

diverges.

§2.9 Historical Epilogue

After his stay in Paris, where he supported himself by giving German lessons until 1827, Dirichlet returned to Prussia. On the recommendation of Fourier, the physical geographer Alexander von Humboldt secured a teaching position for him at the University of Breslau in Silesia—modern Wroclaw in Poland. Then he became a professor at the General Military School in Berlin in 1828 and later at the University of Berlin in 1839. Finally, in 1855, four years before his death of heart trouble, he succeeded Gauss in his chair at the University of Göttingen in the Germanic Confederation.

While Riemann's definition of integral is applicable to some functions with infinitely many discontinuities, there are heavily discontinuous functions that fail to be Riemann integrable. The following is a classical example given, rather reluctantly, by Dirichlet at the end of his 1829 paper on trigonometric series. If $f : (0, 1) \to \mathbb{R}$ is defined by

$$f(x) = \begin{cases} 1 & \text{if } x \text{ is rational} \\ 0 & \text{if } x \text{ is irrational,} \end{cases}$$

then the sum $\sum_{i=1}^{n} f(t_i)(x_i - x_{i-1})$ has value one if each t_i is chosen to be a rational number and value zero if each t_i is chosen to be irrational. Therefore, this sum has no limit as $n \to \infty$, and f is not Riemann integrable. This strange function, which is discontinuous at every point, has become known as the *Dirichlet function*.

His work on trigonometric series notwithstanding, Dirichlet's greatest achievements were in the theory of numbers, to which he contributed several important theorems—such as the fact that there are infinitely many prime numbers ending in 1, 3, 7, and 9, respectively. His great work on the subject, *Vorlesungen über Zahlentheorie*, appeared posthumously in 1863 and was reedited four times.

In 1859 Riemann succeeded Dirichlet in his chair at Göttingen, which had already become one of the foremost mathematical centers of the time. A timid and unassuming man, in fact a candid soul, Riemann can be regarded as the first modern mathematician of the nineteenth century, and his work is not easily described in elementary terms. His contributions and his influence on modern mathematics are incomparable. Already in his thesis of 1851, written under Gauss and dealing with complex function theory, he proved the existence of a conformal mapping of a simply connected domain in the plane onto any other simply connected domain—the *Riemann mapping theorem*—which led him to the creation of *Riemann surfaces* or analytic manifolds. We have already mentioned his 1853 presentation on Fourier series and his discovery of the Riemann integral, a piece of work which was only part of the requirements to obtain a position as *Privatdozent*. To qualify for this position he also delivered a report—*Habilitationsvortrag*—on the foundations of geometry in June of 1854, introducing some n-dimensional topological manifolds with an associated metric: the *Riemann manifolds* of non-Euclidean geometry. On the recommendation of Dirichlet he started receiving a stipend from the university in 1855, and in 1857 he became a lecturer. This was a welcome financial relief because, after the death of his brother that very same year and of a younger sister soon after, he was the sole guardian of his remaining three sisters. In 1859, the year he became Dirichlet's successor, he made a remarkable contribution to number theory by applying the methods of complex function theory, in particular through his use of the so called *Riemann zeta function*. On July 20, 1866, he succumbed to tubercolosis in Italy, where he spent the last few years of his life for health reasons, shortly after his election to the *Académie des Sciences* of Paris and to the Royal Society of London.

★   Another mathematician interested in trigonometric series—the particular series of events leading to this interest will be explored in Chapter 3—was Georg Ferdinand Louis Philippe Cantor (1845–1918), born in Saint Petersburg of Danish descent. His father had come to Russia from Copenhagen, but the family moved to Wiesbaden in the Germanic Confederation when Cantor was still a child. Originally a student of engineering to please his father, soon, like Riemann, he turned to mathematics. In 1869 he became a *Privatdozent* and in 1879 a professor at the University of Halle, where he remained until his retirement in 1905. His interest in trigonometric series was centered on the problem of uniqueness, that is, whether or not two such series that converge to the same function on $(-\pi, \pi)$ are necessarily identical. More precisely, if

$$(18) \qquad \tfrac{1}{2}a_0 + \sum_{n=1}^{\infty}(a_n \cos nx + b_n \sin nx) = 0$$

for all points x in $(-\pi, \pi)$ except, possibly, those of some small set P, then are all the coefficients a_n and b_n equal to zero? He answered this question in the affirmative in 1870 provided that (18) holds for all x, and then in 1871, the year of the unification of Germany with the formation of the German Empire under Kaiser Wilhelm I, he was able to show that uniqueness still holds if the convergence of (18) is given up at a finite number of points. But Cantor was ambitious and found these results short of what he wanted to do, namely to allow the convergence to fail at an infinite number of points and still be able to conclude uniqueness. But then, what kind of infinite set of points should this be? Surely, not every infinite set is a likely candidate.

In 1872, in order to construct such an infinite set, Cantor found that he needed to develop first a theory of the real numbers. Having accomplished this and stated as an axiom that to every real number there corresponds a unique point on a straight line whose coordinate is equal to the given number, he started the construction of his set as follows. First he defined the important concept of limit point:

> Given a set of points P, if there are an infinite number of points of P in every neighborhood, no matter how small, of a point p, then p is said to be a limit point of the set P.

By a neighborhood of p Cantor meant an open interval containing p. Then he defined the *derived set* P' of P as the set of all limit points of P, the second derived set P'' of P as the derived set of P', and so on until, after n iterations, the nth derived set $P^{(n)}$ of P is the derived set of $P^{(n-1)}$. Later he would refer to a set P whose nth derived set is empty for some n as a set of the *first species*, all other sets being of the *second species*. Cantor's 1872 paper on trigonometric series is concerned with sets of the first species exclusively. He proved the existence of sets of the first species by starting with an arbitrary number and, using the axiom stated above, identifying it with a point p on the real line. Then he took $P^{(n)} = \{p\}$ and worked backward from here. Using his newly constructed theory of the real numbers, he then considered the points corresponding to a sequence of rational numbers determining p as the elements of $P^{(n-1)}$, then a sequence of rationals determining each point of $P^{(n-1)}$, and so on. The set of points corresponding to all these rationals is of the first species. On this basis, he proved his most general uniqueness theorem: that, if (18) holds for all values of x in $[-\pi, \pi]$ except for those corresponding to a subset of the first species, then $a_0 = 0$ and $a_n = b_n = 0$ for all $n \geq 1$.

Having found his motivation on questions about trigonometric series, Cantor had just laid the foundations on which he would then build his acclaimed and controversial theory of sets. He defined many of the concepts familiar to anyone acquainted with mathematical analysis. Here is a sample: an *infinite* set is one that can be put in a one-to-one correspondence with a subset of itself, a *closed* set is one that contains all its limit points, a set is *open* if each of its points is the center of an interval contained in the set, a set is *countable* if it can be put in a one-to-one correspondence with the positive integers, and many more. He introduced the concepts of *union* and *intersection* of sets and proved that the set of rational numbers is countable and that the set of real numbers is not. But his main work on set theory and transfinite numbers was yet to begin and would be developed in a series of papers in the *Mathematische Annalen* beginning in 1879 and concluding with two major works of 1895 and 1897 that the interested reader can consult in an edited English translation.[1]

This is, then, the way it was: in 1870 Cantor gave the first steps toward the theory of sets by investigating the set of points at which (18) may fail to hold and still imply that $a_n = b_n = 0$. This is, instead, the way it could have been: in 1870 Hermann Hankel (1839–1873) could have given the first steps toward the theory of sets when he investigated the set of points at which a function may be discontinuous and still Riemann integrable. But it was not to be. In fact, his premature death prevented him from even becoming acquainted with Cantor's greatest achievements. A professor at the University of Tübingen, Hankel had been a student of Riemann at Göttingen and was seeking a necessary and sufficient condition for integrability. True, Riemann had already given such a condition but, in view of his example of a highly discontinuous integrable function, Hankel wanted to characterize integrability in terms of the set of points at which a function is discontinuous and started by defining the *jump* of f at a point x_0 to be the largest of all positive numbers σ such that in any interval containing x_0 there is an x for which $|f(x) - f(x_0)| > \sigma$. Then, if S_σ denotes the set of points at which the jump of f is greater than σ, Hankel came to the conclusion that a bounded function is integrable if and only if, for every $\sigma > 0$, the set S_σ can be enclosed in a finite collection of intervals of arbitrarily small total length. When a set can be so enclosed we shall say that it has *content zero* and, otherwise, that it has *positive content*. Therefore, a bounded function is integrable if and only if for every $\sigma > 0$ the set S_σ has content zero.

But instead of developing these ideas, Hankel next made a mistake and stated the wrong theorem. First he defined a set to be *nowhere dense*—this is the modern termi-

[1] *Contributions to the Founding of the Theory of Transfinite Numbers*, with a long introduction by P. E. B. Jourdain, Dover, New York, 1955.

nology, due to Cantor—if between any two of its points there is an entire interval that contains no points of the set. And then, erroneously thinking that a set has content zero if and only if it is nowhere dense, he stated that a bounded function is integrable if and only if for every $\sigma > 0$ the set S_σ is nowehere dense. Half of this statement is true, for it is quite clear that, if S_σ has content zero, then it is nowhere dense. But the converse is not true, although it was widely believed in large segments of the mathematical community for more than a decade. However, Henry John Stephen Smith (1826–1883), Savilian Professor of Geometry at Oxford, carefully read Hankel's paper and found the error. Furthermore, in a largely ignored paper, *On the Integration of Discontinuous Functions*, of 1875, he gave several methods to construct nowhere dense sets of positive content (see Exercise 9.1). The interesting thing is that it is easy to exhibit functions discontinuous on such sets that are not Riemann integrable (see Exercise 9.2).

In fact, quite a number of these functions can be exhibited, but they are all sufficiently weird and their mere existence does not detract from the wide generality and applicability of the Riemann integral. However, there is more. In 1881 Vito Volterra (1860–1940), then a student at the *Scuola Normale Superiore di Pisa*, used a nowhere dense set of positive content—which he discovered independently of Smith—to construct a nonconstant function f on $[0, 1]$ such that f' exists and is bounded at every point, but is not integrable.[1] The meaning of such an example is that, while f' always has an integral in the sense of prederivative, it may not have an integral in Riemann's sense. It can then be said that Riemann's definition is beginning to show some rough edges. And there are more, because it was known, at least since 1875, that the equality

$$(19) \qquad \int_a^b \left[\lim_{n \to \infty} f_n(x) \right] dx = \lim_{n \to \infty} \int_a^b f_n(x)\, dx.$$

need not hold. Here is a simple example due to René Louis Baire (1874–1932). Define $f_n : [0, 1] \to \mathbb{R}$ as follows: $f_n(x) = 1$ if $x = p/q$, where p and q are integers with no common factors and $q \leq n$, and $f_n(x) = 0$ otherwise. Each f_n is integrable because it is zero except at a finite number of points. But it is easy to see that the f_n converge to the Dirichlet function, known not to be integrable.

What all this means is that the definition of integrability must come up for review, although this was far from clear in the 1880s. In view of Hankel's characterization of this concept in terms of sets of content zero, the first thing to do is learn how to measure the content of a set. Significant contributions in this direction were made in France by Marie Ennemond Camille Jordan (1838–1922) and Émile Felix Edouard Justin Borel (1871–1956). But it was Henri Léon Lebesgue (1875–1941) who, building on their efforts, was able to give a definition of integral superior to Riemann's. He accomplished this in his doctoral dissertation *Intégrale, longueur, aire*, written in 1902 at the Sorbonne, and later, already a professor at the *Collège de France*, incorporated it in his *Leçons sur l'intégration et la recherche des fonctions primitives* of 1904.

---

[1]This construction, whose length does not allow its inclusion here, can be seen in T. Hawkins, *Lebesgue's Theory of Integration: Its Origins and Development*, 2nd ed., Chelsea, New York, 1975, page 57.

HENRI LÉON LEBESGUE, *circa* 1904
From Lebesgue's *Œuvres Scientifiques en Cinq Volumes*, vol. I.
Courtesy of *l'Enseignement Mathématique*, Université de Genève.

Lebesgue proceeded as follows. If S is a set of points contained in an interval $[a, b]$ and if $\{I_n\}$ is a sequence of open intervals in $\mathbb{R}$ such that S is contained in their union, then add the lengths of these intervals. If this is done for all possible sequences of such intervals, the sums so obtained form a set of positive numbers whose infimum—that is, the largest real number that is not larger than any of the sums—is called the *outer measure* of S. Its *inner measure* is defined to be $b - a$ minus the outer measure of $[a, b] - S$. Then S is said to be *measurable* when its outer and inner measures have the same value, and this value, denoted by $m(S)$, is called the *measure* of S.[1]

Using Lebesgue's definition of measure it is possible to show that

1. the measure of a set is nonnegative,

2. a set is measurable if and only if its complement is measurable,

3. the empty set and any finite set are measurable and have measure zero,

4. any interval is measurable and has its length as measure,

---

[1]It is possible to extend the definition of measure to unbounded sets, but we shall not do so here.

5. if $\{S_i\}$ is a (possibly finite) sequence of measurable sets—points in particular—then their union $\cup S_i$ is measurable with $m(\cup S_i) \leq \sum m(S_i)$, and equality holds if the S_i are disjoint, and

6. if S and T are measurable sets then $S - T$ is measurable and if, in addition, $T \subset S$, then $m(S - T) = m(S) - m(T)$, which implies that $m(T) \leq m(S)$ when $T \subset S$.[1]

Next, if S is a measurable set, Lebesgue defined a function $f : S \to \mathbb{R}$ to be *measurable* if and only if the set $\{x \in S : f(x) > r\}$ is measurable for any $r \in \mathbb{R}$.

Turning now to integration, Lebesgue's basic idea, in order to properly account for any discontinuities of the function, was to partition its range into subintervals, not its domain. More specifically, let S be a measurable subset of $[a, b]$, let $f : S \to \mathbb{R}$ be a measurable bounded function, and let m and M be numbers such that $m < f(x) < M$ for every x in S. Then let $m = y_0 < y_1 < \cdots < y_n = M$ be a partition of the interval $[m, M]$ and define the set

$$S_i = \{x \in S : y_{i-1} < f(x) \leq y_i\},$$

represented by the union of the darkened intervals in the example of Figure 2.11. Note

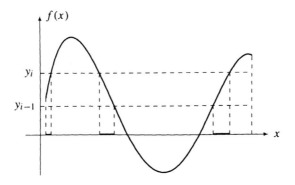

Figure 2.11

that each S_i is measurable because of the sixth property of measure since

$$S_i = \{x \in S : f(x) > y_{i-1}\} - \{x \in S : f(x) > y_i\},$$

and that these sets are disjoint. Then it is easy to see that there is a unique number I such that

$$\sum_{i=1}^{n} y_{i-1} m(S_i) \leq I \leq \sum_{i=1}^{n} y_i m(S_i)$$

---

[1] For a proof of these properties see, for instance, H. L. Royden, *Real Analysis*, 2nd ed., Macmillan, New York, 1968, Chapter 3.

for every partition $m = y_0 < y_1 < \cdots < y_n = M$ of $[m, M]$. In fact, given $\epsilon > 0$ let the partition be so fine that $y_i - y_{i-1} < \epsilon$ for $i = 1, \ldots, n$. Then the difference of the two sums above is

$$\sum_{i=1}^{n}(y_i - y_{i-1})m(S_i) < \epsilon \sum_{i=1}^{n} m(S_i) = \epsilon \, m(\cup S_i) \le \epsilon \, m([a, b]) = \epsilon \, (b - a),$$

which approaches zero as $\epsilon \to 0$. This unique number I is called the *Lebesgue integral* of f on S and is denoted by $\int_S f$ or, if $S = [a, b]$, by $\int_a^b f$. Note that it always exists if S and f are measurable.

Even a function as heavily discontinuous as the Dirichlet function defined at the beginning of the section is Lebesgue integrable. Indeed, if $0 = y_0 < y_1 < \cdots < y_n = 1$ is a partition of $[0, 1]$, then each S_i with $0 \le i < n$ is empty and S_n consists of the rational numbers in $[0, 1]$, which was shown to be countable by Cantor. Then the third and fifth properties of measure imply that $m(S_i) = 0$ for $0 \le i < n$ on $[0, 1]$ and that its integral is zero. It can also be shown that, if a bounded function is Riemann integrable, then it is Lebesgue integrable and that the two integrals have the same value. Thus, the class of Lebesgue integrable functions is larger than the class of Riemann integrable functions.

Furthermore, the Lebesgue integral is free from the defects of the Riemann integral that were pointed out above. First, if $\{f_n\}$ is a sequence of measurable functions that are uniformly bounded, that is, if $|f_n(x)| < M$ on $[a, b]$ for some $M > 0$ and all n, as is the case for the functions in Baire's example, then (19) always holds (a more general version of this result is stated in Exercise 9.13). That is, the order of the operations of passage to the limit and integration can be interchanged. As an immediate application of this fact, Lebesgue showed that a bounded derivative—which need not be Riemann integrable—is always Lebesgue integrable. Indeed, if F is a function that has a bounded derivative f, we have

$$\int_a^x f(s)\, ds = \lim_{h \to 0} \frac{1}{h} \int_a^x [F(s + h) - F(s)]\, ds$$

$$= \lim_{h \to 0} \frac{1}{h}\left[\int_x^{x+h} F - \int_a^{a+h} F \right]$$

$$= F(x) - F(a).$$

In this application we have just a sample of the power of the Lebesgue integral. It should not be thought that its only virtue is to turn some pretty strange functions into integrable objects. That is, actually, neither here nor there. The point is that, by allowing us to make very general statements and by eliminating any concern that a particular argument may be spoiled in the end by the presence of one single function that is not *comme il faut*, the Lebesgue integral becomes a very powerful tool both for theoretical purposes and in applications.[1]

---

[1] For proofs of the results stated above and a good exposition of the theory of Lebesgue integration see, for instance, H. L. Royden, *op. cit.*, Chapter 4. A much briefer development is contained in Exercises 9.3 to 9.16.

This is particularly so in the case of Fourier analysis, the field that Lebesgue chose to show the applicability of his new integral and, in particular, of those new features that are not valid for the Riemann integral. He wrote several papers on the subject and even a book, *Leçons sur les séries trigonométriques*, in 1906. The previous year Lebesgue had closed a full circle by giving a sufficient condition for the convergence of a Fourier series that contained all previously known sufficient conditions. Namely, if we put $g_x(t) = f(x + t) + f(x - t) - f(x+) - f(x-)$, then the Fourier series of f on $(-\pi, \pi)$ converges to the value

$$\tfrac{1}{2}[f(x+) + f(x-)]$$

at every point x for which

$$\frac{1}{\delta} \int_0^\delta |g_x(t)| \, dt \to 0$$

as $\delta \to 0$ and

$$\int_\eta^\pi \frac{|g_x(t) - g_x(t + \eta)|}{t} \, dt \to 0$$

as $\eta = \pi/n \to 0$.[1] Then, in his book, he closed a wider circle by showing that the Fourier series of an integrable function, convergent or divergent, can always be integrated term by term. It was this integration that allowed us to determine the Fourier coefficients at the start of this chapter.

EXERCISES

Find the Fourier series on $(-\pi, \pi)$ of the functions in Exercises 2.1 through 2.5 below. In each case make a sketch of the function.

2.1 $f(x) = \begin{cases} -\pi, & -\pi < x \leq 0 \\ x, & 0 < x < \pi \end{cases}$

2.2 $f(x) = \begin{cases} -x + \dfrac{\pi}{2}, & -\pi < x \leq 0 \\ x - \dfrac{\pi}{2}, & 0 < x < \pi \end{cases}$

2.3 $f(x) = \begin{cases} x + \pi, & -\pi < x \leq 0 \\ x, & 0 < x < \pi \end{cases}$

2.4 $f(x) = e^{ax}$, where a is a nonnegative constant.

---

[1] See, for instance, Zygmund, *op. cit.*, page 65.

2.5 $f(x) = \frac{1}{2}(\pi - x)\sin x$.

2.6 Determine whether each of the following functions is even or odd, and then find its Fourier series on $(-\pi, \pi)$

$(i)\ f(x) = \begin{cases} -1, & -\pi < x \le 0 \\ 1, & 0 < x < \pi \end{cases}$
$\qquad (ii)\ f(x) = \begin{cases} -e^{-x}, & -\pi < x \le 0 \\ e^x, & 0 < x < \pi \end{cases}$

$(iii)\ f(x) = x|x|$
$\qquad\qquad\qquad (iv)\ f(x) = |\sin x|$.

2.7 Find the Fourier series of the following functions on $(-\pi, \pi)$, where δ is a positive constant smaller than π, and also find the limiting values of their Fourier coefficients as $\delta \to 0$

$(i)\ f(x) = \begin{cases} 0, & -\pi < x < -\delta \\ 1/\delta, & -\delta \le x \le \delta \\ 0, & \delta < x < \pi \end{cases}$
$\qquad (ii)\ f(x) = \begin{cases} 0, & -\pi < x < -\delta \\ \dfrac{\delta + x}{2\delta^2}, & -\delta \le x \le 0 \\ \dfrac{\delta - x}{2\delta^2}, & 0 < x \le \delta \\ 0, & \delta < x < \pi \end{cases}$

2.8 Show that, if $f, g : (-\pi, \pi) \to \mathbb{R}$ are functions whose Fourier coefficients are a_n, b_n and α_n, β_n, respectively, and if r is a real number, then the Fourier coefficients of
(i) $f + g$ are $a_n + \alpha_n$ and $b_n + \beta_n$ and of
(ii) rf are ra_n and rb_n.

2.9 Find the Fourier series on $(-\pi, \pi)$ of
(i) $f(x) = \sinh x$ and
(ii) $f(x) = \cosh x$.

2.10 Show that any function $f : (-\pi, \pi) \to \mathbb{R}$ can be written in a unique way as the sum of an even function and an odd function, called its *even* and *odd components*.

2.11 Write the function

$$f(x) = \begin{cases} 0, & -\pi < x \le 0 \\ \sin x, & 0 < x < \pi \end{cases}$$

as the sum of its even and odd components, and then find its Fourier series on $(-\pi, \pi)$.

2.12 Write the function $f(x) = \frac{1}{2}(\pi - x)\sin x$ as the sum of its even and odd components, and then find the Fourier series of its even component on $(-\pi, \pi)$.

2.13 Write the function

$$f(x) = \begin{cases} x + \pi, & -\pi < x \le 0 \\ x, & 0 < x < \pi \end{cases}$$

as the sum of its even and odd components, and then find the Fourier series of its odd component on $(-\pi, \pi)$.

2.14 A function $f : (-\pi, \pi) \to \mathbb{R}$ is said to be *alternating* if and only if $f(x + \pi) = -f(x)$ for $-\pi < x < 0$. Show that, if f is alternating, then its Fourier series on $(-\pi, \pi)$ contains only terms of the form $a_n \cos nx + b_n \sin nx$, where n is odd.

2.15 Show that each of the following functions is alternating in the sense of Exercise 2.14 and then find its Fourier series on $(-\pi, \pi)$.

(i) $f(x) = \begin{cases} x(\pi + x), & -\pi < x \le 0 \\ x(\pi - x), & 0 < x < \pi \end{cases}$ (ii) $f(x) = \dfrac{\pi}{8}(\pi - 2|x|)$.

(iii) $f(x) = \begin{cases} -1, & -\pi < x < -\pi/2 \\ 1, & -\pi/2 \le x \le \pi/2 \\ -1, & \pi/2 < x < \pi \end{cases}$

2.16 Show that, if $0 < \delta < \pi$, the nth Fourier coefficient on $(-\pi, \pi)$ of the function shown in Figure 2.12 is

$$b_n = \frac{2 \sin n\delta}{\delta(\delta - \pi)n^2}.$$

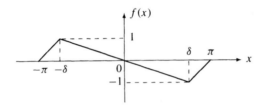

Figure 2.12

2.17 For which value of δ is the function in Exercise 2.16 alternating (see Exercise 2.14)? Find its Fourier series on $(-\pi, \pi)$ for this value of δ and also for $\delta = 0$ and $\delta = \pi$. Then show that in the last two cases these coefficients can be obtained by letting $\delta \to 0$ and $\delta \to \pi$ in the expression for b_n in Exercise 2.16.

2.18 Prove the following formula due to Leopold Kronecker. If f is a continuous function and p is a polynomial function of degree m, then

$$\int pf = pf_1 - p'f_2 + p''f_3 - \cdots + (-1)^m p^{(m)} f_{m+1} + C,$$

where C is a constant, $p', p'', \ldots, p^{(m)}$ are the first m derivatives of p, f_1 is a prederivative of f, and, for $1 \le k \le m$, f_{k+1} is a prederivative of f_k.

2.19 Use the formula in Exercise 2.18 to evaluate the following integrals for $m = 1, 2, 3, 4$:

$$\int_0^\pi x^m \cos nx \quad \text{and} \quad \int_0^\pi x^m \sin nx.$$

2.20 Use the result of Exercise 2.19 to find the Fourier series on $(-\pi, \pi)$ of

(i) $f(x) = x^2$ (ii) $f(x) = x^3$ (iii) $f(x) = x^4$

(iv) $f(x) = x^3 - \pi^2 x$ (v) $f(x) = x^4 - 2\pi^2 x^2$.

3.1   Let $f : (a, b) \to \mathbb{R}$ be a bounded function that is either nonincreasing or nondecreasing on (a, b). Prove that f is Riemann integrable.

3.2   Show that, if a function $f : (-\pi, \pi) \to \mathbb{R}$ has no negative values, then the its Fourier coefficients on $(-\pi, \pi)$ satisfy the inequalities $|a_n| \leq a_0$ and $|b_n| \leq a_0$. If, in addition, f is odd, then $|b_n| \leq n b_1$.

3.3   Prove or disprove that there a Riemann integrable function whose Fourier series on $(-\pi, \pi)$ is

$$\tfrac{1}{2} + \sum_{n=1}^{\infty} (\cos nx + \sin nx).$$

3.4   Consider the odd function $f : (-\pi, \pi) \to \mathbb{R}$ defined by $f(x) = \tfrac{1}{2} \cot \tfrac{1}{2} x$. Prove that $b_n = 1$ for all $n \geq 1$. Does this contradict the Riemann-Lebesgue theorem (**Hint:** if $n > 1$, show that $b_n - b_{n-1} = 0$)?

3.5   Let $f : (-\pi, \pi) \to \mathbb{R}$ be a piecewise continuous function such that $f(-\pi+) = f(\pi-)$ and such that f' is piecewise continuous on $(-\pi, \pi)$. Prove the following limits for its Fourier coefficients on $(-\pi, \pi)$:

$$\lim_{n \to \infty} na_n = 0 \qquad \text{and} \qquad \lim_{n \to \infty} nb_n = 0.$$

Which of these two limits is still valid if the hypothesis $f(-\pi+) = f(\pi-)$ is deleted? Show by example that the other limit need not hold.

3.6   Show that, if a trigonometric series of the form

$$\tfrac{1}{2}a_0 + \sum_{n=1}^{\infty} (a_n \cos nx + b_n \sin nx)$$

converges uniformly [1] on $(-\pi, \pi)$ to a function $f : (-\pi, \pi) \to \mathbb{R}$, then the constants a_0, a_n, and b_n must be the Fourier coefficients of f on $(-\pi, \pi)$.

3.7   Show that, if $f, g : [-\pi, \pi] \to \mathbb{R}$ are continuous functions and if there is a point x_0 in $[-\pi, \pi]$ such that $f(x_0) \neq g(x_0)$, then f and g cannot have the same Fourier coefficients on $(-\pi, \pi)$. (**Hint:** complete the details of the following proof by Lebesgue:

(*i*) If f and g have the same Fourier coefficients on $(-\pi, \pi)$, m is a positive integer, and $T_m(x)$ is a trigonometric polynomial of the form

$$T_m(x) = A_0 + \sum_{n=1}^{m} (A_n \cos nx + B_n \sin nx),$$

where A_0, A_n, and B_n are arbitrary constants, then show that

$$\int_{-\pi}^{\pi} (f - g)T_m = 0.$$

(*ii*) Show that x_0 can be assumed to be in $(-\pi, \pi)$ without loss of generality. If $f(x_0) - g(x_0) = c > 0$ (otherwise interchange the roles of f and g) let $\delta > 0$ be so

---

[1] Refer to Appendix A for the necessary definitions of and results for uniform convergence.

small that the interval $I_\delta = [x_0 - \delta, x_0 + \delta]$ is in $(-\pi, \pi)$ and $f(x) - g(x) \geq c/2$ if x is in I_δ. Then define

$$T_m(x) = [1 - \cos(x - x_0) - \cos \delta]^m$$

and show, by induction on m, that $T_m(x)$ is a trigonometric polynomial of the form given in (i).

(iii) Show that $|T_m(x)| \geq 1$ for x in I_δ and $|T_m(x)| < 1$ for x in $(-\pi, \pi)$ but outside I_δ. Then, by a suitable partition of the interval of integration, show that

$$\int_{-\pi}^{\pi} (f - g)T_m > 0.$$

This contradicts the result of (i) and proves the desired result.)

3.8 Use the results of Exercises 3.6 and 3.7 to prove that, if $f : [-\pi, \pi] \to \mathbb{R}$ is a continuous function whose Fourier series converges uniformly on $[-\pi, \pi]$, then the sum of this series is $f(x)$ at every point x in $[-\pi, \pi]$ and $f(-\pi) = f(\pi)$.

4.1 Verify the following properties of the Dirichlet kernel:

(i) $D_N(\pi) = (-1)^N$,

(ii) the smallest positive number t such that $D_N(t) = 0$ is $t = 2\pi/(2N + 1)$,

(iii) if n is an integer, the value of $\int_{n\pi}^{(n+1)\pi} D_N$ is independent of n, and

(iv) there is a constant C such that

$$\int_0^{\pi} |D_N| \geq C \log N$$

(**Hint:** for (iv) show first that

$$1 + \frac{1}{2} + \cdots + \frac{1}{N} > \int_1^N \frac{1}{t} \, dt \,).$$

4.2 Show that the Fourier series on $(-\pi, \pi)$ of the function given by $f(x) = |\sin x|$ converges to $f(x)$ at each point x in $[-\pi, \pi]$. Then, by substituting $x = 0$ and $x = \pi/2$ in this series (see Exercise 2.6 (iv), show that

$$\sum_{n=1}^{\infty} \frac{1}{4n^2 - 1} = \frac{1}{2} \quad \text{and} \quad \sum_{n=1}^{\infty} \frac{(-1)^{n+1}}{4n^2 - 1} = \frac{\pi - 2}{4}.$$

4.3 Prove that

$$|x| = \frac{\pi}{2} - \frac{4}{\pi} \sum_{n=1}^{\infty} \frac{\cos(2n - 1)x}{(2n - 1)^2}$$

on $[-\pi, \pi]$. What equality results by substituting either $x = \pi$ or $x = -\pi$ above?

4.4 Prove that, if a is a nonintegral constant, then

$$\cos ax = \frac{2a \sin a\pi}{\pi} \left[\frac{1}{2a^2} + \sum_{n=1}^{\infty} \frac{(-1)^n}{a^2 - n^2} \cos nx \right]$$

for every x in $[-\pi, \pi]$. Then derive the following formulas due to Euler

$$\frac{a\pi}{\sin a\pi} = 1 + 2a^2 \sum_{n=1}^{\infty} \frac{(-1)^n}{a^2 - n^2}$$

and

$$a\pi \cot a\pi = 1 + 2a^2 \sum_{n=1}^{\infty} \frac{1}{a^2 - n^2}.$$

4.5 Using the Fourier series on $(-\pi, \pi)$ of the function in Exercise 2.4, show that

$$a\pi \coth a\pi = 1 + 2a^2 \sum_{n=1}^{\infty} \frac{1}{a^2 + n^2}.$$

for $a \neq 0$.

4.6 Using the Fourier series on $(-\pi, \pi)$ of the functions in Exercise 2.20 (i), (iii) and (iv), show that

(i) $\displaystyle\sum_{n=1}^{\infty} \frac{1}{n^2} = \frac{\pi^2}{6}$ (ii) $\displaystyle\sum_{n=1}^{\infty} \frac{(-1)^{n+1}}{n^2} = \frac{\pi^2}{12}$

(iii) $\displaystyle\sum_{n=1}^{\infty} \frac{1}{n^4} = \frac{\pi^2}{90}$ (iv) $\displaystyle\sum_{n=1}^{\infty} \frac{(-1)^{n+1}}{(2n-1)^3} = \frac{\pi^3}{32}.$

4.7 If N is a positive integer, prove the following formula for the Nth partial sum of a Fourier series

$$|s_N(x)| \leq A \log N + B,$$

where A and B are positive constants and log is the natural logarithm. Thus, whether or not $\{s_N(x)\}$ converges, the sequence of absolute values cannot grow faster than $\log N$ (**Hint:** use the substitution $\left(N + \frac{1}{2}\right)t = u$ and the inequalities $\sin\frac{1}{2}t \geq t$ on $[0, \pi]$ and $\sin t \leq t$ for $t \geq 0$).

4.8 Show that the Fourier series of the function given by

$$f(x) = \begin{cases} -\sqrt{-x}, & -\pi < x \leq 0 \\ \sqrt{x}, & 0 < x < \pi \end{cases}$$

on $(-\pi, \pi)$ converges to $f(x)$ at every point x in $[-\pi, \pi]$. This shows that the existence of the one-sided derivatives in Theorems 2.2 or 2.3 is not necessary for convergence.

4.9 Prove *Dini's convergence test*: if $f : \mathbb{R} \to \mathbb{R}$ is a periodic function of period 2π that is integrable on $(-\pi, \pi)$ and if x and r are real numbers such that

$$\frac{f(x+t) + f(x-t) - 2r}{t}$$

is integrable on $[0, \delta]$ for an arbitrarily small $\delta > 0$, then the Fourier series of f on $(-\pi, \pi)$ converges to r at x. Then deduce the Lipschitz test from Dini's test.

4.10 Assume that a function f is equal to its Fourier series, with coefficients a_n and b_n, on some interval. In practical applications one simply takes the first few terms of the series to approximate the function. Thus, the faster the Fourier coefficients decrease as n increases, the fewer the number of terms that are needed in the approximation. This suggests the following method of improving the convergence: find constants α_n, A_n, β_n, and B_n such that

(a) $a_n = \alpha_n + A_n$ and $b_n = \beta_n + B_n$,

(b) $|\alpha_n|$ and $|\beta_n|$ decrease faster than $|a_n|$ and $|b_n|$ as n increases, and

(c) the sum of the series

$$\tfrac{1}{2}A_0 + \sum_{n=1}^{\infty}(A_n \cos nx + B_n \sin nx)$$

is a known function $g(x)$.

Then

$$f(x) = g(x) + \tfrac{1}{2}\alpha_0 + \sum_{n=1}^{\infty}(\alpha_n \cos nx + \beta_n \sin nx).$$

Apply this method to improve the convergence in

(i) $f(x) = \sum_{n=1}^{\infty}(-1)^{n+1}\dfrac{1+n^5}{n^6}\sin nx$ (ii) $f(x) = \sum_{n=1}^{\infty}(-1)^n\dfrac{n^3-1}{n^5}\cos nx$

(**Hint:** use the series in Examples 2.2 and 2.14).

5.1 Find the Fourier series on $(-1, 1)$ of each of the functions in Figure 2.5a–2.5c.

5.2 Find the Fourier series on $(-2, 2)$ of the function in Figure 2.5d.

5.3 Find the Fourier series of the function $f(x) = x$ on $(\pi, 2\pi)$.

5.4 Show that the Fourier series of the function $f(x) = ax^2 + bx$ on $(0, 2\pi)$ is

$$\frac{4a\pi^2}{3} + b\pi + \sum_{n=1}^{\infty}\left(\frac{4a}{n^2}\cos nx - \frac{4a\pi + 2b}{n}\sin nx\right).$$

Use this result to find the sum of the following series on $(0, 2\pi)$:

(i) $\sum_{n=1}^{\infty}\dfrac{\sin nx}{n}$ (ii) $\sum_{n=1}^{\infty}\dfrac{\cos nx}{n^2}$ (ii) $\sum_{n=1}^{\infty}\dfrac{1}{n^2}(\cos nx - n\pi \sin nx)$.

5.5 If a is a real number such that $a > \pi/2$ and $a \neq n\pi$ for any positive integer n, find the Fourier series on $(-a, a)$ of

$$f(x) = \begin{cases} 0, & -a < x < -\pi/2 \\ \cos x, & -\pi/2 \leq x \leq \pi/2 \\ 0, & \pi/2 < x < a \end{cases}$$

5.6 Prove that for x in $(0, \pi)$

$$\cos x = \frac{8}{\pi}\sum_{n=1}^{\infty}\frac{n}{4n^2-1}\sin 2nx.$$

5.7 Find a Fourier series expansion of $f(x) = e^x$ valid on $(0, 1)$ that contains

(i) no cosine terms,

(ii) no sine terms, and

(iii) both sine and cosine terms.

5.8 Prove that for x in $[\pi, 2\pi]$

$$x = \frac{3\pi}{2} + \frac{4}{\pi} \sum_{n=1}^{\infty} \frac{\cos(2n-1)x}{(2n-1)^2}.$$

Compare this result with the result of Exercise 5.3 and explain.

5.9 If a and δ are positive numbers such that $2\delta \leq a$, find the Fourier sine series on $(0, a)$ of the function f shown in Figure 2.13. Then find the limit of the nth Fourier coefficient as $\delta \to 0$. Is this limiting value the nth Fourier coefficient of the function in Exercise 2.6 (i)?

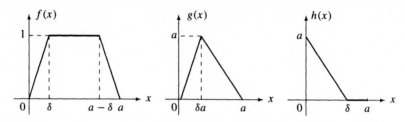

Figure 2.13

5.10 If a and δ are positive numbers and $\delta \leq 1$, find the Fourier sine series on $(0, a)$ of the function g shown in Figure 2.13. Then find the limit of the nth Fourier coefficient as $\delta \to 1$. Is this limiting value the nth Fourier coefficient of the function in Example 2.2 when $a = \pi$?

5.11 If a and δ are positive numbers such that $\delta < a$, find the Fourier sine series and the Fourier cosine series on $(0, a)$ of the function h shown in Figure 2.13. Show that the limit as $\delta \to a$ of the nth coefficient of the Fourier sine series found here is equal to the limit as $\delta \to 0$ of the nth Fourier coefficient found in Exercise 5.10.

5.12 Using the Fourier series found in some of the exercises in this section, find the sum of each of the following numerical series:

(i) $\displaystyle\sum_{n=1}^{\infty} \frac{\sin n}{n}$ $\qquad$ (ii) $\displaystyle\sum_{n=1}^{\infty} (-1)^{\frac{n+3}{2}} \frac{\sin n}{n}.$

6.1 Find the derivative of the function

$$\varphi(x) = 2 \int_0^{\left(N + \frac{1}{2}\right)} x \frac{\sin u}{u} \, du + \pi$$

at $x = 0$.

6.2 Prove that the Fourier series of a function that has a jump discontinuity cannot converge uniformly to the function in any interval containing the point of discontinuity.

6.3 Use the expression for $\int_0^x D_N$ developed in this section to show that

$$\lim_{x \to \infty} \int_0^x \frac{\sin t}{t} \, dt = \frac{\pi}{2}.$$

6.4 Demonstrate the existence of the Gibbs phenomenon for the function defined on $(-\pi, \pi)$ by

$$f(x) = \tfrac{1}{2}(\pi - x).$$

6.5 Demonstrate the existence of the Gibbs phenomenon for the sum of the Fourier series

$$2 \sum_{n=1}^{\infty} \frac{(-1)^{n-1}}{n} \sin nx$$

(**Hint:** use a translation along the x-axis).

7.1 If $\sigma_N(x)$ denotes the Nth Fejér sum of a Fourier series on $(-\pi, \pi)$ as in Definition 2.1, show that for $N \geq 1$

$$\sigma_N(x) = \tfrac{1}{2}a_0 + \sum_{n=1}^{N} \frac{N + 1 - n}{N + 1}(a_n \cos nx + b_n \sin nx).$$

7.2 At which point would the proof of Theorem 2.7 fail, and why, if it is attempted with D_N in place of F_N (**Hint:** refer to Exercise 4.1 (iv))?

7.3 This exercise is devoted to proving one of the classic results of mathematical analysis.

(i) Show that, if an interval $[a, b]$ is contained in $(-\pi, \pi)$ and if $f : [a, b] \to \mathbb{R}$ is a continuous function, then, for every $\epsilon > 0$, there is a trigonometric polynomial,

$$T(x) = \sum_{n=0}^{N}(A_n \cos nx + B_n \sin nx),$$

such that $|f(x) - T(x)| < \epsilon$ on $[a, b]$.

(ii) Make a change of variable mapping $[a, b]$ into a subset of $(-\pi, \pi)$ and then use (i) to prove the *Weierstrass second approximation theorem*: if $f : [a, b] \to \mathbb{R}$ is a continuous function, then, for every $\epsilon > 0$, there is a trigonometric polynomial $T(x)$ such that $|f(x) - T(x)| < \epsilon$ on $[a, b]$

(iii) Prove the *Weierstrass approximation theorem*: if $f : [a, b] \to \mathbb{R}$ is a continuous function, then, for every $\epsilon > 0$ there is a polynomial $P(x)$ such that $|f(x) - P(x)| < \epsilon$ on $[a, b]$ (**Hint:** use (ii) to obtain a trigonometric polynomial $T(x)$ such that $|f(x) - T(x)| < \epsilon/2$ and then use the fact that the Taylor series for the sine and cosine functions converge uniformly on $\mathbb{R}$ to find a polynomial $P(x)$ such that $|T(x) - P(x)| < \epsilon/2$).

7.4 Show that, if a Fourier series is such that $s_N(x) \to \infty$ for some x, then $\sigma_N(x) \to \infty$. Do the same for the case in which ∞ is replaced by $-\infty$.

7.5 Show that, if $m \leq f(x) \leq M$ on $(-\pi, \pi)$, then $m \leq \sigma_N(x) \leq M$ for all N.

7.6 Let (a, b) be a subinterval of $(-\pi, \pi)$ and assume that $m \leq f(x) \leq M$ on (a, b). Show that for every $\epsilon > 0$ there is a positive integer N such that $m - \epsilon \leq \sigma_n(x) \leq M + \epsilon$ on $(a + \epsilon, b - \epsilon)$ for $n > N$ and then use this result to show that the Gibbs phenomenon cannot occur for Fejér sums.

8.1 Carry out the term by term integration of the Fourier series of

(i) $f(x) = x^3 - \pi^2 x$ obtained in Exercise 2.20(iv) and

(ii) $f(x) = |x|$ given in Exercise 4.3.

8.2 Use the identity in Exercise 5.6 to obtain the Fourier series expansion

$$\sin x = \frac{2}{\pi} + \frac{4}{\pi} \sum_{n=1}^{\infty} \frac{1}{1 - 4n^2} \cos 2nx.$$

8.3 Let $f : \mathbb{R} \to \mathbb{R}$ be a periodic function of period 2π that is piecewise continuous on $(-\pi, \pi)$ and has a Fourier series

$$\tfrac{1}{2}a_0 + \sum_{n=1}^{\infty} (a_n \cos nx + b_n \sin nx)$$

on $(-\pi, \pi)$. Show that, if a and b are any two points in $\mathbb{R}$, then

$$\int_a^b f = \tfrac{1}{2}a_0(b - a) + \sum_{n=1}^{\infty} \int_a^b (a_n \cos nx + b_n \sin nx).$$

8.4 State and prove the equivalent of Theorems 2.8 and 2.9 for Fourier series on an arbitrary interval (**Hint:** it is possible to use Theorem 2.8 in conjunction with a change of variable, but it is simpler to define

$$F(x) = \int_c^x f - \tfrac{1}{2}a_0 x,$$

where c is the center of the given interval, and imitate the proof of Theorem 2.8).

8.5 Obtain three new Fourier series on $(0, 2\pi)$ by term-by-term integration of

$$\frac{\pi - x}{2} = \sum_{n=1}^{\infty} \frac{\sin nx}{n}$$

(**Hint:** use some of the series in Exercise 4.6).

8.6 Let $f : (0, 2\pi) \to \mathbb{R}$ be a piecewise continuous function that is continuous on $(\pi, 2\pi)$ and has Fourier series

$$\sum_{n=1}^{\infty} \frac{\sin(2n - 1)x}{2n - 1}$$

on $(0, 2\pi)$. Find f (**Hint:** use the identity

$$x = \frac{3\pi}{2} + \frac{4}{\pi} \sum_{n=1}^{\infty} \frac{\cos(2n - 1)x}{(2n - 1)^2},$$

proved in Exercise 5.8 for x in $[\pi, 2\pi]$).

8.7 Prove that, under the hypotheses of Theorem 2.8,

$$\int_0^x f = \sum_{n=1}^{\infty} \frac{b_n}{n} + \sum_{n=1}^{\infty} \frac{1}{n} \left[(a_n + (-1)^{n+1} a_0) \sin nx - b_n \cos nx \right]$$

(**Hint:** use the Fourier series of Example 2.2).

8.8 Let $f : (-\pi, \pi) \to \mathbb{R}$ be a differentiable function such that

$$f(x) = \frac{\sinh \pi}{\pi} \left[1 + 2 \sum_{n=1}^{\infty} \frac{(-1)^n}{1 + n^2} \cos nx \right]$$

on $(-\pi, \pi)$. Find $\int_0^x f$ as a Fourier series, and show that $f(x) = \cosh x$.

8.9 If $f : (-\pi, \pi) \to \mathbb{R}$ is a differentiable function such that

$$f(x) = \frac{\sin \pi}{\pi} \left[1 + 2 \sum_{n=1}^{\infty} \frac{(-1)^n}{1 - n^2} \cos nx \right]$$

on $(-\pi, \pi)$, find $f(x)$ in closed form.

8.10 Show that, if f is an integrable function on $(-\pi, \pi)$, then the function F defined in the proof of Theorem 2.8 satisfies the conditions of the Lipschitz test and equals its Fourier series on $(-\pi, \pi)$.

9.1 Let I be a closed interval in $\mathbb{R}$ of length $2l$, where $l > 0$ is fixed. From I remove the open central subinterval of length $l/3$; from each of the two remaining subintervals remove the open central subinterval of length $l/3^2$; from each of the four remaining subintervals remove the open central subinterval of length $l/3^3$; and so on indefinitely. Show that the remaining subset S of I

    (i) has total length l,

    (ii) contains its derived set, and

    (iii) does not contain any nonempty open subinterval

(Hint: for (i) observe that the total length of the removed subintervals is

$$\frac{l}{3} + \frac{2l}{3^2} + \frac{4l}{3^3} + \frac{8l}{3^4} + \cdots).$$

9.2 Let I and S be as in Exercise 9.1 and define $f : I \to \mathbb{R}$ by

$$f(x) = \begin{cases} 1, & x \text{ in } S \\ 0, & x \text{ in } I - S \end{cases}$$

Show that f is not Riemann integrable on I but that it is Lebesgue integrable.

9.3 In this exercise we interpret the Lebesgue integral as a Riemann integral. Let S be a measurable set in $[a, b]$ and $f : S \to \mathbb{R}$ a measurable function such that $0 \leq f(x) \leq M$ for all x in S, and define a new function $\mu_f : [0, M] \to \mathbb{R}$ by

$$\mu_f(y) = m\{ x \in S : f(x) > y \},$$

where m denotes Lebesgue measure. Show that, if $\{y_0, \ldots, y_n\}$ is a partition of $[0, M]$, then

$$\sum_{i=1}^{n} y_i m(S_i) = \sum_{i=1}^{n} \mu_f(y_{i-1})(y_i - y_{i-1})$$

and

$$\sum_{i=1}^{n} y_{i-1} m(S_i) = \sum_{i=1}^{n} \mu_f(y_i)(y_i - y_{i-1}).$$

Then use Definition 2.2 to show that μ_f is Riemann integrable on $(0, M)$ and that the Lebesgue integral of f on S is equal to the Riemann integral

$$\int_0^M \mu_f.$$

If $f : S \to \mathbb{R}$ is not restricted to be bounded or nonnegative, then define the *measure function* of f, $\mu_f : \mathbb{R} - \{0\} \to \mathbb{R}$, by [1]

$$\mu_f(y) = \begin{cases} m\{x \in S : f(x) > y\}, & y > 0 \\ -m\{x \in S : f(x) < y\}, & y < 0. \end{cases}$$

When $|f| \le M$ arguments similar to those in Exercise 9.3 lead to the fact that

$$\int_S f = \mathcal{R} \int_{-M}^{M} \mu_f,$$

where the $\mathcal{R}$ in front of an integral indicates that it is a Riemann integral. Then, when the restriction $|f| \le M$ is removed, it is natural to define the *Lebesgue integral* of f on S to be

$$\int_S f = \mathcal{R} \int_{-\infty}^{\infty} \mu_f$$

if the improper Riemann integral on the right exists—for the definition and properties of this type of integral refer to Appendix B. If, in addition, it has a finite value, we say that f is *Lebesgue integrable* on S and we write $f \in \mathcal{L}(S)$.

It is possible to construct a rather simple theory of Lebesgue integration from the preceding definition and the properties of the Lebesgue measure listed in §2.9. This is done in the remaining exercises. It is assumed at all times that S is a measurable subset of an interval and that $f : S \to \mathbb{R}$ is a measurable function.

9.4   Show that $f(x) = 1/\sqrt{x}$ is Lebesgue integrable on $(0, 1)$ and evaluate its Lebesgue integral on this interval.

9.5   Show that the product of Lebesgue integrable functions need not be Lebesgue integrable (**Hint:** consider the square of the function in Exercise 9.4).

9.6   Prove the following properties of the measure function:
   (*i*) $\mu_f \le 0$ on $(-\infty, 0)$ and $\mu_f \ge 0$ on $(0, \infty)$,
   (*ii*) μ_f is nonincreasing on $(-\infty, 0)$ and on $(0, \infty)$,
   (*iii*) if $f \le g$, then $\mu_f \le \mu_g$, and
   (*iv*) if c is a nonzero constant, then

$$\mu_{cf}(y) = \frac{c}{|c|} \mu_f\left(\frac{y}{c}\right)$$

   for all $y \ne 0$.

9.7   We shall write μ_f^S instead of μ_f when it is necessary to specify the domain S of f. Let $\{S_i\}$ be a—possibly finite—sequence of disjoint, measurable sets and let $S = \cup S_i$. Show that
   (*i*) $\mu_f^S = \Sigma \mu_f^{S_i}$ and
   (*ii*) if $T \subset S$, then $\mu_f^T \ge \mu_f^S$ on $(-\infty, 0)$ and $\mu_f^T \le \mu_f^S$ on $(0, \infty)$
   (**Hint:** $S = T \cup (S - T)$ when $T \subset S$).

---

[1] The development of Lebesgue's integration theory outlined in Exercises 9.3 to 9.16 is adapted from the author's "The Lebesgue Integral as a Riemann Integral," *Int. J. Math. Math. Sci.*, **10** (1987) 693–706.

9.8   Use Exercises 9.6 and 9.7 to prove the following properties of the Lebesgue integral:

(*i*) if $f \leq g$, then $\int_S f \leq \int_S g$,

(*ii*) if f has a constant value c, then $f \in \mathcal{L}(S)$ and $\int_S f = cm(S)$,

(*iii*) if $m(S) = 0$, then $f \in \mathcal{L}(S)$ and $\int_S f = 0$,

(*iv*) if there are constants a and b such that $a \leq f \leq b$, then $f \in \mathcal{L}(S)$ and $am(S) \leq \int_S f \leq bm(S)$,

(*v*) if c is a nonzero constant, we have $\int_S cf = c \int_S f$ and then $f \in \mathcal{L}(S)$ implies that $cf \in \mathcal{L}(S)$,

(*vi*) if $\{S_i\}$ is a finite collection of disjoint measurable sets and $S = \cup S_i$, then $\int_S f = \Sigma \int_{S_i} f$, and

(*vii*) if $T \subset S$ and $f \in \mathcal{L}(S)$, then $f \in \mathcal{L}(T)$.

9.9   Prove that, if $f \in \mathcal{L}(S)$, then $|f| \in \mathcal{L}(S)$ and

$$\left| \int_S f \right| \leq \int_S |f|.$$

9.10   A property is said to hold *almost everywhere* on S if and only if the set of points at which it does not hold has measure zero. Show that, if $f = g$ almost everywhere on S, then $\int_S f = \int_S g$ when the integrals exist.

9.11   Let $f \in \mathcal{L}(S)$ and let $T \subset S$. Show that for any $\epsilon > 0$ there is a $\delta > 0$ such that

$$\left| \int_T f \right| < \epsilon$$

if $m(T) < \delta$; that is, integrals of integrable functions over small sets have small values (**Hint:** in view of Exercise 9.9 it is necessary only to consider the case $f \geq 0$).

9.12   Let $\{f_n\}$ be a sequence of functions such that $f_n \in \mathcal{L}(S)$ for each n and assume that there is a function $f \in \mathcal{L}(S)$ such that $f_n \to f$ uniformly on S. Prove that

$$\int_S f_n \to \int_S f$$

(**Hint:** given $\epsilon > 0$ there is a positive integer N such that $f - \epsilon \leq f_n \leq f + \epsilon$ for $n > N$, and then $\mu_{f-\epsilon} \leq \mu_{f_n} \leq \mu_{f+\epsilon}$ by Exercise 9.6 (*iii*). Show that if $n > N$ we have $\mu_f(y + \epsilon) \leq \mu_{f_n}(y)$ for $y > 0$ and $\mu_{f_n}(y) \leq \mu_f(y - \epsilon)$ for $y > \epsilon$. Then use these inequalities to show that

$$\mathcal{R} \int_0^\infty \mu_f - \epsilon m(S) \leq \mathcal{R} \int_0^\infty \mu_{f_n} \leq \mathcal{R} \int_0^\infty \mu_f + \epsilon m(S)$$

for $n > N$).

9.13   For this exercise we shall accept two results on measurable functions. In either case let $\{f_n\}$ be a sequence of measurable functions that converges to a limit function f on S. Then

(*a*) f is measurable and

(*b*) for any $\delta > 0$ there is a subset $T \subset S$ with $m(T) < \delta$ such that $f_n \to f$ uniformly on $S - T$ (*Egorov's theorem*, 1911).[1]

---

[1] For proofs see, for instance, Royden, *op. cit.*, pages 67 and 72.

Prove the *Lebesgue dominated convergence theorem* (1902): Let $\{f_n\}$ be a sequence of measurable functions that converges to a limit function f on S. If there is a function $g \in \mathcal{L}(S)$ such that $|f_n| \le g$ for all n, then

$$\int_S f_n \to \int_S f.$$

(**Hint:** first show that $f_n \in \mathcal{L}(S)$ for each n. Then, given $\epsilon > 0$ use Exercise 9.11 to choose a $\delta > 0$ such that

$$\left| \int_T g \right| < \frac{\epsilon}{4}$$

if $T \subset S$ and $m(T) < \delta$. Now use Egorov's theorem to conclude that there is a subset $T \subset S$ with $m(T) < \delta$ such that $f_n \to f$ uniformly on $S - T$. Next use Exercise 9.12 to show that there is an N such that

$$\left| \int_{S-T} f_n - \int_{S-T} f \right| < \frac{\epsilon}{2}$$

for $n > N$.)

9.14 Let f be a nonnegative measurable function. Show that,

(*i*) if $f_n \in \mathcal{L}(S)$, then for any $\delta > 0$ there is a bounded function $g \in \mathcal{L}(S)$ such that $0 \le g \le f$ and

$$\int_S f - \int_S g < \delta$$

and

(*ii*) if $f \notin \mathcal{L}(S)$, then for any $M > 0$ there is a bounded function $g \in \mathcal{L}(S)$ such that $0 \le g \le f$ and

$$\int_S g > M$$

(**Hint:** let $c > 0$ be so large that either

$$\int_S f - \mathcal{R} \int_0^c \mu_f < \delta$$

or

$$\mathcal{R} \int_0^c \mu_f > M$$

and then define $g = \min\{f, c\}$.)

9.15 Prove *Fatou's lemma* (1906): if $\{f_n\}$ is a sequence of nonnegative measurable functions that converges to a limit function f on S, then for any $\epsilon > 0$ there is a positive integer N such that

$$\int_S f \le \int_S f_n + \epsilon$$

for $n > N$ (**Hint:** let g be the function of Exercise 9.14 corresponding to $\delta = \epsilon/2$ if $f \in \mathcal{L}(S)$ or to an arbitrary M if $f \notin \mathcal{L}(S)$ and apply the Lebesgue dominated convergence theorem to $g_n = \min\{f_n, g\}$ to show that

$$\int_S g \le \int_S f_n + \frac{\epsilon}{2}$$

for $n > N$).

9.16 Prove the *Levi monotone convergence theorem* (1906): if $\{f_n\}$ is a sequence of nonnegative measurable functions that converges to a function f on S and such that $f_m \leq f_n$ on S for $m < n$, then

$$\int_S f_n \to \int_S f$$

(**Hint:** use Fatou's lemma).

3

RETURN TO THE HEATED BAR

§3.1 Existence of a Solution

In Chapter 1 we considered a heated bar or slab located along the x-axis between $x = 0$ and $x = a$, and, after defining the sets

and
$$D = \{ (x, t) \in \mathbb{R}^2 : 0 < x < a, \ t > 0 \}$$
$$\bar{D} = \{ (x, t) \in \mathbb{R}^2 : 0 \le x \le a, \ t \ge 0 \},$$

we posed the following boundary value problem: find a continuous function $u : \bar{D} \to \mathbb{R}$ that is of class C^2 (that is, u has continuous first- and second-order partial derivatives), satisfies the heat equation in D, vanishes at $x = 0$ and $x = a$ for all $t \ge 0$, and is such that $u(x, 0) = f(x)$ for $0 \le x \le a$, where f is a given continuous function representing the initial temperature distribution on the bar.

However, in some applications it is convenient to model a given physical problem using a function f that is only piecewise continuous (see Exercise 1.6), and for this reason we shall modify this problem as follows. Find a function $u : \bar{D} \to \mathbb{R}$ that is of class C^2 in D, is continuous on the set $D_0 = \{ (x, t) \in \bar{D} : t > 0 \}$, and satisfies

(1)	$u_t = k u_{xx}$	in D
(2)	$u(0, t) = 0$	$t > 0$
(3)	$u(a, t) = 0$	$t > 0$
(4)	$u(x, 0) = f(x)$	$0 \le x \le a$
(5)	$u(x, t) \to f(x)$	as $t \to 0$, $\quad 0 \le x \le a$,

where f is a given piecewise continuous function and (4) and (5) are required to hold only at any x at which f is continuous. Note that u cannot be required to be continuous on $\bar{D}$ because now f may not be continuous on $[0, a]$. Instead we require that u be continuous on D_0, and the meaning of the new condition (5) is that u should be a continuous function of t at $t = 0$ for each fixed x in $(0, a)$ at which f is continuous.

In this section we shall solve first the modified boundary value problem and then the original one. We have already shown in Chapter 1 that each term of the series

(6)
$$\sum_{n=1}^{\infty} c_n e^{-n^2\pi^2 kt/a^2} \sin \frac{n\pi}{a}x,$$

where the c_n are arbitrary constants, satisfies the heat equation and the endpoint conditions. If we assume that (6) converges and denote its sum by $u(x, t)$, then the initial condition becomes

(4)
$$\sum_{n=1}^{\infty} c_n \sin \frac{n\pi}{a}x = f(x).$$

It was shown in Chapter 2 that this equality will hold at every x where f is continuous if we:

1. choose the c_n to be the coefficients of the Fourier sine series of f on $(0, a)$ as described in Definition 2.6, that is,

(7)
$$c_n = \frac{2}{a} \int_0^a f(x) \sin \frac{n\pi}{a}x \, dx,$$

   and

2. require f to have one-sided derivatives at every x in $(0, a)$ and then apply Theorem 2.5 to its odd, piecewise continuous, periodic extension to $\mathbb{R}$ of period $2a$ to conclude that (4) holds at every x in $(0, a)$ at which f is continuous.

The preceding work suggests that, under these conditions on f, the sum of (6), where the c_n are given by (7), is the solution of our boundary value problem. However, all we know about the series (6) at this point is that

1. each of its terms is continuous in $\bar{D}$, is of class C^2 in D, and satisfies (1)–(3) and

2. it converges at each point of the form $(x, 0)$ in $\bar{D}$ and its sum satisfies (4) at every x at which f is continuous.

What we do not know about the series (6) is whether or not

1. it converges for all (x, t) in $\bar{D}$,

2. its sum, if it converges, is continuous on D_0 and of class C^2 in D, and

3. its sum, if it converges, satisfies (1)–(3) and (5).

It may seem very discouraging at this moment that we have worked so much to advance so little. However, this is only an apparent setback because it turns out that the most difficult part of the problem is to prove (4), and we have done that. To deal with the rest it would be helpful to answer some more general questions, such as (*i*) is the sum of a series of continuous functions itself continuous and (*ii*) is the sum of a series of differentiable functions also differentiable and, if so, can we obtain its derivative by differentiating term by term? These questions are taken up in Appendix A, in which, in order to provide some affirmative answers, we shall find it necessary to replace the ordinary concept of convergence with that of uniform convergence. With the results developed in this appendix we are in a position to solve our boundary value problem in full and rather quickly. We start with two partial results. In both cases assume that f and the c_n are as in 1 and 2 above.

Proposition 3.1 *The series* (6) *converges to a function that is continuous on the set* $D_0 = \{(x, t) \in \bar{D} : t > 0\}$ *and satisfies* (2)–(4).

Proof. By the choice of the c_n and Theorem 2.5, the series (6) converges for $t = 0$ and its sum satisfies (4).

By the Riemann-Lebesgue theorem $c_n \to 0$ as $n \to \infty$, and then there is a positive constant M such that $|c_n| < M$ for all n. If D_{t_0} denotes the set $\{(x, t) \in D_0 : t \geq t_0\}$ for each $t_0 > 0$, it follows that

$$\left| c_n e^{-n^2\pi^2 kt/a^2} \sin \frac{n\pi}{a} x \right| < M e^{-n^2\pi^2 kt_0/a^2}$$

for (x, t) in D_{t_0} and all n. But the series of numbers

$$\sum_{n=1}^{\infty} M e^{-n^2\pi^2 kt_0/a^2}$$

converges by the root test, and then (6) converges uniformly on D_{t_0} by the Weierstrass M-test (Theorem A.2). Since $t_0 > 0$ is arbitrary it converges on D_0 and since we showed above that it also converges for $t = 0$, we have proved that (6) converges on $\bar{D}$.

The fact that (6) is a series of continuous functions that converges uniformly on D_{t_0} implies that its sum is a continuous function on D_{t_0} by Theorem A.4 and then on D_0 since $t_0 > 0$ is arbitrary.

It is clear that (6) satisfies (2) and (3) because all its terms vanish for $x = 0$ and $x = a$, Q.E.D.

Proposition 3.2 *The sum of* (6) *is a function of class* C^2 *in* D *and satisfies* (1).

Proof. Consider the series

$$(8) \qquad \sum_{n=1}^{\infty} \left(-\frac{n^2\pi^2}{a^2} k \right) c_n e^{-n^2\pi^2 kt/a^2} \sin \frac{n\pi}{a} x,$$

obtained by differentiating (6) term by term with respect to t. Using the Weierstrass M-test as in Proposition 3.1 shows that it is uniformly convergent on the set

$$D_{t_0} = \{(x, t) \in D_0 : t \geq t_0\}$$

for any $t_0 > 0$. In particular, for a given x in $(0, a)$ it is uniformly convergent on the interval $[t_0, \infty)$ for any $t_0 > 0$. Then, if $u(x, t)$ denotes the sum of (6), Theorem A.6 shows that the sum of (8) is $u_t(x, t)$ for $t \geq t_0$. Since t_0 and x are arbitrary, (8) converges to $u_t(x, t)$ for any (x, t) in D.

Analogously, for a given $t > 0$, differentiating (6) term by term with respect to x twice gives

(9)
$$\sum_{n=1}^{\infty} \left(-\frac{n^2 \pi^2}{a^2} \right) c_n e^{-n^2 \pi^2 kt/a^2} \sin \frac{n\pi}{a} x.$$

A similar argument shows that (9) is uniformly convergent on D_{t_0} for any t_0 and that its sum is $u_{xx}(x, t)$ for any (x, t) in D. Then a comparison of (8) and (9) makes it clear that $u_t = k u_{xx}$ in D.

Since (8) and (9) are uniformly convergent on D_{t_0}, their sums are continuous there by Theorem A.4 and then, since $t_0 > 0$ is arbitrary, u_t and u_{xx} are continuous on D. Similarly, u_{tx} and u_{tt} are continuous on D, Q.E.D.

We can now state and prove the complete solution of our boundary value problem.

Theorem 3.1 Let $f : (0, a) \to \mathbb{R}$ be a piecewise continuous function that has one-sided derivatives on $(0, a)$, and let c_n be the nth coefficient of its Fourier sine series on $(0, a)$. Then the series

(6)
$$\sum_{n=1}^{\infty} c_n e^{-n^2 \pi^2 kt/a^2} \sin \frac{n\pi}{a} x$$

converges to a function $u : \overline{D} \to \mathbb{R}$ that is of class C^2 in D, is continuous on the set $D_0 = \{ (x, t) \in \overline{D} : t > 0 \}$, and satisfies the boundary value problem (1)–(5).

Proof. In view of the preceding propositions it remains only to prove (5). Fix a point x in $(0, a)$ at which f is continuous and define

$$a_n = c_n \sin \frac{n\pi}{a} x \qquad \text{and} \qquad g_n(t) = e^{-n^2 \pi^2 kt/a^2}.$$

Then $\sum_{n=1}^{\infty} a_n$ is a convergent Fourier series, $g_{n+1}(t) \le g_n(t)$ for each n and all t in $[0, \infty)$, and $|g_n(t)| \le 1$ for all t in $[0, \infty)$. Hence, by the corollary of Theorem A.3, the series (6) converges uniformly with respect to t on the interval $[0, \infty)$, and its sum is continuous by Theorem A.4. Then $u(x, t) \to u(x, 0)$ as $t \to 0$ and, because of (4), $u(x, t) \to f(x)$ as $t \to 0$, Q.E.D.

Example 3.1 A bar of steel 10 cm long has an initial temperature distribution given by $f(x) = x$ if $0 < x < 10$. The ends of the bar are maintained at $0°C$ for $t > 0$. With a value of $k = 0.128 \, cm^2/sec$ for steel and $a = 10$, (6) is a solution of this problem if the c_n are computed using (7). We have

$$c_n = \frac{2}{10} \int_0^{10} x \sin \frac{n\pi}{10} x \, dx = \frac{20}{n\pi} (-1)^{n+1}$$

and then

$$u(x, t) = \frac{20}{\pi} \sum_{n=1}^{\infty} \frac{(-1)^{n+1}}{n} e^{-0.128 n^2 \pi^2 t/100} \sin \frac{n\pi}{10} x.$$

Because of the exponentials, most of the terms in this series are negligible for t large enough. In fact, if we define $\alpha = e^{-0.00128\pi^2} \approx 0.987$, we have

$$u(x, t) \approx \frac{20}{\pi} \left(\alpha^t \sin \frac{\pi}{10}x - \frac{1}{2}\alpha^{4t} \sin \frac{\pi}{5}x + \frac{1}{3}\alpha^{9t} \sin \frac{3\pi}{10}x \right)$$

for t large enough. This is represented in Figure 3.1 for $t = 10, 30, 60,$ and 120 seconds.

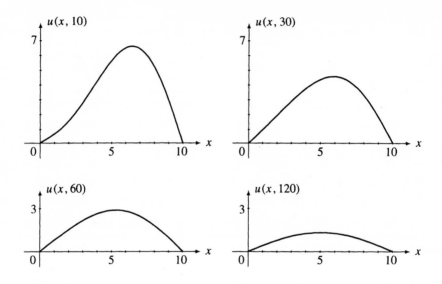

Figure 3.1

We now return to the boundary value problem originally posed. That is, the original temperature distribution f is assumed to be continuous on $[0, a]$, and we want to show that the series (6) has a continuous sum on $\bar{D}$ that satisfies the initial and endpoint conditions. In view of the preceding work, all that remains to be shown is that the sum of (6) is continuous on $\bar{D}$. This would be an immediate consequence of Theorem A.4 if we show that (6) is uniformly convergent on $\bar{D}$, which, in particular, requires its uniform convergence for $t = 0$. Thus, the question that we face now is: does the Fourier series of a continuous function that also has one-sided derivatives converge uniformly?

Clearly, this question could not have been answered in 1829, when Dirichlet first proved the convergence of Fourier series, as the concept of uniform convergence had not yet been discovered. As it is, the same question was posed by Heinrich Eduard Heine (1821–1881), a professor at the University of Halle. He was not concerned with any kind of heat problem, but rather with the following theoretical question. Knowing that

$$f(x) = \tfrac{1}{2}a_0 + \sum_{n=1}^{\infty}(a_n \cos nx + b_n \sin nx)$$

for a continuous function $f : (-\pi, \pi) \to \mathbb{R}$ that satisfies the conditions of Theorem 2.2 and has Fourier coefficients a_n and b_n, he asked whether or not this equality is also valid for the same function but replacing the Fourier coefficients with a different set of constants a_n and b_n. Heine was at this time far from the Weierstrassian world at Berlin, but he had been a student of Weierstrass and may have learned about uniform convergence before leaving Berlin. Or he may have heard about it from a new arrival from Berlin, Georg Ferdinand Louis Philippe Cantor (1845–1918), who had become a *Privatdozent* in 1869 at Halle. Whatever the case, Heine knew about uniform convergence and about Weierstrass' proof that a uniformly convergent series of integrable functions can be integrated term by term. He also knew, from Fourier's work—as we have shown in §2.2—that if the equation above holds and if the series on the right can be integrated term by term, then the coefficients are uniquely determined and have the values given by Definition 2.1. Heine made the connection between these ideas and stated that if a function $f : (-\pi, \pi) \to \mathbb{R}$ is the sum of a uniformly convergent trigonometric series of the form given above, then its coefficients are uniquely determined and happen to be the Fourier coefficients of f on $(-\pi, \pi)$. It was in this manner that Heine posed himself the question: when does the Fourier series of a function converge uniformly? He provided the answer in a paper of 1870, a particular case of which, sufficient for our purposes, is stated below in Theorem 3.2. Heine did not work any further on this type of problem, but it was he who encouraged Cantor to pursue the question of the uniqueness of the coefficients of a trigonometric series when the requirement of uniform convergence is dropped. Cantor's solution of this problem and his consequent discovery of the theory of sets have already been described in §2.9.

Theorem 3.2 *If a and α are real numbers with $a > 0$, let $f : [\alpha - a, \alpha + a] \to \mathbb{R}$ be a continuous function that has a piecewise continuous derivative and is such that $f(\alpha - a) = f(\alpha + a)$. Then its Fourier series on $(\alpha - a, \alpha + a)$ converges uniformly to f on $[\alpha - a, \alpha + a]$.*

We shall postpone the proof until §6.3, in which new mathematical tools will allow us to give a rather quick proof.[1] For the time being, Heine's result can be used to prove the following theorem.

---

[1] It is interesting to mention that, in order to complete the research in his 1870 paper, Heine introduced two concepts that are now indispensable in modern analysis. The first was uniform continuity, more precisely defined in a follow-up paper of 1872. The second was the observation that, if a closed and bounded interval is contained in a sequence of open intervals, then it is contained in a finite number of these intervals. This fact was formally proved in 1895 by Émile Borel (1871–1938) and named the *Heine-Borel theorem* in 1900 by Arthur Schönflies. The theorem currently known by this name does not require the original cover of a closed and bounded interval by open intervals to be countable, a generalization proved by Lebesgue in 1904. Eventually, this covering property of a closed and bounded set by open sets became the modern definition of compactness as given by Pavel Sergeevich Alexandrov (1896–1985) and Pavel Samuelovich Urysohn (1898–1924) in 1924.

Theorem 3.3 *Let* $f : [0, a] \to \mathbb{R}$ *be a continuous function that has a piecewise continuous derivative and is such that* $f(0) = f(a) = 0$, *and let* c_n *be nth coefficient of its Fourier sine series on* $(0, a)$. *Then the series*

$$(6) \qquad \sum_{n=1}^{\infty} c_n e^{-n^2\pi^2 kt/a^2} \sin \frac{n\pi}{a} x$$

converges to a continuous function $u : \overline{D} \to \mathbb{R}$ *such that*

$$
\begin{aligned}
u_t &= k u_{xx} & &\text{in } D \\
u(0, t) &= 0 & &t \geq 0 \\
u(a, t) &= 0 & &t \geq 0 \\
u(x, 0) &= f(x) & &0 \leq x \leq a.
\end{aligned}
$$

Proof. In view of Theorem 3.1, it remains to be proved that u is continuous on $\overline{D}$. The condition $f(0) = f(a) = 0$ implies that the odd extension of f to $[-a, a]$ is continuous and that $f(-a) = f(a)$, and then, by Theorem 3.2, its Fourier series

$$\sum_{n=1}^{\infty} c_n \sin \frac{n\pi}{a} x$$

converges uniformly on $[0, a]$. Then (6) converges uniformly on $\overline{D}$ by Theorem A.3 with

$$f_n(x, t) = c_n \sin \frac{n\pi}{a} x \qquad \text{and} \qquad g_n(x, t) = e^{-n^2\pi^2 kt/a^2},$$

and its sum is continuous on $\overline{D}$ by Theorem A.4, Q.E.D.

Note that the condition $f(0) = f(a) = 0$ does not only allow us to use Theorem 3.2, but is actually necessary for the conclusion of Theorem 3.3 to hold. Without it, u could not be continuous on $\overline{D}$ and satisfy $u(0, t) = u(a, t) = 0$ for $t \geq 0$.

§3.2 Uniqueness and Stability of the Solution

Having exhibited a solution of the boundary value problem in Theorem 3.3, it remains, in order to show that this problem is well posed in the sense of Hadamard, only to prove its uniqueness and stability or continuous dependence on initial values. These facts will be easily established from what is called the *maximum principle*, which is a mathematical expression of the fact that heat flows from a hotter to a cooler spot. Unless the temperature is the same at all points of the bar, it cannot be highest at a point x_0 with $0 < x_0 < a$ at a time $t_0 > 0$. If this were so the temperature would have been increasing at x_0 for some time prior to the instant t_0. But the heat necessary to increase this temperature must have come from at least one nearby point $x \neq x_0$ that was hotter at some time $t < t_0$. In short, the temperature cannot be highest at (x_0, t_0). This was formally proved by Eugenio Elia Levi (1883–1917) in 1907.

Theorem 3.4 (**The Maximum Principle for the Heat Equation**) *Let D and $\bar{D}$ be as those in §3.1, let T be an arbitrary positive number, and define the sets*

$$D_T = \{ (x, t) \in D : t \leq T \}$$

$$\bar{D}_T = \{ (x, t) \in \bar{D} : t \leq T \},$$

and $\gamma_T = \bar{D}_T - D_T$, as shown in Figure 3.2. If $u : \bar{D}_T \to \mathbb{R}$ is a continuous function that satisfies the heat equation in D_T, then u attains its maximum and minimum values on γ_T.[1]

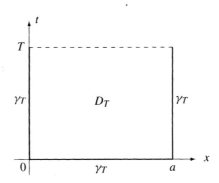

Figure 3.2

Proof. Assume that $u : \bar{D}_T \to \mathbb{R}$ is continuous, satisfies the heat equation on D_T, has a minimum value m at a point (x_0, t_0) in D_T, and that, contrary to what we wish to prove, its minimum value on γ_T is $M > m$. Then define a new function $v : \bar{D}_T \to \mathbb{R}$ by

$$v(x, t) = u(x, t) + \frac{M - m}{2T}(t - t_0).$$

Now, $t - t_0 \geq -T$ and if (x, t) in γ_T, where $u(x, t) \geq M$, we have

$$v(x, t) \geq M - \frac{M - m}{2} = \frac{M + m}{2} > m,$$

while

$$v(x_0, t_0) = u(x_0, t_0) = m.$$

Hence, v also attains its minimum at a point (x_1, t_1) in D_T, but then

$$v_{xx}(x_1, t_1) \geq 0 \qquad \text{and} \qquad v_t(x_1, t_1) \leq 0$$

---

[1] Weierstrass proved in his lectures at Berlin that a continuous function on a closed and bounded set has a maximum value and a minimum value on that set. For a proof see W. Rudin, *Principles of Mathematical Analysis*, 3rd ed., McGraw-Hill, New York, 1976, page 89.

(in fact, $v_t(x_1, t_1) = 0$ if $t_1 < T$ but $v_t(x_1, t_1)$ can be negative if $t_1 = T$). This contradicts the fact that

$$v_t - kv_{xx} = u_t - ku_{xx} + \frac{M - m}{2T} = \frac{M - m}{2T} > 0$$

in D_T, and we must conclude that u attains its minimum value on γ_T.

For the same reasons, $-u$ attains its minimum value on γ_T, so that u attains its maximum value on γ_T, Q.E.D.

Using the maximum principle we can conclude the study of our boundary value problem rather quickly.

Theorem 3.5 *The boundary value problem of Theorem 3.3 is well posed in the sense of Hadamard and, if u is the solution, then $u \geq 0$ if $f \geq 0$.*

Proof. The existence of a solution was already proved in Theorem 3.3. Assume now that u_1 and u_2 are solutions corresponding to initial temperature distributions given by continuous functions f and g on $[0, a]$. Then $u_1 - u_2$ is a solution corresponding to the initial temperature distribution given by $f - g$, and $u_1 - u_2$ is continuous on $\bar{D}_T$ because u_1 and u_2 are. Then, if M_{f-g} denotes the maximum value of $|f - g|$ on $[0, a]$, the fact that $u_1 - u_2$ vanishes on the lines $x = 0$ and $x = a$ and the maximum principle imply that

$$|u_1(x, t) - u_2(x, t)| \leq M_{f-g}$$

for all (x, t) in $\bar{D}_T$. Since T is arbitrary, this shows that a small change in the initial values results in a small change in the solution. If $f \equiv g$, then $u_1 \equiv u_2$ for all (x, t) in $\bar{D}_T$, and, since T is arbitrary, this shows uniqueness of solutions.

Finally, if $f \geq 0$, then the maximum principle and the fact that u vanishes on the lines $x = 0$ and $x = a$ imply that the minimum value of u is zero, Q.E.D.

Note that this theorem goes beyond stating that the problem is well posed and asserts that the solution is nonnegative. In fact, if we assume, as we shall from now on for convenience, that temperatures are measured from absolute zero, then f must be a nonnegative function and u must be shown to be nonnegative. If, instead, a different scale of temperature were used then both f and u should not be smaller than a certain constant. For instance, this constant is -273.15 if the temperature is measured in degrees Celsius.

To conclude this section we discuss the asymptotic behavior of the solution. It is expected on purely physical grounds that if heat is allowed to escape at the endpoints of the bar, and, if their temperatures are kept at zero degrees, then the temperature at every point of the bar will naturally decay from $f(x)$ to zero as time passes. Mathematically, this is strongly suggested by the decaying exponentials in (6) and can be proved as follows. If M is such that $|c_n| < M$ for all n and if we define $\alpha = \pi^2 k / a^2$ and $r = e^{-\alpha t}$, we obtain the following bound for the nth term of (6)

$$\left| c_n e^{-n^2 \alpha t} \sin \frac{n\pi}{a} x \right| < M e^{-n^2 \alpha t} \leq M e^{-n\alpha t} = M r^n.$$

Then, since $0 < r < 1$ for $t > 0$ and since the sum of the geometric series $\sum_{n=1}^{\infty} r^n$ with $|r| < 1$ is $r/(1 - r)$, we have

$$|u(x, t)| \leq M \frac{e^{-\alpha t}}{1 - e^{-\alpha t}}.$$

This shows that $u(x, t) \to 0$ as $t \to \infty$. Moreover, the quotient $u(x, t)/e^{-\alpha t}$ remains bounded as $t \to \infty$.

The problem that we have just solved provides the basis for the study of similar problems in which the boundary conditions are modified. In the remaining sections we shall briefly indicate how the basic procedure should be altered to accommodate some of these changes. In some cases this will lead to a complete solution, but in other cases it will lead to new, open mathematical problems that will be explored in Chapter 4.

§3.3 Nonzero Temperature at the Endpoints

We now pose the boundary value problem

(10)	$u_t = k u_{xx}$	in D
(11)	$u(0, t) = T_1$	$t \geq 0$
(12)	$u(a, t) = T_2$	$t \geq 0$
(13)	$u(x, 0) = f(x)$	$0 \leq x \leq a,$

where T_1 and T_2 are nonnegative constants and f is a continuous, nonnegative function with a piecewise continuous derivative and such that $f(0) = T_1$ and $f(a) = T_2$. The new endpoint conditions (11) and (12) are not linear since the sum of solutions that satisfy (11) and (12) does not satisfy (11) and (12), and this problem cannot be solved by just using the principle of superposition of solutions.

However, some physical considerations will help us out of this trouble. It is expected that after a sufficiently long time the effects of the initial temperature distribution will dissipate and that the temperature of the bar will become, for all practical purposes, independent of time. If U denotes this temperature distribution for very large t, called the *steady-state solution* of our problem, then U is a function of x only and must satisfy (10)–(12) but not the initial condition (13). That is,

$$0 = kU''$$
$$U(0) = T_1$$
$$U(a) = T_2.$$

From the first of these equations we obtain $U(x) = Ax + B$, where A and B are constants, and these are easily evaluated from the second and third equations. We obtain

$$U(x) = \frac{T_2 - T_1}{a} x + T_1.$$

Now, if u is a solution of (10)–(13), then the function $v = u - U$ satisfies

$$
\begin{aligned}
v_t &= k v_{xx} && \text{in } D \\
v(0, t) &= 0 && t \geq 0 \\
v(a, t) &= 0 && t \geq 0 \\
v(x, 0) &= f(x) - U(x) && 0 \leq x \leq a.
\end{aligned}
$$

Since U has a continuous derivative, a solution of this problem is found as in Theorem 3.3, and then $u = U + v$ is a solution of (10)–(13).

If u_1 and u_2 are solutions of (10)–(13) corresponding to initial temperature distributions f and g and endpoint temperatures T_{1f}, T_{2f} and T_{1g}, T_{2g}, let M_{f-g} denote the maximum value of $|f - g|$ on $[0, a]$. The maximum principle implies that

$$
|u_1(x, t) - u_2(x, t)| \leq \max\{M_{f-g}, |T_{1f} - T_{1g}|, |T_{2f} - T_{2g}|\}.
$$

From this inequality the given problem is seen to be well posed in the sense of Hadamard exactly as in the proof of Theorem 3.5. The maximum principle also shows that $u \geq 0$ if $f \geq 0$.

Since v is the solution of a problem with zero endpoint values, it was shown at the end of §3.2 that $v(x, t) \to 0$ as $t \to \infty$. For this reason this function v is called the *transient solution* of the problem (10)–(13), whose complete solution is thus seen to be the sum of the steady-state solution and the transient solution.

This problem is solved in exactly the same way when f is only piecewise continuous and has one-sided derivatives and this whether or not it satisfies the endpoint conditions $f(0) = T_1$ and $f(a) = T_2$, but then the complete solution cannot be expected to be continuous on $\overline{D}$. Instead, at every point at which f is continuous the limit $u(x, t) \to f(x)$ as $t \to 0$ can be proved as in §3.1.

Example 3.2 A bar 20 cm long has a constant temperature of $400°$K. For $t \geq 0$ the ends of the bar are maintained at 300 and $500°$K. We have $a = 20$, $T_1 = 300°$K, and $T_2 = 500°$K, and then the steady-state solution is

$$
U(x) = \frac{500 - 300}{20} x + 300 = 10x + 300.
$$

The transient solution satisfies the initial condition

$$
v(x, 0) = 400 - 10x - 300 = 100 - 10x,
$$

and, using (7), it is easily verified that

$$
c_n = \frac{2}{20} \int_0^{20} (100 - 10x) \sin \frac{n\pi}{20} x \, dx =
\begin{cases}
0 & \text{if } n \text{ is odd} \\
\dfrac{400}{n\pi} & \text{if } n \text{ is even.}
\end{cases}
$$

The complete solution is then given by

$$
u(x, t) = U(x) + v(x, t) = 10x + 300 + \frac{200}{\pi} \sum_{n=1}^{\infty} \frac{1}{n} e^{-n^2 \pi^2 k t / 100} \sin \frac{n\pi}{10} x.
$$

If the bar is made of aluminum, for which $k = 0.84\,\mathrm{cm}^2/\mathrm{sec}$, then $e^{-\pi^2 k/100} \approx 0.92$ and

$$u(x, t) \approx 10x + 300 + \frac{200}{\pi} 0.92^t \sin \frac{\pi}{10} x$$

for t large enough. It is represented in Figure 3.3 for $t = 10$ seconds.

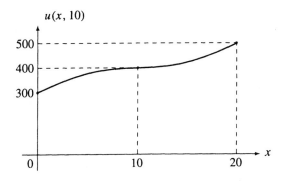

Figure 3.3

§3.4 Bar Insulated at the Endpoints

The bar is now totally insulated, along its lateral surface and on its faces at $x = 0$ and $x = a$. Since, according to Fourier's law, the heat flows through the cross section of the bar at x at the rate $-\kappa A u_x(x, t)$ and since there is no heat flow through the faces at the ends, we have the boundary value problem

$$\begin{aligned}
u_t &= k u_{xx} & &\text{in } D \\
u_x(0, t) &= 0 & &t \geq 0 \\
u_x(a, t) &= 0 & &t \geq 0 \\
u(x, 0) &= f(x) & &0 \leq x \leq a,
\end{aligned}$$

where f is continuous, has a piecewise continuous derivative and $f'(0) = f'(a) = 0$.

A constant is always a solution of the heat equation that satisfies the endpoint conditions. Further solutions can be found by the method of separation of variables. If we assume that the heat equation has a solution of the form $u(x, t) = X(x)T(t)$, then, as in Chapter 1, X and T satisfy the ordinary differential equations

$$kX'' + \lambda X = 0 \qquad \text{and} \qquad T' + \lambda T = 0,$$

where λ is a constant and the first equation is subject to the boundary conditions $X'(0) = X'(a) = 0$.

It is easily seen that for $\lambda \leq 0$ only a constant solution of the first equation satisfies the endpoint conditions. If $\lambda > 0$, its general solution is

$$X(x) = A \sin \sqrt{\frac{\lambda}{k}} x + B \cos \sqrt{\frac{\lambda}{k}} x,$$

and the conditions $X'(0) = X'(a) = 0$ imply that

$$A = 0 \qquad \text{and} \qquad \lambda = \frac{n^2 \pi^2}{a^2} k,$$

where n is an integer. Corresponding to each such value of λ and incorporating the solution of $T' + \lambda T = 0$, we have a product solution of the heat equation of the form

$$c_n e^{-n^2 \pi^2 kt/a^2} \cos \frac{n\pi}{a} x,$$

where c_n is an arbitrary constant, and this solution satisfies the endpoint conditions.

As in §3.1 and using the same theorems on uniform convergence from Appendix A, it can be shown that the sum of the series

$$\frac{1}{2} c_0 + \sum_{n=1}^{\infty} c_n e^{-n^2 \pi^2 kt/a^2} \cos \frac{n\pi}{a} x,$$

where

$$c_n = \frac{2}{a} \int_0^a f(x) \cos \frac{n\pi}{a} x \, dx,$$

is a solution of our boundary value problem that is continuous on $\bar{D}$. On the other hand, this problem cannot be shown to be well posed in the sense of Hadamard from the maximum principle alone, because the values of u are not specified for $x = 0$ or $x = a$. However, under the given endpoint conditions u is expected to attain its maximum and minimum values for $t = 0$. This is clear on physical grounds because, if an end face remains insulated for all time, then heat cannot enter the bar through this end. Then the temperature cannot attain a maximum there at some time $t_0 > 0$, for the heat necessary to bring this temperature to a maximum must have come from at least one nearby point that was hotter at a time $t < t_0$. This new type of maximum principle, which we shall prove in a more general setting in §3.10, can then be used to show that our boundary value problem is well posed in the sense of Hadamard and that $u \geq 0$ if $f \geq 0$.

As in previous cases, if f is just required to be piecewise continuous, u may not be continuous on $\bar{D}$, and only the limit $u(x, t) \to f(x)$ as $t \to 0$ can be guaranteed at every point x at which f is continuous.

Finally, note that the bar is now completely insulated and that the total amount of heat in the bar remains constant. It is then expected that any initial temperature differences will disappear as time passes and that the temperature at any point of the bar will approach the average value of the initial temperature distribution as time increases. Indeed, an analysis identical to that at the end of §3.2 shows that

$$u(x, t) \to \frac{c_0}{2} = \frac{1}{a} \int_0^a f(x) \, dx$$

as $t \to \infty$ for any x in $[0, a]$.

★ ## §3.5 Mixed Endpoint Conditions

Assume now that the face at $x = 0$ is kept at zero temperature while the face at $x = a$ is insulated to prevent heat flow. We are dealing with the boundary value problem

$$
\begin{aligned}
u_t &= k u_{xx} && \text{in } D \\
u(0, t) &= 0 && t \geq 0 \\
u_x(a, t) &= 0 && t \geq 0 \\
u(x, 0) &= f(x) && 0 \leq x \leq a,
\end{aligned}
$$

where f is continuous, is nonnegative, and has a piecewise continuous derivative and $f(0) = f'(a) = 0$.

Separation of variables leads to the ordinary differential equations

$$
k X'' + \lambda X = 0 \qquad \text{and} \qquad T' + \lambda T = 0.
$$

For $\lambda \leq 0$ only the trivial solution of the first equation satisifes $X(0) = X'(a) = 0$. For $\lambda > 0$ its general solution is

$$
X(x) = A \sin \sqrt{\frac{\lambda}{k}}\, x + B \cos \sqrt{\frac{\lambda}{k}}\, x,
$$

and it is readily seen that the endpoint conditions $X(0) = X'(a) = 0$ imply that

$$
B = 0 \qquad \text{and} \qquad \lambda = \frac{(2n - 1)^2 \pi^2}{4a^2} k.
$$

Hence, for each positive integer N, the sum

$$
\sum_{n=1}^{N} c_n e^{-(2n-1)^2 \pi^2 kt / 4a^2} \sin \frac{(2n - 1)\pi}{2a} x,
$$

where the c_n are arbitrary constants, is a solution of the heat equation that satisfies the endpoint conditions.

To satisfy the initial condition we try an infinite series and ask if the constants c_n can be determined so that, for $t = 0$,

$$
f(x) = \sum_{n=1}^{\infty} c_n \sin \frac{(2n - 1)\pi}{2a} x.
$$

The difficulty here is that the number $(2n - 1)/2$ is not an integer, and then the right-hand side above cannot be a Fourier sine series on $(0,a)$. Therefore, the theory developed in Chapter 2 is not directly applicable to determine the constants c_n or to prove the convergence of the series.

We could, of course, think of developing a similar theory for the new type of trigonometric series that has arisen in this problem. However, a moment's reflection will show that there is a simpler way out of this trouble. The series above can be the Fourier sine series of a function on the interval $(0, 2a)$, since $2n - 1$ is indeed an integer. Now, nothing prevents us from imagining that our bar extends from $x = 0$ to $x = 2a$, and then, in order to pose a boundary value problem for the extended bar, we proceed as follows:

1. Extend f to the interval $[0, 2a]$ in such a manner that its graph is symmetric with respect to the vertical line $x = a$; that is, we define

$$\tilde{f}(x) = \begin{cases} f(x) & \text{if} \quad 0 \le x \le a \\ f(2a - x) & \text{if} \quad a < x \le 2a. \end{cases}$$

2. Require that the endpoint at $x = 2a$ be kept at zero temperature for all time.

Note that $\tilde{f}$ is continuous by construction and that $\tilde{f}'$ is piecewise continuous because f' is. This extension of f makes the situation entirely symmetric with respect to the vertical line $x = a$, and no heat can flow through the cross section of the extended bar at $x = a$. Thus, if u is a solution of the boundary value problem

$$\begin{aligned} u_t &= ku_{xx} & 0 &< x < 2a, \quad t > 0 \\ u(0, t) &= 0 & t &\ge 0 \\ u(2a, t) &= 0 & t &\ge 0 \\ u(x, 0) &= \tilde{f}(x) & 0 &\le x \le 2a, \end{aligned}$$

then this solution satisfies $u_x(a, t) = 0$ for $t \ge 0$ and is also a solution of the original problem for $0 \le x \le a$.

By Theorem 3.3, the unique solution of the extended problem is

$$\sum_{n=1}^{\infty} c_n e^{-n^2\pi^2 kt/4a^2} \sin \frac{n\pi}{2a} x,$$

where

$$\begin{aligned} c_n &= \frac{2}{2a} \int_0^{2a} \tilde{f}(x) \sin \frac{n\pi}{2a} x \, dx \\ &= \frac{1}{a} \int_0^a f(x) \sin \frac{n\pi}{2a} x \, dx + \frac{1}{a} \int_a^{2a} f(2a - x) \sin \frac{n\pi}{2a} x \, dx \\ &= \frac{1}{a} \int_0^a f(x) \sin \frac{n\pi}{2a} x \, dx - \frac{1}{a} \int_a^0 f(s) \sin \frac{n\pi}{2a} (2a - s) \, ds \\ &= \frac{1}{a} \int_0^a f(x) \sin \frac{n\pi}{2a} x \, dx + \frac{1}{a} \int_0^a (-1)^{n+1} f(s) \sin \frac{n\pi}{2a} s \, ds \\ &= \begin{cases} 0 & \text{if} \quad n \quad \text{is even} \\ \dfrac{2}{a} \int_0^a f(x) \sin \dfrac{n\pi}{2a} x \, dx & \text{if} \quad n \quad \text{is odd} \end{cases} \end{aligned}$$

Thus, the sum $u(x, t)$ of the series

$$\sum_{n=1}^{\infty} c_{2n-1} e^{-(2n-1)^2\pi^2 kt/4a^2} \sin \frac{(2n - 1)\pi}{2a} x,$$

where

$$c_{2n-1} = \frac{2}{a} \int_0^a f(x) \sin \frac{(2n-1)\pi}{2a} x \, dx,$$

is a solution of our original boundary value problem.

It is easily seen that u is continuous on $\bar{D}$ (although only the limit $u(x, t) \to f(x)$ as $t \to 0$ can be guaranteed at the points of continuity of f when f is just required to be piecewise continuous) and then that the problem is well posed in the sense of Hadamard and that $u \geq 0$ if $f \geq 0$ (Theorem 3.8). Since u satisfies a problem of the type discussed in §3.1 and §3.2, it is finally clear that $u(x, t) \to 0$ as $t \to \infty$.

§3.6　Heat Convection at One Endpoint

While the face of the bar at $x = 0$ is to be kept now at a constant temperature $T_1 \geq 0$, we shall assume that there is heat transfer through the face at $x = a$ into the surrounding medium. According to Newton's law of cooling, the temperature at this point decreases at a rate that is proportional to the temperature difference between the bar and its surroundings; that is,

$$u_x(a, t) = -h[u(a, t) - T_2],$$

where h is a positive constant that depends on the material of the bar and $T_2 \geq 0$ is the temperature of the medium outside the bar. This type of heat interchange is called *convection*. Therefore, our new boundary value problem is

$$
\begin{aligned}
u_t &= k u_{xx} &&\text{in } D \\
u(0, t) &= T_1 &&t \geq 0 \\
u_x(a, t) &= h[T_2 - u(a, t)] &&t \geq 0 \\
u(x, 0) &= f(x) &&0 \leq x \leq a,
\end{aligned}
$$

where f is a nonnegative, continuous function with a piecewise continuous derivative and such that $f(0) = T_1$ and $f'(a) = h[T_2 - f(a)]$.

Since the endpoint conditions are not linear, we shall follow the procedure outlined in §3.3 and look first for the steady-state solution U. This satisfies

$$
\begin{aligned}
0 &= kU'' \\
U(0) &= T_1 \\
U'(a) &= h[T_2 - U(a)],
\end{aligned}
$$

and we easily obtain $U(x) = Ax + B$, where the endpoint conditions give

$$
\begin{aligned}
B &= U(0) = T_1 \\
A &= U'(a) = h[T_2 - U(a)] = h(T_2 - Aa - B)
\end{aligned}
$$

so that

$$A = \frac{h(T_2 - B)}{1 + ha} = \frac{h(T_2 - T_1)}{1 + ha}.$$

This determines the steady-state solution

$$U(x) = \frac{h(T_2 - T_1)}{1 + ha}x + T_1.$$

If we now define a function $v = u - U$, where u and U are as above, then the endpoint condition at a implies that $v_x(a, t) = u_x(a, t) - U'(a) = h[T_2 - u(a, t)] - h[T_2 - U(a)] = -h[u(a, t) - U(a)] = -hv(a, t)$. Then it is easily seen that v is a solution of the boundary value problem

$$
\begin{array}{ll}
v_t = kv_{xx} & \text{in } D \\
v(0, t) = 0 & t \geq 0 \\
v_x(a, t) = -hv(a, t) & t \geq 0 \\
v(x, 0) = f(x) - U(x) & 0 \leq x \leq a,
\end{array}
$$

and $U + v$ is a complete solution of the original problem.

To find this new function v, which we shall call the transient solution of the original problem, we try separation of variables, and this leads to

$$kX'' + \lambda X = 0 \qquad \text{and} \qquad T' + \lambda T = 0,$$

subject to the conditions $X(0) = 0$ and $X'(a) = -hX(a)$.

As in previous cases, it is possible to see that, if $\lambda < 0$, only the trivial solution of the first equation satisfies the endpoint conditions. If $\lambda > 0$ and if we put $\mu = \lambda/k$, the general solution for X is of the form

$$X(x) = A \sin \sqrt{\mu}\, x + B \cos \sqrt{\mu}\, x.$$

Then $X(0) = 0$ and $X'(a) = -hX(a)$ imply that $B = 0$ and that

$$A\sqrt{\mu} \cos \sqrt{\mu}\, a = -hA \sin \sqrt{\mu}\, a.$$

Since $h > 0$, it follows that $\cos \sqrt{\mu}\, a \neq 0$, and then

$$\tan a\sqrt{\mu} = -\frac{\sqrt{\mu}}{h}.$$

There are infinitely many solutions of this transcendental equation in μ, as can be seen graphically in Figure 3.4. For each positive integer n there is a unique solution μ_n of the given equation with

$$\frac{(2n - 1)\pi}{2a} < \sqrt{\mu_n} < \frac{n\pi}{a}.$$

Therefore, using the principle of superposition of solutions and incorporating for each n the corresponding solution of $T' + \lambda_n T$ with $\lambda_n = \mu_n k$, every finite sum of the form

$$\sum_{n=1}^{N} c_n e^{-\mu_n kt} \sin \sqrt{\mu_n}\, x,$$

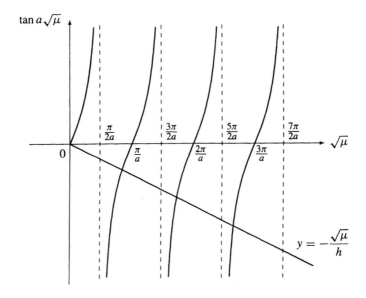

Figure 3.4

where N is a positive integer and the c_n are arbitrary constants, is a solution of the heat equation that satisfies the endpoint conditions for v.

The values of the μ_n can be evaluated by numerical methods, but there is no way to exactly solve the transcendental equation involved. One thing is certain, that the value of $\sqrt{\mu_n}$ is not a constant times n. Thus, when we replace the finite sum with an infinite series and put $t = 0$ to satisfy the initial condition, the series so obtained is not a Fourier series of the type studied in Chapter 2. By making the assumption that this new series converges uniformly to $f - U$ we can then find the coefficients c_n and the transient solution, as outlined in Exercises 6.1 to 6.3. But, in any particular case, a solution so obtained must be verified *a posteriori*, since there is no guarantee of the validity of this assumption. The nature of the new mathematical problems that we must solve to avoid unjustified assumptions will be made clearer by considering some more general cases in the remaining sections.

§3.7   Time-Independent Problems

In the preceding sections we have presented just a few boundary value problems for the heat equation. A variety of other problems can be obtained by choosing either a more general equation of heat propagation or more general endpoint conditions. For instance, if the cross section A of the bar remains constant but the specific heat capacity c, the density ρ, and the thermal conductivity κ of the material are functions of x rather than constants, if there is internal heat generated at a rate q that is also a function of x, and if we write c instead of $c\rho$ for simplicity, then the equation of heat propagation is

$$(14) \qquad\qquad cu_t = (\kappa u_x)_x + q$$

in $D = \{(x, t) \in \mathbb{R}^2 : 0 < x < a,\ t > 0\}$ (Exercise 2.7 of Chapter 1). We assume that c, κ, and q are nonnegative, continuous functions on $[0, a]$, that c and κ are positive, and that κ has a continuous derivative.

Assuming again, without loss of generality, that temperatures are measured from absolute zero, we shall consider the problem of finding a continuous, nonnegative, bounded solution of (14) subject to the boundary conditions

$$(15) \qquad a_1 u(0, t) - b_1 u_x(0, t) = c_1 \qquad t \geq 0$$

$$(16) \qquad a_2 u(a, t) + b_2 u_x(a, t) = c_2 \qquad t \geq 0$$

$$(17) \qquad u(x, 0) = f(x) \qquad 0 \leq x \leq a.$$

We shall work under the following hypotheses:

1. a_1, a_2, b_1, b_2, c_1, and c_2 are nonnegative constants,

2. $a_1 + b_1 > 0$ and $a_2 + b_2 > 0$,

3. if $a_1 = a_2 = 0$, then $c_1 = c_2 = 0$ and $q \equiv 0$, and

4. f is a nonnegative, continuous function with a piecewise continuous derivative such that $a_1 f(0) - b_1 f'(0) = c_1$ and $a_2 f(a) + b_2 f'(a) = c_2$.

Since these hypotheses may seem excessive, and even somewhat mysterious, a word of explanation is in order.

As for the first hypothesis, it is clear that we can always take $a_1 \geq 0$ or else (15) can be multiplied throughout by -1. But then if, for example, $a_1 > 0$, $b_1 = 0$ and $c_1 < 0$, we would have $u(0, t) < 0$, contrary to our agreement that temperatures should be nonnegative. Similarly, if $a_1 > 0$, $b_1 < 0$, $c_1 > 0$ and if $u(0, t) > c_1/a_1$, we would have $u_x(0, t) < 0$; that is, heat would flow into the bar at $x = 0$ when the temperature at this point, $u(0, t)$, is higher than that of the surrounding medium, c_1/a_1, which is absurd. Thus, the first hypotheses will rule out this type of situation.

Unfortunately, the first hypothesis will also rule out some well-posed problems, as, for example, when $a_1 = 0$, $b_1 < 0$ and $c_1 > 0$, for this choice simply means that heat leaves the bar at $x = 0$ at a constant rate. This is permissible, but such types of problem should be posed with extreme caution. For instance, if, in addition, $a_2 = c_2 = 0$, the bar is insulated at $x = a$, and it can lose heat at a constant rate at the other end only if heat is generated internally at a sufficiently high rate. Even this may not be enough if the amount of heat initially stored in the bar is insufficient. Therefore, we reaffirm our decision to work under the first hypothesis, but in some of the exercises, in this section and in §3.8 and §3.10, the case in which $b_1 < 0$ or $b_2 < 0$ will be explored further.

The second hypothesis is included to make sure that none of the endpoint conditions is missing, since we cannot expect the problem to be well posed without one of these.

For instance, if $a_1 \geq 0$ and $b_1 \geq 0$, the inequality $a_1 + b_1 > 0$ guarantees that these constants are not both zero, as this would imply that $c_1 = 0$, and then there would be no endpoint condition at $x = 0$.

The case in which $a_1 = a_2 = 0$ is very delicate. When $c_1 = c_2 = 0$ the bar is insulated at both endpoints and the temperature is expected to rise without bound unless $q \equiv 0$. This is also true, even if $q \equiv 0$, if either $c_1 > 0$ or $c_2 > 0$, because now, according to the second hypothesis, we have $b_1 > 0$ and $b_2 > 0$, and then heat is entering the bar at a constant rate at one or both endpoints. To verify our physical intuition we shall investigate how the total amount of heat in the bar varies with time. We have

$$(18) \qquad \frac{d}{dt} \int_0^a cAu(s,t)\,ds = A \int_0^a cu_t(s,t)\,ds$$

$$= A \int_0^a [(\kappa u_x)_x(s,t) + q(s)]\,ds$$

$$= A\left[(\kappa u_x)(a,t) - (\kappa u_x)(0,t) + \int_0^a q \right]$$

$$= A\left[\kappa(a)\frac{c_2}{b_2} + \kappa(0)\frac{c_1}{b_1} + \int_0^a q \right].$$

Hence, the heat will increase without bound unless this derivative is zero. Since A, κ, b_1, and b_2 are positive, this is possible only when $c_1 = c_2 = 0$ and $q \equiv 0$. This justifies the third hypothesis if a bounded solution is desired.

Note next that this general boundary value problem contains all the previous ones. When c and κ are constant, $q \equiv 0$, and $b_1 = b_2 = 0$ we have the problem of §3.1 or §3.3. If, instead, $a_1 = a_2 = c_1 = c_2$, then the bar is insulated at the endpoints. Finally, with the choice of signs in (15) and (16) we have convection at $x = 0$ when $a_1 > 0$ and $b_1 > 0$ or convection at $x = a$ when $a_2 > 0$ and $b_2 > 0$. What we have learned in these particular cases allows us to outline the following general procedure to solve the boundary value problem (14)–(17) and to verify that it is well posed in the sense of Hadamard.

1. Find the steady-state solution. This is a nonnegative solution U that is independent of time and satisfies the endpoint conditions (15) and (16), but need not satisfy the initial condition (17). It is expected to be the limiting value of the complete solution as $t \to \infty$.

2. Find the transient solution. This is a solution v corresponding to the case $q \equiv 0$, $c_1 = c_2 = 0$, and the new initial condition $v(x, 0) = f(x) - U(x)$. Since both the equation and the endpoint conditions are now linear, it may be possible to find v by the superposition of solutions obtained by the method of separation of variables. As in §3.2, it is expected that $v(x, t) \to 0$ as $t \to \infty$.

3. Find the complete solution. If both the steady-state and the transient solutions, U and v, have been found, then the function $u = U + v$ is clearly a solution of the original problem. It must be shown that $u \geq 0$.

4. Verify the uniqueness and stability of the solution. If a solution has been found, we must verify that it is unique and continuously dependent on initial and endpoint values.

These four aspects of the problem will be studied in the next three sections.

§3.8   The Steady-State Solution

It is expected that in practice the transient solution will become negligible after a short amount of time. For this reason, it is the steady-state solution that is most important in many applications, and it can always be found as shown below. What we want is to find a nonnegative solution U of (14)–(16) that is independent of time, that is, a function U such that

$$(19) \qquad 0 = (\kappa U')' + q$$

$$(20) \qquad a_1 U(0) - b_1 U'(0) = c_1$$

$$(21) \qquad a_2 U(a) + b_2 U'(a) = c_2.$$

We start with the particular case $a_1 = a_2 = 0$. Then $c_1 = c_2 = 0$ and $q \equiv 0$ as required by the third hypothesis. But then (19) implies that $\kappa U'$ is constant, and this constant is zero because of (20) or (21). Since $\kappa > 0$, $U' \equiv 0$ and then U is also constant. Its value cannot be found from (19)–(21), but if a steady-state solution is to represent the limit of $u(x, t)$ as $t \to \infty$ and if, according to (18), the total heat stored in the bar remains constant, then it must retain its original value. This implies that

$$\int_0^a cU = \int_0^a cf$$

and, since U is a constant,

$$U \equiv \frac{\displaystyle\int_0^a cf}{\displaystyle\int_0^a c}.$$

Note that when c is also constant this is simply the average value of the original temperature distribution, in agreement with the result of §3.4.

In the remaining case, when $a_1 + a_2 > 0$, it is usually simple to integrate (19) twice for any given κ and q and then evaluate the constants of integration from (20) and (21).

Example 3.3 Find the steady-state solution of the boundary value problem

$$\begin{array}{ll}
u_t = u_{xx} + \sin x & 0 < x < \pi, \quad t > 0 \\
u(0, t) = 1 & t \geq 0 \\
u_x(\pi, t) = 2 & t \geq 0 \\
u(x, 0) = 1 + \sin 2x & 0 \leq x \leq \pi.
\end{array}$$

Then (19)–(21) become

$$0 = U'' + \sin x$$
$$U(0) = 1$$
$$U'(\pi) = 2.$$

Integrating the first equation twice gives

$$U'(x) = \cos x + A$$
$$U(x) = \sin x + Ax + B,$$

where A and B are integration constants. Putting $x = \pi$ yields $A - 1 = U'(\pi) = 2$, so that $A = 3$, and putting $x = 0$ yields $B = U(0) = 1$. Hence,

$$U(x) = \sin x + 3x + 1.$$

Clearly, it is nonnegative for $0 \le x \le \pi$.

★   It is also possible to find a general expression for the steady-state solution.[1] Such an expression will be useful in §3.10 in order to show that the boundary value problem (14)–(17) is well posed in the sense of Hadamard. Two consecutive integrations of (19) yield

$$U'(x) = -\frac{1}{\kappa(x)}\left(\int_0^x q - A\right)$$

and

$$U(x) = -\int_0^x \frac{1}{\kappa(s)}\left(\int_0^s q - A\right) ds + B,$$

where A and B are constants. At this point we shall find it convenient, for the sake of brevity, to define

$$Q(x) = \int_0^x q \qquad \text{and} \qquad K(x) = \int_0^x \frac{1}{\kappa}.$$

In this notation our solution is

(22) $$U(x) = -\int_0^x \frac{Q}{\kappa} + AK(x) + B.$$

The constants A and B are uniquely determined by (20) and (21). We omit the purely algebraic details (Exercise 8.5) and simply assert that these values are

(23) $$A = \frac{\kappa(0)}{\Delta}\left[\kappa(a)\left(a_1 c_2 - a_2 c_1 + a_1 a_2 \int_0^a \frac{Q}{\kappa}\right) + a_1 b_2 Q(a)\right]$$

---

[1]The rest of this section is based on the author's "The Existence of a Steady-State Solution for a Type of Parabolic Boundary Value Problem," *Int. J. Math. Educ. Sci. Technol.*, **19** (1988) 413–419.

and

$$(24) \quad B = \frac{1}{\Delta}\Bigg[b_1c_2\kappa(a) + b_2c_1\kappa(0) + b_1b_2Q(a)$$
$$+ a_2c_1\kappa(0)\kappa(a)K(a) + a_2b_1\kappa(a)\int_0^a \frac{Q}{\kappa}\Bigg],$$

where

$$\Delta = [a_1\kappa(0)K(a) + b_1]a_2\kappa(a) + a_1b_2\kappa(0).$$

Our hypotheses on the constants and the fact that $a_1 + a_2 > 0$ in this case imply that $\Delta > 0$ and $B \geq 0$.

It remains to show that U is nonnegative. We already know that $U(0) = B \geq 0$. Then a short computation (Exercise 8.5) gives

$$U(a) = \frac{1}{\Delta}\Bigg[b_1c_2\kappa(a) + b_2c_1\kappa(0) + b_1b_2Q(a)$$
$$+ a_1c_2\kappa(0)\kappa(a)K(a) + a_1b_2\kappa(0)\Big(Q(a)K(a) - \int_0^a \frac{Q}{\kappa}\Big)\Bigg].$$

Since Q is an increasing function we have

$$\int_0^x \frac{Q}{\kappa} \leq \int_0^x \frac{Q(x)}{\kappa} = Q(x)K(x)$$

for any $0 \leq x \leq a$, which, with $x = a$, shows that $U(a) \geq 0$. Thus, the only way for U to have negative values is to have a negative minimum at some point x in $(0, a)$. But this is impossible for at such a point

$$U'(x) = -\frac{1}{\kappa(x)}[Q(x) - A] = 0,$$

so that $A = Q(x)$ and then, from (22),

$$U(x) = -\int_0^x \frac{Q}{\kappa} + Q(x)K(x) + B \geq 0.$$

Example 3.4 The conductivity of a thin bar located between $x = 0$ and $x = 9$ is given by $\kappa(x) = 1 + x$ and the rate of internal heat generation, by $q(x) = 2(1+x)$. The left endpoint is kept insulated and the right endpoint, at $300°$K. Find the steady-state solution.

We are dealing with a boundary value problem of the form

$$\begin{aligned}
cu_t &= [(1+x)u_x]_x + 2(1+x) & 0 < x < 9, \quad t > 0 \\
u_x(0, t) &= 0 & t \geq 0 \\
u(9, t) &= 300 & t \geq 0 \\
u(x, 0) &= f(x) & 0 \leq x \leq 9.
\end{aligned}$$

Recall that knowledge of c and f is not necessary to determine the steady-state solution.

In this case $a = 9$, $a_1 = c_1 = b_2 = 0$, $b_1 = a_2 = 1$ and $c_2 = 300$. Also,

$$Q(x) = 2 \int_0^x (1+s)\, ds = 2x + x^2$$

$$K(x) = \int_0^x \frac{1}{1+s}\, ds = \log(1+x)$$

and

$$\int_0^x \frac{Q}{\kappa} = \int_0^x \frac{2s + s^2}{1+s}\, ds = \frac{x^2}{2} + x - \log(1+x).$$

With these values (23) and (24) become $A = 0$ and

$$B = \frac{b_1 c_2 \kappa(9) + a_2 b_1 \kappa(9) \int_0^9 \frac{Q}{\kappa}}{b_1 a_2 \kappa(9)} = \frac{c_2 + a_2 \int_0^9 \frac{Q}{\kappa}}{a_2} = \frac{699}{2} - \log 10.$$

Therefore, using (22) we obtain

$$U(x) = \log \frac{1+x}{10} + \frac{699 - x^2}{2} - x.$$

Example 3.5 The conductivity of a thin bar located between $x = 0$ and $x = 10$ is given by $\kappa(x) = e^{-x}$, and there is no internal heat generation. Instead, there is heat transfer by convection at the endpoints and, at each of them, the temperature varies at a rate equal to the temperature difference between the bar and the medium outside. If this medium is kept at $320°$K, find the steady-state solution.

Our boundary value problem is of the form

$$\begin{aligned}
cu_t &= (e^{-x}u_x)_x & 0 < x < 10, \quad t > 0 \\
u(0,t) - u_x(0,t) &= 320 & t \geq 0 \\
u(10,t) + u_x(10,t) &= 320 & t \geq 0 \\
u(x,0) &= f(x) & 0 \leq x \leq 10.
\end{aligned}$$

Since $a_1 c_2 = a_2 c_1$ and $Q \equiv 0$ it follows that $A = 0$ and that $U \equiv B$. There is no need to use (24) to evaluate B, for a moment's reflection shows that, if U is constant, then $U'(0) = 0$, and $U(x) = U(0) = 320$ from the left endpoint condition.

Note that the same constant solution is obtained if we allow the convection proportionality constants b_1 and b_2 to assume any arbitrary values instead of unity. For the steady-state solution to be other than constant either internal heat must be introduced or the temperatures of the medium outside the bar must be different at the endpoints.

Example 3.6 If the temperature of the medium at the right endpoint of the bar in Example 3.5 is lowered to $300°$K, then, since $Q \equiv 0$, the steady-state solution is of the form $U(x) = AK(x) + B$. But this time $K(x) = e^x - 1$ and $\Delta = 2$, and then (23), (24) and (22) become

$$A = \frac{\kappa(0)\kappa(10)(a_1c_2 - a_2c_1)}{2} = -10e^{-10}$$

$$B = \frac{b_1c_2\kappa(10) + b_2c_1\kappa(0) + a_2c_1\kappa(0)\kappa(10)K(10)}{2} = 320 - 10e^{-10}$$

and

$$U(x) = -10e^{-10}(e^x - 1) + 320 - 10e^{-10} = 320 - 10e^{x-10}.$$

In the preceding work it has been convenient to assume that κ has a continuous derivative so that U has two. However, when modeling a concrete physical situation, it may be more convenient to assume that κ' is only piecewise continuous. For example, when two bars made of different materials are put in contact end to end to form a longer bar, the resulting κ is piecewise constant. We can accept a piecewise continuously differentiable κ in the model if we assume that it is just an approximation of a continuously differentiable physical κ, the two being exactly alike except that the latter has rounded corners. If this rounding is assumed to take place over arbitrarily small intervals about the points at which the κ in the model is not differentiable, then using this κ in (22) will give us a good approximation of the steady-state solution. We shall adopt this point of view and still refer to (22) as the steady-state solution in cases in which κ' is only piecewise continuous (see Exercise 10.3).

Finally, the analysis carried out in this section excludes some valid boundary value problems when either $b_1 < 0$ or $b_2 < 0$. A more complete study covering those cases is contained in Exercises 8.6 to 8.12.

§3.9 The Transient Solution

If we define a new function $v = u - U$, where u is a solution of (14)–(17) and U is a solution of (19)–(21), then v is a solution of the boundary value problem

(25) $\qquad cv_t = (\kappa v_x)_x \qquad\qquad 0 < x < a \quad t > 0$

(26) $\qquad a_1v(0, t) - b_1v_x(0, t) = 0 \qquad t \geq 0$

(27) $\qquad a_2v(a, t) + b_2v_x(a, t) = 0 \qquad t \geq 0$

(28) $\qquad v(x, 0) = f(x) - U(x) \qquad\qquad 0 \leq x \leq a$

and $U + v$ is a complete solution of the original problem (14)–(17).

Since both the equation and the endpoint conditions for v are linear, we shall attempt to solve this problem by separation of variables. If there is a solution of the form $v(x, t) = X(x)T(t)$, then substitution into (25) gives

$$cXT' = \kappa X''T + \kappa'X'T.$$

After dividing by cXT, the left-hand side becomes a function of t only and the right-hand side a function of x only. Thus, each side must be equal to a constant that we can denote by $-\lambda$, and this leads to the ordinary differential equations $T' + \lambda T = 0$ and

(29) $$\kappa X'' + \kappa' X' + \lambda c X = 0.$$

The general solution of the first is well known, but that of the second is unknown except in some special cases—as when c and κ are constants. Even if it were known, it is not immediately clear that there are any values of λ for which there is a solution X that satisfies the endpoint conditions $a_1 X(0) - b_1 X'(0) = 0$ and $a_2 X(a) + b_2 X'(a) = 0$.

However, let us assume that in a given particular case we can find a sequence of real numbers $\lambda_1, \lambda_2, \ldots, \lambda_n, \ldots$ and, corresponding to each value $\lambda = \lambda_n$, a solution $X_n \not\equiv 0$ of (29) that satisfies the endpoint conditions. Then, for each n, the function

$$v_n(x, t) = c_n e^{-\lambda_n t} X_n(x),$$

where c_n is an arbitrary constant, is a solution of (25)–(27). Because of the linearity, any finite sum of the v_n also satisfies (25)–(27). Then, in order to satisfy an arbitrary initial condition, we try an infinite series of the form

(30) $$\sum_{n=1}^{\infty} c_n e^{-\lambda_n t} X_n(x)$$

as a possible solution. If it converges to a continuous function $v : \overline{D} \to \mathbb{R}$, then, putting $t = 0$ in (30) and using (28), we obtain the equation

(31) $$\sum_{n=1}^{\infty} c_n X_n(x) = f(x) - U(x).$$

It can be used to determine the constants c_n if we assume, as we did in Chapter 2 for Fourier series, that the series on the left converges, that it can integrated term by term—which is the case if the convergence is uniform—that equality holds, and that there is an integrable function $r \not\equiv 0$ such that

$$\int_0^a r X_m X_n = 0$$

for $m \neq n$. Indeed, if we multiply each term of (31) by $r X_m$, integrate term by term, and solve for c_n we obtain

$$c_n = \frac{\displaystyle\int_0^a r(f - U) X_n}{\displaystyle\int_0^a r X_n^2}.$$

The function r, which was identically equal to 1 in the case of Fourier series, is introduced here for the sake of flexibility. In general, we cannot expect the integral $\int_0^a X_m X_n$ to be zero, but it may be zero if the integrand is multiplied by some suitable factor $r \not\equiv 0$.

Therefore, to solve the problem in a given practical situation, one must be able to find specific values of λ_n and corresponding solutions X_n of (29) with the properties described above. This can be done in many cases, but is not always possible.

Regardless of whether or not the transient solution v can be found in a given particular case, can we just show its existence in the general case? For this to be possible a number of questions must be provided with affirmative answers, and the preceding discussion has already posed the following.

(i) Are there any values of λ for which (29) has a solution $X_n \not\equiv 0$ that satisfies the endpoint conditions $a_1 X(0) - b_1 X'(0) = 0$ and $a_2 X(a) + b_2 X'(a) = 0$?

(ii) Are they real numbers?

(iii) Do they form a sequence $\{\lambda_n\}$; that is, can they be put in a one-to-one correspondence with the positive integers?

(iv) Is there a unique—up to a multiplicative constant—solution $X_n \not\equiv 0$ of (29) for each λ_n?

(v) Does $m \neq n$ imply that $\int_0^a r X_m X_n = 0$ for some integrable function r?

(vi) With the values of c_n given above, does the series in (31) converge to $f(x) - U(x)$ uniformly on $[0, a]$?

Assuming that these questions have affirmative answers, we still have to show that (30) converges, that its sum is continuous on $\overline{D}$, and that it solves the boundary value problem (25)–(28).[1] The proof given in §3.1 can be adapted to the general case as follows. If we use the notation

$$X_n(x) = \sin \frac{n\pi}{a} x \qquad \text{and} \qquad g_n(t) = e^{-n^2 \pi^2 kt / a^2},$$

then the proofs of Propositions 3.1 and 3.2 are based on the inequalities $|c_n X_n| < M$ on $[0, a]$ for some $M > 0$ and all n;

$$g_n(t) \le g_n(t_0)$$

for $t \ge t_0$, any $t_0 > 0$, and all n; and

$$\lim_{n \to \infty} [g_n(t_0)]^{\frac{1}{n}} < 1.$$

Also, the proofs of Theorems 3.1 and 3.3 rest on the additional inequalities

$$g_n(t) \le 1 \qquad \text{and} \qquad g_{n+1}(t) \le g_n(t)$$

for all n and all t in $[0, \infty)$.

In the general case, when the X_n are solutions of (29) and $g_n(t) = e^{-\lambda_n t}$, the inequality $|c_n X_n| < M$ is a consequence of the uniform convergence of the series

---

[1] The rest of this section assumes knowledge of the proofs of Theorems 3.1 and 3.2.

in (31), assumed in answer to our last question (use Theorem A.1 with $\epsilon = 1$ and $m = n$ and take M to be any number larger than 1 and than any of the numbers

$$\max_{x \in [0,a]} |c_n X_n(x)|$$

for $n = 1, \ldots, N$). Also, the four inequalities stated above for the g_n remain valid if the next two questions have affirmative answers.

(vii) Is $0 \le \lambda_n < \lambda_{n+1}$ for all n?

(viii) Is $\lim_{n \to \infty} \dfrac{\lambda_n}{n} > 0$?

If they do, then (30) converges to a continuous solution $v : \overline{D} \to \mathbb{R}$ of the boundary value problem (25)–(28). For this v to be the transient solution it remains to show that $v(x, t) \to 0$ as $t \to \infty$. This is true if $\lambda_1 > 0$, for then the inequalities in (vii) imply that $\lambda_n / n > 0$ for all n and the limit in (viii) implies that $\lambda_n / n \ge \alpha$ for some $\alpha > 0$ and all n. Then,

$$|v(x, t)| \le \sum_{n=1}^{\infty} \left| c_n e^{-\lambda_n t} X_n(x) \right| \le \sum_{n=1}^{\infty} M e^{-\lambda_n t} \le M \sum_{n=1}^{\infty} (e^{-\alpha t})^n = \frac{e^{-\alpha t}}{1 - e^{-\alpha t}},$$

which apprarches zero as $t \to \infty$. If $\lambda_1 = 0$, the inequality $\lambda_n \ge n\alpha$ is still valid for $n \ge 2$ and a similar proof shows that $v(x, t) - c_1 X_1(x) \to 0$ as $t \to \infty$. Hence, v is still the transient solution if $c_1 = 0$.

The preceding discussion motivates our last question.

(ix) Is $c_1 = 0$ when $\lambda_1 = 0$?

The careful study of all these questions will be postponed until Chapter 4, in which we shall provide affirmative answers for all of them, and the existence of the continuous transient solution will be firmly established by Theorem 4.7.

★   ## §3.10 The Complete Solution

Once the existence of both the steady-state and the transient solutions, U and v, has been proved, then the function $u = U + v$ satisfies

(14)	$cu_t = (\kappa u_x)_x + q$	in D
(15)	$a_1 u(0, t) - b_1 u_x(0, t) = c_1$	$t \ge 0$
(16)	$a_2 u(a, t) + b_2 u_x(a, t) = c_2$	$t \ge 0$
(17)	$u(x, 0) = f(x)$	$0 \le x \le a,$

where the hypotheses of §3.7 are still applicable. That u is nonnegative and its uniqueness and continuous dependence on initial and boundary values remain to be proved.[1]

---

[1] This section is based on the author's "Uniqueness and Stability of Solutions for a Type of Parabolic Boundary Value Problem," *Int. J. Math. Math. Sci.*, **12** (1989) 735–740.

While such matters were decided on the basis of the maximum principle in the particular cases in §3.2 to §3.5, the maximum principle need not hold when there is internal heat generation (Exercise 2.1). However, the following two statements about the minimum of a solution, the first for the equation of heat propagation and the second for the boundary value problem, will be sufficient for our purposes.

Theorem 3.6 *Let D and $\bar{D}$ be as in §3.1, let T be an arbitrary positive number, and define the sets*

$$D_T = \{\, (x, t) \in D : t \leq T \,\}$$
$$\bar{D}_T = \{\, (x, t) \in \bar{D} : t \leq T \,\},$$

and $\gamma_T = \bar{D}_T - D_T$. If $u : \bar{D}_T \to \mathbb{R}$ is a continuous function that satisfies (14) in D_T, then it attains its minimum value on γ_T.

Proof. The proof is identical to that of Theorem 3.4 to the point at which it is established that v has its minimum at $(x_1, t_1) \in D_T$. But then $v_t(x_1, t_1) \leq 0$, $v_x(x_1, t_1) = 0$, and $v_{xx}(x_1, t_1) \geq 0$, so that $cv_t - (\kappa v_x)_x = cv_t - \kappa'v_x - \kappa v_{xx} \leq 0$. Since this contradicts the fact that

$$cv_t - (\kappa v_x)_x = cu_t - (\kappa u_x)_x + c\frac{M-m}{2T} = q + c\frac{M-m}{2T} > 0$$

in D_T, we must conclude that u attains its minimum value on γ_T, Q.E.D.

Theorem 3.7 *Let D_T, $\bar{D}_T$, γ_T, and u be as in Theorem 3.6 and define the sets*

$$\gamma_0 = \{\, (x, t) \in \gamma_T : x = 0, \ t > 0 \,\}$$

and

$$\gamma_a = \{\, (x, t) \in \gamma_T : x = a, \ t > 0 \,\}.$$

If u has a negative minimum on $\bar{D}_T$ and
  (i) *satisfies (15), then it attains its minimum on $\gamma_T - \gamma_0$, or*
  (ii) *satisfies (16), then it attains its minimum on $\gamma_T - \gamma_a$.*

Proof. By Theorem 3.6, u attains its minimum value on γ_T. We now consider two cases.

1. Assume that u satisfies (15) but that, contrary to the desired conclusion, $u(0, t_0) = m < 0$ for some $t_0 > 0$ while $u \geq M > m$ on $\gamma_T - \gamma_0$. Since the minimum is at $(0, t_0)$ we must have $u_x(0, t_0) \geq 0$. In fact, $u_x(0, t_0) = 0$, or else the hypothesis $a_1 + b_1 > 0$ implies that $a_1 u(0, t_0) - b_1 u_x(0, t_0) < 0$, contradicting (15) because $c_1 \geq 0$.
   Now define a function $v : \bar{D}_T \to \mathbb{R}$ by

$$v(x, t) = u(x, t) - \frac{M-m}{2K(a)} K(x),$$

where $K(x) = \int_0^x (1/\kappa)$. Then v is continuous and satisfies (14) since the derivative of

$$\kappa \frac{M-m}{2K(a)} K'(x) = \frac{M-m}{2K(a)}$$

with respect to x is zero. Hence, by Theorem 3.6, v attains its minimum value on γ_T. Now, $K(0) = 0$ implies that $v \equiv u$ on γ_0, but, if (x, t) is in $\gamma_T - \gamma_0$, the inequalities $u(x, t) \geq M$ and $K(x) \leq K(a)$ imply that

$$v(x, t) \geq u(x, t) - \frac{M - m}{2} \geq M - \frac{M - m}{2} = \frac{M + m}{2} > m.$$

This shows that v has its minimum at $(0, t_0)$, contradicting the fact that

$$v_x(0, t_0) = u_x(0, t_0) - \frac{M - m}{2K(a)\kappa(0)} = -\frac{M - m}{2K(a)\kappa(0)} < 0.$$

Therefore, u cannot have a negative minimum at a point in γ_0.

2. The proof that, if u satisfies (16), then it attains its minimum on $\gamma_T - \gamma_a$ is analogous to the one above, but now based on the choice

$$v(x, t) = u(x, t) - \frac{M - m}{2K(a)}[K(a) - K(x)],$$

Q.E.D.

These theorems have two corollaries that are applicable to solutions of

(25) $cu_t = (\kappa u_x)_x$ in D

(26) $a_1 u(0, t) - b_1 u_x(0, t) = 0$ $t \geq 0$

(27) $a_2 u(a, t) + b_2 u_x(a, t) = 0$ $t \geq 0,$

and we shall use them to complete the discussion of our boundary value problem.

Corollary 1 *A continuous function* $u : \bar{D}_T \to \mathbb{R}$ *that satisfies* (25) *in* D_T *attains its maximum value on* γ_T. *If this maximum value is positive and if*
  (i) *u satisfies* (26), *then it attains its maximum on* $\gamma_T - \gamma_0$, *or*
  (ii) *u satisfies* (27), *then it attains its maximum on* $\gamma_T - \gamma_a$.

Proof. If u satisfies any of the last three equations, so does $-u$, and thus $-u$ satisfies the hypotheses of Theorems 3.6 or 3.7 with $q \equiv 0$, $c_1 = c_2 = 0$. Accordingly, the statements made there about the minimum of $-u$ translate into the statements made here about the maximum of u, Q.E.D.

Corollary 2 *Let* $u : \bar{D} \to \mathbb{R}$ *be a continuous function that satisfies* (25)–(27). *If* M_0 *and* m_0 *denote the maximum and minimum values of* $u(x, 0)$, *then*

$$\min\{0, m_0\} \leq u(x, t) \leq \max\{0, M_0\}$$

for all (x, t) *in* $\bar{D}$.

Proof. Let $T > 0$ be arbitrary. By Corollary 1, the restriction of u to $\bar{D}_T$ attains its maximum value M on γ_T. Either $M \leq 0$, and then $u(x, t) \leq 0$ for $t \leq T$, or $M > 0$. In the second case, and again by Corollary 1, the restriction of u to $\bar{D}_T$ attains its maximum for $t = 0$, and thus $u(x, t) \leq M = M_0$ for $t \leq T$. Since T is arbitrary, this proves the inequality on the right. The one on the left is proved analogously but using the theorems instead of Corollary 1, Q.E.D.

With all this we are ready to prove the desired result.

Theorem 3.8 *If the boundary value problem* (14)–(17) *has a continuous solution* $u : \bar{D} \rightarrow \mathbb{R}$, *then it is well posed in the sense of Hadamard and* $u \geq 0$ *if* $f \geq 0$.

Proof. If u_1 and u_2 are solutions of (14)–(17) corresponding to initial temperature distributions f and g, then $u_1 - u_2$ satisfies (25)–(27) and the new initial condition $(u_1 - u_2)(x, 0) = f(x) - g(x)$. If M_0 and m_0 denote the maximum and minimum values of $f - g$, then Corollary 2 implies that

$$\min\{0, m_0\} \leq (u_1 - u_2)(x, t) \leq \max\{0, M_0\}$$

for all (x, t) in $\bar{D}$. This proves the continuity of solutions with respect to initial conditions and, taking $f \equiv g$, proves uniqueness.

To prove the continuity of a solution u with respect to both endpoint and initial conditions, note that the steady-state solution U, which exists and was found in §3.8, depends continuously on c_1, c_2 and f as is shown by the equations giving U. Then, if we define $v = u - U$, this function satisfies (25)–(27) and the initial condition $v(x, 0) = f(x) - U(x)$. As above, v depends continuously on $f - U$ which, in turn, depends continuously on c_1, c_2, and f. Therefore, so does $u = U + v$.

Finally, if u has negative values, then it must attain a negative minimum on $\bar{D}_T$ for some $T > 0$. By Theorem 3.7, this minimum is attained for $t = 0$, which is impossible because $f \geq 0$, Q.E.D.

§3.11   Time-Dependent Problems

The last heat problem ever studied by Fourier, in December of 1828, was that of a bar whose endpoint temperatures depend on time. We shall consider now a more general situation in which both the rate of internal heat generation and the endpoint conditions depend on time, that is, a boundary value problem of the form

$$
\begin{aligned}
cu_t &= (\kappa u_x)_x + q(x, t) && \text{in } D \\
a_1 u(0, t) - b_1 u_x(0, t) &= g(t) && t \geq 0 \\
a_2 u(a, t) + b_2 u_x(a, t) &= h(t) && t \geq 0 \\
u(x, 0) &= f(x) && 0 \leq x \leq a,
\end{aligned}
$$

where, in addition to the usual hypotheses:

1. a_1, a_2, b_1, and b_2 are nonnegative,

2. $a_1 + b_1 > 0$ and $a_2 + b_2 > 0$, and

3. if $a_1 = a_2 = 0$, then $g \equiv h \equiv 0$ and $q \equiv 0$,

we shall assume that g and h are nonnegative, differentiable functions and that f is nonnegative, is continuous, has a piecewise continuous derivative, and satisfies $a_1 f(0) - b_1 f'(0) = g(0)$ and $a_2 f(a) + b_2 f'(a) = h(0)$.

In view of the third hypothesis, the case $a_1 = a_2 = 0$ is actually time independent and has already been considered. Therefore, we assume that $a_1 + a_2 > 0$, and the first thing to note is that there is no such thing as a steady-state solution because the whole situation is dependent on time. However, it is still possible to reduce this problem to one with linear endpoint conditions if we find a function U of x and t that satisfies the given ones. For then the function $v = u - U$ would satisfy the same type of problem but with $g \equiv h \equiv 0$. The simplest approach is to investigate whether or not a function of the form $U(x, t) = A(t)x + B(t)$ would do the job. Indeed, such a function is easily found by substitution of this expression into the given endpoint conditions. This gives two equations from which the unknown functions A and B are readily obtained. The result is

$$(32) \qquad U(x, t) = \frac{a_1 h(t) - a_2 g(t)}{(aa_1 + b_1)a_2 + a_1 b_2} x + \frac{b_1 h(t) + (aa_2 + b_2)g(t)}{(aa_1 + b_1)a_2 + a_1 b_2}.$$

It should be no surprise that when $q \equiv 0$, $\kappa \equiv 1$, and g and h are constant, U becomes the steady-state solution of §3.8.

It is now easy to see that the function $v = u - U$ must be such that

$$cv_t = (\kappa v_x)_x + q(x, t) + \kappa' U_x(x, t) - cU_t(x, t) \qquad \text{in } D$$
$$(26) \qquad a_1 v(0, t) - b_1 v_x(0, t) = 0 \qquad t \geq 0$$
$$(27) \qquad a_2 v(a, t) + b_2 v_x(a, t) = 0 \qquad t \geq 0$$
$$v(x, 0) = f(x) - U(x, 0) \qquad 0 \leq x \leq a.$$

The next thing to observe regarding this problem is that the method of separation of variables is no longer applicable because $q + \kappa' U_x - cU_t$ is a function of both x and t. This rules out the procedure outlined in §3.9, but it is legitimate to ask whether or not it is possible to modify that procedure to yield a solution in this case, and we have nothing to lose in doing so. Assume then that there is a sequence of real constants λ_n and functions X_n as described in §3.9. That is, assume that for each n there is a solution $X_n \not\equiv 0$ of

$$(29) \qquad \kappa X'' + \kappa' X' + \lambda_n cX = 0$$

that satisfies $a_1 X_n(0) - b_1 X_n'(0) = 0$ and $a_2 X_n(a) + b_2 X_n'(a) = 0$, and assume that there is an integrable function r such that $\int_0^a r X_m X_n = 0$ (we shall learn in Chapter 4 that r must be precisely the specific heat capacity c of the material). Our question now, inspired by Lagrange's method of variation of constants, is whether or not our problem has a solution of the form

$$(33) \qquad v(x, t) = \sum_{n=1}^{\infty} c_n(t) X_n(x),$$

where the c_n are unknown functions to be determined, rather than constants as in §3.9. The advantage of this type of series is that, if it converges, it automatically satisfies (26) and (27).

Now, the initial condition will be satisfied if

$$\sum_{n=1}^{\infty} c_n(0)X_n(x) = f(x) - U(x, 0),$$

and then multiplication of both sides of this equation by $rX_m(x)$ and integration term by term from 0 to a, which we shall assume to be permissible, give

$$(34) \qquad c_n(0) = \frac{\displaystyle\int_0^a r[f(x) - U(x, 0)]X_n(x)\,dx}{\displaystyle\int_0^a rX_n^2(x)\,dx}.$$

It remains to determine the values of c_n for $t > 0$ so that v satisfies the desired partial differential equation. This is not difficult if the series in (33) can be differentiated term by term as necessary, for then

$$cv_t = \sum_{n=1}^{\infty} cc_n' X_n \qquad \text{and} \qquad \kappa v_x = \sum_{n=1}^{\infty} \kappa c_n X_n',$$

and, using (29),

$$(\kappa v_x)_x = \sum_{n=1}^{\infty} c_n(\kappa X''_n + \kappa' X_n') = \sum_{n=1}^{\infty} c_n(-\lambda_n c X_n).$$

Substitution into $cv_t - (\kappa v_x)_x = q + \kappa' U_x - cU_t$ requires that

$$(35) \qquad \sum_{n=1}^{\infty} (c_n' + \lambda_n c_n)X_n = \frac{q + \kappa' U_x}{c} - U_t.$$

If we define

$$\tilde{q} = \frac{q + \kappa' U_x}{c} - U_t;$$

if, for each fixed t, $\tilde{q}$ has a series expansion in terms of the X_n of the form

$$(36) \qquad \tilde{q}(x, t) = \sum_{n=1}^{\infty} q_n(t)X_n(x);$$

and if this series can be integrated term by term from 0 to a, we obtain

$$(37) \qquad q_n(t) = \frac{\displaystyle\int_0^a r\tilde{q}(x, t)X_n(x)\,dx}{\displaystyle\int_0^a rX_n^2(x)\,dx}.$$

Therefore, (35) and (36) imply that, for each n,

$$c_n'(t) + \lambda_n c_n(t) = q_n(t).$$

This is a first-order ordinary differential equation whose solution is known to be

$$(38) \qquad c_n(t) = e^{-\lambda_n t} \left[\int_0^t q_n(\tau) e^{\lambda_n \tau} \, d\tau + c_n(0) \right].$$

This completes the determination of the function v.

When, in a practical situation, the steps outlined here lead to a series of the form (33), because of the unjustified assumptions made in the derivation of the formulas given above, it must be verified *a posteriori* that its sum is indeed a solution as desired.

We shall consider only very simple examples, both to shorten the computations and because of the difficulty in determining the λ_n, r, and X_n in more complicated cases.

Example 3.7 Find a solution of the boundary value problem

$$\begin{array}{ll}
u_t = u_{xx} + e^{-t}(x - 1 + \sin \pi x) & 0 < x < 1, \quad t > 0 \\
u(0, t) = e^{-t} & t \geq 0 \\
u(1, t) = 2 & t \geq 0 \\
u(x, 0) = x + 1 & 0 \leq x \leq 1.
\end{array}$$

Here $a = a_1 = a_2 = 1$, $b_1 = b_2 = 0$, $g(t) = e^{-t}$, and $h(t) = 2$ for all t. From (32) we obtain

$$U(x, t) = [h(t) - g(t)]x + g(t) = (2 - e^{-t})x + e^{-t} = e^{-t}(1 - x) + 2x.$$

Since $c \equiv \kappa \equiv 1$, (29) becomes $X'' + \lambda X = 0$, and (26) and (27) require that $X(0) = X(1) = 0$. Then

$$X_n(x) = \sin n\pi x \qquad \text{for} \qquad \lambda = \lambda_n = n^2\pi^2$$

and $r \equiv 1$. Now, $f(x) - U(x, 0) = 0$ for all x and (34) shows that $c_n(0) = 0$ for all n. Also,

$$\tilde{q}(x, t) = q(x, t) - U_t(x, t) = e^{-t}(x - 1 + \sin \pi x) + e^{-t}(1 - x) = e^{-t} \sin \pi x,$$

and then (37) gives

$$q_n(t) = \frac{e^{-t} \displaystyle\int_0^1 \sin \pi x \sin n\pi x \, dx}{\displaystyle\int_0^1 \sin^2 n\pi x \, dx} = \begin{cases} 0 & \text{if} \quad n > 1 \\ e^{-t} & \text{if} \quad n = 1. \end{cases}$$

Finally, (38) implies that $c_n(t) = 0$ for all $t \geq 0$ if $n > 1$ and, with $\lambda_1 = \pi^2$,

$$c_1(t) = e^{-\pi^2 t} \int_0^t e^{-\tau} e^{\pi^2 \tau} \, d\tau = e^{-\pi^2 t} \frac{e^{(\pi^2 - 1)t} - 1}{\pi^2 - 1} = \frac{e^{-t} - e^{-\pi^2 t}}{\pi^2 - 1}.$$

The complete solution of our problem is

$$u(x, t) = U(x, t) + \sum_{n=1}^{\infty} c_n(t) \sin n\pi x = e^{-t}(1 - x) + 2x + \frac{e^{-t} - e^{-\pi^2 t}}{\pi^2 - 1} \sin \pi x.$$

It is easily verified that this is indeed a solution of our problem.

Example 3.8 Find a solution of the boundary value problem

$$
\begin{aligned}
u_t &= u_{xx} + q(x, t) & 0 &< x < a, \quad t > 0 \\
u(0, t) &= 0 & t &\geq 0 \\
u(a, t) + u_x(a, t) &= 0 & t &\geq 0 \\
u(x, 0) &= f(x) & 0 &\leq x \leq a
\end{aligned}
$$

In this problem the endpoint conditions are linear and $U \equiv 0$. Again (29) becomes $X'' + \lambda X = 0$, but now we must find solutions of this equation such that $X(0) = 0$ and $X(a) + X'(a) = 0$. This was done in §3.6, in which we found that $r \equiv 1$ and, putting $\mu_n = \lambda_n$ because $k = 1$ in this particular case, the solutions $X_n(x) = \sin \sqrt{\lambda_n} \, x$, where each λ_n is such that

$$\tan a\sqrt{\lambda_n} = -\sqrt{\lambda_n} \quad \text{and} \quad \frac{(2n - 1)\pi}{2a} < \sqrt{\lambda_n} < \frac{n\pi}{a}.$$

Then

$$
\begin{aligned}
\int_0^a X_n^2 &= \int_0^a \sin^2 \sqrt{\lambda_n} \, x \, dx \\
&= \int_0^a \left(\frac{1}{2} - \frac{\cos 2\sqrt{\lambda_n} \, x}{2} \right) dx \\
&= \frac{a}{2} - \frac{\sin 2a\sqrt{\lambda_n}}{4\sqrt{\lambda_n}} \\
&= \frac{a}{2} - \frac{\sin a\sqrt{\lambda_n} \cos a\sqrt{\lambda_n}}{2\sqrt{\lambda_n}} \\
&= \frac{a}{2} - \frac{\tan a\sqrt{\lambda_n} \cos^2 a\sqrt{\lambda_n}}{2\sqrt{\lambda_n}} \\
&= \frac{a + \cos^2 a\sqrt{\lambda_n}}{2}.
\end{aligned}
$$

The complete solution can now be assembled using (34), (37) with $\tilde{q} \equiv q$, (38), and (33), in this order. The result is

$$u(x, t) = \sum_{n=1}^{\infty} \frac{2e^{-\lambda_n t}}{a + \cos^2 a\sqrt{\lambda_n}} \left[\int_0^t \left(\int_0^a q(s, \tau) \sin \sqrt{\lambda_n} \, s \, ds \right) e^{\lambda_n \tau} d\tau \right.$$
$$\left. + \int_0^a f(s) \sin \sqrt{\lambda_n} \, s \, ds \right] \sin \sqrt{\lambda_n} \, x.$$

★ **Example 3.9** Find a solution of the boundary value problem

$$u_t = u_{xx} - x \sin t \qquad\qquad 0 < x < 2\pi, \quad t > 0$$
$$u(0, t) = 1 \qquad\qquad\qquad\quad t \geq 0$$
$$u_x(2\pi, t) = \cos t \qquad\qquad\;\; t \geq 0$$
$$u(x, 0) = 1 \qquad\qquad\qquad\;\; 0 \leq x \leq 2\pi.$$

In this case $a = 2\pi$, $a_1 = b_2 = 1$, $b_1 = a_2 = 0$, $g(t) = 1$ for all t, and $h(t) = \cos t$. Then, from (32),

$$U(x, t) = h(t)x + g(t) = x \cos t + 1.$$

Now (29) becomes $X'' + \lambda X = 0$, and (26) and (27) require that $X(0) = 0$ and $X'(2\pi) = 0$. It was found in §3.5 that the solutions are

$$X_n(x) = \sin \frac{2n - 1}{4} x \qquad \text{for} \qquad \lambda = \lambda_n = \left(\frac{2n - 1}{4}\right)^2$$

for $n \geq 1$ with $r \equiv 1$. Since $f(x) - U(x, 0) = -x$, (34) yields

$$
\begin{aligned}
c_n(0) &= \frac{\displaystyle\int_0^{2\pi} -x \sin \frac{2n-1}{4} x \, dx}{\displaystyle\int_0^{2\pi} \sin^2 \frac{2n-1}{4} x \, dx} \\[3mm]
&= \frac{1}{\pi}\left[\frac{4x}{2n-1} \cos \frac{2n-1}{4} x \Big|_0^{2\pi} - \frac{4}{2n-1} \int_0^{2\pi} \cos \frac{2n-1}{4} x \, dx\right] \\[3mm]
&= -\frac{1}{\pi}\left(\frac{4}{2n-1}\right)^2 \sin \frac{2n-1}{4} x \Big|_0^{2\pi} \\[3mm]
&= \frac{16(-1)^n}{(2n-1)^2 \pi}.
\end{aligned}
$$

Next we compute

$$\tilde{q}(x, t) = q(x, t) - U_t(x, t) = -x \sin t + x \sin t = 0$$

for all x and t, so that $q_n \equiv 0$ for all n. Therefore,

$$c_n(t) = c_n(0)e^{-\lambda_n t} = \frac{16(-1)^n}{(2n-1)^2 \pi} e^{-(2n-1)^2 t/16},$$

and the complete solution is

$$u(x, t) = x \cos t + 1 + \frac{16}{\pi} \sum_{n=1}^{\infty} \frac{(-1)^n}{(2n-1)^2} e^{-(2n-1)^2 t/16} \sin \frac{2n-1}{4} x.$$

As in §3.1, it is easy to verify that this series can be differentiated term by term and that u is a solution of our problem. The initial condition is satisfied because, for $t = 0$, the series on the right adds up to $-x$ on $[0, 2\pi]$ (Exercise 11.2).

Note that this method of solution for time-dependent problems emphasizes again the need for affirmative answers to the questions posed in §3.9. Their study is taken up in Chapter 4.

The results of §3.10 remain valid when q, g, and h depend on time, as only the fact that q, c_1, and c_2 are nonnegative was used to establish them. The new function U, although not a steady-state solution, depends continuously on g and h. Therefore, if our problem has a solution, then it is well posed in the sense of Hadamard by arguments identical to those used for time-independent problems.

EXERCISES

1.1   Find a solution of the boundary value problem (1)–(5) if

    (i) $f(x) = C(a - x)$ (ii) $f(x) = \cos \dfrac{5\pi}{a} x,$

    where C is a constant.

1.2   Let δ be a constant such that $0 < \delta < a$. Find a solution of the boundary value problem (1)–(5) if

$$(i)\ f(x) = \begin{cases} a - \dfrac{x}{\delta}, & 0 \le x \le \delta \\ 0, & \delta < x \le a \end{cases} \qquad (ii)\ f(x) = \begin{cases} 0, & 0 \le x \le a/2 \\ \sin \dfrac{\pi x}{a}, & a/2 < x \le a. \end{cases}$$

1.3   A bar of glass ($k = 0.0036\,\text{cm}^2/\text{sec}$) 20 cm long is initially at a temperature of 100°C and its ends are maintained at 0°C for $t > 0$.

    (i) By taking a finite number of terms of (6), find a good approximation to the temperature distribution on the bar for $t > 0$.

    (ii) Make a graph of such a temperature distribution for $t = 15, 30$, and 60 sec.

    (iii) Express the total amount of heat stored in the bar as a function of time.

1.4   For the glass bar of Exercise 1.3 ($c = 0.39\,\text{cal/cm}^3\,°\text{C}$) compute the amount of heat remaining in the bar after every hour for 5 hr, and show that it comes down to approximately for five hours, and show that it comes down to approximately 16.38% of the amount initially stored. If the bar were made of steel ($c = 0.858\,\text{cal/cm}^3\,°\text{C}$ and $k = 0.128\,\text{cm}^2/\text{sec}$), show that

    (i) the amount of heat initially stored in the bar is more than double the amount initially stored in the glass bar, and

    (ii) it takes less than 13 min for the amount of heat remaining in the bar to equal the amount remaining in the glass bar at the end of 5 hr.

1.5 If $T > 0$ and $0 < \delta < a$ are constants, show that the boundary value problem (1)–(5) with

$$f(x) = \begin{cases} T & \text{if } 0 \le x \le \delta \\ 0 & \text{if } \delta < x \le a \end{cases}$$

has the solution

$$u(x, t) = \frac{4T}{\pi} \sum_{n=1}^{\infty} \frac{1}{n} \sin^2 \frac{n\pi\delta}{2a} e^{-n^2\pi^2 kt/a^2} \sin \frac{n\pi}{a} x.$$

Prove or disprove that if $k \ge 0.1\,\mathrm{cm}^2/\mathrm{sec}$, $a \le 10\,\mathrm{cm}$ and $t \ge 120\,\mathrm{sec}$, then all the terms in this series are negligible—less than 3% in amplitude—compared with the one for $n = 1$. Prove or disprove the same statement for the series (6).

1.6 Two cylindrical bars of steel ($k = 0.128\,\mathrm{cm}^2/\mathrm{sec}$), each 5 cm long, are kept at temperatures 50 and 0°C, respectively. At a certain time they are placed end to end in perfect contact so as to form a single bar 10 cm long. Their lateral surfaces are insulated and their exposed ends are kept at 0°C for $t \ge 0$.

(*i*) Compute the approximate temperature at the junction every 10 sec for 1 min after contact, and show that the junction cools down to approximately 15°C at the end of 1 min.

(*ii*) Sketch the temperature distribution on the composite steel bar for $t = 10, 30$, and 60 sec.

(*iii*) How long would it take for the junction to cool down to this temperature if the bars were made of copper ($k = 1.1\,\mathrm{cm}^2/\mathrm{sec}$)?

2.1 Prove or disprove that the conclusion of the maximum principle holds if the heat equation is replaced by $u_t = ku_{xx} + q$, where q is a nonnegative, continuous function such that $q \not\equiv 0$ (**Hint:** of which boundary value problem is $u(x, t) = (1 - e^{-t}) \sin x$ a solution?).

In Exercises 2.2 to 2.4 solve the boundary value problem

$$\begin{aligned} u_t &= ku_{xx} & 0 < x < a, \quad t > 0 \\ u(0, t) &= 0 & t \ge 0 \\ u(a, t) &= 0 & t \ge 0 \\ u(x, 0) &= f(x) & 0 \le x \le a \end{aligned}$$

for the conditions given.

2.2 $f(x) = \sin^2 \dfrac{\pi}{a} x$.

2.3 δ is a constant such that $0 < \delta < a/2$ and f is the function of Exercise 5.9 of Chapter 2.

2.4 $a = \pi$ and $f(x) = x \sin x$.

2.5 Show that, if $f \ge 0$ and $f \not\equiv 0$, then $c_1 > 0$ in (6). Then, if we define $\alpha = \pi^2 k/a^2$ and $0 < x < a$, rewrite the sum of (6) as

$$u(x, t) = c_1 e^{-\alpha t} \sin \frac{\pi}{a} x \left[1 + \frac{e^{\alpha t}}{c_1 \sin \frac{\pi}{a} x} \sum_{n=2}^{\infty} c_n e^{-n^2 \alpha t} \sin \frac{n\pi}{a} x \right].$$

Prove that as $t \to \infty$ the term in the brackets approaches 1 and then that

$$\lim_{t \to \infty} \frac{\log u(x, t)}{t} = -\frac{\pi^2}{a^2} k.$$

Thus, while $u(x,t) \to 0$ and $\log u(x,t) \to -\infty$ as $t \to \infty$, $\log u(x,t)/t$ has a finite limit independent of f.

2.6 Is the boundary value problem

$$
\begin{aligned}
u_t &= ku_{xx} & 0 < x < a, \quad t < 0 \\
u(0,t) &= 0 & t \leq 0 \\
u(a,t) &= 0 & t \leq 0 \\
u(x,0) &= f(x) & 0 \leq x \leq a
\end{aligned}
$$

for negative time well posed in the sense of Hadamard? (**Hint:** what is the solution corresponding to

$$
f(x) = \frac{1}{m} \sin \frac{m\pi}{a} x,
$$

where m a positive integer?)

3.1 If $T > 0$ and $0 < \delta < a$ are constants, find a solution of the boundary value problem

$$
\begin{aligned}
u_t &= ku_{xx} & 0 < x < a, \quad t > 0 \\
u(0,t) &= T & t \geq 0 \\
u(a,t) &= 0 & t \geq 0 \\
u(x,0) &= \begin{cases} T, & 0 \leq x \leq \delta \\ 0, & \delta < x \leq a. \end{cases}
\end{aligned}
$$

3.2 If the left endpoint of the bar in Exercise 1.6 is kept at 50°C for $t \geq 0$,

(*i*) compute the temperature at the junction every 10 sec for 1 min after contact, for the cases of both steel and copper, and

(*ii*) sketch the temperature distribution on the composite bar for $t = 10, 30$, and 60 sec.

3.3 Solve the boundary value problem

$$
\begin{aligned}
u_t &= ku_{xx} & 0 < x < a, \quad t > 0 \\
u(0,t) &= T_1 & t \geq 0 \\
u(a,t) &= T_2 & t \geq 0 \\
u(x,0) &= 0 & 0 \leq x \leq a.
\end{aligned}
$$

3.4 Solve the boundary value problem

$$
\begin{aligned}
u_t &= ku_{xx} & 0 < x < a, \quad t > 0 \\
u(0,t) &= T_1 & t \geq 0 \\
u(a,t) &= T_2 & t \geq 0 \\
u(x,0) &= T_1 + (T_2 - T_1)\left(\frac{x}{a}\right)^2 & 0 \leq x \leq a.
\end{aligned}
$$

3.5 Solve the boundary value problem

$$
\begin{aligned}
u_t &= ku_{xx} & 0 < x < 3, \quad t > 0 \\
u(0,t) &= 1 & t \geq 0 \\
u(3,t) &= 4 & t \geq 0 \\
u(x,0) &= (x-1)^2 & 0 \leq x \leq 3.
\end{aligned}
$$

3.6   Solve the boundary value problem

$$u_t = ku_{xx} \qquad\qquad\qquad 0 < x < 10, \quad t > 0$$
$$u(0, t) = 300 \qquad\qquad\quad t \geq 0$$
$$u(10, t) = 300 \qquad\qquad t \geq 0$$
$$u(x, 0) = 300 + x(10 - x) \qquad 0 \leq x \leq 10.$$

3.7   If δ is a constant such that $0 < \delta < a$, solve the boundary value problem

$$u_t = ku_{xx} \qquad\qquad 0 < x < a, \quad t > 0$$
$$u(0, t) = 0 \qquad\qquad\quad t \geq 0$$
$$u(a, t) = a \qquad\qquad\quad t \geq 0$$
$$u(x, 0) = \begin{cases} ax/\delta, & 0 \leq x < \delta \\ a, & \delta \leq x \leq a. \end{cases}$$

3.8   Prove or disprove that the boundary value problem

$$u_t = ku_{xx} \qquad\qquad\qquad\quad 0 < x < a, \quad t > 0$$
$$u(x, t) \to T \quad \text{as} \quad x \to 0 \qquad t > 0$$
$$u(x, t) \to 0 \quad \text{as} \quad t \to 0 \qquad 0 < x < a$$

is correctly posed in the sense of Hadamard (**Hint:** refer to Example 1.2 of Chapter 1).

3.9   Assume that the bar is not insulated along its lateral surface and that it loses heat through it at a rate per unit length proportional to the difference $u(x, t) - T$, where T is the temperature of the surrounding medium. The equation of heat propagation is now

$$u_t = ku_{xx} - h(u - T),$$

where h is a positive constant (see Exercise 2.5 of Chapter 1). If the endpoints of the bar are maintained at a temperature T for $t \geq 0$ and the initial temperature distribution is $u(x, 0) = f(x)$, use the function

$$v = e^{ht}(u - T)$$

to reduce this problem to one already solved.

3.10   Use the method of Exercise 3.9 to solve the boundary value problem

$$u_t = ku_{xx} - h(u - 1) \qquad 0 < x < a, \quad t > 0$$
$$u(0, t) = 1 \qquad\qquad\qquad\quad t \geq 0$$
$$u(a, t) = 1 \qquad\qquad\qquad\quad t \geq 0$$
$$u(x, 0) = 1 - \sin \frac{\pi}{a}x \qquad 0 \leq x \leq a,$$

where h is a positive constant.

In Exercises 4.1 to 4.5 solve the boundary value problem corresponding to a bar insulated at the endpoints for the conditions given.

4.1   (i) $f(x) = \sin^2 \dfrac{\pi}{a}x$ \qquad\qquad\qquad (ii) $f(x) = 1 + \cos \dfrac{3\pi}{a}x.$

4.2 C and δ are constants with $0 < \delta < a$ and

$$f(x) = \begin{cases} Cx, & 0 \le x \le \delta \\ \dfrac{C\delta}{a - \delta}(a - x), & \delta < x \le a. \end{cases}$$

4.3 T_1 and T_2 are constants and

$$f(x) = T_1 + (T_2 - T_1)\left(\frac{x}{a}\right)^2.$$

4.4 T_1, T_2, and δ are constants with $0 < \delta < a$ and

$$f(x) = \begin{cases} T_1, & 0 \le x \le \delta \\ T_2, & \delta < x \le a. \end{cases}$$

4.5 $a = \pi$ and

$$f(x) = \begin{cases} 0, & 0 \le x \le \pi/2 \\ -\cos x, & \pi/2 < x \le 3\pi/2 \\ 0, & 3\pi/2 < x \le \pi. \end{cases}$$

4.6   Show that the boundary value problem for a bar insulated at the left endpoint cannot have a bounded solution if the endpoint condition at $x = a$ is $u_x(a, t) = \epsilon > 0$ for $t \ge 0$ (**Hint:** examine how the total amount of heat in the bar varies with time).

4.7   Assume that the bar is insulated at the endpoints only and that it loses heat through its lateral surface at a rate per unit length proportional to the difference $u(x, t) - T$, where T is the temperature of the medium surrounding the bar. The equation of heat propagation is now

$$u_t = ku_{xx} - h(u - T),$$

where h is a positive constant (see Exercise 2.5 of Chapter 1). Use the function

$$v = e^{ht}(u - T)$$

to reduce this problem to one already solved.

4.8   Repeat Exercise 4.4 if the heat equation is replaced by that in Exercise 4.7.

4.9   Consider a thin piece of wire of length $2a$, $a > 0$, insulated along its lateral surface and bent into the shape of a circular ring. If s denotes arc length from its midpoint, the temperature in the ring satisfies the heat equation

$$u_t = ku_{ss} \qquad -a < s < a, \quad t > 0$$

subject to the conditions

$$\begin{aligned} u(-a, t) &= u(a, t) & t \ge 0 \\ u_s(-a, t) &= u_s(a, t) & t \ge 0. \end{aligned}$$

If the initial temperature distribution is given by

$$u(s, 0) = f(s) \qquad -a \le s \le a,$$

use separation of variables to find a formal solution of this boundary value problem. Find the particular form of the solution if $a = \pi$, δ is a constant with $0 < \delta < \pi$ and f is the function of Exercise 2.16 of Chapter 2.

4.10 If the function f in Exercise 4.9 is continuous, has a piecewise continuous derivative, and is such that $f(-a) = f(a)$ and $f'(-a) = f'(a)$, verify that the formal solution of Exercise 4.9 is continuous on $\{(s, t) \in \mathbb{R}^2 : -a \leq s \leq a, t \geq 0\}$ and satisfies the stated boundary value problem. Prove the uniqueness of the solution (**Hint:** for the uniqueness proof let u_1 and u_2 be two solutions, define $u = u_1 - u_2$ and

$$I(t) = \int_{-a}^{a} u^2(s, t) \, ds,$$

and show that $I'(t) \leq 0$).

In Exercises 5.1 and 5.2 solve the boundary value problem

$$
\begin{aligned}
u_t &= ku_{xx} & 0 < x < a, \quad t > 0 \\
u(0, t) &= 0 & t \geq 0 \\
u_x(a, t) &= 0 & t \geq 0 \\
u(x, 0) &= f(x) & 0 \leq x \leq a
\end{aligned}
$$

for the conditions given.

5.1 T and δ are constants with $0 \leq \delta < a$ and

$$(i) \ f(x) = \begin{cases} 0, & 0 \leq x \leq \delta \\ T, & \delta < x \leq a \end{cases} \qquad (ii) \ f(x) = \begin{cases} Tx, & 0 \leq x \leq \delta \\ T\delta, & \delta < x \leq a. \end{cases}$$

5.2 $(i) \ f(x) = \sin \dfrac{\pi}{2a} x$ $\qquad\qquad\qquad (ii) \ f(x) = \dfrac{x}{a} - \dfrac{1}{2\pi} \sin \dfrac{2\pi}{a} x.$

5.3 If $h > 0$ and T are constants, solve the boundary value problem

$$
\begin{aligned}
u_t &= ku_{xx} - h(u - T) & 0 < x < a, \quad t > 0 \\
u(0, t) &= T & t \geq 0 \\
u_x(a, t) &= 0 & t \geq 0 \\
u(x, 0) &= f(x) & 0 \leq x \leq a
\end{aligned}
$$

(**Hint:** refer to Exercise 3.9).

5.4 Solve the boundary value problem

$$
\begin{aligned}
u_t &= ku_{xx} & 0 < x < a, \quad t > 0 \\
u_x(0, t) &= 0 & t \geq 0 \\
u(a, t) &= 0 & t \geq 0 \\
u(x, 0) &= f(x) & 0 \leq x \leq a.
\end{aligned}
$$

5.5 If T is a constant, solve the boundary value problem

$$
\begin{aligned}
u_t &= ku_{xx} & 0 < x < a, \quad t > 0 \\
u(0, t) &= T & t \geq 0 \\
u_x(a, t) &= 0 & t \geq 0 \\
u(x, 0) &= f(x) & 0 \leq x \leq a
\end{aligned}
$$

(**Hint:** first find a solution, U, of the first three equations that is independent of t, and consider the function $v = u - U$).

5.6   Solve the boundary value problem of Exercise 5.5 if δ is a constant such that $0 < \delta < a$ and

$$(i) \ f(x) = \begin{cases} T, & 0 \le x \le \delta \\ 0, & \delta < x \le a \end{cases} \qquad (ii) \ f(x) = \begin{cases} \dfrac{T}{\delta}(\delta - x), & 0 \le x \le \delta \\ 0, & \delta < x \le a. \end{cases}$$

What limiting values are obtained as $\delta \to 0$ and as $\delta \to a$?

In Exercises 6.1 to 6.3 assume that μ_m and μ_n are solutions of the equation

$$\tan a \sqrt{\mu} = -\frac{\sqrt{\mu_n}}{h},$$

where a and h are positive constants.

6.1   Verify that

$$\int_0^a \sin^2 \sqrt{\mu_n}\, x = \frac{ah + \cos^2 a \sqrt{\mu_n}}{2h}$$

and that

$$\int_0^a \sin \sqrt{\mu_m}\, x \, \sin \sqrt{\mu_n}\, x \, dx = 0, \qquad m \neq n.$$

Then show that, if $g : [0, a] \to \mathbb{R}$ is a piecewise continuous function and if

$$\sum_{n=1}^{\infty} c_n \sin \sqrt{\mu_n}\, x = g(x)$$

uniformly on $[0, a]$,

$$c_n = \frac{2h}{ah + \cos^2 a} \int_0^a g(x) \sin \sqrt{\mu_n}\, x \, dx.$$

Compute c_n in the particular cases $g \equiv 1$ and $g(x) = x$.

6.2   If c_n is as in Exercise 6.1, show that $\mu_n/n \to \infty$ and that $c_n \to 0$ as $n \to \infty$. Then prove that

$$\sum_{n=1}^{\infty} c_n e^{-\mu_n k t} \sin \sqrt{\mu_n}\, x$$

converges to a function v that satisfies

$$\begin{aligned} v_t &= k v_{xx} & 0 < x < a, \quad t > 0 \\ v(0, t) &= 0 & t \ge 0 \\ v_x(a, t) &= -h v(a, t) & t \ge 0. \end{aligned}$$

6.3   Let U be the steady-state solution of the boundary value problem

$$\begin{aligned} u_t &= k u_{xx} & 0 < x < a, \quad t > 0 \\ u(0, t) &= T_1 & t \ge 0 \\ u_x(a, t) &= h[T_2 - u(a, t)] & t \ge 0 \\ u(x, 0) &= f(x) & 0 \le x \le a. \end{aligned}$$

Assume that for $t = 0$ the series in Exercise 6.2 converges uniformly on $[0, a]$, and let the c_n be as in Exercise 6.1 with $g = f - U$. Show that the function u defined by

$$u(x, t) = U(x) + \sum_{n=1}^{\infty} c_n e^{-\mu_n kt} \sin \sqrt{\mu_n}\, x$$

is a solution of this boundary value problem and that $u(x, t) \to U(x)$ as $t \to \infty$.

6.4 Use the constants evaluated at the end of Exercise 6.1 to show that, if $f \equiv T_3$, the function u of Exercise 6.3 takes the form

$$u(x, t) = \frac{h(T_2 - T_1)}{1 + ah} x + T_1 + 2h \sum_{n=1}^{\infty} \frac{T_3 - T_1 - (T_2 - T_3) \cos a\sqrt{\mu_n}}{\sqrt{\mu_n}\,(ah + \cos^2 a\sqrt{\mu_n}\,)} e^{-\mu_n kt} \sin \sqrt{\mu_n}\, x.$$

If $a = 20$, $k = 0.128$, $h = 1$, $T_1 = T_3 = 300$, $T_2 = 510$, $\mu_1 = 0.17496875$, and $\mu_2 = 0.701273437$, compute and sketch the sum of the first four terms of the series above for $t = 60$ and $t = 300$ sec.

6.5 Use the steady-state solution and separation of variables to show that the boundary value problem

$$\begin{aligned}
u_t &= ku_{xx} & 0 &< x < a, \quad t > 0 \\
u_x(0, t) &= 0 & t &\geq 0 \\
u_x(a, t) &= h[T - u(a, t)] & t &\geq 0 \\
u(x, 0) &= f(x) & 0 &\leq x \leq a
\end{aligned}$$

has a formal solution of the form

$$u(x, t) = T + \sum_{n=1}^{\infty} c_n e^{-\mu_n kt} \cos \sqrt{\mu_n}\, x,$$

where μ_n are the solutions of the equation

$$\tan a\sqrt{\mu} = \frac{h}{\sqrt{\mu}}.$$

Then, as in Exercise 6.1, arrive at the formula

$$c_n = \frac{2h}{ah + \sin^2 a\sqrt{\mu_n}} \left[\int_0^a f(x) \cos \sqrt{\mu_n}\, x\, dx - \frac{T}{\sqrt{\mu_n}} \sin a\sqrt{\mu_n} \right].$$

6.6 Show that, if C is a constant and $f(x) = T + C$ in Exercise 6.5, then

$$u(x, t) = T + 2hC \sum_{n=1}^{\infty} \frac{\sin a\sqrt{\mu_n}}{\sqrt{\mu_n}\,(ah + \sin^2 a\sqrt{\mu_n}\,)} e^{-\mu_n kt} \cos \sqrt{\mu_n}\, x.$$

6.7 Retrace the steps in §3.6 to arrive at a proposed solution of the given boundary value problem when $h < 0$. Consider separately the cases $ah < -1$, $ah = -1$, and $ah > -1$.

6.8 Let $f : [-a, a] \to \mathbb{R}$ be an odd function and assume that the function v of Exercise 6.2 satisfies $v(x, 0) = f(x)$ for $0 \leq x \leq a$. Show that v also satisfies the boundary value problem

$$\begin{aligned}
v_t &= kv_{xx} & -a &< x < a, \quad t > 0 \\
v_x(-a, t) &= hv(-a, t) & t &\geq 0 \\
v_x(a, t) &= -hv(a, t) & t &\geq 0 \\
v(x, 0) &= f(x) & -a &\leq x \leq a
\end{aligned}$$

6.9   Let $f : [-a, a] \to \mathbb{R}$ be an even function, put $T = 0$ in the boundary value problem of Exercise 6.5, and assume that the function u given there is a solution of that problem. Show that u also solves the boundary value problem of Exercise 6.8.

6.10   Use Exercises 6.8 and 6.9 to sketch a method of solution of the same boundary value problem if f is arbitrary (**Hint:** write f as the sum of an odd function and an even function).

7.1   Show that any solution of the boundary value problem

$$
\begin{aligned}
u_t &= ku_{xx} & 0 &< x < 1, \quad t > 0 \\
u(0, t) &= 0 & t &\geq 0 \\
u_x(1, t) &= -1 & t &\geq 0 \\
u(x, 0) &= x(1 - x) & 0 &\leq x \leq 1
\end{aligned}
$$

must have negative values (**Hint:** what boundary value problem does the function v defined by $v(x, t) = u(x, t) + x$ satisfy?).

7.2   Find a continuous, nonnegative, bounded solution of the boundary value problem

$$
\begin{aligned}
u_t &= ku_{xx} & 0 &< x < 1, \quad t > 0 \\
u(0, t) &= 0 & t &\geq 0 \\
u(1, t) - u_x(1, t) &= 0 & t &\geq 0 \\
u(x, 0) &= x & 0 &\leq x \leq 1.
\end{aligned}
$$

7.3   Show that the boundary value problem

$$
\begin{aligned}
u_t &= u_{xx} + 2 & 0 &< x < a, \quad t > 0 \\
u_x(0, t) &= 2 & t &\geq 0 \\
u_x(a, t) &= 0 & t &\geq 0 \\
u(x, 0) &= f(x) & 0 &\leq x \leq a
\end{aligned}
$$

cannot have a bounded solution unless $a = 1$. In such a case, find a solution when

$$
f(x) = \frac{4}{\pi} \sin \frac{\pi}{2} x.
$$

7.4   For what constant value of q does the boundary value problem

$$
\begin{aligned}
u_t &= u_{xx} + q & 0 &< x < 2, \quad t > 0 \\
u_x(0, t) &= 2 & t &\geq 0 \\
u_x(2, t) &= 0 & t &\geq 0 \\
u(x, 0) &= f(x) & 0 &\leq x \leq 2
\end{aligned}
$$

have a bounded solution? Find such a solution when

$$
f(x) = \begin{cases} 2x, & 0 \leq x \leq 1 \\ 2, & 1 < x \leq 2. \end{cases}
$$

Find the steady-state solutions of the boundary value problems in Exercises 8.1 to 8.4.

8.1

$$
\begin{aligned}
cu_t &= \kappa u_{xx} + Ce^{-mx} & 0 &< x < a, \quad t > 0 \\
u(0, t) &= 0 & t &\geq 0 \\
u(a, t) &= 0 & t &\geq 0 \\
u(x, 0) &= f(x) & 0 &\leq x \leq a,
\end{aligned}
$$

where κ, C, and m are positive constants.

8.2
$$cu_t = (e^{-x}u_x)_x + \cos \tfrac{1}{2}\pi x \qquad 0 < x < 1, \quad t > 0$$
$$u(0, t) - u_x(0, t) = 0 \qquad t \geq 0$$
$$u_x(1, t) = 300 \qquad t \geq 0$$
$$u(x, 0) = f(x) \qquad 0 \leq x \leq 1.$$

8.3
$$u_t = u_{xx} + \frac{1}{x + 10}u_x + 1 \qquad 0 < x < 10, \quad t > 0$$
$$u_x(0, t) = 0 \qquad t \geq 0$$
$$u(10, t) = 300 \qquad t \geq 0$$
$$u(x, 0) = f(x) \qquad 0 \leq x \leq 10.$$

8.4
$$cu_t = \kappa u_{xx} - h(u - T) \qquad 0 < x < a, \quad t > 0$$
$$u(0, t) = T_1 \qquad t \geq 0$$
$$u(a, t) = T_2 \qquad t \geq 0$$
$$u(x, 0) = f(x) \qquad 0 \leq x \leq a,$$

where $\kappa, h, T, T_1 - T$, and $T_2 - T$ are positive constants.

8.5   Derive Equations (23) and (24) and the equation giving $U(a)$.

For Exercises 8.6 to 8.12[1] replace the stated hypotheses on the constants with the conditions
  1. a_1, a_2, c_1, and c_2 are nonnegative,
  2. $a_1^2 + b_1^2 > 0$ and $a_2^2 + b_2^2 > 0$, and
  3. $a_1b_1 \geq 0$ and $a_2b_2 \geq 0$.
Then define a *steady-state solution* to be a nonnegative function that is independent of time, satisfies (14)–(16), and, when $a_1 = a_2 = 0$, also satisfies

$$\int_0^a cU = \int_0^a cf.$$

8.6   Consider the boundary value problem

$$cu_t = (\kappa u_x)_x + q \qquad 0 < x < a, \quad t > 0$$
$$-b_1 u_x(0, t) = c_1 \qquad t \geq 0$$
$$b_2 u_x(a, t) = c_2 \qquad t \geq 0$$
$$u(x, 0) = f(x) \qquad 0 \leq x \leq a.$$

Let $Q(x)$ and $K(x)$ be as given and define the constant

$$C = b_1 c_2 \kappa(a) + b_2 c_1 \kappa(0) + b_1 b_2 Q(a).$$

Show that this problem has a solution independent of time if and only if $C = 0$ and that this solution is of the form

$$U(x) = -\int_0^x \frac{Q}{\kappa} - \frac{c_1}{b_1}\kappa(0)K(x) + B,$$

where B is a constant.

---

[1]These exercises are based on the author's "The Existence of a Steady-State Solution for a Type of Parabolic Boundary Value Problem," *Int. J. Math. Educ. Sci. Technol.*, **19** (1988) 413–419.

8.7   Show that, if u is a solution of the boundary value problem of Exercise 8.6, then

$$\int_0^a cu = \int_0^a cf$$

if and only if $C = 0$.

8.8   Prove that the boundary value problem of Exercise 8.6 has a steady-state solution if and only if $C = 0$ and none of the following two conditions is true.

(i) $b_1 < 0, c_1 > 0$ and

$$\int_0^a cf < -\frac{c_1}{b_1}\kappa(0) \int_0^a cK - \int_0^a c(s)\left(\int_0^s \frac{Q}{\kappa}\right) ds.$$

(ii) $b_2 < 0, c_2 > 0$ and

$$\int_0^a cf < \frac{c_2}{b_2}\kappa(a) \int_0^a c[K - K(a)] + \int_0^a c(s)\left(Q(a)[K(s) - K(a)] - \int_0^s \frac{Q}{\kappa}\right) ds.$$

Then evaluate the constant B in Exercise 8.6.

8.9   Find the constant value of q and the smallest value of the constant M for which the boundary value problem

$$\begin{aligned}
u_t &= u_{xx} + q & 0 < x < 1, \quad t > 0 \\
u_x(0, t) &= 1 & t \geq 0 \\
-u_x(1, t) &= 3 & t \geq 0 \\
u(x, 0) &= x(1 - 2x) + M & 0 \leq x \leq 1
\end{aligned}$$

has a steady-state solution and then find this solution (**Hint:** do not use Exercise 8.8).

8.10   Consider the boundary value problem

$$\begin{aligned}
cu_t &= (\kappa u_x)_x + q & 0 < x < a, \quad t > 0 \\
a_1 u(0, t) - b_1 u_x(0, t) &= c_1 & t \geq 0 \\
a_2 u(a, t) + b_2 u_x(a, t) &= c_2 & t \geq 0 \\
u(x, 0) &= f(x) & 0 \leq x \leq a
\end{aligned}$$

and define the constant

$$C = b_1 c_2 \kappa(a) + b_2 c_1 \kappa(0) + b_1 b_2 Q(a).$$

Prove that, if $a_1 + a_2 > 0$, this problem has a steady-state solution, which is given by (23) and (24), except in the following cases.

(i) $b_1 < 0, c_1 > 0$ and

$$C > \min\left\{0, -a_2\kappa(a)\left(c_1\kappa(0)K(a) + b_1 \int_0^a \frac{Q}{\kappa}\right)\right\}.$$

(ii) $b_2 < 0, c_2 > 0$ and

$$C > \min\left\{0, -a_1\kappa(0)\left[c_2\kappa(a)K(a) + b_2\left(Q(a)K(a) - \int_0^a \frac{Q}{\kappa}\right)\right]\right\}.$$

8.11   Use the result of Exercise 8.10 to prove or disprove that the boundary value problem

$$
\begin{aligned}
u_t &= u_{xx} & 0 < x < \pi, \quad t > 0 \\
u_x(0, t) &= c_1 > 0 & t \geq 0 \\
u(\pi, t) &= 0 & t \geq 0 \\
u(x, 0) &= c_1 \sin x & 0 \leq x \leq \pi
\end{aligned}
$$

has a steady-state solution.

8.12   For what values of ϵ does the boundary value problem

$$
\begin{aligned}
u_t &= u_{xx} + \sin x & 0 < x < \pi, \quad t > 0 \\
u(0, t) &= 1 & t \geq 0 \\
-u_x(\pi, t) &= 1 + \epsilon & t \geq 0 \\
u(x, 0) &= f(x) & 0 \leq x \leq \pi
\end{aligned}
$$

have a steady-state solution? Find this solution (**Hint:** use the result of Exercise 8.10).

10.1   Solve the boundary value problem of Exercise 8.1 when $c = a = \kappa = C = m = 1$ and

$$
f(x) = \sin \pi x - \frac{e(e^x - 1) - xe^x(e - 1)}{ee^x}.
$$

10.2   Solve the boundary value problem

$$
\begin{aligned}
u_t &= u_{xx} + 1 & 0 < x < \pi, \quad t > 0 \\
u(0, t) &= 1 & t \geq 0 \\
u_x(\pi, t) &= \pi & t \geq 0 \\
u(x, 0) &= 1 + \frac{x^2}{2} & 0 \leq x \leq \pi.
\end{aligned}
$$

10.3   Solve the boundary value problem

$$
\begin{aligned}
cu_t &= (\kappa u_x)_x & 0 < x < 10, \quad t > 0 \\
u(0, t) &= 300 & t \geq 0 \\
u(10, t) &= 320 & t \geq 0 \\
u(x, 0) &= 300 + 2x & 0 \leq x \leq 10,
\end{aligned}
$$

where

$$
\kappa(x) = \begin{cases} k, & 0 \leq x \leq 5 \\ 3k, & 5 < x \leq 10 \end{cases}
$$

and k is a constant (**Hint:** find two solutions, one for $0 < x < 5$ and one for $5 < x < 10$, that match up at $x = 5$).

10.4   Solve the boundary value problem of Exercise 8.4 if $T_1 = T_2 = T_0$ and $f \equiv T_0$, where T_0 is a given constant.

10.5   Solve the boundary value problem

$$
\begin{aligned}
u_t &= u_{xx} + \sin x & 0 < x < \pi, \quad t > 0 \\
u_x(0, t) &= 1 & t \geq 0 \\
-u_x(\pi, t) &= 1 & t \geq 0 \\
u(x, 0) &= \sin x + \cos x + 1 & 0 \leq x \leq \pi
\end{aligned}
$$

(**Hint:** this problem is of the type described in Exercises 8.6 to 8.8).

10.6   Solve the boundary value problem

$$
\begin{aligned}
u_t &= u_{xx} + \sin x & 0 < x < \pi, \quad t > 0 \\
u(0, t) &= 1 & t \geq 0 \\
-4u_x(\pi, t) &= 5 & t \geq 0 \\
u(x, 0) &= 1 + \frac{5}{4}\sin x & 0 \leq x \leq \pi
\end{aligned}
$$

(**Hint:** this problem is of the type described in Exercise 8.12).

11.1   Verify Equation (36).

11.2   Show that for $t = 0$ the series in Example 3.9 adds up to $-x$ on $[0, 2\pi]$.

Solve the boundary value problems in Exercises 11.3 to 11.6 and, assuming that the series solutions obtained are differentiable term by term, verify these solutions.

11.3
$$
\begin{aligned}
u_t &= u_{xx} + e^{-t}(x - 1 + \sin \pi x) & 0 < x < 1, \quad t > 0 \\
u(0, t) &= e^{-t} & t \geq 0 \\
u(1, t) &= 3 & t \geq 0 \\
u(x, 0) &= x + 2 & 0 \leq x \leq 1.
\end{aligned}
$$

11.4
$$
\begin{aligned}
u_t &= u_{xx} - xe^{-t/4} & 0 < x < \pi, \quad t > 0 \\
u(0, t) &= 0 & t \geq 0 \\
u_x(\pi, t) &= 0 & t \geq 0 \\
u(x, 0) &= 0 & 0 \leq x \leq \pi.
\end{aligned}
$$

11.5
$$
\begin{aligned}
u_t &= u_{xx} + 3e^{-t}\cos t + 3 & 0 < x < \pi, \quad t > 0 \\
u_x(0, t) &= 0 & t \geq 0 \\
u_x(\pi, t) &= 0 & t \geq 0 \\
u(x, 0) &= \cos x & 0 \leq x \leq \pi.
\end{aligned}
$$

11.6
$$
\begin{aligned}
u_t &= u_{xx} + e^{-t}\sin x & 0 < x < \pi/2, \quad t > 0 \\
u(0, t) &= 0 & t \geq 0 \\
u_x(\pi/2, t) &= e^{-t} & t \geq 0 \\
u(x, 0) &= x & 0 \leq x \leq \pi/2.
\end{aligned}
$$

11.7   A bar of length $a > 0$, initially at a temperature $T_0 > 0$, is submerged in a medium whose temperature is $T < T_0$. While the ends of the bar are maintained at temperature T_0 for $t \geq 0$, there is heat loss by convection along the lateral surface, and the equation of heat propagation is

$$
u_t = ku_{xx} - h(u - T),
$$

where h is a positive constant. State and solve the corresponding boundary value problem and verify your solution (**Hint:** use the function $v = e^{ht}(u - T)$).

11.8   Repeat Exercise 11.7 if $a = 10$, $h = k = 1$, $T = 300$, and the endpoint and initial temperatures are modified as follows:

$$
\begin{aligned}
u(0, t) &= 310 & t \geq 0 \\
u(10, t) &= 320 & t \geq 0 \\
u(x, 0) &= 310 + x & 0 \leq x \leq 10.
\end{aligned}
$$

4

GENERALIZED FOURIER SERIES

§4.1 Sturm-Liouville Problems

In §3.6 we encountered for the first time a boundary value problem whose solution by separation of variables involves $\sin rx$ and $\cos rx$, where r is not an integer, and this precludes the use of Fourier series as developed in Chapter 2. It was Fourier who first came across this kind of problem when studying the cooling of a sphere.[1] But already in his early work—in articles 116 to 139 of his 1807 memoir, later incorporated into his book—he had to deal with an even more general type of situation.[2] Trigonometric functions turned out to be of no use in solving the problem of heat propagation in a solid cylinder, and Fourier found it necessary to use another kind of function, known today as the *Bessel function of the first kind of order zero*. We shall postpone the presentation of this type of problem and the development of the properties and applications of Bessel functions until Chapter 10.

One of Fourier's biggest detractors, Siméon Denis Poisson (1781–1840), was, ironically, one of the men most influenced by Fourier's work. Poisson, first a student and then a professor at the *École Polytechnique* and later also at the *Faculté des Sciences*, had been working on heat conduction problems since 1815, and in 1835 he published a book on the subject: *Théorie Mathématique de la Chaleur*. Many of Poisson's problems also require the use of functions other than sines and cosines in their series solutions, including Bessel functions and even newer kinds.

Jacques Charles François Sturm (1803–1855), originally from Geneva, was a member of a small group of young physicists and mathematicians—including Dirichlet and later Liouville—that gathered around Fourier in the 1820s in some sort of "scientific salon" (there was number of such groups in Paris at the time, and young scientists usually joined several). Sturm developed an interest in heat conduction problems since the beginning of his professional life, had read Fourier's original papers when still in

---

[1] *The Analytical Theory of Heat*, pages 268–278.
[2] *The Analytical Theory of Heat*, pages 291–310.

CHARLES FRANÇOIS STURM
Drawing by the author from a sketch by
Sturm's contemporary Daniel Colladon.

manuscript form, and was also well aware of Poisson's work. He made the development
of a general theory that would cover all the cases studied by Fourier and Poisson one of
his mathematical goals. In particular, Sturm devoted himself to the study of heat propagation in a bar of variable density, in which case the heat capacity c and the diffusivity κ
depend on x. Using separation of variables led him to consider a general type of second
order differential equation, depending on a parameter λ and subject to some endpoint
conditions on an interval. By 1836 he had proved the existence of a sequence of real
values of the parameter $\lambda_0 < \lambda_1 < \cdots < \lambda_n < \cdots$, where $\lambda_n \to \infty$, for which the
ordinary differential equation has solutions that satisfy the endpoint conditions. In some
simple cases these solutions are sines and cosines, but in general they need not be.

Even more generally, we shall consider the partial differential equation

$$u_t = P u_{xx} + R u_x + Q u,$$

where P, R, and Q are continuous functions and P does not vanish. This includes all the
equations of Chapter 3. Assuming there is a solution of the form $u(x, t) = X(x)T(t)$
we obtain

$$XT' = P X''T + R X'T + Q XT,$$

that is,

$$\frac{T'}{T} = \frac{PX'' + RX'}{X} + Q = -\lambda,$$

where λ is a constant. It should be said, for the benefit of those readers getting weary of heat propagation problems, that the second equality above is obtained regardless of whether the left-hand side of the given partial differential equation is u_t or, for example, u_{tt}. The latter equation arises in many problems involving vibrations, whether acoustical, mechanical, or electromagnetic. Thus, the theory about to be developed is more general than it might seem in the present context. We then concentrate on the equation

$$PX'' + RX' + (\lambda + Q)X = 0$$

and assume, without loss of generality, that P has only positive values—or else multiply the equation by -1 and replace λ by $-\lambda$.

It will be easier to study this equation if we multiply it by a new function p defined by

$$p = e^{\int R/P}$$

and divide it by P. Then, if we notice that $pR = p'P$ and if we define

$$q = -\frac{pQ}{P} \qquad \text{and} \qquad r = \frac{p}{P},$$

our equation becomes

(1) $$(pX')' + (\lambda r - q)X = 0,$$

where $p > 0$ and $r > 0$.

This is the equation studied by Sturm, except that he used a function G in place of the factor $\lambda r - q$. Joseph Liouville (1809–1882), an alumnus of the *École Polytechnique* and now a professor at the *École Centrale des Arts et Manufactures*, became acquainted with this type of problem through Sturm and considered it in three papers of 1836 and 1837, already writing $\lambda r - q$ explicitly and using endpoint conditions.[1] We shall refer to the following as a *Sturm-Liouville problem*:

---

[1] Sturm's 1836 papers and Liouville's three-part "Mémoire sur le développement des fonctions ou parties de fonctions en séries, dont les divers termes sont assujettis à satisfaire a une même équation différentielle du second ordre, contenant un paramètre variable," as well as the extract of a joint paper with almost the same title—their only joint paper even though they collaborated closely on the elaboration of what is now called the *Sturm-Liouville theory*—appeared in the first two volumes of the *Journal de Mathématiques Pures et Appliquées*, just founded by Liouville and popularly known as Liouville's *Journal*.

JOSEPH LIOUVILLE
Courtesy of Springer-Verlag Archives.

Let p and r be continuous on an interval $[a, b]$, let q be continuous and p continuously differentiable on (a, b), and let $p > 0$ and $r > 0$ on (a, b). If λ is a parameter, find solutions $X : [a, b] \to \mathbb{R}$ of (1) subject to the following conditions:

    1. if $p(a) > 0$ then there are constants a_1 and b_1, not both zero, such that

(2) $$a_1 X(a) - b_1 X'(a) = 0$$

and

    2. if $p(b) > 0$ then there are constants a_2 and b_2, not both zero, such that

(3) $$a_2 X(b) + b_2 X'(b) = 0.$$

The implication is that the endpoint condition (2) is not present when $p(a) = 0$ and that the endpoint condition (3) is not present when $p(b) = 0$. The reason for this, which can be understood only by an examination of the proofs of Theorems 4.1 and 4.3 below, is to make these results as general as possible. For the same reason the hypotheses on the constants are as weak as possible. Additional conditions will be imposed in Theorem 4.3 below.

Definition 4.1 *Each real value of λ for which a Sturm-Liouville problem has a solution $X \not\equiv 0$ defined on $[a, b]$ is called an* **eigenvalue** *and X an* **eigenfunction**.

Example 4.1 Consider the differential equation

$$x^2 X'' + x X' + \lambda X = 0$$

subject to the endpoint conditions $X'(1) = 0$ and $X(b) = 0$, where $b > 1$. We have $P(x) = x^2$, $R(x) = x$, and $Q \equiv 0$. The given equation can be put in the form (1) with

$$p(x) = e^{\int \frac{1}{x} \, dx} = e^{\log x} = x$$

$$r(x) = \frac{1}{x}$$

and $q \equiv 0$. This yields the Sturm-Liouville problem

$$(x X')' + \frac{\lambda}{x} X = 0$$
$$X'(1) = 0$$
$$X(b) = 0.$$

In order to solve the differential equation for X we transform it by using the change of variable $x = e^s$ (see Exercise 1.5). If we define $Y(s) = X(e^s)$, we obtain

$$Y'(s) = X'(e^s) e^s$$

and, using the given differential equation,

$$Y''(s) = X''(e^s)(e^s)^2 + X'(e^s) e^s = -\lambda X(e^s) = -\lambda Y(s).$$

Then the original Sturm-Liouville problem is transformed into

$$Y'' + \lambda Y = 0$$
$$Y'(0) = 0$$
$$Y(\log b) = 0.$$

For $\lambda < 0$ the general solution for Y is

$$Y(s) = A e^{\sqrt{-\lambda} s} + B e^{-\sqrt{-\lambda} s},$$

where A and B are arbitrary constants, and the endpoint conditions imply that

$$A - B = 0$$
$$A e^{\sqrt{-\lambda} \log b} + B e^{-\sqrt{-\lambda} \log b} = 0,$$

a system of equations that has no real solutions for A and B. For $\lambda = 0$ we obtain $Y \equiv 0$, which is not an eigenfunction.

Finally, for $\lambda > 0$ the general solution for Y is

$$Y(s) = A \sin \sqrt{\lambda} s + B \cos \sqrt{\lambda} s,$$

where A and B are arbitrary constants, and the endpoint conditions imply that $A = 0$ and then that $\cos(\sqrt{\lambda} \log b) = 0$. This leads to the solutions

$$Y_n(s) = \cos \sqrt{\lambda_n}\, s,$$

where

$$\lambda_n = \frac{(2n-1)^2 \pi^2}{4 \log^2 b}$$

and n is a positive integer. These are, then, the eigenvalues of the original problem for X, and the corresponding eigenfunctions are

$$X_n(x) = Y_n(\log x) = \cos\left(\frac{2n-1}{2\log b}\, \pi \log x\right).$$

Example 4.2 If $a > 0$, consider the Sturm-Liouville problem on $(0, a)$

$$X'' + \lambda X = 0$$
$$hX(0) - X'(0) = 0$$
$$X'(a) = 0,$$

where h is a positive constant.

As in the previous example, only the trivial solution is obtained for $\lambda \le 0$. For $\lambda > 0$ the general solution for X is

$$X(x) = A \sin \sqrt{\lambda}\, x + B \cos \sqrt{\lambda}\, x,$$

where A and B are arbitrary constants, and the endpoint conditions imply that

$$hB - A\sqrt{\lambda} = 0$$
$$A \cos \sqrt{\lambda}\, a - B \sin \sqrt{\lambda}\, a = 0.$$

The first equation shows that neither A nor B can be zero, and then it follows from the second that $\cos \sqrt{\lambda}\, a \ne 0$; that is, λ is an eigenvalue if and only if it is a solution of the transcendental equation

$$\tan \sqrt{\lambda}\, a = \frac{h}{\sqrt{\lambda}}.$$

An eigenfunction corresponding to an eigenvalue λ_n is

$$X_n(x) = \sin \sqrt{\lambda_n}\, x + \frac{\sqrt{\lambda_n}}{h} \cos \sqrt{\lambda_n}\, x = \frac{\cos \sqrt{\lambda_n}\,(a-x)}{\sin a\sqrt{\lambda_n}}.$$

§4.2 The Eigenvalues and Eigenfunctions

The work we do in this section shall, in particular, justify some of the steps in the method of solution proposed in §3.9. We start with a small technical result.

Lemma 4.1 *Let X_1 and X_2 be eigenfunctions of a Sturm-Liouville problem. Then,*
(i) *if $p(a) > 0$, there is a constant C such that*

$$(4) \qquad X_2(a) = C X_1(a) \qquad and \qquad X'_2(a) = C X'_1(a),$$

and,
(ii) *if $p(b) > 0$, there is a constant C such that*

$$(5) \qquad X_2(b) = C X_1(b) \qquad and \qquad X'_2(b) = C X'_1(b).$$

Proof. If $p(a) > 0$, then (2) implies that

$$a_1 X_1(a) - b_1 X'_1(a) = 0$$
$$a_1 X_2(a) - b_1 X'_2(a) = 0,$$

and, since a_1 and b_1 are not both zero, this system of equations has a solution for a_1 and b_1 only if the determinant of the system is zero; that is, if its second row is a constant multiple of the first. This proves (4), and (5) is obtained analogously, Q.E.D.

The first theorem was discovered in 1823 by Poisson, who used it to prove that the eigenvalues are necessarily real valued (see Exercise 2.8). It is the key to computing the coefficients of the series in Equation (31) of Chapter 3.

Theorem 4.1 *If X_1 and X_2 are eigenfunctions of a Sturm-Liouville problem corresponding to distinct eigenvalues λ_1 and λ_2, then*

$$\int_a^b r X_1 X_2 = 0.$$

Proof. Multiplying the equations

$$(p X'_1)' + (\lambda_1 r - q) X_1 = 0$$
$$(p X'_2)' + (\lambda_2 r - q) X_2 = 0$$

by X_2 and X_1, respectively, and subtracting we obtain

$$(\lambda_1 - \lambda_2) r X_1 X_2 = X_1 (p X'_2)' - X_2 (p X'_1)' = [p(X_1 X'_2 - X_2 X'_1)]'.$$

Since all functions are integrable, integration from a to b gives

$$(\lambda_1 - \lambda_2) \int_a^b r X_1 X_2 = p(b)[X_1(b) X'_2(b) - X_2(b) X'_1(b)]$$
$$- p(a)[X_1(a) X'_2(a) - X_2(a) X'_1(a)].$$

It follows from Lemma 4.1 that the right-hand side is zero, and the result follows because $\lambda_1 \neq \lambda_2$, Q.E.D.

Definition 4.2 *Any two integrable functions* X_1 *and* X_2 *for which*

$$\int_a^b r X_1 X_2 = 0$$

are said to be **orthogonal** *with* **weight function** r *on the interval* (a, b).

Therefore, Theorem 4.1 simply says that eigenfunctions corresponding to distinct eigenvalues are orthogonal with weight function r on (a, b).

Next we show that if at least one of the endpoint conditions is present then each eigenvalue has a unique eigenfunction up to a multiplicative constant.

Theorem 4.2 *If* λ *is an eigenvalue of a Sturm-Liouville problem, if* X_1 *and* X_2 *are eigenfunctions corresponding to* λ, *and if* $p(a) + p(b) > 0$, *then there is a constant* C *such that* $X_2 = C X_1$.

Proof. If $p(a) > 0$ and if C is the constant in Lemma 4.1(i), then the function $\varphi = X_2 - C X_1$ is a solution of (1) because it is a linear combination of solutions, and (4) implies that $\varphi(a) = \varphi'(a) = 0$. But it is known from the theory of ordinary differential equations that the only solution of (1) that, together with its derivative, vanishes at $x = a$ is the trivial solution $\varphi \equiv 0$. Similarly if $p(b) > 0$, Q.E.D.

The conclusion of this theorem need not hold if $p(a) = p(b) = 0$, that is, if both endpoint conditions are missing (see Exercise 2.2). By imposing some additional conditions on q and the coefficients it is possible to show that the eigenvalues are nonnegative or even positive.

Theorem 4.3 *Let* λ *be an eigenvalue of a Sturm-Liouville problem and assume that* q *is continuous and nonnegative on* $[a, b]$ *and that,*
  (i) *if* $p(a) > 0$, *then* a_1 *and* b_1 *are nonnegative and,*
  (ii) *if* $p(b) > 0$, *then* a_2 *and* b_2 *are nonnegative.*
Then $\lambda \geq 0$. *Moreover,* $\lambda = 0$ *is an eigenvalue if and only if* $q \equiv 0$ *and* $p(a)a_1 = p(b)a_2 = 0$, *and then the corresponding eigenfunction is a constant.*

Proof. Let X be an eigenfunction corresponding to an eigenvalue λ. If we multiply (1) by X and rearrange terms, we obtain

$$\lambda r X^2 = q X^2 - (p X')' X,$$

and then integration from a to b gives

(6) $\quad \lambda \int_a^b r X^2 = \int_a^b q X^2 - \int_a^b (p X')' X = \int_a^b q X^2 - p X' X \Big|_a^b + \int_a^b p (X')^2.$

Now,

$$-p X' X \Big|_a^b = p(a) X'(a) X(a) - p(b) X'(b) X(b),$$

and this is always nonnegative. In fact, either $p(a) = 0$ or the endpoint condition (2) implies that $X(a) = 0$ when $b_1 = 0$ or that

$$X'(a) = \frac{a_1}{b_1} X(a)$$

when $b_1 > 0$. In either case $p(a)X'(a)X(a) \geq 0$. Similarly, $p(b)X'(b)X(b) \leq 0$. Since $p > 0, r > 0$ and $q \geq 0$ on (a, b), (6) shows that $\lambda \geq 0$.

If $\lambda = 0$ is an eigenvalue then (6) implies that $q \equiv X' \equiv 0$, and then X is constant. Either $p(a) = 0$ or, since X cannot be identically zero, condition (2) shows that $a_1 = 0$. Similarly, $p(b)a_2 = 0$. Conversely, if $q \equiv 0$ and $p(a)a_1 = p(b)a_2 = 0$ then the Sturm-Liouville problem (1)–(3) consists of the equation $(pX')' + \lambda rX = 0$ and, possibly, the endpoint conditions $X'(a) = 0$ and $X'(b) = 0$, and it has a constant nonzero solution for $\lambda = 0$, Q.E.D.

A Sturm-Liouville problem may have negative eigenvalues if $a_1 b_1 \leq 0$ (see Exercise 2.6) or if q has negative values.

Example 4.3 The Sturm-Liouville problem

$$X'' + (\lambda + 1)X = 0$$
$$X'(a) = 0$$
$$X'(b) = 0$$

has a constant solution corresponding to the eigenvalue $\lambda = -1$.

Since the equation $\kappa X'' + \kappa' X + \lambda cX = (\kappa X')' + \lambda cX = 0$ of §3.9 is of the form (1), the preceding work shows, in particular, that it may be reasonable to seek a solution to the problem (25)–(28) of Chapter 3 in the form of a series with terms $c_n e^{-\lambda_n t} X_n(x)$. Reasonable to the extent that, according to the theorems in this section, there is essentially one X_n for each λ_n, that the exponentials are bounded—and all but at most one are decreasing as t increases—and that the coefficients c_n may be obtained from the orthogonality of the eigenfunctions if term by term integration is possible.

Specifically, our results give affirmative answers to (iv), (v), and the first inequality in (vii), posed in §3.9.

§4.3 The Existence of the Eigenvalues

It is customary to classify Sturm-Liouville problems into two broad categories. Our main reason for this is that we want to single out in one the type of problem for which the method of solution proposed in §3.9 can be fully justified. Guided by the hypotheses in Theorems 4.2 and 4.3, we shall focus our attention on the following type.

Definition 4.3 *A Sturm-Liouville problem is called* **regular** *if and only if $p(a) > 0$, $p(b) > 0$, q is continuous on $[a, b]$, and the constants $a_1, b_1, a_2,$ and b_2 are nonnegative. A Sturm-Liouville problem that is not regular is called* **singular**.

Therefore, the endpoint conditions (2) and (3) are always present in the case of a regular Sturm-Liouville problem.

Example 4.4 The Sturm-Liouville problem

$$[(1 - x^2)X']' + \lambda X = 0$$

on $[-1, 1]$, with no endpoint conditions, is singular because $p(x) = 1 - x^2$ vanishes at $x = \pm 1$.

Example 4.5 The Sturm-Liouville problem

$$(xX')' + \left(\lambda x - \frac{n^2}{x}\right)X = 0$$
$$X(a) = 0$$
$$X(b) = 0$$

is regular on an interval (a, b) if $a > 0$ but is singular if $a = 0$ and if the endpoint condition at a is removed, because $p(x) = x$ vanishes at $x = 0$ and $q(x) = n^2/x$ is not continuous on $[0, b]$.

All the problems in Chapter 3 are regular except for the one in Exercise 4.9. In fact, this is not a Sturm-Liouville problem at all, but since this type of situation appears in some applications we shall introduce the following terminology.

Definition 4.4 *Let p, q, r, and λ be as in the definition of a Sturm-Liouville problem, and assume, in addition, that $p(a) = p(b)$. The problem of finding solutions $X: [a, b] \to \mathbb{R}$ of (1) subject to the endpoint conditions*

$$X(a) = X(b) \qquad \text{and} \qquad X'(a) = X'(b)$$

is called a **periodic Sturm-Liouville problem**.

Example 4.6 The *Hill differential equation*

$$X'' + (\lambda - q)X = 0$$

subject to the endpoint conditions

$$X(0) = X(\pi) \qquad \text{and} \qquad X'(0) = X'(\pi)$$

gives a periodic Sturm-Liouville problem on $(0, \pi)$. This equation was introduced by George William Hill (1838–1914), of the Nautical Almanac Office in Washington, in his 1877 study on the motion of the Moon: *Researches on Lunar Theory*.

It is simpler to study regular Sturm-Liouville problems, which represent the most frequent type in applications, and we shall do so in the rest of this section. We shall prove, in particular, the existence of eigenvalues, that these form an infinite, increasing sequence $\{\lambda_n\}$, and that $\lambda_n/n \to \infty$ as $n \to \infty$. Together with the work done in §4.2, this justifies the method proposed in §3.9, except for the convergence of the series in (30) and (31). This will be done in §4.4. Some singular problems will be seen in later chapters and will be treated as individual cases. There is, however, a general theory that was developed by Claude Hugo Hermann Weyl (1885–1955) in his doctoral dissertation of 1909 at Göttingen.[1]

The study of a second-order ordinary differential equation is often easier after transforming it into a system of two first-order equations. In the case of (1), define a new function $Y = pX'$. This definition and (1) give the first order system

(7)
$$X' = \frac{Y}{p}$$

(8)
$$Y' = (q - \lambda r)X.$$

We shall start our study with the following question: does an eigenfunction X have zeros? This is a natural question since these zeros were needed in order to satisfy the endpoint conditions in some of the problems of Chapter 3, and this is also the case in many other boundary value problems. Now, the answer to this question is yes for $X'' + \lambda X = 0$, as we have seen repeatedly, and the same answer may be expected for large λ in the general case because p and r are approximately constant in a very small neighborhood of any given point of (a, b) and then, if $\lambda r \gg q$, (1) can be approximated by $pX'' + \lambda rX = 0$ in that neighborhood. This equation has sinusoidal solutions with zeros in this neighborhood for λ sufficiently large. Figure 4.1 represents such an

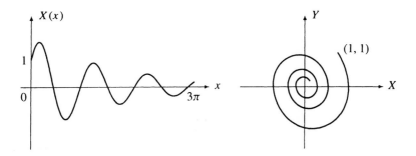

Figure 4.1

---

[1] For a general treatment of singular and periodic Sturm-Liouville problems see, for instance, E. A. Coddington and N. Levinson, *Theory of Ordinary Differential Equations*, McGraw-Hill, New York, 1955, Chapter 9.

eigenfunction X with zeros and the corresponding parametric representation in the XY-plane. The right-hand side of this figure suggests that polar coordinates might be best to continue our investigation. In 1926 Heinz Prüfer (1896–1934) chose coordinates R and θ such that $X = R\sin\theta$ and $Y = R\cos\theta$, a choice slightly different from the usual. Then the pair of equations

$$X = R\sin\theta \quad \text{and} \quad pX' = R\cos\theta$$

is known as *Prüfer's substitution*. Taking these expressions for X and Y to (7) and (8) yields

(9) $$R'\sin\theta + R\theta'\cos\theta = \frac{R}{p}\cos\theta$$

(10) $$R'\cos\theta - R\theta'\sin\theta = (q - \lambda r)R\sin\theta$$

If we multiply (9) by $\sin\theta$, multiply (10) by $\cos\theta$, and add the results we obtain

(11) $$R' = \frac{R}{2}\left(\frac{1}{p} + q - \lambda r\right)\sin 2\theta.$$

Similarly, multiplying (9) by $\cos\theta$, multiplying (10) by $\sin\theta$, and subtracting yield

(12) $$\theta' = \frac{1}{p}\cos^2\theta + (\lambda r - q)\sin^2\theta.$$

Since (12) does not contain R then, for any α in $\mathbb{R}$ and each λ, it has a unique solution θ_λ such that $\theta_\lambda(a) = \alpha$. Taking this solution to (11), denoting the factor $\frac{1}{2}(1/p + q - \lambda r)$ by F_λ, and denoting the solution corresponding to a fixed value of λ by R_λ, one obtains

$$R_\lambda(x) = R_\lambda(a)e^{\int_a^x F_\lambda(s)\sin 2\theta_\lambda(s)\,ds}.$$

Note that $R_\lambda(a) \neq 0$ if X_λ is an eigenfunction corresponding to λ, or else

$$0 = R_\lambda^2(a) = X_\lambda^2(a) + Y_\lambda^2(a) = X_\lambda^2(a) + p^2(a)[X_\lambda'(a)]^2$$

implies that $X_\lambda(a) = X_\lambda'(a) = 0$ and then that $X_\lambda \equiv 0$, which is a contradiction. Therefore, $R_\lambda > 0$ for all x, and we conclude that $X_\lambda = R_\lambda\sin\theta_\lambda$ has zeros if and only if $\theta_\lambda = k\pi$ for some integer k.

Before we study these zeros we must prove the existence of eigenvalues and eigenfunctions. This will involve comparing the solutions of (12) for different values of λ, and, in this respect, the following version of one of Sturm's comparison theorems is useful.

Lemma 4.2 *Let p, q, and r be continuous on $[a, b]$ with $p > 0$ and continuously differentiable, and let θ_{λ_1} and θ_{λ_2} be solutions of (12) corresponding to $\lambda = \lambda_1$ and $\lambda = \lambda_2$, respectively. If $\lambda_1 < \lambda_2$ and if $\theta_{\lambda_1}(a) \leq \theta_{\lambda_2}(a)$, then $\theta_{\lambda_1} < \theta_{\lambda_2}$ on $(a, b]$.*

Proof. For the sake of brevity we shall use the notation

$$\omega = \theta_{\lambda_2} - \theta_{\lambda_1},$$

$$f = \left(\lambda_1 r - q - \frac{1}{p}\right)\frac{\sin^2 \theta_{\lambda_2} - \sin^2 \theta_{\lambda_1}}{\theta_{\lambda_2} - \theta_{\lambda_1}},$$

$$g = (\lambda_2 - \lambda_1) r \sin^2 \theta_{\lambda_2},$$

and then using (12) we obtain

$$\omega' = \frac{1}{p}(\cos^2 \theta_{\lambda_2} - \cos^2 \theta_{\lambda_2}) + (\lambda_2 r - q)\sin^2 \theta_{\lambda_2} - (\lambda_1 r - q)\sin^2 \theta_{\lambda_1}$$

$$= \frac{1}{p}(-\sin^2 \theta_{\lambda_2} + \sin^2 \theta_{\lambda_1}) + (\lambda_2 r - q)\sin^2 \theta_{\lambda_2} - (\lambda_1 r - q)\sin^2 \theta_{\lambda_1}$$

$$= \left(\lambda_1 r - q - \frac{1}{p}\right)(\sin^2 \theta_{\lambda_2} - \sin^2 \theta_{\lambda_1}) + (\lambda_2 - \lambda_1) r \sin^2 \theta_{\lambda_2}$$

$$= f\omega + g.$$

The mean value theorem shows that

$$\sin^2 \theta_{\lambda_2} - \sin^2 \theta_{\lambda_1} = (\sin 2\theta)(\theta_{\lambda_2} - \theta_{\lambda_1})$$

for some θ between θ_{λ_1} and θ_{λ_2}, and this shows that, if $\theta_{\lambda_2} - \theta_{\lambda_1}$ approaches zero at some point x in $[a, b]$, then f has a limit at x. Hence, f is or can be redefined as a continuous function on $[a, b]$. Integration of $\omega' = f\omega + g$ now gives

$$\omega(x) = e^{\int_a^x f}\left[\int_a^x g(s)e^{-\int_a^s f} ds + \omega(a)\right],$$

as shown for any $x \geq a$. Now, $g \not\equiv 0$ on $[a, x]$ or else $\theta_{\lambda_2} \equiv k\pi$ for some integer k, contradicting (12). Since g is continuous and nonnegative it follows that $g > 0$ on some subinterval of $[a, x]$. This and $\omega(a) \geq 0$ imply that $\omega(x) > 0$ and then, since x is arbitrary, that $\omega = \theta_{\lambda_2} - \theta_{\lambda_1} > 0$ on $[a, b]$, Q.E.D.

Next we prove the existence of the eigenvalues and eigenfunctions using Prüfer's method rather than Sturm's.

Theorem 4.4 *A regular Sturm-Liouville problem,* (1)–(3), *has an infinite collection of eigenvalues such that*
   (i) *they form an increasing sequence,* $\lambda_1 < \lambda_2 < \cdots < \lambda_n < \cdots$, *and*
   (ii) *an eigenfunction* X_n *corresponding to an eigenvalue* λ_n *has exactly* $n - 1$ *zeros on* (a, b), *and each of them is simple.*

Proof. If θ_λ is a solution of (12) corresponding to a given value of the parameter λ, then λ would be an eigenvalue and $X_\lambda = R_\lambda \sin \theta_\lambda$ would be an eigenfunction of the Sturm-Liouville problem (1)–(3) if X_λ satisfies (2) and (3), that is, if

$$0 = a_1 X_\lambda(a) - b_1 X_\lambda'(a) = R_\lambda(a)\left[a_1 \sin \theta_\lambda(a) - \frac{b_1}{p(a)} \cos \theta_\lambda(a)\right]$$

and

$$0 = a_2 X_\lambda(b) + b_2 X_\lambda'(b) = R_\lambda(b)\left[a_2 \sin \theta_\lambda(b) + \frac{b_2}{p(b)} \cos \theta_\lambda(b)\right].$$

To satisfy the condition at $x = a$ we choose, for each λ, the solution θ_λ of (12) such that $\theta_\lambda(a) = \alpha$, where $\alpha = \pi/2$ if $a_1 = 0$ and

$$0 \le \alpha < \frac{\pi}{2} \qquad \text{with} \qquad \tan \alpha = \frac{b_1}{a_1 p(a)} \ge 0$$

if $a_1 \ne 0$. Then, to satisfy the condition at $x = b$, $\theta_\lambda(b)$ must be of the form $\theta_\lambda(b) = \beta + (n-1)\pi$ for some integer n, where $\beta = \pi/2$ if $a_2 = 0$ and

$$\frac{\pi}{2} < \beta \le \pi \qquad \text{with} \qquad \tan \beta = -\frac{b_2}{a_2 p(b)} \le 0$$

if $a_2 \ne 0$. In fact, $n > 0$ or else $\theta_\lambda(b) \le 0$, which together with $\theta_\lambda(a) \ge 0$ implies that $\theta_\lambda(x) = 0$ and $\theta_\lambda'(x) \le 0$ for some $x \le b$, and this contradicts (12).

To show that $\theta_\lambda(b)$ actually attains each of the desired values for some specific values of λ, recall that $\theta_\lambda(b)$ is a continuous function of λ, as is shown in the theory of ordinary differential equations,[1] and Lemma 4.2 shows that this function is strictly increasing. Thus, if we show that $\theta_\lambda(b) < \pi/2$ for some λ and that $\theta_\lambda(b) \to \infty$ as $\lambda \to \infty$—this will be done below—it follows that for each positive integer n there is a unique λ_n such that

$$\theta_{\lambda_n}(b) = \beta + (n-1)\pi$$

and that $\lambda_n < \lambda_{n+1}$. This proves (i).

An eigenfunction $X_n = R_{\lambda_n} \sin \theta_{\lambda_n}$ has a zero at a point x in (a, b) if and only if $\theta_{\lambda_n}(x) = k\pi$ for some integer k, and then $\theta_{\lambda_n}'(x) > 0$ according to (12), so that θ_{λ_n} assumes the value $k\pi$ at most once in (a, b). Now, $0 < \beta \le \pi$ implies that

$$(n-1)\pi < \theta_{\lambda_n}(b) = \beta + (n-1)\pi \le n\pi.$$

Since $0 \le \theta_{\lambda_n}(a) \le \pi/2$, it follows that θ_{λ_n} assumes each value $k\pi$ for $k = 1, \ldots, n-1$ and each value exactly once in (a, b). Therefore, X_n has exactly $n-1$ zeros in (a, b). Furthermore, each such zero is simple because, if $\theta_{\lambda_n}(x) = k\pi$, then

$$p(x)X_n'(x) = Y_n(x) = R_{\lambda_n}(x) \cos \theta_{\lambda_n}(x) \ne 0.$$

This proves (ii).

---

[1] See, for instance, E. A. Coddington and N. Levinson, *op. cit.*, page 29.

We now turn to the remaining technical details of the proof of (i). To see that $\theta_\lambda(b) < \pi/2$ for some λ, choose a positive number $\epsilon < \pi/2$ and let m be the slope of the line segment L from $(a, \alpha + \epsilon)$ to $(b, \pi/2 - \epsilon)$. If, contrary to what we wish to prove, $\theta_\lambda(b) \geq \pi/2$ for all λ, then, for each λ, the graph of θ_λ intersects L; that is, for each λ there is a point x in (a, b) such that $\theta_\lambda(x)$ is in L and $\theta_\lambda'(x) \geq m$. But this is impossible because $\sin^2 \theta$ has a positive minimum for $\alpha + \epsilon \leq \theta \leq \pi/2 - \epsilon$, and then we can choose a $\lambda < 0$ such that the right-hand side of (12) is less than m for $\alpha + \epsilon \leq \theta \leq \pi/2 - \epsilon$.

It remains to be shown that $\theta_\lambda(b) \to \infty$ as $\lambda \to \infty$. First note that, if λ is so large that $\lambda r - q > 0$ on $[a, b]$, (12) implies that $\theta_\lambda' > 0$ and then θ_λ is strictly increasing on $[a, b]$. Assume now that for any such λ, and contrary to what we wish to prove, $\theta_\lambda(b) < K\pi$ for some positive integer K. Then for each such λ there are at most K points x in $[a, b]$ such that $\theta_\lambda(x)$ is an integral multiple of π, and then $\sin \theta_\lambda(x) = 0$. Corresponding to each such x, and if $0 < \epsilon < 1$, there is a largest interval I_x in $[a, b]$ containing x, where $-\epsilon \leq \sin \theta_\lambda \leq \epsilon$. If we denote the length of I_x by l_x the mean value theorem implies the existence of a point c in I_x such that

$$2\epsilon \geq \theta_\lambda'(c) \cos \theta_\lambda(c) \, l_x.$$

Then, if p_M is the maximum value of p, if we use (12) and if we note that $\cos^2 \theta_\lambda(c) \geq 1 - \epsilon^2$, we have

$$2\epsilon \geq \frac{1}{p(c)} \cos^3 \theta_\lambda(c) \, l_x \geq \frac{1}{p_M} (1 - \epsilon^2)^{3/2} l_x > 0,$$

so that $l_x \to 0$ as $\epsilon \to 0$ for each x. Thus, ϵ can be chosen so small that the sum of the lengths of all the I_x is less than $(b - a)/2$. Since

$$\theta_\lambda' \geq (\lambda r - q) \sin^2 \theta_\lambda > (\lambda r - q) \epsilon^2$$

on the remaining portion of (a, b), λ can be chosen so large that $\theta_\lambda' > 2M/(b - a)$ on this portion, and then

$$\theta_\lambda(b) - \theta_\lambda(a) = \int_a^b \theta_\lambda' > M,$$

contradicting the original assumption. Therefore, $\theta_\lambda(b) \to \infty$ as $\lambda \to \infty$, Q.E.D.

It is, of course, possible for an eigenfunction to have additional zeros at $x = a$ and $x = b$.

Example 4.7 If $a > 0$, then the Sturm-Liouville problem $X'' + \lambda X = 0$, $X(0) = X(a) = 0$ has eigenvalues $\lambda_n = n^2 \pi^2 / a^2$ and eigenfunctions

$$X_n(x) = \sin \frac{n\pi}{a} x,$$

as seen in §1.4. Then X_n has zeros at $x = 0$, $x = a$ and, if $n \geq 2$, also at $x_k = ka/n$, $k = 1, \ldots, n - 1$, in $(0, a)$.

Example 4.8 If $a > 0$, then the Sturm-Liouville problem $X'' + \lambda X = 0$, $X'(0) = X'(a) = 0$ has eigenvalues

$$\lambda_1 = 0 \quad \text{and} \quad \lambda_n = \frac{(n-1)^2 \pi^2}{a^2}, \quad n > 1$$

with corresponding eigenfunctions

$$X_1 \equiv 1 \quad \text{and} \quad X_n(x) = \cos \frac{(n-1)\pi}{a} x, \quad n > 1,$$

as shown in §3.4. X_1 has no zeros. If $n > 1$, X_n has zeros at

$$x_k = \frac{(2k-1)a}{2(n-1)}, \quad k = 1, \dots, n-1$$

in $(0, a)$, but no extra zeros at the endpoints.

Example 4.9 If a and h are positive constants, consider the Sturm-Liouville problem on $[0, a]$

$$X'' + \lambda X = 0$$
$$hX(0) - X'(0) = 0$$
$$X'(a) = 0.$$

As shown in Example 4.2, its eigenvalues λ_n are the solutions of

$$\tan \sqrt{\lambda}\, a = \frac{h}{\sqrt{\lambda}}$$

and corresponding eigenfunctions are

$$X_n(x) = \frac{\cos \sqrt{\lambda_n}\, (a - x)}{\sin a \sqrt{\lambda_n}}.$$

It is shown graphically in Figure 4.2 that there is an increasing sequence of positive eigenvalues. The nth eigenvalue λ_n satisfies the inequalities

$$\frac{(n-1)\pi}{a} < \sqrt{\lambda_n} < \frac{(2n-1)\pi}{2a}.$$

Then X_1 has no zeros in $(0, a)$ because the inequalities $0 < x < a$ and $0 < \sqrt{\lambda_1} < \pi/2a$ imply that

$$0 < \sqrt{\lambda_1}\, (a - x) < \frac{\pi}{2}.$$

For $n > 1$, however, X_n has a zero at each point x in $(0, a)$ such that $\sqrt{\lambda_n}\, (a - x)$ is an odd integral multiple of $\pi/2$, that is, at each x_k such that

$$x_k = a - \frac{(2k-1)\pi}{2\sqrt{\lambda_n}},$$

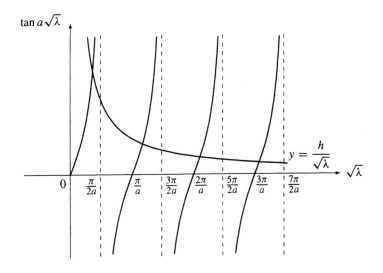

Figure 4.2

$k = 1, \ldots, n - 1$. Indeed, x_k is a decreasing function of k that has value a for $k = \frac{1}{2}$ and value zero for

$$k = \frac{2a\sqrt{\lambda_n} + \pi}{2\pi},$$

which, in view of the inequalities for $\sqrt{\lambda_n}$, is between $n - \frac{1}{2}$ and n. Thus, the only integral values of k for which x_k is in $(0, a)$ are $k = 1, \ldots, n - 1$, and X_n has exactly $n - 1$ zeros in $(0, a)$.

The next theorem, due to Liuoville, provides some needed information about the distribution of large eigenvalues.

Theorem 4.5 *If* $\{\lambda_n\}$ *is the sequence of eigenvalues of a regular Sturm-Liouville problem,* (1)–(3), *then*

$$\frac{\lambda_n}{n} \to \infty$$

as $n \to \infty$.

Proof. It was shown in the proof of Theorem 4.4 that, if $n > 1$ is large enough, then θ_{λ_n} is strictly increasing and that there are exactly $n - 1$ points $x_1 < x_2 < \cdots < x_{n-1}$ in (a, b) such that $\theta_{\lambda_n}(x_k) = k\pi$.

If h is the inverse function of θ_{λ_n}, differentiating the identity $\theta = \theta_{\lambda_n}(h(\theta))$, valid for $\pi \leq \theta \leq (n-1)\pi$, yields $1 = \theta'_{\lambda_n}(h(\theta))h'(\theta)$, and then

$$b - a > x_{n-1} - x_1 = h[(n-1)\pi] - h(\pi) = \int_{\pi}^{(n-1)\pi} h'(\theta)\, d\theta = \int_{\pi}^{(n-1)\pi} \frac{1}{\theta'_{\lambda_n}(h(\theta))}\, d\theta.$$

Now, if p_m, q_m, and r_M denote the minimum values of p and q and the maximum value of r on $[a, b]$ and if we recall that $\theta_{\lambda_n}(h(\theta)) = \theta$, (12) implies that

$$\theta'_{\lambda_n}(h(\theta)) \leq \frac{1}{p_m}\cos^2\theta + (\lambda_n r_M - q_m)\sin^2\theta.$$

Since the right-hand side is a periodic function of θ of period π, it follows that

$$b - a > (n - 2)\int_0^\pi \frac{1}{\dfrac{1}{p_m}\cos^2\theta + (\lambda_n r_M - q_m)\sin^2\theta}\, d\theta.$$

From tables of integrals we obtain

$$\int_0^\pi \frac{1}{s^2\cos^2\theta + t^2\sin^2\theta}\, d\theta = \frac{\pi}{st},$$

and then

$$b - a > \frac{\pi(n-2)}{\sqrt{\dfrac{1}{p_m}(\lambda_n r_M - q_m)}} = \frac{\pi(n-2)\sqrt{p_m}}{\sqrt{\lambda_n r_M - q_m}};$$

that is, if λ_n is large enough,

$$\lambda_n > \frac{q_m}{r_M} + \left(\frac{\pi}{b-a}\right)^2 \frac{p_m}{r_M}(n-2)^2.$$

It follows that $\lambda_n/n \to \infty$ as $n \to \infty$, Q.E.D.

Example 4.10 It is clear that the nth eigenvalue of each of the Sturm-Liouville problems in Examples 4.7 to 4.9 satisfies the inequality

$$\lambda_n \geq \frac{\pi^2(n-1)^2}{a^2}.$$

It follows, in particular, that $\lambda_n/n \to \infty$ as $n \to \infty$.

The preceding results complete a set of affirmative answers to all the questions posed in §3.9, except (vi) and (ix). This will be done in the next section.

§4.4 Generalized Fourier Series

Building on the preceding results, Liouville then considered the problem of expanding an arbitray function $f : [a, b] \to \mathbb{R}$ in a series involving the eigenfunctions X_n of a given regular Sturm-Liouville problem. More precisely, the problem of finding constants c_n such that

$$f = \sum_{n=1}^{\infty} c_n X_n.$$

Assuming for the time being, as was done in the case of Fourier series, that the series on the rigth-hand side converges, that its sum equals f, and that term-by-term integration is possible after multiplication of its terms by $r X_m$, we have

$$\int_a^b r f X_m = \sum_{n=1}^{\infty} c_n \int_a^b r X_n X_m,$$

and we deduce from Theorem 4.1, as Liouville did, that

$$c_n = \frac{\displaystyle\int_a^b r f X_n}{\displaystyle\int_a^b r X_n^2}.$$

Definition 4.5 *If* $f : [a, b] \to \mathbb{R}$ *is an integrable function and* X_n *is the nth eigenfunction of a regular Sturm-Liouville problem,* (1)–(3), *then the constants* c_n *given above are called the* **Fourier coefficients** *of* f *with respect to the* X_n, *and the series* $\sum_{n=1}^{\infty} c_n X_n$ *is called the* **Fourier series** *of* f *with respect to the* X_n.

Note that the Fourier coefficients of f depend on the choice of the eigenfunctions, that is, that each c_n changes if X_n is replaced by $C_n X_n$, where C_n is an arbitrary constant. However, each product $c_n X_n$ and then the Fourier series remain unchanged.

The square root of the denominator in the expression defining c_n will be denoted by $\|X_n\|_r$ or, simply, by $\|X_n\|$ when $r = 1$. The functions

$$Y_n = \frac{X_n}{\|X_n\|_r}$$

are then referred to as the *normalized eigenfunctions*. Their use is sometimes advantageous in that $\|Y_n\|_r = 1$, yielding the simpler expression

$$c_n = \int_a^b r f Y_n$$

for the Fourier coefficients with respect to the Y_n.

Once in possession of the Fourier coefficients of a function with respect to the eigenvalues of a Sturm-Liouville problem, Liouville gave an incomplete proof of the convergence of its Fourier series to the given function. A complete, correct proof was first given by Vladimir Andreevich Steklov (1864–1926) in 1898 for twice differentiable functions that satisfy the boundary conditions. An independent proof of this fact, along entirely different lines, was also obtained in 1904 by David Hilbert (1862–1943) under some restrictive assumptions. These restrictions were shown to be superfluous the next year by Erhardt Schmidt (1876–1959), a student of Hilbert, in his doctoral dissertation. Although the methods of Hilbert and Schmidt would be more influential in subsequent developments of mathematics (especially after the introduction of L^2 spaces, as related in §6.6), a correct proof of the convergence of generalized Fourier series for piecewise continuous functions was first given by Julius Carl Chr. Adolph Kneser (1862–1930), of Breslau, in 1904. The proof of this theorem is beyond the scope of this book, so we limit ourselves to give its statement, which we combine with the equivalent of Heine's result on uniform convergence.[1]

Theorem 4.6 *If f and f' are piecewise continuous on $[a, b]$, then the Fourier series of f with respect to the eigenfunctions of a regular Sturm-Liouvillle problem, (1)–(3), converges at each x in (a, b) to the value*

$$\tfrac{1}{2}[f(x+) + f(x-)].$$

If, in addition, f is continuous and satisfies the boundary conditions

$$a_1 f(a) - b_1 f'(a) = 0$$
$$a_2 f(b) + b_2 f'(b) = 0,$$

then it converges absolutely and uniformly to f on $[a, b]$.

Example 4.11 Find the Fourier series of the function defined by $f(x) = Kx$, where K is a constant, with respect to the eigenfunctions of the Sturm-Liouville problem

$$X'' + \lambda X = 0$$
$$hX(0) - X'(0) = 0$$
$$X'(a) = 0$$

of Example 4.9. The eigenfunctions are

$$X_n(x) = \frac{\cos \sqrt{\lambda_n}\,(a - x)}{\sin a \sqrt{\lambda_n}}$$

with $\tan a \sqrt{\lambda_n} = h/\sqrt{\lambda_n}$. The Fourier coefficients of f with respect to the X_n are

$$c_n = \frac{\displaystyle\int_0^a K x X_n(x)\,dx}{\displaystyle\int_0^a X_n^2(x)\,dx}.$$

---

[1] For a proof see, for instance, F. V. Atkinson, *Discrete and Continuous Boundary Value Problems*, Academic Press, New York, 1965.

Straightforward computations show that

$$\int_0^a x X_n(x)\, dx = \frac{1}{\sin a\sqrt{\lambda_n}} \int_0^a x \cos \sqrt{\lambda_n}\,(a-x)\, dx$$

$$= \frac{1}{\sin a\sqrt{\lambda_n}} \left[\frac{x \sin \sqrt{\lambda_n}\,(a-x)}{\sqrt{\lambda_n}} \Big|_0^a - \int_0^a \frac{\sin \sqrt{\lambda_n}\,(a-x)}{\sqrt{\lambda_n}}\, dx \right]$$

$$= \frac{1}{\sin a\sqrt{\lambda_n}} \frac{\cos \sqrt{\lambda_n}\,(a-x)}{\lambda_n} \Big|_0^a$$

$$= \frac{1 - \cos a\sqrt{\lambda_n}}{\lambda_n \sin a\sqrt{\lambda_n}}$$

and that

$$\int_0^a X_n^2 = \frac{1}{\sin^2 a\sqrt{\lambda_n}} \int_0^a \cos^2 \sqrt{\lambda_n}\,(a-x)\, dx$$

$$= \frac{1}{\sin^2 a\sqrt{\lambda_n}} \int_0^a \frac{1 + \cos 2\sqrt{\lambda_n}\,(a-x)}{2}\, dx$$

$$= \frac{1}{2\sin^2 a\sqrt{\lambda_n}} \left[a + \frac{\sin 2\sqrt{\lambda_n}\,(a-x)}{2\sqrt{\lambda_n}} \Big|_0^a \right]$$

$$= \frac{2a\sqrt{\lambda_n} + \sin 2a\sqrt{\lambda_n}}{4\sqrt{\lambda_n} \sin^2 a\sqrt{\lambda_n}}.$$

Thus,

$$c_n = \frac{4 \sin a\sqrt{\lambda_n}\,(1 - \cos a\sqrt{\lambda_n})}{2a\lambda_n + \sqrt{\lambda_n} \sin 2a\sqrt{\lambda_n}}$$

and, therefore,

$$f(x) \sim 4K \sum_{n=1}^{\infty} \frac{1 - \cos a\sqrt{\lambda_n}}{2a\lambda_n + \sqrt{\lambda_n} \sin 2a\sqrt{\lambda_n}} \cos \sqrt{\lambda_n}\,(a-x).$$

★ As for the general boundary value problem of §3.7 and the questions raised in §3.9, Theorem 4.6 provides an affirmative answer to question (*vi*). Therefore, a proof analogous to those of Theorems 3.1 and 3.2 is now valid in the general case. Finally, the transient solution v, which satisfies the initial condition $v(x, 0) = f(x) - U(x)$, has the limit $v(x, t) \to 0$ as $t \to \infty$. To see this note that if $\lambda_1 = 0$ is an eigenvalue then, according to Theorem 4.3, $a_1 = a_2 = 0$ and any corresponding eigenfunction is constant. But then, with $r \equiv c$, we have

$$c_1 = \frac{\int_0^a c(f-U)X_1}{\int_0^a cX_1^2} = \frac{X_1}{\|X_1\|_c^2} \left(\int_0^a cf - \int_0^a cU \right),$$

which is zero by the construction of the steady state solution in the case $a_1 = a_2 = 0$. This answers the last question in §3.9 and shows that v is indeed the transient solution. Together with the work done in §3.10, this allows us to state the following result for a time-independent boundary value problem.

Theorem 4.7 *Consider the boundary value problem*

$$
\begin{aligned}
cu_t &= (\kappa u_x)_x + q & &\text{in } D \\
a_1 u(0, t) - b_1 u_x(0, t) &= c_1 & &t \geq 0 \\
a_2 u(a, t) + b_2 u_x(a, t) &= c_2 & &t \geq 0 \\
u(x, 0) &= f(x) & &0 \leq x \leq a,
\end{aligned}
$$

where the set D; the functions c, κ, q, and f; and the constants a, a_1, b_1, c_1, a_2, b_2, and c_2 are as in §3.7. Then this boundary value problem is well posed in the sense of Hadamard. Furthermore, let U be its steady-state solution, let λ_n and X_n be the eigenvalues and corresponding eigenfunctions of the regular Sturm-Liouville problem

$$
\begin{aligned}
(\kappa X')' + \lambda c X &= 0 \\
a_1 X(0) - b_1 X'(0) &= 0 \\
a_2 X(a) + b_2 X'(a) &= 0,
\end{aligned}
$$

and let c_n be the nth Fourier coefficient of $f - U$ with respect to the X_n. Then the series

$$
\sum_{n=1}^{\infty} c_n e^{-\lambda_n t} X_n(x)
$$

converges to a continuous function $v : \overline{D} \to \mathbb{R}$ such that $v(x, t) \to 0$ as $t \to \infty$, and $u = U + v$ is the solution of our boundary value problem.

The study of time-dependent problems carried out in §3.11 remains valid as is, but note that $r \equiv c$ in (34) and (37).

★   ## §4.5 Approximations

The theory presented in the preceding sections has shown the existence of the eigenvalues for regular problems and exhibited some of their essential properties. But, from a practical point of view, it has not given us something very crucial: a method to compute them. In fact, Equation (1) can be solved in very few cases, such as when it has constant coefficients or is a *Cauchy-Euler* equation. However, we can prove a general result that can be used to estimate the first eigenvalue, the most interesting one in applications.

If λ_n is an eigenvalue with eigenfunction X_n, we start by rewriting Equation (1) in the form

$$
\lambda_n r X_n = q X_n - (p X_n')'.
$$

If we multiply both sides by X_n, integrate from a to b, and solve for λ_n, we obtain

$$
\lambda_n = \frac{\displaystyle\int_a^b [q X_n - (p X_n')'] X_n}{\|X_n\|_r^2}.
$$

The right-hand side of this equation is called the *Rayleigh quotient* in honor of John William Strutt, Lord Rayleigh, of Cambridge (1842–1919), who introduced it in his studies on the theory of sound.

Note that this quotient cannot be used to evaluate λ_n because X_n is not known until λ_n is. However, we can ask whether or not replacing X_n by a function f that is close to X_n would give an approximate value for λ_n. Since X_n is twice differentiable and satisfies the endpoint conditions (2) and (3), it is natural to require the same of f. However, f should not be required to be a solution of (1). We have the following result.

Theorem 4.8 *The first eigenvalue λ_1 of a regular Sturm-Liouville problem, (1)–(3), is equal to the minimum value of*

$$R(f) = \frac{\int_a^b [qf - (pf')']f}{\|f\|_r^2}$$

over all continuous functions f that have two piecewise continuous derivatives and satisfy the endpoint conditions $a_1 f(a) - b_1 f'(a) = 0$ and $a_2 f(b) + b_2 f'(b) = 0$, and this minimum value is attained for $f = X_1$.

Proof. Note first that, if f is as above, then, according to Theorem 4.6, it is the sum of its Fourier series with respect to the X_n

(13) $$f = \sum_{n=1}^{\infty} c_n X_n,$$

where

$$\int_a^b r f X_n = c_n \|X_n\|_r^2.$$

Next we use integration by parts twice to compute the general term in the numerator of $R(f)$. This yields

$$\int_a^b [qf - (pf')']X_n = \int_a^b qfX_n - pf'X_n \Big|_a^b + \int_a^b pf'X_n'$$

$$= \int_a^b qfX_n - pf'X_n \Big|_a^b + pfX_n' \Big|_a^b - \int_a^b (pX_n')'f$$

$$= \int_a^b [qX_n - (pX_n')']f + p(a)[f'(a)X_n(a) - f(a)X_n'(a)]$$

$$- p(b)[f'(b)X_n(b) - f(b)X_n'(b)].$$

Both f and X_n satisfy the endpoint condition at a, that is, the system of equations

$$a_1 f(a) - b_1 f'(a) = 0$$
$$a_1 X_n(a) - b_1 X_n'(a) = 0.$$

It has a nontrivial solution for a_1 and b_1 only if its determinant is zero, that is, only if

$f'(a)X_n(a) - f(a)X'_n(a) = 0$. Similarly, $f'(b)X_n(b) - f(b)X'_n(b) = 0$. It follows that the last two integrated terms are zero, and then

$$\int_a^b [qf - (pf')']X_n = \int_a^b [qX_n - (pX'_n)']f = \lambda_n \int_a^b rfX_n = \lambda_n c_n \|X_n\|_r^2.$$

To complete the computation of the numerator of $R(f)$ recall that, by Theorem 4.6, the convergence in (13) is uniform so that term-by-term integration is possible. Then

$$\int_a^b [qf - (pf')']f = \int_a^b [qf - (pf')'] \sum_{n=1}^{\infty} c_n X_n$$

$$= \sum_{n=1}^{\infty} c_n \int_a^b [qf - (pf')']X_n$$

$$= \sum_{n=1}^{\infty} \lambda_n c_n^2 \|X_n\|_r^2$$

$$\geq \lambda_1 \sum_{n=1}^{\infty} c_n^2 \|X_n\|_r^2.$$

The last inequality is due to the fact that λ_1 is the smallest eigenvalue, and equality holds only if $c_n = 0$ for all $n \geq 2$, that is, if $f = X_1$.

To evaluate the infinite series on the right we multiply (13) by rf and integrate term by term. We obtain

$$\|f\|_r^2 = \int_a^b rf^2 = \int_a^b rf \sum_{n=1}^{\infty} c_n X_n = \sum_{n=1}^{\infty} c_n \int_a^b rfX_n = \sum_{n=1}^{\infty} c_n^2 \|X_n\|_r^2.$$

We conclude that $\lambda_1 \leq R(f)$ and that equality holds if $f = X_1$, Q.E.D.

Theorem 4.8 is valuable in practice because there are techniques that work well to estimate the minimum of the right-hand side, but they are beyond the scope of this book. What we can do, instead, is to compute $R(f)$ for some function f that satisfies the properties listed in the theorem, which will always yield an upper bound for λ_1. Since $R(f)$ has its minimum for $f = X_1$ and since X_1 has no zeros between a and b by Theorem 4.4, it seems reasonable to select f with no zeros between a and b, hoping that the closer f is to X_1 the closer $R(f)$ will be to its minimum value. It turns out that in practice this often gives a pretty good estimate of λ_1.

Example 4.12 Consider the Sturm-Liouville problem

$$X'' + \lambda X = 0$$
$$X(0) = 0$$
$$X(10) = 0,$$

whose first eigenvalue is known to be $\lambda_1 = \pi^2/100$. To see how this value is confirmed by the method outlined above note that, in this case,

$$R(f) = \frac{-\int_0^{10} f'' f}{\int_0^{10} f^2}.$$

If we choose $f(x) = x(10 - x)$, which satisfies the endpoint conditions and has no zeros on $(0, 10)$, we obtain

$$R(f) = \frac{\int_0^{10} (20x - 2x^2)\, dx}{\int_0^{10} (10x - x^2)^2\, dx} = 0.1,$$

which is a good approximation of the true value $\pi^2/100$.

Example 4.13 Find a good approximation of the first eigenvalue of the Sturm-Liouville problem

$$X'' + \lambda X = 0$$
$$X(0) - 4X'(0) = 0$$
$$X'(\pi) = 0.$$

We have

$$R(f) = \frac{-\int_0^{\pi} f'' f}{\int_0^{\pi} f^2},$$

and if we choose $f(x) = 2 + \sin \frac{1}{2}x$, which satisfies the endpoint conditions and has no zeros on $(0, \pi)$, we obtain

$$R(f) = \frac{\int_0^{\pi} \left(\frac{1}{2} \sin \frac{1}{2}x + \frac{1}{4} \sin^2 \frac{1}{2}x \right) dx}{\int_0^{\pi} \left(4 + 4 \sin \frac{1}{2}x + \sin^2 \frac{1}{2}x \right) dx} = \frac{1 + \frac{1}{8}\pi}{4\pi + 8 + \frac{1}{2}\pi} \approx 0.063.$$

§4.6 Historical Epilogue

Sturm and Liouville, who had become friends in the late 1820s and remained friends for life, went on to have distinguished careers in mathematics.[1] Sturm, who had changed to mathematics from an early interest in fluids, had to overcome the difficulties of being a

---

[1] Most of this section is based on J. Lützen, *Joseph Liouville 1809–1882: Master of Pure and Applied Mathematics*, Springer-Verlag, New York, 1990.

foreigner in Paris (although naturalized in 1833) and a Protestant, but could always count on the support of Liouville. In 1836 both of them and three others were candidates for a position at the *Académie des Sciences*. Liouville addressed the *Académie* and praised Sturm's work, comparing some of the latter's memoirs to those of Lagrange. As a result, Sturm was ranked number one by the geometry section and was elected by a majority of 46 votes out of 52 when—to everyone's surprise—Liouville and Duhamel, one of the other applicants of whose work we shall learn in Chapter 9, withdrew in Sturm's favor.

Liouville's opportunity came in 1839, when a position became available in the astronomy section. This prompted Liouville, who had previously done some work on celestial mechanics, to quickly present several papers in astronomy to the *Académie*. He was elected on June 3 and officially appointed on June 10. The previous year Liouville had resigned his post at the *École Centrale* and become Professor of Analysis and Mechanics at the *École Polytechnique*, nominating Sturm as his assistant (*répétiteur*). Thus, Liouville had become a professor at a prestigious institution and an academician by his thirtieth year. For the older Sturm, who had neglected the search for a permanent job in favor of research, this assistantship was his first full-time position. Two years later, with Liouville as a member of a committee in charge of suggesting candidates for new vacancies, Sturm became a second Professor of Analysis and Mechanics at the *École Polytechnique*. For the next ten years, Liouville taught in the odd-numbered years and Sturm in the even-numbered years. In 1840, after the death of Poisson, Sturm became his successor as Professor of Mechanics at the Paris *Faculté des Sciences*.[1]

With the revolution of 1848, which resulted in the second republic, Liouville became involved in politics and was elected a *représentant du peuple* to the National Constituting Assembly. Although he served for only one year—for as long as the republic stayed actually democratic—this new activity implied a noticeable decrease in his mathematical output (a state of affairs that he would later describe in a letter to Dirichlet, when the problem resurfaced for personal reasons, as a neglect of the x's and the y's), and just when it started to pick up again something happened. In 1850 idle politicians decided to tamper with the teaching at the *École Polytechnique*. It was found that the school had deviated from its original mission to train military engineers and seemed to be devoted to producing future professors. The prescribed cure was the elimination of abstract mathematical theories and the introduction of practical topics. This was distressful to both Sturm and Liouville, but they reacted very differently to these demands.

For Liouville it meant that he had to find another job, a task in which he succeeded with his election to a professorship at the *Collège de France*, a higher level institution,

---

[1] The French educational system, as revamped by Napoléon Bonaparte, included a few special schools, such as the *École Polytechnique* and the *Collège de France*, but otherwise entrusted all basic education to the *Université Impériale de France*. This was a nationwide institution divided into local branches, each of them including at least one of five *Facultés* (science, letters, medicine, law, and theology). Professors were chosen from other major institutions and their salaries were added to those of their regular positions. In order to make ends meet, multiemployment—called the *cumul*—was then the norm rather than the exception in the French academic world. Bright students who desired further education would enter, upon graduation from the *Faculté* or from the *École Polytechnique*, a specialist *école d'application*, of which the most noteworthy was the *École des Ponts et Chaussées*, Liouville's own, and twenty years earlier Cauchy's.

at the start of 1851. His teaching had always been found to be most illuminating by his best students but to go over the heads of the average ones, and in this respect his move to the *Collège* was just what the doctor ordered. For Sturm the pressures of the new situation proved to be too much. He felt persecuted and suffered a nervous breakdown. In 1851 he resigned his position at the *École Polytechnique* and took a leave of absence from the *Faculté des Sciences*. Although he resumed his lectures here two years later, he never recovered from his nervous ailment, and he is said to have died insane on December 15, 1855. A grief-stricken Liouville delivered a moving eulogy at the grave of his friend, to whose work he devoted his own lectures over the coming semester, and Liouville and his family always kept in touch with Sturm's widow.

Two years later, the necessities of life being what they were, Liouville succeeded Sturm in his chair at the *Faculté des Sciences*. Once again, his creativity decreased, in part because of his increased teaching duties (at some point in his life he taught about 35 hours a week, including Saturdays and Sundays) and in part because of his failing health. By 1861—already twice a grandfather—he considered himself an old man at the age of 52, the age at which Sturm died. From this point on, afflicted by gout in a very serious way, he spent the rest of his life in pain, and starting in 1872 he had first Darboux and then Tisserand permanently substitute for him at the *Faculté des Sciences*.

In the 1860s he developed a new mathematical interest in analytic number theory, a subject on which he published numerous notes but always without proofs or justifications. Perhaps he felt that he was running out of time in his life and decided to write just the bare facts of his discoveries, but this behavior hurt his reputation. His constant pain made him miserable and even mean toward some people. Such was his suffering that in 1876—the year of his election to the Berlin Academy—he wrote in his diary: *Ah! death, what a relief*. It was not to be yet, but by a cruel twist of fate he lost both his wife Marie Louise in a terrible accident and his son Ernest in 1880. Although he had stopped publishing in 1874, Joseph Liouville continued lecturing at the *Collège de France* to the end of his days. In fact, it was only during these lectures—and he had to be helped to the blackboard—that he regained a measure of his former energy and wit. His release from life did not come until September 8, 1882, and he rests since then in the *Cimetière du Montparnasse*, next to his wife.

EXERCISES

1.1 Transform the following equations to the form (1).

(*i*) The *Laguerre* differential equation $x X'' + (1 - x)X' + \lambda X = 0, \quad x > 0$.

(*ii*) The *Hermite* differential equation $X'' - 2x X' + 2\lambda X = 0, \quad -\infty < x < \infty$.

(*iii*) The *Chebyshev* differential equation $(1 - x^2)X'' - x X' + \lambda X = 0, \quad -1 \leq x \leq 1$.

(*iv*) The *Legendre* differential equation $(1 - x^2)X'' - 2x X' + \lambda X = 0, \quad -1 \leq x \leq 1$.

1.2 Identify all the Sturm-Liouville problems arising from the study of the heated bar in §3.1 to §3.6. In each case identify the eigenvalues and the eigenfunctions.

1.3   Determine which of the following are Sturm-Liouville problems and find all their eigenvalues and eigenfunctions. Give reasons why the remaining problems are not of the
Sturm-Liouville type.

(i) $X'' + \lambda X = 0$, $X'(0) = 0$, $X(a) = 0$, $a > 0$.

(ii) $X'' - \lambda X = 0$, $X(0) = 0$, $X(\pi) = 0$.

(iii) $X'' + \lambda X = 0$, $X(0) = 1$, $X(1) = 0$.

(iv) $X'' + \lambda X = 0$, $X(0) - 2X(1) = 0$, $X'(1) = 0$.

(v) $x^2 X'' + xX' + \lambda X = 0$, $X(1) = 0$, $X(e) = 0$.

1.4   Show that π^2 is an eigenvalue of the Sturm-Liouville problem

$$X'' + \lambda X = 0$$
$$X(0) - X'(0) = 0$$
$$\pi^2 X(1/2) + X'(1/2) = 0$$

and find a corresponding eigenfunction.

1.5   Show that for $x > 0$ the second-order *Cauchy-Euler* differential equation

$$c_2 x^2 X'' + c_1 x X' + c_0 X = 0,$$

where c_2, c_1, and c_0 are constants, can be transformed into the constant coefficients
differential equation

$$c_2 Y'' + (c_1 - c_2)Y' + c_0 Y = 0,$$

by the substitution $x = e^s$, $Y(s) = X(e^s)$.

1.6   Find the eigenvalues and eigenfunctions of the following Sturm-Liouville problems.
Use Exercise 1.5 where applicable.

(i) $(x^2 X')' + \lambda X = 0$, $X(1) = 0$, $X(b) = 0$, $b > 1$.

(ii) $(e^{2x} X')' + e^{2x}(\lambda + 1)X = 0$, $X(0) = 0$, $X(\pi) = 0$.

(iii) $(e^{2x} X')' + e^{2x}(\lambda + 1)X = 0$, $X'(0) = 0$, $X'(\pi) = 0$.

2.1   Verify the orthogonality of the eigenfunctions for the Sturm-Liouville problems of Exercise 1.6 directly by integration.

2.2   Verify that, if $p(x) = \sin x$ then the Sturm-Liouville problem

$$(pX')' + \lambda X = 0$$

on $[0, \pi]$, which has no endpoint conditions, has the eigenvalue $\lambda = 0$ and find two
linearly independent eigenfunctions.

2.3   Show that solutions of the problem

$$X'' + \lambda X = 0$$
$$X(0) = 0$$
$$\lambda X(\pi) + X'(\pi) = 0$$

corresponding to different values of $\lambda > 0$ are not orthogonal.

2.4   Show that the problem

$$X'' + X = 0$$
$$X(0) - X(2\pi) = 0$$
$$X'(0) - X'(2\pi) = 0$$

has two solutions X_1 and X_2 such that $X_1 \neq C X_2$ for any value of the constant C.

2.5 Show that the problem

$$X'' + \lambda X = 0$$
$$X(0) = 0$$
$$2X(1) = 3$$

has a solution for some $\lambda < 0$. Determine such values of λ and the corresponding solutions.

2.6 Show that the Sturm-Liouville problem

$$X'' + \lambda X = 0$$
$$X'(0) = 0$$
$$3X(\log \tfrac{1}{2}) + 5X'(\log \tfrac{1}{2}) = 0$$

has a nontrivial solution for $\lambda = -1$.

2.7 Do the results of Exercises 2.3 to 2.6 contradict Theorems 4.1 to 4.3, respectively? Explain fully.

2.8 Use Theorem 4.1 to show that, if a Sturm-Liouville problem has a nonzero solution for some value of λ, then λ is necessarily a real number.

3.1 Prove that Theorems 4.1 and 4.3 are valid for periodic Sturm-Liouville problems but Theorem 4.2 is not.

3.2 Consider the regular Sturm-Liouville problem

$$X'' + \lambda X = 0$$
$$a_1 X(0) - b_1 X'(0) = 0$$
$$a_2 X(a) + b_2 X'(a) = 0$$

on an interval $[0, a]$, where $a_1 + a_2 > 0$ and $b_1 + b_2 > 0$ (this only excludes the well-known cases of §3.1 and §3.4). Show that the eigenvalues of this problem are $a_1 a_2 / b_1 b_2$, if and only if $b_1 b_2 \neq 0$ and

$$\sqrt{\frac{a_1 a_2}{b_1 b_2}} = \frac{(2N - 1)\pi}{2a}$$

for some positive integer N, and the positive roots of the transcendental equation

$$\tan a\sqrt{\lambda} = \frac{(a_1 b_2 + a_2 b_1)\sqrt{\lambda}}{b_1 b_2 \lambda - a_1 a_2}.$$

Show that corresponding eigenfunctions are

$$X(x) = b_1\sqrt{\lambda} \cos \sqrt{\lambda}\, x + a_1 \sin \sqrt{\lambda}\, x.$$

3.3 Verify graphically that, if $b_1 b_2 \neq 0$ and if

$$\frac{(2N - 1)\pi}{2a} < \sqrt{\frac{a_1 a_2}{b_1 b_2}} < \frac{(2N + 1)\pi}{2a}$$

for some positive integer N, then the eigenvalues of the Sturm-Liouville problem of Exercise 3.2 form an increasing sequence $\{\lambda_n\}$ such that

$$\frac{(2n - 1)\pi}{2a} < \sqrt{\lambda_n} < \frac{n\pi}{a} \qquad \text{for} \quad n = 1, \ldots, N$$

and
$$\frac{(n-1)\pi}{2a} < \sqrt{\lambda_n} < \frac{(2n-1)\pi}{2a} \qquad \text{for} \quad n = N+1, \dots .$$
Find similar inequalities for the eigenvalues in the cases

(i) $0 < \sqrt{\dfrac{a_1 a_2}{b_1 b_2}} < \dfrac{\pi}{2a}$,

(ii) $\sqrt{\dfrac{a_1 a_2}{b_1 b_2}} = \dfrac{(2N-1)\pi}{2a}$ for some positive integer N, and

(iii) $b_1 b_2 = 0$.

3.4 If $\{\lambda_n\}$ is as in Exercise 3.3, verify that
$$\lim_{n \to \infty} \frac{\lambda_n}{n} = \infty.$$

What is the value of
$$\lim_{n \to \infty} \frac{\lambda_n}{n^2}$$

in each case?

3.5 Show that the eigenvalues of the Sturm-Liouville problem
$$x^2 X'' + x X' + \lambda X = 0$$
$$X'(1) = 0$$
$$a_2 X(b) + b_2 X'(b) = 0,$$

where $b > 1$ and $a_2 + b_2 > 0$, are the positive roots of the transcendental equation
$$b_2 \sqrt{\lambda} \sin(\sqrt{\lambda} \log b) = a_2 b \cos(\sqrt{\lambda} \log b).$$

Determine the corresponding eigenfunctions and evaluate
$$\lim_{n \to \infty} \frac{\lambda_n}{n^2}.$$

Find the eigenvalues and eigenfunctions of this Sturm-Liouville problem if $a_2 = 0$ and if $b_2 = 0$ (**Hint:** use Exercise 1.5).

3.6 Show that the eigenvalues of the Sturm-Liouville problem
$$(e^{-2x} X')' + \lambda e^{-2x} X = 0$$
$$X(0) - X'(0) = 0$$
$$X'(a) = 0$$

on an interval $[0, a]$ are the solutions $\lambda_n > 1$ of the transcendental equation
$$\tan a\sqrt{\lambda_n - 1} = \frac{1}{\sqrt{\lambda_n - 1}}$$

and have corresponding eigenfunctions
$$X_n(x) = e^x \cos \sqrt{\lambda_n - 1}\, x.$$

If the endpoint condition $X'(a) = 0$ is replaced by $X(a) + X'(a) = 0$, how are the eigenfunctions and the equation for the eigenvalues to be modified?

3.7   Verify directly that the nth eigenfunction of the Sturm-Liouville problem described in Exercises 3.2 and 3.3 has exactly $n - 1$ zeros in the interval $[0, a]$.

3.8   Find the zeros of the eigenfunction X_n of Exercise 3.6 in the interval $[0, a]$.

Solve the boundary value problems of Exercises 4.1 to 4.6.

4.1

$$e^{2x}u_t = (e^{2x}u_x)_x \qquad 0 < x < \pi, \qquad t > 0$$
$$u(0, t) = 0 \qquad t \geq 0$$
$$u(\pi, t) = 0 \qquad t \geq 0$$
$$u(x, 0) = \sin x \qquad 0 \leq x \leq \pi$$

(**Hint:** refer to Exercise 1.6 (*ii*)).

4.2

$$e^{-2x}u_t = (e^{-2x}u_x)_x \qquad 0 < x < \pi, \qquad t > 0$$
$$u(0, t) - u_x(0, t) = 0 \qquad t \geq 0$$
$$u_x(\pi, t) = 0 \qquad t \geq 0$$
$$u(x, 0) = \begin{cases} e^x, & 0 \leq x \leq \pi/2 \\ e^{\pi/2}, & \pi/2 < x \leq \pi \end{cases}$$

(**Hint:** refer to Exercise 3.6).

4.3

$$u_t = u_{xx} - 2u_x \qquad 0 < x < \pi, \qquad t > 0$$
$$u(0, t) - u_x(0, t) = 0 \qquad t \geq 0$$
$$u(\pi, t) + u_x(\pi, t) = 0 \qquad t \geq 0$$
$$u(x, 0) = 1 + \sin x \qquad 0 \leq x \leq \pi$$

(**Hint:** refer to Exercise 3.6).

4.4

$$e^{2x}u_t = (e^{2x}u_x)_x + e^{x-t} \qquad 0 < x < \pi, \qquad t > 0$$
$$u(0, t) = 0 \qquad t \geq 0$$
$$u(\pi, t) = 0 \qquad t \geq 0$$
$$u(x, 0) = \sin x \qquad 0 \leq x \leq \pi$$

(**Hint:** refer to Exercise 4.1).

4.5

$$u_t = [(x + 1)^2 u_x]_x + \frac{e^{-t}}{\sqrt{x + 1}} \qquad 0 < x < 9, \qquad t > 0$$
$$u(0, t) = 0 \qquad t \geq 0$$
$$u(9, t) = 0 \qquad t \geq 0$$
$$u(x, 0) = 0 \qquad 0 \leq x \leq 9$$

(**Hint:** make the substitution $s = x + 1$ and refer to Exercise 1.6 (*i*)).

4.6

$$u_t = u_{xx} - hu \qquad 0 < x < a, \qquad t > 0$$
$$u_x(0, t) = 0 \qquad t \geq 0$$
$$u(a, t) = T \qquad t \geq 0$$
$$u(x, 0) = \frac{T}{a^2}x^2 \qquad 0 \leq x \leq a,$$

where h and T are positive constants (**Hint:** make the substitution $v = e^{ht}u$ and refer to Exercise 1.3 (*i*)).

4.7   Repeat Exercise 4.6 if the equation $u_t = u_{xx} - hu$ is replaced with $u_t = u_{xx} - h(u - T)$.

4.8   A bar of length a, initially at a temperature T_0, is submerged in a medium whose temperature is $T < T_0$. There is heat loss by convection along the lateral surface and at both ends. Assuming that the proportionality constant is a rational number, state and solve the corresponding boundary value problem (**Hint:** refer to Exercise 3.2).

Estimate the first eigenvalue for each of the following Sturm-Liouville problems.

5.1
$$X'' + \lambda X = 0$$
$$X(0) = 0$$
$$X(10) = 0$$

using the function

$$f(x) = \begin{cases} x, & 0 \le x \le 5 \\ 10 - x, & 5 < x \le 10. \end{cases}$$

5.2
$$X'' + \lambda X = 0$$
$$X(0) - 4X'(0) = 0$$
$$X'(\pi) = 0$$

using a function of the form $f(x) = ax^2 + bx + c$, where a, b, and c are constants.

5.3
$$X'' + \lambda X = 0$$
$$X(0) = 0$$
$$X(3) + X'(3) = 0$$

(**Hint:** $\lambda_1 \approx 0.67002076$).

5.4
$$(xX')' + \frac{\lambda}{x}X = 0$$
$$X'(1) = 0$$
$$X(9) = 0$$

(**Hint:** it was shown in Example 4.1 that $\lambda_1 = \pi^2 / \log^2 9$).

5.5
$$X'' + \frac{4\lambda}{x^4}X = 0$$
$$X(2) = 0$$
$$X(3) = 0.$$

Also verify that the general solution of the given differential equation is

$$X(x) = Ax \cos \frac{2\sqrt{\lambda}}{x} + Bx \sin \frac{2\sqrt{\lambda}}{x},$$

where A and B are arbitrary constants, and find λ_1 exactly.

5.6
$$X'' + (\lambda - x)X = 0$$
$$X(0) - X'(0) = 0$$
$$X(3\pi/2) = 0.$$

5.7
$$X'' + (\lambda - x^2)X = 0$$
$$X(0) = 0$$
$$X'(2) = 0,$$

first using the function $f(x) = \sin(\pi x/4)$ and then using another function that gives a better estimate.

5.8
$$X'' + 2X' + (\lambda + 1)X = 0$$
$$X(0) = 0$$
$$X(\pi) = 0$$

(**Hint:** refer to Exercise 1.6 (*ii*)).

5

THE WAVE EQUATION

§5.1 Introduction

In the small hours of November 17, 1717, a Parisian *gendarme* walking his beat came across a small basket that had been left on the front steps of the church of Saint Jean Le Rond and contained a recently born infant. This little baby, the illegitimate son of Mme. de Tencin and the Chevalier Destouches, was destined to become the greatest mathematician of France in his times. Jean Le Rond d'Alembert, as he was appropriately named, was placed by his father in the foster home of Mme. Rousseau, in the Rue Michel-Ange, where he was to remain for the first 48 years of his life, the years that would witness all of his creative mathematical activity. The path that led d'Alembert to mathematics was not straight and narrow, for he was a man with a wide range of intellectual interests. His mathematical career may have come to an end at the age of 48, but not his intellectual output. As it happened, he first became a lawyer in 1738 and then took one year of medicine. But with the publication of his first mathematical paper in 1739, at the age of 21, the direction of his main area of research became firmly established. So was his daily routine, consisting of study and research in the morning, in a small garret room at Mme. Rousseau's, while his afternoons and evenings were entirely devoted to Parisian society life. Witty, charming, a brilliant conversationalist, he became indispensable in the *salons*—he regularly attended those of Mme. Geoffrin and Mme. la Marquise du Deffand—where everyone was interested in what d'Alembert had to say. Music and dance were then, as had already been for the longest time, an integral part of society life, and it was natural for any intelligent man, and in particular for a man of science, to wonder about such things as the consonance of musical sounds such as, for instance, those produced by a violin string.

When a taut piece of string, tied down at both ends, is either plucked or hit it emits audible vibrations. Their frequency, that is, the number of vibrations per second, was first determined by none other than Brook Taylor (1685–1731), of *Taylor's Theorem* fame, in his 1713 paper *De motu nervi tensi*. For a string of length a, mass per unit

JEAN LE ROND D'ALEMBERT
David Eugene Smith Collection.
Rare Book and Manuscript Library.
Columbia University.
Reprinted with permission.

length m and subject to a tension force τ, this frequency is

$$f = \frac{1}{2a}\sqrt{\frac{\tau}{m}}.$$

As for the shape of the string at any given instant of its motion, Taylor believed that it was a sine wave. But there is usually more than one audible frequency of vibration when such a string is in motion. The one that Taylor had computed is called *fundamental* because it is the frequency of the loudest vibration. The other component vibrations that make up the motion of the string are called *harmonics*, and in 1726 the composer Jean-Philippe Rameau explained that the reason that musical sounds are pleasing to the ear is that the frequencies of the harmonics are integral multiples of the fundamental frequency.

Some other researchers also become involved in the study of the vibrating string. For the time being, suffice it to mention that Johann Bernoulli (1667–1748) arrived independently at the same formula that Taylor had obtained for the fundamental frequency, a fact that he communicated to his son Daniel in a letter of 1727 and that he published the year after. Bernoulli also thought that the shape of the string at any instant was sinusoidal. His son Daniel went a step farther in a paper of 1732, in which he stated that the vibrating string has nodes in its motion, that is, stationary points, points that do not ever move. This is, of course, compatible with the idea of a sinusoidal shape for the moving string since a sine function has plenty of zeros. What eluded these and other investigators was a complete description of the motion of the string, that is, of its exact shape at each instant. They could not know that such an accomplishment was impossible using the tools at their disposal, namely, ordinary differential equations.

Success would be reserved for d'Alembert, and it was based on the fact that in 1747 he derived a partial differential equation, the now well-known *wave equation*, that accurately describes the phenomenon in question and then solved it. While partial differential equations were not entirely new at that time—they appeared earlier in 1734 in some work of Euler and in d'Alembert's own *Traité de Dynamique* of 1743—this was, however, the first instance in which their use led to the solution of a worthwhile mathematical problem. In his papers *Recherches sur la courbe que forme une corde tendüe mise en vibration*, d'Alembert was able to follow the matter to its conclusion and to show that, earlier predictions notwithstanding, the shape of the vibrating string at any given time need not be sinusoidal.

§5.2 The Vibrating String

We shall now derive the partial differential equation describing the motion of the vibrating string, but using a method different from the original one by d'Alembert. Specifically, we shall work under the following hypotheses (refer to Exercises 2.1 and 2.4 for alternate derivations with less restrictive hypotheses).

1. The string is uniform so that the mass per unit length, m, is constant.

2. The string is elastic and so flexible that it offers no resistance to bending, and then the tension on the string is a force tangent to the string at every point. This is shown in Figure 5.1 for the piece of string between two points of abscissas x and $x + h$.

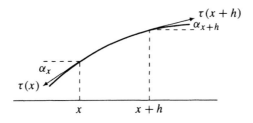

Figure 5.1

3. The angles involved are so small that, at each point, the string moves only vertically.

4. If the displacement of the string at a point of abscissa x and at a time t such that
$0 \le x \le a$ and $t \ge 0$ is denoted by $u(x, t)$, then u is a function of class C^2 on
the set

$$D = \{ (x, t) \in \mathbb{R}^2 : 0 < x < a, \, t > 0 \}.$$

Note that the third hypothesis actually means that the horizontal components of the
tension on each piece of string balance each other out. Thus, if $\tau(x)$ and $\tau(x + h)$ are
the tensile forces and α_x and α_{x+h} are the angles shown in Figure 5.1, we have

$$-\tau(x) \cos \alpha_x + \tau(x + h) \cos \alpha_{x+h} = 0$$

for every x and arbitrary h. In other words, there is a positive constant τ such that

$$\tau(x) \cos \alpha_x = \tau(x + h) \cos \alpha_{x+h} = \tau.$$

The vertical component of the tensile force on the piece of string from x to $x + h$ is

$$-\tau(x) \sin \alpha_x + \tau(x + h) \sin \alpha_{x+h} = -\tau \tan \alpha_x + \tau \tan \alpha_{x+h}.$$

If $u_{tt}(x, t)$ is the acceleration of the point of the string of abscissa x at time t, then
Newton's second law of motion states that

$$m \int_x^{x+h} u_{tt}(s, t) \, ds = -\tau \tan \alpha_x + \tau \tan \alpha_{x+h} - mhg,$$

approximately, where g is the acceleration of gravity. Since $\tan \alpha_x$ and $\tan \alpha_{x+h}$ are
the slopes of the string at x and $x + h$, respectively, and since these are given by the
derivatives of u with respect to x, we obtain

$$m \int_0^{x+h} u_{tt}(s, t) \, ds - \int_0^x u_{tt}(s, t) \, ds = \tau[u_x(x + h, t) - u_x(x, t)] - mhg.$$

Dividing by mh, defining $c^2 = \tau/m$, and letting $h \to 0$,

$$u_{tt}(x, t) = c^2 u_{xx}(x, t) - g.$$

When g is considered negligible because c^2 is very large—lightweight, very tight
string—the resulting equation

$$u_{tt} = c^2 u_{xx}$$

is called the *wave equation*. This equation, derived here for the motion of a string,
is also important in the description of many other physical phenomena, including the
propagation of acoustic, fluid, and electromagnetic waves. Our object is to solve it in D
or, more precisely, to solve a boundary value problem for the wave equation in D. We
start with a very simple one in order to get a feeling of what is involved.

Example 5.1 If $a = \pi$, find a function $u \in C^2(D)$ that is continuous on

$$\bar{D} = \{ (x, t) \in \mathbb{R}^2 : 0 \le x \le \pi, \quad t \ge 0 \}$$

and satisfies

$$
\begin{array}{ll}
u_{tt} = c^2 u_{xx} & \text{in } D \\
u(0, t) = 0 & t \ge 0 \\
u(\pi, t) = 0 & t \ge 0 \\
u_t(x, 0) = 0 & 0 \le x \le \pi \\
u(x, 0) = M \sin x & 0 \le x \le \pi,
\end{array}
$$

where M is a constant. That is, the string is tied down at the endpoints, its initial velocity is zero, and it is initially displaced in the shape of a sine wave.

Using the familiar method of separation of variables, if the solution that we seek is of the form $u(x, t) = X(x)T(t)$, then we obtain

$$X'' + \lambda X = 0 \qquad \text{and} \qquad T'' + \lambda c^2 T = 0,$$

where λ is a constant, subject to the endpoint and initial conditions

$$X(0) = X(\pi) = 0 \qquad \text{and} \qquad T'(0) = 0.$$

According to Theorem 4.3, the Sturm-Liouville problem for X has only positive eigenvalues. It is easy to see that the eigenvalues and eigenfunctions are

$$\lambda_n = n^2 \qquad \text{and} \qquad X_n(x) = \sin nx$$

for any positive integer n. Then all the solutions of the initial value problem

$$T'' + n^2 c^2 T = 0, \qquad T'(0) = 0$$

are of the form

$$T(t) = C \cos nct,$$

where C is an arbitrary constant. Therefore,

$$u(x, t) = C \sin nx \cos nct.$$

This satisfies the initial condition $u_t(x, 0) = 0$, and then $u(x, 0) = M \sin x$ implies that $C = M$ and $n = 1$, so that the complete solution is

$$u(x, t) = M \sin x \cos ct = \tfrac{1}{2} M \, [\sin(x - ct) + \sin(x + ct)].$$

We have written the solution in a way that allows us to graph the shape of the string at time t as the sum of two curves. Both of them are sine waves whose amplitudes are one half the amplitude of the original, but the first is displaced to the right by the amount ct while the second is displaced to the left by the same amount. This is represented in Figure 5.2 for values of $ct = k\pi/4$, $k = 0$, 1, 2, and 3. To sum up, the displacement of the string at time t is the sum of two waves that—except for their amplitudes—are replicas of the original curve and are moving in opposite directions with speed c. It is this observation that will be used next to obtain the solution of a more general boundary value problem.

§5.3 D'Alembert's Solution

What d'Alembert proposed to do in his first paper of 1747 is to show that there is an infinity of curves, other than the sine, that the vibrating string can adopt in its motion. After putting down four hypotheses and deriving the wave equation, he found, by a most ingenious method, what in modern terminology is known as its *general solution*.

We shall use a simpler method and start by remarking that in Example 5.1 the solution was written in terms of the variables $x + ct$ and $x - ct$. Hoping that this may be possible in the general case, we denote these new variables by

$$r = x + ct \qquad \text{and} \qquad s = x - ct$$

and then determine the form adopted by the wave equation when written in terms of r and s. To this end, we introduce a new function v defined by

$$v(r, s) = u(x, t)$$

and use the chain rule in making the change of variables. This yields

$$
\begin{aligned}
u_x &= v_r r_x + v_s s_x = v_r + v_s \\
u_{xx} &= v_{rr} r_x + v_{rs} s_x + v_{sr} r_x + v_{ss} s_x = v_{rr} + 2v_{rs} + v_{ss}
\end{aligned}
$$

where we have used the fact that, if v is of class C^2, then $v_{rs} = v_{sr}$.[1] Also,

$$
\begin{aligned}
u_t &= v_r r_t + v_s s_t = cv_r - cv_s \\
u_{tt} &= cv_{rr} r_t + cv_{rs} s_t - cv_{sr} r_t - cv_{ss} s_t \\
&= c^2 v_{rr} - c^2 v_{rs} - c^2 v_{sr} + c^2 v_{ss} \\
&= c^2 (v_{rr} - 2v_{rs} + v_{ss})
\end{aligned}
$$

Taking these expressions for u_{xx} and u_{tt} to the wave equation $u_{tt} = c^2 u_{xx}$ we arrive at

$$c^2(v_{rr} - 2v_{rs} + v_{ss}) = c^2(v_{rr} + 2v_{rs} + v_{ss}),$$

that is,

$$v_{rs} = 0.$$

This is the simpler form that the wave equation adopts in the new variables r and s. It is now very easy to solve this equation. Indeed, $v_{rs} \equiv 0$ implies that v_r is a function of r only:

$$v_r(r, s) = f(r).$$

Then, if F is a prederivative of f, it is clear that $(v - F)_r \equiv 0$ so that $v - F$ is a function of s only:

$$v(r, s) - F(r) = G(s).$$

---

[1] See, for instance, W. Rudin, *Principles of Mathematical Analysis*, 3rd ed., McGraw-Hill, New York, 1976, page 236.

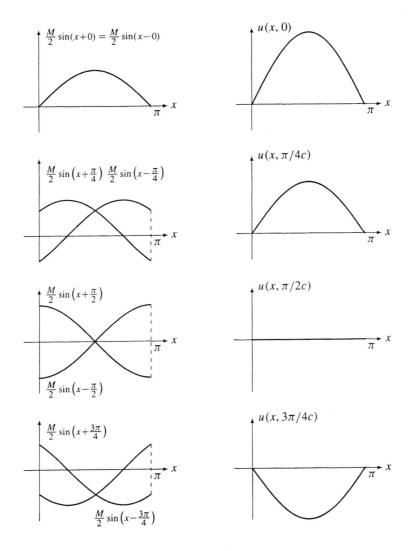

Figure 5.2

Note that both F and G are of class C^2 because v is. Finally, undoing the change of variables,

$$u(x, t) = v(r, s) = F(r) + G(s) = F(x + ct) + G(x - ct).$$

This tells us that any solution of the wave equation must be of this form. Conversely, any function of this form in which F and G are of class C^2 is a solution of the wave equation because, using the chain rule again,

$$u_{xx} = F''(x + ct) + G''(x - ct)$$
$$u_t = F'(x + ct)c + G'(x - ct)(-c)$$
$$u_{tt} = F''(x + ct)c^2 + G''(x - ct)(-c)^2$$
$$= c^2 [F''(x + ct) + G''(x - ct)]$$
$$= c^2 u_{xx}$$

We have, therefore, proved the following theorem.

Theorem 5.1 (D'Alembert, 1747) *The general solution of the wave equation is of the form*

$$u(x, t) = F(x + ct) + G(x - ct),$$

where F and G are arbitrary functions of class C^2 in $\mathbb{R}$.

The expression $F(x + ct)$ is called a *traveling wave* moving to the left with speed c because the graph of $F(x + ct)$ is the graph of $F(x)$ displaced ct units to the left, as shown in Figure 5.3. Similarly, $G(x - ct)$ is called a *traveling wave* moving to the right

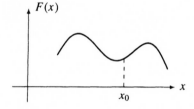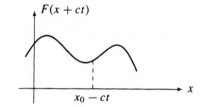

Figure 5.3

with speed c.

What remains, once a boundary value problem is posed for the wave equation, is to select from the wealth of solutions given by Theorem 5.1 the one that satisfies any given boundary and initial conditions. The first problem that we pose is that in which the string is initially displaced from its equilibrium position, its shape being the graph of a given function f, and then let go. That is, the initial velocity is zero. Our problem, then, consists of finding a function u of class C^2 in D, continuous on

$$\bar{D} = \{ (x, t) \in \mathbb{R}^2 : 0 \leq x \leq a, \quad t \geq 0 \},$$

such that

(1)	$u_{tt} = c^2 u_{xx}$	in D
(2)	$u(0, t) = 0$	$t \geq 0$
(3)	$u(a, t) = 0$	$t \geq 0$
(4)	$u(x, 0) = f(x)$	$0 \leq x \leq a$
(5)	$u_t(x, 0) = 0$	$0 \leq x \leq a,$

where f is continuous, and such that $f(0) = f(a) = 0$.

Following d'Alembert and Euler, we shall make some general observations that will lead to a unique determination of the particular solution of this problem from the general solution in Theorem 5.1. Note, however, that F and G cannot be determined uniquely because, if C is an arbitrary constant, $F + C$ and $G - C$ lead to the same solution as F and G. For this reason there is no loss of generality in imposing the additional condition

$$(6) \qquad\qquad\qquad\qquad F(0) = 0,$$

and we shall do so. Furthermore, this problem involves only those values of x and t such that $0 \le x \le a$ and $t \ge 0$. Consequently, only the values of F on $[0, \infty)$ and the values of G on $(-\infty, a]$ are relevant in the solution. It will then be these values that will be uniquely determined by our boundary value problem.

From the initial conditions (4) and (5) and the equation in Theorem 5.1 we obtain

$$(7) \qquad\qquad F(x) + G(x) = f(x), \qquad 0 \le x \le a$$

$$(8) \qquad\qquad cF'(x) - cG'(x) = 0, \qquad 0 \le x \le a.$$

The first of these gives $F(0) + G(0) = f(0) = 0$, and then (6) implies that $G(0) = 0$. Then, integrating (8) from 0 to x yields

$$(9) \qquad\qquad\qquad F(x) - G(x) = 0, \qquad 0 \le x \le a,$$

and solving (7) and (9) we arrive at

$$F(x) = G(x) = \tfrac{1}{2} f(x), \qquad 0 \le x \le a.$$

This determines F and G uniquely on $[0, a]$.

To find their values beyond this interval, we turn to the endpoint conditions. From $u(0, t) = 0$ for $t \ge 0$ we obtain $F(ct) + G(-ct) = 0$ for $t \ge 0$, that is

$$(10) \qquad\qquad\qquad F(x) + G(-x) = 0, \qquad x \ge 0.$$

Note that (9) and (10) imply that

$$G(-x) = -G(x), \qquad 0 \le x \le a,$$

so that G is an odd function on the interval $[-a, a]$.

Next, from $u(a, t) = 0$ for $t \ge 0$ we obtain $F(a + ct) + G(a - ct) = 0$ for $t \ge 0$, that is

$$(11) \qquad\qquad F(a + x) + G(a - x) = 0, \qquad x \ge 0,$$

and then (11) and (10) imply that

$$G(a - x) = -F(a + x) = G(-a - x), \qquad x \ge 0,$$

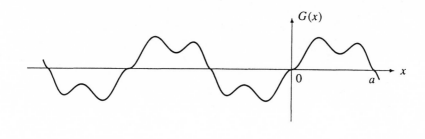

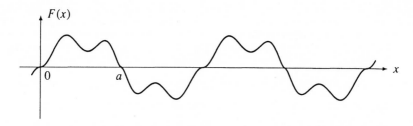

Figure 5.4

which means that G is the restriction to $(-\infty, a]$ of a periodic function of period $2a$. Its graph is as shown in the top part of Figure 5.4. Finally, (10) implies that $F(x) = -G(-x)$ for $x \geq 0$, so that the graph of F on $[0, \infty)$, represented in the bottom part of Figure 5.4, is obtained from the restriction of the graph of G to $(-\infty, 0]$ by symmetry with respect to the origin.

To sum up, if $\tilde{f}$ denotes the odd function on $\mathbb{R}$ of period $2a$ that coincides with f on $[0, a]$, we have found that

$$F(x) = \tfrac{1}{2}\tilde{f}(x), \qquad x \geq 0$$
$$G(x) = \tfrac{1}{2}\tilde{f}(x), \qquad x \leq a,$$

and our boundary value problem has the unique solution

$$u(x, t) = \tfrac{1}{2}[\tilde{f}(x + ct) + \tilde{f}(x - ct)].$$

Note that $\tilde{f}$ must be of class C^2 because F and G are, and then, since $\tilde{f}$ is odd, differentiating $\tilde{f}(x) = -\tilde{f}(-x)$ twice yields $\tilde{f}''(x) = -\tilde{f}''(-x)$ for all x in $\mathbb{R}$. In particular, $\tilde{f}''(0) = -\tilde{f}''(-0)$ and $\tilde{f}''(a) = -\tilde{f}''(-a) = -\tilde{f}''(a)$, the last equality being a consequence of the fact that $\tilde{f}$ is periodic of period $2a$. We conclude that

$$\tilde{f}''(0) = \tilde{f}''(a) = 0.$$

It is very easy to see that the solution obtained above depends continuously on the initial values. In fact, if v is the solution such that $v(x, 0) = g(x)$ and if M_{f-g} denotes the maximum value of $|f - g|$ on $[0, a]$, then M_{f-g} is also the maximum value of $|\tilde{f} - \tilde{g}|$ on $\mathbb{R}$ and then

$$|u(x, t) - v(x, t)| \leq \tfrac{1}{2} [M_{f-g} + M_{f-g}] = M_{f-g}.$$

This completes the proof of the following theorem.

Theorem 5.2 *The boundary value problem*

$$
\begin{aligned}
u_{tt} &= c^2 u_{xx} &&\text{in } D \\
u(0, t) &= 0 &&t \geq 0 \\
u(a, t) &= 0 &&t \geq 0 \\
u(x, 0) &= f(x) &&0 \leq x \leq a \\
u_t(x, 0) &= 0 &&0 \leq x \leq a
\end{aligned}
$$

where f is of class C^2 on $[a, b]$ and such that $f(0) = f(a) = f''(0) = f''(a) = 0$, is correctly posed in the sense of Hadamard. Then, if $\tilde{f}$ is the odd periodic extension of f to $\mathbb{R}$ of period $2a$, the solution is

$$u(x, t) = \tfrac{1}{2} [\tilde{f}(x + ct) + \tilde{f}(x - ct)].$$

Note that all the listed conditions on f, which may appear excessive at first glance, are actually necessary for the solution to exist as a function of class C^2. D'Alembert himself was not too precise about these conditions in his own statement of the theorem, but it seems that he thought of f as having derivatives of all orders.

To interpret d'Alembert's result physically consider the case, illustrated in Figure 5.5, in which only a small portion of the string about a point $x_0 > a/2$ is initially displaced in the shape of a hump symmetric about its center. As the string is released and according to the interpretation of $\tilde{f}(x + ct)$ and $\tilde{f}(x - ct)$ as traveling waves, the originally displaced portion splits into the sum of two waves with an amplitude one-half of the original amplitude, one moving to the left and the other to the right, with the same speed c. An observer located at a point $p > x_0$ will see the traveling wave to the right centered at p at time $t = (p - x_0)/c$. Since $\tilde{f}$ is odd and periodic of period $2a$, its graph initially exhibits another hump at $-x_0$ and another at $-x_0 + 2a$, both identical to the one at x_0 but in the opposite vertical direction. These are, of course, virtual humps. They have no physical existence because they are located beyond the endpoints of the string, and for this reason they are represented by a dashed curve. Mathematically, however, each of them splits at $t = 0$ into two traveling waves, one to the right and one to the left, both moving with speed c. Now, let time pass and the observer at p keep alert. Both the real hump from x_0 and the virtual one from $-x_0 + 2a$ reach the point $x = a$ at time $t = (a - x_0)/c$. At this time they annihilate each other, both physically and mathematically, but moments later, at time $t = (2a - p - x_0)/c$, the observer at p sees the inverted hump pass toward the left with speed c. Now this is not a virtual hump but a real hump. What has happened physically is that the original wave traveling to the right, the one that passed by p at time $t = (p - x_0)/c$, unable to spend its energy on arrival at the endpoint $x = a$, has been

Figure 5.5

reflected and is now traveling to the left. This is not where the matter rests, however. At time $t = x_0/c$ the original wave traveling to the left is reflected at the endpoint $x = 0$ and starts moving to the right. The two inverted humps between 0 and a will eventually coincide and double up at $x = a - x_0$ at time $t = a/c$. Were the observer at p patient enough, an unending number of reflected waves will pass by this spot in both directions. The first of these will be the original wave traveling to the left after reflection at $x = 0$.

It is possible to obtain a graphical representation of the motion of these reflected waves in space-time. In fact, the original wave traveling to the right with speed c is centered at a point x at time $t = (x - x_0)/c$, and then $x - ct = x_0$. Hence, the points (x, t) in space-time that represent the position of the wave top are those on the line $x - ct = x_0$. Similarly, those for the wave traveling to the left are on the line $x + ct = x_0$. When the wave moving to the right is reflected at $x = a$ at time $t = (a - x_0)/c$, it starts moving to the left, and the points (x, t) representing the position of the wave are now on a line of the form $x + ct =$ constant. The precise value of the constant is found from the fact that $x = a$ when $t = (a - x_0)/c$, and it is easy to see that the equation of this line is $x + ct = 2a - x_0$. A similar argument applies to any further reflections of these traveling waves, a few of which are represented in Figure 5.6. These straight line

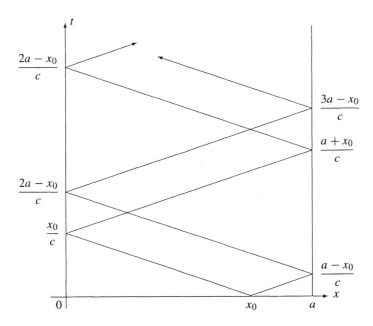

Figure 5.6

segments are called *characteristics* or *wave fronts*.

Returning to Theorem 5.2, it is quite obvious that the differentiability conditions on f are very heavy. Shortly after the publication of d'Alembert's papers containing this solution, Leonhardt Euler came up with one of his own in which he mostly repeated what d'Alembert had already said except for one important point. Euler rejected the differentiability requirements on f and accepted functions with discontinuous derivatives, as the one represented in Figure 5.7, as the initial displacement of the string. This type

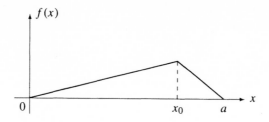

Figure 5.7

of function is a better model for a plucked string than a twice differentiable function, and Euler was more interested in a good physical model than in a carefully formulated theorem. D'Alembert would not accept such functions, and this disagreement marked the beginning of a lively mathematical argument between the two men. For the time being let us simply say that the conditions stated in Theorem 5.2 are necessary for u to be of class C^2 and to satisfy the wave equation in D. However, if f fails to be of class C^2 at a single point x_0, then u fails to be of class C^2 at any (x, t) in D such that

$$x + ct = \begin{cases} x_0 + 2ka, & \text{if } k = 0, 1, 2, \ldots \\ -x_0 + 2ka, & \text{if } k = 1, 2, \ldots \end{cases}$$

or $x - ct = \pm x_0 - 2ka, k = 0, 1, 2, \ldots$. Thus, if the string is initially displaced as in Figure 5.7, u is of class C^2 in D except on the characteristics or wave fronts associated with x_0. It can be regarded as a kind of generalized solution, the only kind possible when f is not as in Theorem 5.2.

§5.4   A Struck String

So far we have considered only the case of a plucked string, that is, a string that is initially displaced and then released. This may be the way to play the guitar or the harp, but it is no way to play the piano. In this latter case the strings are struck, and this means that an initial velocity is prescribed instead of or in addition to an initial displacement. The solution of this boundary value problem was found in 1760–1761 by Joseph Louis de Lagrange Tournier (1736–1813), then a professor of mathematics at the artillery school in Turin, where he was born of French-Italian ancestry.

We shall not follow Lagrange's original method, which was not entirely correct, but rather use the general solution as in §5.3 for the plucked string, and then we can proceed very quickly. As before, we can assume without loss of generality that $F(0) = 0$. If

JOSEPH LOUIS LAGRANGE
From W. B. Ford and C. Ammerman, *Second Course
in Algebra*, Macmillan, New York, 1926.

the initial displacement and velocity are represented by given functions f and g with
$f(0) = f(a) = g(0) = g(a) = 0$, then, putting $t = 0$ in the expression for $u(x, t)$ in
Theorem 5.1,

$$F(x) + G(x) = f(x), \qquad 0 \le x \le a,$$

so that $G(0) = f(0) - F(0) = 0$, and then $u_t(x, 0) = g(x)$ implies that $cF'(x) - cG'(x) = g(x)$. Integrating the last equation,

$$F(x) - G(x) = \frac{1}{c} \int_0^x g, \qquad 0 \le x \le a.$$

We conclude that

(12) $$F(x) = \frac{1}{2} f(x) + \frac{1}{2c} \int_0^x g, \qquad 0 \le x \le a$$

(13) $$G(x) = \frac{1}{2} f(x) - \frac{1}{2c} \int_0^x g, \qquad 0 \le x \le a.$$

An argument identical to that in §5.3 shows that the endpoint conditions imply that

(10) $$F(x) + G(-x) = 0, \qquad x \ge 0$$

(11) $$F(x + a) + G(a - x) = 0, \qquad x \ge 0$$

According to (10), $G(x) = -F(-x)$ for $x \le 0$, and this means that G can be constructed on $[-a, 0]$ as the odd extension of the restriction of F to $[0, a]$, as represented in

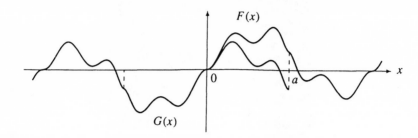

Figure 5.8

Figure 5.8—note that $F(a)$ need not be zero because the integral of g over $(0, a)$ need not be zero. More precisely, using (10) and (12) we obtain

$$(14) \qquad G(x) = -F(-x) = -\frac{1}{2}f(-x) - \frac{1}{2c}\int_0^{-x} g, \qquad -a \le x \le 0.$$

Now, let $\tilde{f}$ and $\tilde{g}$ be the extensions of f and g to $\mathbb{R}$ as odd periodic functions of period $2a$. If $x \le 0$, we have $\tilde{f}(x) = -f(-x)$ and $\tilde{g}(x) = -g(-x)$, and then

$$\int_0^{-x} g(\sigma)\,d\sigma = -\int_0^{-x} \tilde{g}(-\sigma)\,d\sigma = \int_0^x \tilde{g}(s)\,ds,$$

so that (13) and (14) yield

$$(15) \qquad G(x) = \frac{1}{2}\tilde{f}(x) - \frac{1}{2c}\int_0^x \tilde{g}, \qquad -a \le x \le a.$$

Finally, (11) and (10) imply that

$$G(a - x) = -F(a + x) = G(-a - x), \qquad x \ge 0,$$

so that G is the restriction to $(-\infty, a]$ of a periodic function of period $2a$. Then it follows from (10) that the graph of F on $[0, \infty)$ is obtained from the graph of G on $(-\infty, 0]$ by symmetry with respect to the origin, so that F is the restriction of a periodic function of period $2a$ to $[0, \infty)$, as shown in Figure 5.8. Finally, $\int_0^x \tilde{g}$ is also a periodic function of period $2a$ because $\tilde{g}$ is odd, and then (12), (15), and the preceding discussion allow us to conclude that

$$F(x) = \frac{1}{2}\tilde{f}(x) + \frac{1}{2c}\int_0^x \tilde{g}, \qquad x \ge 0$$

$$G(x) = \frac{1}{2}\tilde{f}(x) - \frac{1}{2c}\int_0^x \tilde{g}, \qquad x \le a.$$

Here $\tilde{f}$ must be of class C^2 and $\tilde{g}$ of class C^1 so that F and G are of class C^2, and then, as in the proof of Theorem 5.2, $f''(0) = f''(a) = 0$. Since F and G have been determined uniquely on $[0, \infty)$ and $(-\infty, a]$, respectively, the unique solution of our boundary value problem is

$$u(x, t) = \frac{1}{2}[\tilde{f}(x + ct) + \tilde{f}(x - ct)] + \frac{1}{2c}\int_{x-ct}^{x+ct} \tilde{g}.$$

This solution depends continuously on initial values, for, if u_1 and u_2 are solutions corresponding to initial conditions f_1, g_1 and f_2, g_2, respectively, and if $M_{f_1-f_2}$ and $M_{g_1-g_2}$ denote the maximum values of $|f_1 - f_2|$ and $|g_1 - g_2|$ on $[0, a]$, the fact that $\tilde{g}_1 - \tilde{g}_2$ is an odd function implies that

$$\left| \int_{x-ct}^{x+ct} (\tilde{g}_1 - \tilde{g}_2) \right| \le \int_{-a}^{a} |\tilde{g}_1 - \tilde{g}_2| \le 2aM_{g_1-g_2},$$

and then

$$|u_1(x, t) - u_2(x, t)| \le M_{f_1-f_2} + \frac{a}{c}M_{g_1-g_2}.$$

Putting $f_1 = f_2$ and $g_1 = g_2$ proves uniqueness. We have therefore completed the proof of the following result.

Theorem 5.3 *The boundary value problem*

$$\begin{aligned}
u_{tt} &= c^2 u_{xx} & &\text{in } D \\
u(0, t) &= 0 & &t \ge 0 \\
u(a, t) &= 0 & &t \ge 0 \\
u(x, 0) &= f(x) & &0 \le x \le a \\
u_t(x, 0) &= g(x) & &0 \le x \le a
\end{aligned}$$

where f is of class C^2, g is of class C^1, and $f(0) = f(a) = g(0) = g(a) = f''(0) = f''(a) = 0$, is well posed in the sense of Hadamard and, if $\tilde{f}$ and $\tilde{g}$ are the odd periodic extensions of f and g to $\mathbb{R}$ of period $2a$, then its solution is

$$(16) \qquad u(x, t) = \frac{1}{2}[\tilde{f}(x + ct) + \tilde{f}(x - ct)] + \frac{1}{2c}\int_{x-ct}^{x+ct} \tilde{g}.$$

It is possible to study the motion of the string by the method used in Figure 5.5, but the analysis of the term involving the initial velocity is not as straightforward as that of the term involving only the initial displacement. The traveling waves moving to the right and left with speed c are replaced by spreading waves that leave a residual effect (see Exercise 4.10).

§5.5 Bernoulli's Solution

The solution that we have presented gives a complete mathematical description of the motion of the string, in that it specifies the position of each of its points at each time.

Mathematically that is all very well, but where is the musical description of the phenomenon? In fact, the given solution does not even address the question of vibrations, for no study has been made of its possible periodicity in the variable t (except for a passing hint in Exercise 3.1). It was Euler, in his 1748 paper, mentioned above, who brought up the fact that the motion of the string is periodic in time and made up of many individual vibrations. Upon reading d'Alembert's and Euler's papers, Daniel Bernoulli (1700–1782), of Basel, decided to publish his own ideas on the subject, which he did

DANIEL BERNOULLI *circa* 1750
Portrait by Grooth.
Reprinted from *Die Werke von Daniel Bernoulli, Band 2: Analysis Wahrscheinlichkeitsrechnung*, in *Die Gesammelten Werke der Mathematiker und Physiker der Familie Bernoulli*, ed. by David Speiser.
By permission of Birkhäuser-Verlag, Basel.

in 1753. Perhaps there was an element of irritation in the fact that Euler now stated what Bernoulli had known for some time. In a 1740 paper—that did not appear in print until 1750—Bernoulli had already stated that musical modes can coexist; that is, that the shape of the string at a given instant is the superposition of individual vibrations. Now, after having a bit of fun criticizing d'Alembert and Euler—he referred to the former as a great mathematician *in abstractis*—he asserted that the displacement of the string at any given time can be represented by an infinite series of sines. In particular, for $t = 0$,

$$(17) \qquad \sum_{n=1}^{\infty} c_n \sin \frac{n\pi}{a} x.$$

Bernoulli based his assertion on physical considerations alone. He provided no mathematical reasons whatsoever to back it up. Euler pounced on it immediately, the very same year, refusing to accept it. Harping on his earlier idea that the function f representing the initial displacement need not be differentiable at all points, he found it impossible to believe that the sum of a series like (17) could equal $f(x)$, given that the sine functions in the series are all differentiable everywhere. D'Alembert published a similar attack on Bernoulli's paper in 1757, but Bernoulli did not surrender his position for, he said, he had infinitely many coefficients c_n to choose to make the equality true. In 1759 Lagrange decided to put in his ten cents worth in a paper that contained a certain amount of nonsense, in which he sided with Euler about the initial curves that should be allowed. All this created a heated controversy that raged through the 1770s, without any of the participants giving an inch to the others' point of view. It was later revived through Fourier's researches on heat and, as we well know, eventually settled once and for all: the sum of an infinite series of sines can be a function that is not differentiable at all points.

By now we know not only this, but also that this type of series solution is likely to result from the application of the method of separation of variables. Although the idea of separation of variables can be traced back to Daniel Bernoulli's investigation of the hanging chain problem, the method as we know it today was introduced by d'Alembert in a 1750 paper, in which he proceeded exactly as follows. If the wave equation has a solution of the form

$$u(x, t) = X(x)T(t),$$

then

$$XT'' = c^2 X'' T.$$

Dividing by XT and equating each side to a constant $-\lambda$, we obtain

$$T'' + \lambda T = 0 \qquad \text{and} \qquad c^2 X'' + \lambda X = 0.$$

At this point d'Alembert limited himself to conclude that the endpoint conditions imply that X is a sine wave, and so is T because u has to be periodic.

We shall be more precise than either Bernoulli or d'Alembert and arrive at a series expansion for $u(x, t)$. The Sturm-Liouville problem $c^2 X'' + \lambda X = 0$, $X(0) = X(a) = 0$ on $[0, a]$ has eigenvalues and eigenfunctions

$$\lambda_n = \frac{n^2 \pi^2 c^2}{a^2} \qquad \text{and} \qquad X_n(x) = \sin \frac{n\pi}{a} x.$$

Then, the equation $T'' + \lambda_n T = 0$ has the general solution

$$T_n(t) = A_n \cos \frac{n\pi c}{a} t + B_n \sin \frac{n\pi c}{a} t,$$

and in order to satisfy the initial conditions

$$u(x, 0) = f(x) \qquad \text{and} \qquad u_t(x, 0) = g(x)$$

we try a solution in the form of an infinite series

(18) $$u(x, t) = \sum_{n=1}^{\infty} \left(A_n \cos \frac{n\pi c}{a} t + B_n \sin \frac{n\pi c}{a} t \right) \sin \frac{n\pi}{a} x.$$

If f and g have convergent Fourier sine series on $(0, a)$,

$$f(x) = \sum_{n=1}^{\infty} a_n \sin \frac{n\pi}{a} x$$

$$g(x) = \sum_{n=1}^{\infty} b_n \sin \frac{n\pi}{a} x,$$

and if the series in (18) defining u converges, putting $t = 0$ gives $A_n = a_n$. If we differentiate (18) term by term and put $t = 0$, we obtain

$$B_n \frac{n\pi c}{a} = b_n.$$

Thus, our proposed solution is

$$u(x, t) = \sum_{n=1}^{\infty} \left(a_n \cos \frac{n\pi c}{a} t + \frac{a b_n}{n\pi c} \sin \frac{n\pi c}{a} t \right) \sin \frac{n\pi}{a} x.$$

Of course, it must be verified that this is indeed a solution of the wave equation by direct substitution. Unfortunately, this may not be feasible because the last series need not be twice differentiable term by term.

Example 5.2 Consider the vibrations of a string tied down at $x = 0$ and $x = 2$, whose initial displacement and velocity are given by $u(x, 0) = f(x)$, where f is defined by

$$f(x) = \begin{cases} Mx & \text{if } 0 \le x \le 1 \\ M(2 - x) & \text{if } 1 < x \le 2, \end{cases}$$

with M a constant, and $u_t(x, 0) = 0$ for $0 \le x \le 2$.

The coefficients of the Fourier sine series of f on $(0, 2)$ are

$$a_n = \int_0^1 Mx \sin \frac{n\pi}{2} x \, dx + \int_1^2 M(2 - x) \sin \frac{n\pi}{2} x \, dx = \frac{8M}{n^2\pi^2} \sin \frac{n\pi}{2}.$$

Then, since $b_n = 0$ for all n, our proposed solution is

$$\frac{8M}{\pi^2} \sum_{n=1}^{\infty} \frac{1}{n^2} \sin \frac{n\pi}{2} \cos \frac{n\pi c}{2} t \sin \frac{n\pi}{2} x.$$

Differentiating this series term by term twice with respect to t we obtain

$$-\frac{8M}{\pi^2} \sum_{n=1}^{\infty} \frac{\pi^2 c^2}{4} \sin \frac{n\pi}{2} \cos \frac{n\pi c}{2} t \sin \frac{n\pi}{2} x.$$

For $x = 1$ each term contains the factor

$$\sin^2 \frac{n\pi}{2} = \begin{cases} 1 & \text{if } n \text{ is odd} \\ 0 & \text{if } n \text{ is even,} \end{cases}$$

and the differentiated series becomes

$$-2Mc^2 \sum_{n=1}^{\infty} \cos \frac{(2n-1)\pi c}{2} t,$$

which diverges because the general term does not approach zero as $n \to \infty$. The same thing happens if the two differentiations are with respect to x.

To sum up, in this type of problem separation of variables leads to a proposed solution that cannot be verified directly by differentiation. However, we already know its unique solution from Theorem 5.3, and, if we want to write it as an infinite series, we can simply take the Fourier sine series of f and g to equation (16) in Theorem 5.3. Recalling that a Fourier series can integrated term by term (Theorem 2.8), its right-hand side becomes

$$\frac{1}{2} \sum_{n=1}^{\infty} a_n \left[\sin \frac{n\pi}{a}(x+ct) + \sin \frac{n\pi}{a}(x-ct) \right] + \frac{1}{2c} \int_{x-ct}^{x+ct} \left[\sum_{n=1}^{\infty} b_n \sin \frac{n\pi}{a} s \right] ds$$

$$= \sum_{n=1}^{\infty} a_n \sin \frac{n\pi}{a} x \cos \frac{n\pi c}{a} t - \frac{1}{2c} \sum_{n=1}^{\infty} \frac{ab_n}{n\pi} \left[\cos \frac{n\pi}{a}(x+ct) - \cos \frac{n\pi}{a}(x-ct) \right]$$

$$= \sum_{n=1}^{\infty} a_n \sin \frac{n\pi}{a} x \cos \frac{n\pi c}{a} t + \sum_{n=1}^{\infty} \frac{ab_n}{n\pi c} \sin \frac{n\pi}{a} x \sin \frac{n\pi c}{a} t$$

and then

$$u(x, t) = \sum_{n=1}^{\infty} \left(a_n \cos \frac{n\pi c}{a} t + \frac{ab_n}{n\pi c} \sin \frac{n\pi c}{a} t \right) \sin \frac{n\pi}{a} x.$$

This is, of course, the unique solution of the problem as proved by Theorem 5.3, and it is the one found above by separation of variables.

The last expression can be simplified if we now define

$$c_n = \sqrt{a_n^2 + \frac{a^2 b_n^2}{n^2 \pi^2 c^2}},$$

and then θ_n to be uniquely determined in $[0, 2\pi]$ by the equations

$$c_n \cos \frac{n\pi c}{a} \theta_n = \frac{ab_n}{n\pi c}$$

$$c_n \sin \frac{n\pi c}{a} \theta_n = -a_n.$$

The solution takes the final form

$$u(x, t) = \sum_{n=1}^{\infty} c_n \sin \frac{n\pi}{a} x \sin \frac{n\pi c}{a} (t - \theta_n).$$

In this form, for we insist that it is the same solution given by Theorem 5.3, it is known as *Bernoulli's solution*. The advantage of this form is that it is readily amenable to physical interpretation. We see that the vibration of the string is the superposition of infinitely many vibrations. The amplitude of each of these at a point x in $(0, a)$ is

$$c_n \sin \frac{n\pi}{a} x,$$

and its frequency, that is, the number of vibrations per second, is

$$f_n = \frac{nc}{2a} = \frac{n}{2a} \sqrt{\frac{\tau}{m}},$$

where τ is the tensile force on the string and m its mass per unit length. f_1 is called the *fundamental frequency*—which has the value already anticipated by Taylor and Bernoulli—and the remaining f_n the *overtones*. Each frequency increases as τ increases and decreases as the length of the string a or the density of the material increase. The amplitude of each vibration and its *phase lag* θ_n depend on the length of the string and, through the Fourier coefficients, on the initial displacement and velocity.

Figure 5.9 shows the individual vibrations, also called *harmonics*, for $n = 1, 2, 3, 4$.

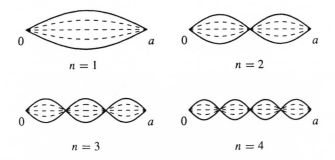

$n = 1$ $n = 2$

$n = 3$ $n = 4$

Figure 5.9

For each harmonic some points on the string remain fixed during its motion. For instance, $x = 0, a/3, 2a/3, a$ for $n = 3$. These are called *nodes*. Note that the displacement of the string has opposite signs on both sides of a node. The vibrations in Figure 5.9 are known as *standing waves*, as they do not travel to the left or right.

§5.6   Time-Independent Problems

Assume now that the mass per unit length, m, and the tensile force, τ, on the string are not constant but, instead, positive, continuous functions of x and that τ is of class C^1. A similar derivation of the equation of motion leads to

$$mu_{tt} = (\tau u_x)_x + q$$

(Exercise 2.1), where q is a continuous function of x that represents external forces. For instance, $q = -mg$ if the only such force is due to gravity.

We shall impose boundary conditions of the form

$$
\begin{array}{ll}
a_1 u(0, t) - b_1 u_x(0, t) = c_1 & t \geq 0 \\
a_2 u(a, t) + b_2 u_x(a, t) = c_2 & t \geq 0 \\
u(x, 0) = f(x) & 0 \leq x \leq a \\
u_t(x, 0) = g(x) & 0 \leq x \leq a,
\end{array}
$$

where a_1, a_2, b_1, and b_2 are nonnegative constants, $a_1 + b_1 > 0$, $a_2 + b_2 > 0$, and $a_1 + a_2 > 0$ (Exercise 6.3 shows that additional conditions may have to imposed on the function q for a solution method similar to the one suggested below to be applicable in the case $a_1 = a_2 = 0$). In addition, f and g are continuous functions and f has a piecewise continuous derivative and it satisfies the endpoint conditions $a_1 f(0) - b_1 f'(0) = c_1$ and $a_2 f(a) - b_2 f'(a) = c_2$.

The endpoint conditions can be physically realized. For instance, if $b_1 = 0$, the string is simply attached at $x = 0$ at a height $u(0, t) = c_1/a_1$. In the general case, the string can be attached to a disc of negligible mass that is allowed to move in a vertical track at the end on an elastic spring of unstretched length l, as shown in Figure 5.10. By Hooke's law of elasticity, the restoring force on the moving disc is proportional to

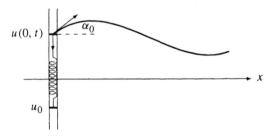

Figure 5.10

$u(0, t) - u_0 - l$, while the vertical component of the tensile force from the string is proportional to $\sin \alpha_0$ and, for small α_0, to $\tan \alpha_0 = u_x(0, t)$. The balance of forces gives

$$k_1[u(0, t) - u_0 - l] = k_2 u_x(0, t),$$

where k_1 and k_2 are positive constants—they cannot have opposite signs or the forces would be in the same direction rather than in balance—that is,

$$k_1 u(0, t) - k_2 u_x(0, t) = c,$$

where $c = k_1(u_0 + l)$ is another constant. This is the general form of the left endpoint condition. Other external forces can be incorporated into c. When the spring is removed, $k_1 = 0$, and this condition becomes $u_x(0, t) = 0$. This simply means that the vertical component of the tensile force at $x = 0$ is zero, as it must be—in the absence of mass, spring, and friction—to prevent an infinite vertical acceleration at this point.

The first thing to do in solving this problem is to deal with the nonlinear equation and endpoint conditions. This is done, as in the case of heat propagation, by finding a solution U that is independent of time and satisfies the endpoint conditions. It is found as in §3.8 but with m in place of c and τ in place of κ. Then, if we define $v = u - U$, where u is the solution of the given boundary value problem, v is a solution of

(19)	$m v_{tt} = (\tau v_x)_x$	$0 < x < a, \quad t > 0$
(20)	$a_1 v(0, t) - b_1 v_x(0, t) = 0$	$t \geq 0$
(21)	$a_2 v(a, t) + b_2 v_x(a, t) = 0$	$t \geq 0$
(22)	$v(x, 0) = f(x) - U(x)$	$0 \leq x \leq a$
(23)	$v_t(x, 0) = g(x)$	$0 \leq x \leq a.$

Separation of variables leads to the differential equation $T'' + \lambda T = 0$ and the regular Sturm-Liouville problem

(24)	$(\tau X')' + \lambda m X = 0$
(25)	$a_1 X(0) - b_1 X'(0) = 0$
(26)	$a_2 X(a) + b_2 X'(a) = 0$

that is well known to us from Chapter 4. By Theorems 4.3 and 4.4, its eigenvalues form a sequence $\{\lambda_n\}$ of positive numbers. If X_n are corresponding eigenfunctions, we propose a series solution of the form

$$v(x, t) = \sum_{n=1}^{\infty} T_n(t) X_n(x),$$

where $T_n(t)$ is a solution of $T'' + \lambda_n T = 0$. Unfortunately, the sum of the series on the right may not be twice differentiable term by term, as shown by the particular case of Example 5.2. This means that we are unable to prove that the proposed v is a solution in the general case. However, we can prove the next best thing: that if our boundary value problem has a solution at all then it has to be of this form.

Theorem 5.4 *Let D and $\bar{D}$ be as usual and let $v : \bar{D} \to \mathbb{R}$ be a continuous solution of the boundary value problem* (19)–(23) *that is of class C^2 in D. Then*

$$(27) \qquad v(x, t) = \sum_{n=1}^{\infty} \left[A_n \cos \sqrt{\lambda_n}\, t + \frac{B_n}{\sqrt{\lambda_n}} \sin \sqrt{\lambda_n}\, t \right] X_n(x),$$

where λ_n and X_n are the nth eigenvalue and eigenfunction of the Sturm-Liouville problem (24)–(26), and A_n and B_n are the Fourier coefficients of $f - U$ and g, respectively, with respect to the X_n.

Proof. By Theorem 4.6 and for each fixed t, v can be expanded in a convergent Fourier series with respect to the X_n

$$(28) \qquad\qquad v(x, t) = \sum_{n=1}^{\infty} T_n(t) X_n(x),$$

and it remains to find the Fourier coefficients $T_n(t)$. We know that this series is independent of the choice of eigenfunctions, and then we assume for simplicity that these are normalized (refer to §4.4). Then, since m takes the place of the function r of §4.4,

$$(29) \qquad\qquad T_n(t) = \int_0^a m(x) v(x, t) X_n(x)\, dx.$$

Since $(\tau X_n')' + \lambda_n m X_n = 0$, we obtain

$$T_n(t) = -\frac{1}{\lambda_n} \int_0^a [\tau(x) X_n'(x)]' v(x, t)\, dx$$

$$= -\frac{1}{\lambda_n} \left[\tau(x) X_n'(x) v(x, t) \Big|_0^a - \int_0^a \tau(x) X_n'(x) v_x(x, t)\, dx \right]$$

$$= -\frac{1}{\lambda_n} \left[\tau(x) X_n'(x) v(x, t) \Big|_0^a - \tau(x) X_n(x) v_x(x, t) \Big|_0^a \right.$$

$$\left. + \int_0^a [\tau(x) v_x(x, t)]_x X_n(x)\, dx \right]$$

$$= -\frac{1}{\lambda_n} \left[\tau(a)[X_n'(a) v(a, t) - X_n(a) v_x(a, t)] \right.$$

$$\left. - \tau(0)[X_n'(0) v(0, t) - X_n(0) v_x(0, t)] + \int_0^a [\tau(x) v_x(x, t)]_x X_n(x)\, dx \right].$$

Now, the system of equations

$$(20) \qquad\qquad a_1 v(0, t) - b_1 v_x(0, t) = 0$$

$$(25) \qquad\qquad a_1 X_n(0) - b_1 X_n'(0) = 0$$

has a nontrivial solution for a_1 and b_1 only if its determinant is zero, that is, only if $X_n'(0) v(0, t) - X_n(0) v_x(0, t) = 0$. Similarly, (21) and (26) imply that $X_n'(a) v(a, t) - X_n(a) v_x(a, t) = 0$, and then

$$(30)\ \ T_n(t) = -\frac{1}{\lambda_n} \int_0^a [\tau(x) v_x(x, t)]_x X_n(x)\, dx = -\frac{1}{\lambda_n} \int_0^a m v_{tt}(x, t) X_n(x)\, dx.$$

Since the integrand in (29) is continuous and has a continuous second partial derivative with respect to t, we can differentiate twice under the integral sign to obtain

$$T_n''(t) = \int_0^a m v_{tt}(x, t) X_n(x) \, dx.$$

This and (30) imply that $T_n'' + \lambda_n T_n = 0$, whose general solution is of the form

$$T_n(t) = A_n \cos \sqrt{\lambda_n} \, t + \frac{B_n}{\sqrt{\lambda_n}} \sin \sqrt{\lambda_n} \, t,$$

where A_n and B_n are arbitrary constants, and taking this to (28) proves that $v(x, t)$ is of the form given by (27).

The assertion that A_n is the nth Fourier coefficient of $f - U$ follows by letting $t \to 0$ in (29). In fact, since T_n and v are continuous,

$$A_n = \lim_{t \to 0} T_n(t) = \lim_{t \to 0} \int_0^a m(x) v(x, t) X_n(x) \, dx = \int_0^a m(x)[f(x) - U(x)] X_n(x) \, dx.$$

Similarly,

$$B_n = \lim_{t \to 0} T_n'(t) = \lim_{t \to 0} \int_0^a m(x) v_t(x, t) X_n(x) \, dx = \int_0^a m(x) g(x) X_n(x) \, dx,$$

Q.E.D.

Note that Theorem 5.4 implies uniqueness of solutions. It also vindicates the method of separation of variables, because if a solution exists at all then it can be found by this method and the associated Sturm-Liouville problem, and is given by (27). This theorem does not show existence of solutions, but in practical applications it is frequently possible to conclude that a solution exists from physical considerations alone.

The right-hand side of (27) may actually fail to be differentiable everywhere, as in the particular case of the wave equation when f is not of class C^2. We can then refer to the sum of this series as a *generalized* or *weak* solution, the best we can hope to find in such a case because Theorem 5.4 tells us that if a solution of class C^2 existed then it would be given by (27). The question is whether such a generalized solution is of any use in applications to physics and engineering, and the answer is yes because we shall see in §6.5, after we introduce a new kind of convergence that is very adequate for work with Fourier series, that it is an approximation to the actual solution.

Finally, it is clear from (27) that $v(x, t)$ does not approach zero as $t \to \infty$, and then that $u(x, t)$ does not approach $U(x)$ as $t \to \infty$. For this reason, we do not refer to $U(x)$ and $v(x, t)$ as the steady state and transient solutions in this case.

Example 5.3 A string is initially at rest, stretched in the shape of a straight line between two points at heights 1 and $1 + h$ meters, respectively, that are 1 meter apart horizontally. Find the motion of the string if its endpoints remain fixed at all times and if it is located in a frictionless vacuum and subject only to the force of gravity.

If g is the acceleration of gravity, the motion of the string is governed by the partial differential equation $u_{tt} = c^2 u_{xx} - g$, as shown in §5.2. Then we are dealing with the boundary value problem

$$
\begin{array}{ll}
u_{tt} = c^2 u_{xx} - g & 0 < x < 1, \quad t > 0 \\
u(0, t) = 1 & t \geq 0 \\
u(1, t) = 1 + h & t \geq 0 \\
u(x, 0) = 1 + hx & 0 \leq x \leq 1 \\
u_t(x, 0) = 0 & 0 \leq x \leq 1.
\end{array}
$$

In this case, a solution independent of time must satisfy $0 = c^2 U'' - g$, and then

$$
U(x) = \frac{g x^2}{2c^2} + Ax + B,
$$

where the constants of integration A and B are easily found from the endpoint conditions $U(0) = 1$ and $U(1) = 1 + h$. We obtain

$$
U(x) = \frac{g}{2c^2} x(x - 1) + hx + 1.
$$

Now, all B_n in Theorem 5.4 are zero because $u_t(x, 0) = 0$. With $\tau/m = c^2$, $a = 1$, and $b_1 = b_2 = 0$ the Sturm-Liouville problem (24)–(26) becomes $c^2 X'' + \lambda X = 0$, $X(0) = 0$, and $X(1) = 0$. It has eigenvalues $\lambda_n = n^2 \pi^2 c^2$ and eigenfunctions $X_n(x) = \sin n\pi x$, and then straightforward computations yield $\|X_n\| = \frac{1}{2}$ and

$$
A_n = -\frac{g}{c^2} \int_0^1 (x^2 - x) \sin n\pi x \, dx = \frac{2g}{n^3 \pi^3 c^2} [1 - (-1)^n].
$$

Therefore, the solution of our boundary value problem is

$$
u(x, t) = \frac{g}{2c^2} x(x - 1) + hx + 1 + \frac{2g}{\pi^3 c^2} \sum_{n=1}^{\infty} \frac{1 - (-1)^n}{n^3} \cos n\pi c t \sin n\pi x.
$$

The presence of undamped vibrations is due to the fact that we have neglected friction, but in practice these vibrations die out (see Exercises 6.5 and 6.6 for the same problem with friction).

§5.7 Time-Dependent Problems

We now consider a boundary value problem of the form

$$
\begin{array}{ll}
m u_{tt} = (\tau u_x)_x + q(x, t) & \text{in } D \\
a_1 u(0, t) - b_1 u_x(0, t) = h(t) & t \geq 0 \\
a_2 u(a, t) + b_2 u_x(a, t) = k(t) & t \geq 0 \\
u(x, 0) = f(x) & 0 \leq x \leq a \\
u_t(x, 0) = g(x) & 0 \leq x \leq a,
\end{array}
$$

where, in addition to the usual hypotheses on the constants and on the functions f and g, we assume that h and k are twice differentiable.

The solution of this problem is very similar to that of §3.11, and so we can proceed rather quickly. First we point out that

$$(31) \qquad U(x,t) = \frac{a_1 k(t) - a_2 h(t)}{(aa_1 + b_1)a_2 + a_1 b_2} x + \frac{b_1 k(t) + (aa_2 + b_2)h(t)}{(aa_1 + b_1)a_2 + a_1 b_2}$$

is a function that satisfies the endpoint conditions. Then the function $v = u - U$ is a solution of

$$
\begin{aligned}
mv_{tt} &= (\tau v_x)_x + q(x,t) + \tau' U_x - mU_{tt}(x,t) && \text{in } D \\
a_1 v(0,t) &- b_1 v_x(0,t) = 0 && t \geq 0 \\
a_2 v(a,t) &+ b_2 v_x(a,t) = 0 && t \geq 0 \\
v(x,0) &= f(x) - U(x,0) && 0 \leq x \leq a \\
v_t(x,0) &= g(x) - U_t(x,0) && 0 \leq x \leq a.
\end{aligned}
$$

If such a solution is found, then $u = U + v$ is the complete solution of the original boundary value problem.

Now let λ_n and X_n be the eigenvalues and eigenfunctions of the Sturm-Liouville problem (24)–(26). We propose to determine functions c_n such that

$$(32) \qquad v(x,t) = \sum_{n=1}^{\infty} c_n(t) X_n(x)$$

is a solution of the preceding boundary value problem. To satisfy $v(x,0) = f(x) - U(x,0)$ we must have

$$\sum_{n=1}^{\infty} c_n(0) X_n(x) = f(x) - U(x,0),$$

and then, since the weight function r of §4.4 is now m,

$$(33) \qquad c_n(0) = \frac{\displaystyle\int_0^a m[f(x) - U(x,0)]X_n(x)\,dx}{\displaystyle\int_0^a mX_n^2(x)\,dx}.$$

Similarly, to satisfy $v_t(x,0) = g(x) - U_t(x,0)$ we must have

$$\sum_{n=1}^{\infty} c_n'(0) X_n(x) = g(x) - U_t(x,0),$$

and then

$$(34) \qquad c_n'(0) = \frac{\displaystyle\int_0^a m[g(x) - U_t(x,0)]X_n(x)\,dx}{\displaystyle\int_0^a mX_n^2(x)\,dx}.$$

To determine the values of c_n for $t > 0$ so that v satisfies the desired partial differential equation we assume that (32) can be differentiated term by term as necessary, and then

$$mv_{tt} = \sum_{n=1}^{\infty} mc''X_n, \qquad \tau v_x = \sum_{n=1}^{\infty} \tau c_n X_n',$$

and, using $(\tau X_n')' + \lambda_n m X_n = 0$,

$$(\tau v_x)_x = \sum_{n=1}^{\infty} c_n (\tau X_n')' = \sum_{n=1}^{\infty} c_n (-\lambda_n m X_n).$$

Substitution into $mv_{tt} - (\tau v_x)_x = q + \tau' U_x - mU_{tt}$ requires that

$$(35) \qquad \sum_{n=1}^{\infty} (c'' + \lambda_n c_n) X_n = \frac{q + \tau' U_x}{m} - U_{tt}.$$

If we define

$$\tilde{q} = \frac{q + \tau' U_x}{m} - U_{tt}$$

and if, for each fixed t, $\tilde{q}$ admits a series expansion in terms of the X_n of the form

$$(36) \qquad \tilde{q}(x, t) = \sum_{n=1}^{\infty} q_n(t) X_n(x),$$

then

$$(37) \qquad q_n(t) = \frac{\displaystyle\int_0^a m\tilde{q}(x, t) X_n(x)\, dx}{\displaystyle\int_0^a m X_n^2(x)\, dx},$$

while (35) and (36) imply that

$$(38) \qquad c''(t) + \lambda_n c_n(t) = q_n(t).$$

The procedure to solve a problem of this kind is to find $U(x, t)$ using (31) and then determine $c_n(0)$, $c_n'(0)$, and $q_n(t)$ from (33), (34), and (37). Finally, solve (38) for $c_n(t)$, if possible, take the solution to (32), and set $u = U + v$. The solution so obtained should be verified *a posteriori*.

Example 5.4 A violin is placed horizontally on a table at rest. An electronic device located nearby generates a pure audio note. Show that each of the violin strings—which we assume to have constant mass per unit length—vibrates, and find the equation of motion.

The vibrations generated by the electronic device produce air pressure waves that reach each violin string and force its displacement. If we assume that these pressure waves vary sinusoidally with time, we are dealing with the boundary value problem

$$
\begin{aligned}
u_{tt} &= c^2 u_{xx} + A \sin \omega t & &\text{in } D \\
u(0, t) &= 0 & &t \geq 0 \\
u(a, t) &= 0 & &t \geq 0 \\
u(x, 0) &= 0 & &0 \leq x \leq a \\
u_t(x, 0) &= 0 & &0 \leq x \leq a.
\end{aligned}
$$

It is clear that in this case $U \equiv 0$, $c_n(0) = c_n'(0) = 0$, and $\tilde{q}(t) = q(t) = A \sin \omega t$. Also (24)–(26) become $c^2 X'' + \lambda X = 0$, $X(0) = 0$, and $X(a) = 0$, a Sturm-Liouville problem whose eigenvalues and eigenfunctions are

$$
\lambda_n = \frac{n^2 \pi^2 c^2}{a^2} \quad \text{and} \quad X_n(x) = \sin \frac{n\pi}{a} x.
$$

Then

$$
q_n(t) = \frac{\displaystyle\int_0^a A \sin \omega t \sin \frac{n\pi}{a} x \, dx}{\displaystyle\int_0^a \sin^2 \frac{n\pi}{a} x \, dx} = \frac{2A}{n\pi} \left[1 - (-1)^n \right] \sin \omega t.
$$

This means that $q_{2m} \equiv 0$ and

$$
q_{2m-1}(t) = \frac{4A}{(2m-1)\pi} \sin \omega t
$$

for $m = 1, 2, 3, \ldots$. Hence, for $n = 2m$, the only solution of (38) subject to zero initial conditions is $c_{2m} \equiv 0$. For each $n = 2m - 1$, assume first that $\omega \neq \sqrt{\lambda_{2m-1}}$ for all m. Then we must look for a particular solution of

$$
(39) \qquad c_{2m-1}''(t) + \lambda_{2m-1} c_{2m-1}(t) = \frac{4A}{(2m-1)\pi} \sin \omega t
$$

of the form $B_m \sin \omega t + C_m \cos \omega t$. If we define $\omega_m = \sqrt{\lambda_{2m-1}}$, we find that

$$
B_m = \frac{4A}{(2m-1)\pi(\omega_m^2 - \omega^2)} \quad \text{and} \quad C_m = 0.
$$

Adding the particular solution to the general solution of $c_{2m-1}'' + \omega_m^2 c_{2m-1} = 0$ and using the zero initial conditions we obtain

$$
c_{2m-1}(t) = \frac{4A}{(2m-1)\pi(\omega_m^2 - \omega^2)} \left(\sin \omega t - \frac{\omega}{\omega_m} \sin \omega_m t \right),
$$

and the motion of the string is given by

$$
u(x, t) = \frac{4A}{\pi} \sum_{m=1}^{\infty} \frac{1}{(2m-1)(\omega_m^2 - \omega^2)} \left(\sin \omega t - \frac{\omega}{\omega_m} \sin \omega_m t \right) \sin \frac{(2m-1)\pi}{a} x.
$$

Suppose next that $\omega = \omega_M = \sqrt{\lambda_{2M-1}}$ for some M, that is, that the frequency of the forcing pressure waves coincides with one of the natural frequencies of the string. Then, for $m = M$, we need a particular solution of (39) of the form $B_M t \sin \omega_M t + C_M t \cos \omega_M t$, and we find that

$$B_M = 0 \qquad \text{and} \qquad C_M = -\frac{2A}{(2M-1)\pi\omega_M}$$

and then that

$$c_{2M-1}(t) = \frac{2A}{(2M-1)\pi\omega_M}\left(\frac{1}{\omega_M}\sin\omega_M t - t\cos\omega_M t\right).$$

In this case the Mth term of the previous series has to be replaced with a term whose amplitude does not remain bounded as time increases. This phenomenon is called *resonance*. If the electronic device is not switched off, the violin string vibrating under resonance may break, much as a crystal goblet may break when it resonates to a singer's voice. But then, of course, the wave equation is valid only for small displacements of the string, so we cannot entirely be sure of what happens under resonance.

§5.8 Historical Epilogue

At the end of §5.3 we mentioned that Euler proposed a function with a discontinuous derivative as the initial displacement of the string and later saw how this caused a big argument among the researchers involved in these matters. The fact is that Euler's proposal represented something very new. The concept of function at the time was that of an analytical expression or formula, such as a power of x, so that when a definite value is plugged in for x the formula gives the value of the function. In fact, this was the year—1748—of the publication of Euler's enormously influential treatise *Introductio in analysin infinitorum*, the standard work on analysis during the second half of the eighteenth century. At the very beginning of this work, in the fourth paragraph, he defined a function of a variable quantity as

> any analytic expression made up in any manner whatever from that variable quantity and numbers and constants.

Euler followed here an earlier definition of Johann Bernoulli, his former teacher at Basel, to which he added the words *analytic expression*. But then, in that very same year, the problem of the vibrating string made him realize that this concept of function was too narrow to fit the needs of applied mathematics. Then he gave a new definition in his *Institutiones calculi differentialis* of 1755 as follows:

> If some quantities depend on other quantities so that they change when the latter are varied, then the former quantities are called functions of the latter.

This represented a radical departure from his previous idea, but would not be the last word. For one thing, it is very vague, lacking the precision that we are used to today, or

LEONHARDT EULER
David Eugene Smith Collection.
Rare Book and Manuscript Library.
Columbia University.
Reprinted with permission.

even the precision already necessary in the first quarter of the nineteenth century after Cauchy's *Cours d'analyse*. For another it was not completely accepted. For instance, in the opening of each edition of Lagrange's *Théorie des fonctions analytiques* of 1797 and 1813, we still have a function defined as any expression of calculation—*expression de calcul*—in which the variable or variables enter in any manner. What definitely won the day was Fourier's work, his use of functions exhibiting not just points of non-differentiability but even discontinuities, and Dirichlet's proof of Fourier's assertion that a trigonometric series could converge to such a function. After this there was no turning back to the purely analytic concept of function. Fourier himself tried his hand at a new definition as follows:

> The function $f(x)$ denotes a function completely arbitrary, that is to say a succession of given values, subject or not to a common law, and answering to all the values of x between 0 and any magnitude X.[1]

But, in spite of this *completely arbitrary* qualifier, it is clear from an examination of his work, that Fourier never had in mind a function with more than a finite number of

---

[1] *The Analytical Theory of Heat*, page 432.

discontinuities.

Dirichlet is sometimes credited with stating the first completely general definition of function in 1837, in the process of extending his convergence theorem to a wider class of functions. This is not correct, because what Dirichlet defined at this time was *continuous* function. However, the comments that he appended to his definition are valuable, for he said that *it is not really necessary for y to depend on x according to the same law on the whole interval, indeed one does not even need to think of expressing this dependence by means of mathematical operations.* This was the first time that anyone said this explicitly.

We shall never know whether or not Dirichlet and Riemann held any private discussions on the concept of function, but such a connection cannot entirely be ruled out in view of the mathematical interests that they shared. Whatever his motivation, the fact is that in the opening paragraph of Riemann's thesis we read:

> If z is a variable quantity that can gradually assume all possible real values, when to each of its values there corresponds a unique value of an undetermined quantity w, then we say that w is a function of z.

And on the same page he added: *obviously this definition does not specify any fixed law between the individual values of the function, because, after this function is defined on a particular interval, the way it can be extended outside remains entirely arbitrary.* Which is what Fourier had been saying all along: no *common law*, and it does not matter how the function is extended beyond $(-\pi, \pi)$. But with Riemann we have precision, we have this correspondence of a unique value of the function to each value of the variable. In short, the first entirely general and modern definition of function.

D'Alembert himself was never a participant in these developments to refine the concept of function, not even at the initial stages, although his papers on the motion of a vibrating string were the immediate cause of all the bustle and stir. For his work in mathematics and mechanics he became well known and admired in his lifetime, and the quality of his contributions made him a very sought-after man. In particular, his friend Frederick the Great of Prussia made him repeated offers to preside over the Berlin Academy. These—as well as similar ones by Catherine the Great of Russia—were rejected because d'Alembert was forever and above all a Parisian and could never think of leaving the city of his dreams. On one occasion, when on a trip to Italy for health reasons, he stopped to visit his friend Voltaire at Ferney. Comforted and refreshed by the companionship of the great man—whose ideas he fully shared—he never made it across the border and, instead, returned to Paris.

In spite of his scientific achievements, d'Alembert was better known in Europe as a *philosophe*, a man of the Age of Reason. It is a fact that he himself thought of the spread of the Enlightenment as one of his most important endeavors, and it was in this capacity that he joined his friend Denis Diderot as coeditor of the famous *Encyclopédie*[1]

---

[1] *Encyclopédie, ou Dictionnaire raisonné des sciences, des arts et des métiers, par une societé de gens de lettres. Mis en ordre et publié par M. Diderot, ... et quant à la partie mathématique, par M. d'Alembert, ...*, 17 volumes, Briasson, David l'aîné, Le Breton, Durand, Paris, 1751–1765, and 18 additional volumes in later years.

to which he contributed the *Discours préliminaire* and numerous articles. It was just at this point in time that he started to find less and less time for mathematics. Not only because of his genuine liking of these literary activities, but also because of pressures put on him by his friends—in particular Mme. du Deffand—who wanted to steer him in this direction, and by his renewed efforts in philosophy when the *Encyclopédie* came under the attack of the Catholic Church. His literary and philosophical achievements were promptly recognized and earned him his election to the *Académie Française* in 1754, and, soon after, his fame was widely spread in Europe, mainly through the *Discours* and through his *Élémens de philosophie* of 1759.

In 1764 Julie de Lespinasse—d'Alembert's lover—was suddenly and summarily banned from the *salon* of her cousin, Mme. du Deffand, when she discovered that Julie had been receiving guests early in her own *pre-salon*. Mlle. de Lespinasse then opened her own fully fledged *salon*, which d'Alembert attended regularly ever since, and where—already an influential man—he started arranging memberships and elections to the academies. But then, in 1765, d'Alembert fell seriously ill. Although restored back to health by Mlle. de Lespinasse—who found this a propitious occasion to convince him to leave Mme. Rousseau and move in with her—this illness left a permanent mark. From now on, as in a letter to Lagrange of 1766, he would frequently complain about his inability to do mathematics due to some weakness of the head. While his ability to carry on in *philosophie* remained unimpaired, his creative mathematical output did indeed come to an end. The unfinished trip to Italy, already mentioned above, took place in 1770, when d'Alembert was taken ill once more. Even as late as 1777—again in a letter to Lagrange—he still complained about being unable to do mathematics. His literary ventures provided him with just *a way to fill the time for the lack of anything better to do.*[1]

And so it was until 1776, when d'Alembert was devastated by the death of Julie and, shortly after, by that of his friend Mme. Geoffrin, the *salonnière*. He suffered an even more devastating blow—in the form of insult added to injury—when he found among Julie's papers some uncontrollably passionate love letters that she had written to the Comte de Guibert and the Marquis de Mora. A broken man, d'Alembert retired to some small, gloomy quarters at the Louvre—to which he was entitled as permanent secretary of the *Académie Française*—until his death in 1783, the same year as Euler's.

EXERCISES

2.1   Assume that the mass per unit length m and the tension τ of the vibrating string are functions of x and that a distributed, time-dependent, external force $q(x, t)$ acts vertically upward on each point of abscissa x at time t. Show that, if the remaining hypotheses

---

[1] Quoted from T. L. Hankins, *Jean d'Alembert: Science and the Enlightenment*, The Clarendon Press, Oxford, 1970.

in §5.2 hold, then the equation of motion for the string is

$$mu_{tt} = (\tau u_x)_x + q(x, t).$$

2.2   Use the equation of Exercise 2.1 to show that, if m and τ are constant and if the vibrating string moves in a medium with friction such that the friction force is proportional to the velocity (the constant of proportionality is called the *friction coefficient*), then its equation of motion is

$$v_{tt} = c^2 v_{xx} + k^2 v,$$

where $c^2 = \tau/m$, k is a positive constant and $v = e^{kt}u$. How is k related to the friction coefficient?

2.3   A string of length a is attached at $x = 0$ to a vertical axis, and the string rotates in a horizontal plane around this axis with constant angular velocity ω. In addition, it vibrates vertically from its equilibrium position in the plane of rotation, with the centrifugal force providing the tension on the string. Show that, if its mass per unit length m is constant and if there are no external forces acting on it, then the equation of motion is

$$u_{tt} = c^2[(a^2 - x^2)u_{xx} - 2xu_x],$$

and find the value of the constant c in terms of ω and m.

2.4   Show that if the third hypothesis in §5.2 is removed, if s denotes the x-coordinate of a point on the string when it is at rest in a horizontal position, if $P(s, t) = (x(s, t), y(s, t))$ is the position of that point at time t, if the mass per unit length m and the tension τ are functions of s, and if an external force $q(x, t)$ acts vertically upward on the point of the string that is at x at time t, then the equations of motion are [1]

$$mx_{tt} = \left[\frac{\tau x_s}{\sqrt{x_s^2 + y_s^2}}\right]_s \quad\text{and}\quad my_{tt} = \left[\frac{\tau y_s}{\sqrt{x_s^2 + y_s^2}}\right]_s + q(x, t)$$

(**Hint:** replace Figure 5.1 with Figure 5.11, use the tangent vectors to the string at $P(s, t)$

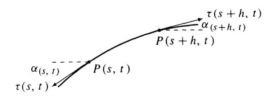

Figure 5.11

and $P(s + h, t)$ to compute the cosines of $\alpha_{(s, t)}$ and $\alpha_{(s+h, t)}$, and use Newton's law to establish the equilibrium of horizontal and vertical force components).

3.1   Consider the boundary value problem of Theorem 5.2 and assume that

$$f\left(\frac{a}{2} - x\right) = f\left(\frac{a}{2} + x\right)$$

---

[1] This is the derivation in H. F. Weinberger, *A First Course in Partial Differential Equations*, Ginn and Company, Waltham, Massachusetts, 1965, pages 1–3.

for all x such that $0 \leq x \leq a/2$. Show that $u(x, a/2c) = 0$ and that $u(x, a/c) = u(x, 0)$. Since the string moves without friction and there are no external forces, can you make a conjecture about its motion for all $t > 0$? If not, compute $u(x, 3a/2c)$ and $u(x, 2a/c)$ and try again. Can you prove your conjecture?

3.2 Consider the equation $u_{tt} = c^2 u_{xx} - g$, where g is a constant.
　　　(*i*) Find all the solutions that are independent of time.
　　　(*ii*) Find the solution independent of time such that $u(0, t) = u(a, t) = 0$.
　　　(*iii*) Can the solution found in (*ii*) be called a steady-state solution? Justify your answer.
　　　(*iv*) Show that the general solution is

$$u(x, t) = \frac{g}{2c^2} x(x - 1) + F(x + ct) + G(x - ct),$$

where F and G are arbitrary functions of class C^2 in $\mathbb{R}$.

3.3 Solve the boundary value problem

$$
\begin{array}{ll}
u_{tt} = c^2 u_{xx} - g & 0 < x < a, \quad t > 0 \\
u(0, t) = 0 & t \geq 0 \\
u(a, t) = 0 & t \geq 0 \\
u(x, 0) = 0 & 0 \leq x \leq a \\
u_t(x, 0) = 0 & 0 \leq x \leq a
\end{array}
$$

(**Hint:** use the solution independent of time found in Exercise 3.2 (*ii*)).

3.4 Use the general solution found in Exercise 3.2 (*iv*) to show that the solution of the boundary value problem

$$
\begin{array}{ll}
u_{tt} = c^2 u_{xx} - g & 0 < x < 1, \quad t > 0 \\
u(0, t) = 1 & t \geq 0 \\
u(1, t) = 1 + h & t \geq 0 \\
u(x, 0) = 1 + hx & 0 \leq x \leq 1 \\
u_t(x, 0) = 0 & 0 \leq x \leq 1
\end{array}
$$

where h is a positive constant, is

$$u(x, t) = \frac{g}{2c^2} x(x - 1) + hx + 1 - \tfrac{1}{2}[\tilde{f}(x + ct) + \tilde{f}(x - ct)],$$

where $\tilde{f}$ is the odd periodic extension of

$$f(x) = \frac{g}{2c^2} x(x - 1)$$

to $\mathbb{R}$ of period 2 (**Hint:** show that

$$\tilde{f}(x) = -\frac{g}{2c^2} x(x + 1)$$

for $-1 \leq x \leq 0$, and assume without loss of generality that $F(0) = \tfrac{1}{2}$).

3.5 Find a simpler way to arrive at the solution of the Exercise 3.4.

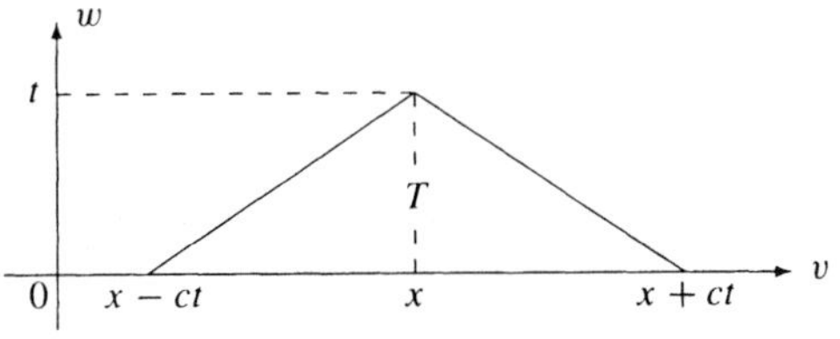

Figure 5.12

3.6 Show that, if m and τ are constant, then the general solution of the equation of Exercise 2.1, $mu_{tt} = \tau u_{xx} + q(x,t)$, is

$$u(x,t) = F(x+ct) + G(x-ct) + \frac{1}{2\sqrt{m\tau}} \int_T q(v,w)\, dv\, dw,$$

where F and G are arbitrary functions of class C^2 in $\mathbb{R}$ and T is the interior of the triangle shown in Figure 5.12. Verify that u is indeed a solution.

3.7 Solve the boundary value problem

$$\begin{aligned}
u_{tt} &= c^2 u_{xx} & x > 0, \quad t > 0 \\
u(0,t) &= h(t) & t \geq 0 \\
u(x,0) &= 0 & x \geq 0 \\
u_t(x,0) &= 0 & x \geq 0
\end{aligned}$$

for an ideal semi-infinite string, and state conditions on the function h for the solution to be of class C^2. Sketch the solution $u(x,t)$ as a function of x for a fixed arbitrary value of t in the following cases.

(i) $h(t) = \begin{cases} \sin \pi t, & 0 \leq t \leq 1 \\ 0, & t > 0. \end{cases}$ (ii) $h(t) = \sin \pi t$, $t \geq 0$.

3.8 Repeat Exercise 3.7 if the endpoint condition $u(0,t) = h(t)$ for $t \geq 0$ is replaced with $u_x(0,t) = h(t)$ for $t \geq 0$.

4.1 The kinetic and potential energies of the vibrating string are given by

$$\frac{1}{2} \int_0^a mu_t^2(s,t)\, ds \quad \text{and} \quad \frac{1}{2} \int_0^a \tau u_x^2(s,t)\, ds,$$

where m is the mass per unit length and τ is the tensile force on the string, which we assume to be constant. Show that if there is no friction or external forces the total energy of the string is independent of time. Use this fact to show that the boundary value problem of Theorem 5.3 has no more than one solution. Show that if there is friction the total energy of the string is a decreasing function of time.

4.2 Solve the boundary value problem

$$\begin{aligned}
u_{tt} &= c^2 u_{xx} + q(x) & 0 < x < 1, \quad t > 0 \\
u(0,t) &= a & t \geq 0 \\
u(1,t) &= b & t \geq 0 \\
u(x,0) &= f(x) & 0 \leq x \leq 1 \\
u_t(x,0) &= g(x) & 0 \leq x \leq 1,
\end{aligned}$$

where q is an integrable function, a and b are constants, and f and g are functions of class C^2 and C^1, respectively. State additional conditions on f and g for the solution to be of class C^2 (**Hint:** find first a solution of the given equation that is independent of time and satisfies the endpoint conditions).

4.3   Find the solution of the equation of Exercise 2.1 subject to the initial and boundary conditions of Theorem 5.3 (**Hint:** use the general solution of Exercise 3.6 and write the integral over T as an iterated integral).

4.4   Solve the boundary value problem

$$
\begin{aligned}
u_{tt} &= c^2 u_{xx} & x > 0, \quad t > 0 \\
u(0, t) &= 0 & t \geq 0 \\
u(x, 0) &= f(x) & x \geq 0 \\
u_t(x, 0) &= g(x) & x \geq 0
\end{aligned}
$$

for an ideal semi-infinite string, where f is of class C^2, g is of class C^1 and $f(0) = 0$. State additional conditions on f and g for the solution to be of class C^2.

4.5   Use linearity to solve the boundary value problem of Exercise 4.4 if the endpoint condition is replaced with that of Exercise 3.7, and show that h must satisfy the conditions $h(0) = f(0)$, $h'(0) = g(0)$, and $h''(0) = c^2 f''(0)$.

4.6   Solve the boundary value problem

$$
\begin{aligned}
u_{tt} &= c^2 u_{xx} & x > 0, \quad t > 0 \\
u_x(0, t) &= 0 & t \geq 0 \\
u(x, 0) &= f(x) & x \geq 0 \\
u_t(x, 0) &= g(x) & x \geq 0
\end{aligned}
$$

for an ideal semi-infinite string, where f is of class C^2, g is of class C^1, and $f'(0) = 0$. Do f and g have to satisfy any additional conditions for the solution to be of class C^2?

4.7   Use linearity to solve the boundary value problem of Exercise 4.6 if the endpoint condition is replaced with that of Exercise 3.8, and show that h must satisfy the conditions $h(0) = f'(0)$ and $h'(0) = g'(0)$.

4.8   Find a solution of the boundary value problem

$$
\begin{aligned}
u_{tt} &= c^2 u_{xx} & -\infty < x < \infty, \quad t > 0 \\
u(x, 0) &= f(x) & -\infty < x < \infty \\
u_t(x, 0) &= g(x) & -\infty < x < \infty
\end{aligned}
$$

for an ideal string of infinite length, where f is of class C^2 and g is of class C^1. Prove or disprove that this problem is well posed in the sense of Hadamard.

4.9   Assume that the motion of an ideal infinite string, stretched with tension τ, is given by $u(x, t) = F(x + ct) + G(x - ct)$, where

$$
F(x + ct) = \begin{cases}
0, & x < -ct \\
\dfrac{Mcv_0}{2\tau}\left(1 - e^{-\tau(x+ct)/Mc^2}\right), & -ct \leq x \leq 0 \\
0, & x > 0
\end{cases}
$$

and

$$G(x - ct) = \begin{cases} 0, & x < 0 \\ \dfrac{Mcv_0}{2\tau}\left(1 - e^{\tau(x-ct)/Mc^2}\right), & 0 \le x \le ct \\ 0, & x > ct. \end{cases}$$

(*i*) Draw the graph of $u(x, t)$ for fixed $t > 0$.

(*ii*) What happened at $t = 0$ to cause this motion of the string?

(*iii*) What is the meaning of the constants M and v_0?

4.10   Consider the ideal infinite string of Exercise 4.8, and let $[a, b]$, be a given interval. Sketch the graph of $u(x, t)$ for fixed arbitrary t and all x if $f \equiv 0$, $g > 0$ on $[a, b]$ and $g \equiv 0$ outside $[a, b]$. Find the limit of $u(x, t)$ for fixed arbitrary x as $t \to \infty$. Compare your results with those for a string with no initial velocity for which $f > 0$ on $[a, b]$ and $f \equiv 0$ outside $[a, b]$ and make appropriate comments. Assume now that the string is finite and tied down at both ends, and describe how the graph of $u(x, t)$ is built from the original disturbance and any reflections at the endpoints.

5.1   Assume that the initial displacement of a string that is tied down at both ends is as represented in Figure 5.7 and that the maximum value of f is M. Find the motion of the string if it has no initial velocity. Choose the value of x_0 so that the nth harmonic and all its integral multiples vanish. Repeat if $f \equiv 0$ and g is as in Figure 5.7.

5.2   A mandolin string 27 cm long is tightened with a tension force of 16 kg. Find the fundamental frequency of vibration if the mass of the string is 1 g. If the initial displacement of the string is

$$f(x) = \begin{cases} Cx, & 0 \le x \le 17 \\ C(27 - x), & 17 < x \le 27, \end{cases}$$

if x is in centimeters, and if the maximum value of f is 1 mm, use the result of Exercise 5.1 to find the amplitude of the fundamental vibration and of the first harmonic.

5.3   Find the motion of a string tied down at the endpoints $x = 0$ and $x = a$ and initially at rest if its initial velocity is

$$u_t(x, 0) = \begin{cases} 4Mx/a, & 0 \le x \le a/4 \\ M, & a/4 \le x \le 3a/4 \\ 4M(a - x)/a, & 3a/4 < x \le a, \end{cases}$$

where M is a constant.

5.4   A string, tied down at the endpoints $x = 0$ and $x = a$ and initially at rest, is struck by a hammer of mass M with velocity v_0. The hammer hits only the points of the string between $a/2 - \epsilon$ and $a/2 + \epsilon$, where ϵ is a small enough positive number, in such a way that the hammer's momentum is evenly distributed over all these points. Find the motion of the string and compute its limit as $\epsilon \to 0$. Show that, if the hammer blow is centered at ka/n, where k and n are positive integers and $k < n$, then the nth harmonic vanishes.

5.5   The equation of motion of a string moving in a medium with friction in which the friction force is proportional to its velocity at each point is $u_{tt} = c^2 u_{xx} - 2ku_t$, where k is a positive constant. Find a formal solution of the boundary value problem

$$
\begin{aligned}
u_{tt} &= c^2 u_{xx} - 2k u_t & 0 < x < a, \quad t > 0 \\
u(0, t) &= 0 & t \geq 0 \\
u(a, t) &= 0 & t \geq 0 \\
u(x, 0) &= f(x) & 0 \leq x \leq a \\
u_t(x, 0) &= g(x) & 0 \leq x \leq a
\end{aligned}
$$

if the friction is so small that $k < \pi c/a$.[1] State the form of the solution in the particular case $g \equiv 0$ (**Hint:** use the substitution $v = e^{kt} u$ to obtain the equation of Exercise 2.2).

5.6   Repeat Exercise 5.5 in the cases $k = \pi c/a$ and $k > \pi c/a$ (**Hint:** note that separation of variables does not lead to a Sturm-Liouville problem).

6.1   Solve the boundary value problem

$$
\begin{aligned}
u_{tt} &= c^2 u_{xx} & 0 < x < a, \quad t > 0 \\
a_1 u(0, t) - b_1 u_x(0, t) &= 0 & t \geq 0 \\
a_2 u(a, t) + b_2 u_x(a, t) &= 0 & t \geq 0 \\
u(x, 0) &= f(x) & 0 \leq x \leq a \\
u_t(x, 0) &= g(x) & 0 \leq x \leq a
\end{aligned}
$$

if

  (*i*) $a_1 = a_2 = 0$ and
  (*ii*) $b_1 = a_2 = 0$.

What can be done in case (*i*) to obtain a bounded solution? How should the length of the string be adjusted in case (*ii*) so that the frequencies of vibration are the same as the odd harmonics in case (*i*)?

6.2   Solve the boundary value problem of Exercise 6.1 (*ii*) if the wave equation is replaced with $u_{tt} = c^2 u_{xx} - g$, where g is the acceleration of gravity. Can you use a similar method to solve Exercise 6.1 (*i*) if the wave equation is replaced with $u_{tt} = c^2 u_{xx} - g$? Explain.

6.3   Consider the boundary value problem

$$
\begin{aligned}
u_{tt} &= c^2 u_{xx} + q(x) & 0 < x < a, \quad t > 0 \\
u_x(0, t) &= 0 & t \geq 0 \\
u_x(a, t) &= 0 & t \geq 0 \\
u(x, 0) &= f(x) & 0 \leq x \leq a \\
u_t(x, 0) &= g(x) & 0 \leq x \leq a.
\end{aligned}
$$

What additional condition must q satisfy for this partial differential equation to have a solution independent of time that satisfies the endpoint conditions? Use the result of Exercise 6.1 to find a bounded solution of the stated boundary value problem if q satisfies this condition.

6.4   The endpoint conditions derived in §5.6 were based on the assumption that discs of negligible mass attached to springs move in vertical tracks at the ends of the string. Show that, if the masses of the discs are not negligible, the endpoint conditions become

$$
\begin{aligned}
u_{tt}(0, t) + a_1 u(0, t) - b_1 u_x(0, t) &= c_1 \\
u_{tt}(a, t) + a_2 u(a, t) + b_2 u_x(a, t) &= c_2.
\end{aligned}
$$

---

[1] The proof that it is a solution is almost identical to the proofs of Propositions 3.1 and 3.2.

Consider now the particular case in which the string is tied down at $x = 0$, so that $u(0, t) = 0$, and in which the spring at $x = a$ is removed, so that $a_2 = c_2 = 0$. Show that the resulting boundary value problem

$$\begin{aligned}
u_{tt} &= c^2 u_{xx} & 0 < x < a, \quad t > 0 \\
u(0, t) &= 0 & t \geq 0 \\
u_{tt}(a, t) + b_2 u_x(a, t) &= 0 & t \geq 0 \\
u(x, 0) &= f(x) & 0 \leq x \leq a \\
u_t(x, 0) &= g(x) & 0 \leq x \leq a
\end{aligned}$$

has a solution of the form

$$u(x, t) = \sum_{n=1}^{\infty} \left(A_n \cos \sqrt{\lambda_n}\, t + \frac{B_n}{\sqrt{\lambda_n}} \sin \sqrt{\lambda_n}\, t \right) \sin \frac{\sqrt{\lambda_n}}{c}\, x,$$

where the λ_n are the positive roots of

$$\tan \frac{\sqrt{\lambda}}{c}\, a = \frac{ab_2}{c\sqrt{\lambda}}.$$

6.5 Solve the boundary value problem of Example 5.3 if the string moves in a medium with friction in which the friction force is proportional to the velocity of the string, that is, if the equation of motion is replaced by

$$u_{tt} = c^2 u_{xx} - 2ku_t - g,$$

where k is a positive constant, and assume that the friction is so small that $k < \pi c/a$ (**Hint:** use the solution of Exercise 5.5).

6.6 Repeat Exercise 6.5 in the cases $k = \pi c/a$ and $k > \pi c/a$ (**Hint:** use the solution of Exercise 5.6).

7.1 A string tied down at the endpoint $x = 0$ and initially at rest horizontally is attached to an oscillator at $x = a$ so that $u(a, t) = A \sin \omega t$ for $t \geq 0$. If the force of gravity is neglected, find the motion of the string for $t > 0$. Consider separately the cases $\omega \neq n\pi c/a$ for any positive integer n and $\omega = N\pi c/a$ for some N.

7.2 Solve the boundary value problem of Example 5.4 if the string is not tied down at $x = a$ but attached to a spring of elastic constant k. Assume that the string is initially at rest horizontally. Consider separately the cases $\omega \neq n\pi c/a$ for any positive integer n and $\omega = N\pi c/a$ for some N.

7.3 Solve the boundary value problem of Example 5.4 if the differential equation is replaced with $u_{tt} = c^2 u_{xx} + A(x) \sin \omega t$. Express your answer in terms of the coefficients of the Fourier sine series of A on $(-a, a)$.

7.4 A string of mass per unit length m is tied down at the endpoints $x = 0$ and $x = a$, stretched with tension τ, and initially at rest. Let x_0 be a point in $(0, a)$ and assume that an external variable force of total magnitude $A \sin \omega t$ acts on the segment between $x_0 - \epsilon$ and $x_0 + \epsilon$, where ϵ is a small enough positive number, in such a way that it is evenly distributed over all these points. Find the motion of the string and, by taking the limit as $\epsilon \to 0$, find also the motion of the string if a concentrated variable force of magnitude $A \sin \omega t$ acts on the point x_0.

7.5 Solve the boundary value problem of Example 5.4 if the string moves in a medium with friction in which the friction force is proportional to the velocity of the string, that is, if the equation of motion is replaced

$$u_{tt} = c^2 u_{xx} - 2ku_t + A \sin \omega t,$$

where k is a positive constant, and assume that the friction is so small that $k < \pi c / a$ (**Hint:** follow the steps in Example 5.4, show that equation (38) is replaced by

$$c_n''(t) + 2kc_n'(t) + \lambda_n c_n(t) = q_n(t),$$

and then solve this equation).

7.6 Repeat Exercise 7.5 in the cases $k = \pi c / a$ and $k > \pi c / a$.

6

ORTHOGONAL SYSTEMS

§6.1 Fourier Series and Parseval's Identity

Generalized Fourier series were used by David Hilbert (1862–1943), of the University of Göttingen, in a series of important papers on integral equations published between 1904 and 1906. But Hilbert took this concept one step further by abandoning any reference to the eigenfunctions of a Sturm-Liouville problem. As we saw in §4.4 the key property of the eigenfunctions in computing the Fourier coefficients is their orthogonality, that is,

$$\int_a^b r X_m X_n = 0 \quad \text{if} \quad m \neq n$$

(Theorem 4.1). What Hilbert did was start with a set of nontrivial functions that satisfy this property, whether or not they are eigenfunctions of a Sturm-Liouville problem.

In order to properly define such a set, we shall adopt the following notation and terminology. If f and g are integrable on (a, b) and r is a fixed function that is continuous and positive on (a, b), the expression

$$(f, g) = \int_a^b r f g$$

will be called the *inner product* of f and g and

$$\|f\| = \sqrt{(f, f)}$$

will be called the *norm* of f. Of course, these definitions depend on the function r, which is called the *weight function*, but in most applications $r \equiv 1$. If f, g, and h are any integrable functions, the basic properties of an inner product are

1. $(f, g) = (g, f)$,

2. $(f + g, h) = (f, h) + (g, h)$, and

3. $(cf, g) = c(f, g)$ for every constant c.

Definition 6.1 *A sequence of integrable functions* $\{\varphi_n\}$ *on an interval* (a, b) *is called an* **orthogonal system** *on* (a, b) *if and only if*

(i) $(\varphi_m, \varphi_n) = 0$ *if* $m \neq n$ *and*

(ii) $\|\varphi_n\| > 0$ *for all* n.

If, in addition, $\|\varphi_n\| = 1$ *for all* n, *then the system is called* **orthonormal**.

In the preceding chapters we have already seen examples of orthogonal systems whose elements are the eigenfunctions of some Sturm-Liouville problem. The following are examples not related to a Sturm-Liouville problem.

Example 6.1 The functions defined by

$$\varphi_n(x) = \cos(n \arccos x)$$

for every nonnegative integer n form an orthogonal system on $(-1, 1)$ with weight function

$$r(x) = \frac{1}{\sqrt{1 - x^2}}$$

because the substitution $t = \arccos x$ shows that

$$\int_{-1}^{1} \frac{1}{\sqrt{1 - x^2}} \varphi_m(x)\varphi_n(x)\, dx = \int_{0}^{\pi} \frac{1}{\sqrt{1 - \cos^2 t}} \cos mt \cos nt \sin t\, dt$$

$$= \int_{0}^{\pi} \cos mt \cos nt\, dt,$$

which is zero if $m \neq n$.[1] These orthogonal functions are called the *Chebyshev polynomials* (see Exercise 1.3 to verify that they are indeed polynomials) because they were introduced by Pafnuti L'vovich Chebyshev (1821–1894), a professor at the University of Saint Petersburg, in answer to the following problem: of all the polynomials $P_n(x)$ of degree n such that $|P(x)| \leq 1$ on $[-1, 1]$, which one has the largest coefficient for x^n?

Example 6.2 Define $\varphi_0 \colon [0, 1) \to \mathbb{R}$ by $\varphi_0 \equiv 1$ and then, for each positive integer m of the form 2^k and each $n = 1, \ldots, m$, $\varphi_{mn} \colon [0, 1) \to \mathbb{R}$ by

$$\varphi_{mn}(x) = \begin{cases} 1 & \text{if } \dfrac{n-1}{m} \leq x < \dfrac{2n-1}{2m} \\[2mm] -1 & \text{if } \dfrac{2n-1}{2m} \leq x < \dfrac{n}{m} \\[2mm] 0 & \text{otherwise} \end{cases}$$

These functions are represented in Figure 6.1 up to $m = 4$, and form an orthogonal

---

[1] In this particular example the integral on the left is improper since the integrand is unbounded near $x = \pm 1$; that is, this integral is defined as the limit as $\epsilon > 0$ approaches zero of the integral from $-1 + \epsilon$ to $1 - \epsilon$.

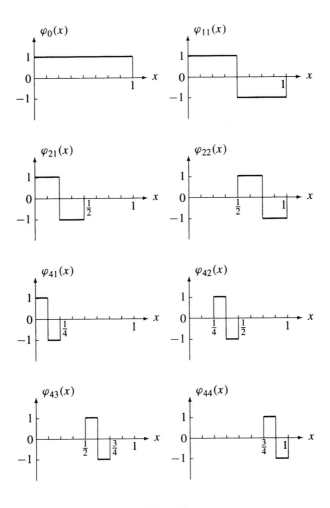

Figure 6.1

system with $r \equiv 1$ by construction. We shall refer to them as the *Haar functions* because they were introduced in 1910 by Alfred Haar (1885–1933), a former student of Hilbert. Although these functions can be listed using a single subscript, the double subscript listing is convenient because $1/m$ is the length of the support of φ_{mn} (defined as the set of all points x in $[0, 1)$ such that $\varphi_{mn}(x) \neq 0$) and n indicates the position of this support in $[0, 1]$.

Every orthogonal system can be *normalized* by replacing each φ_n with the product $\|\varphi_n\|^{-1}\varphi_n$.

Example 6.3 For the Chebyshev polynomials the substitution used in Example 6.1 shows that

$$\|\varphi_0\|^2 = \int_{-1}^{1} \frac{1}{\sqrt{1-x^2}} \, dx = \int_0^{\pi} dt = \pi$$

and, for $n > 0$,

$$\|\varphi_n\|^2 = \int_{-1}^{1} \frac{\cos^2(n \arccos x)}{\sqrt{1-x^2}} \, dx = \int_0^{\pi} \cos^2 nt \, dt = \frac{\pi}{2},$$

and then the corresponding orthonormal system is

$$\frac{1}{\sqrt{\pi}} \varphi_0, \sqrt{\frac{2}{\pi}} \varphi_1, \ldots, \sqrt{\frac{2}{\pi}} \varphi_n, \ldots.$$

For the Haar functions we have $\|\varphi_0\| = 1$ and

$$\|\varphi_{mn}\|^2 = \int_0^1 \varphi_{mn}^2 = \frac{1}{m},$$

and then the normalized Haar functions are $\varphi_0, \varphi_{11}, \ldots, \sqrt{m} \, \varphi_{mn}, \ldots.$

The definition of orthogonal system and much of the theory to be developed below can be generalized in two ways that are often useful. The first is to replace the interval (a, b) by an arbitrary interval I in $\mathbb{R}$, even if I is unbounded, but then care must be taken to ensure that the square of every function involved is integrable on I in the sense of Appendix B. This is necessary for its norm to exist and it does not happen automatically if the function is integrable on I (see Exercise 2 of Appendix B). We shall use unbounded intervals only at the end of §6.4 and in Exercises 2.8 to 2.10. The second generalization is to allow complex valued functions, but then the product fg in the definition of inner product must be replaced with $f\bar{g}$, where the bar denotes complex conjugate, and its first property must be replaced with $(f, g) = \overline{(g, f)}$.

We can now use an arbitrary orthogonal system to further generalize the concept of Fourier series.

Definition 6.2 *If f is an integrable function on (a, b) and $\{\varphi_n\}$ is an orthogonal system on (a, b), then the constants*

$$c_n = \frac{1}{\|\varphi_n\|^2} (f, \varphi_n)$$

are called the **Fourier coefficients** *of f with respect to $\{\varphi_n\}$ and the series*

$$\sum_{n=1}^{\infty} c_n \varphi_n$$

is called its **Fourier series** *with respect to $\{\varphi_n\}$.*

Example 6.4 The system $\{e^{inx}\}$, where n assumes all integral values, is orthogonal on $(-\pi, \pi)$ with weight function $r \equiv 1$ and each function has norm $\sqrt{2\pi}$ because

$$\int_{-\pi}^{\pi} e^{inx} e^{-inx} \, dx = 2\pi.$$

Then, if f is either a real or complex valued function on $(-\pi, \pi)$, its Fourier coefficients with respect to the given system are

$$c_n = \int_{-\pi}^{\pi} f(x) e^{-inx} \, dx$$

and its Fourier series

$$\sum_{n=-\infty}^{\infty} c_n e^{inx}$$

is called the *complex Fourier series* of f on $(-\pi, \pi)$. Note that, if f is real valued, then $c_{-n} = \bar{c}_n$ for all n.

From now on we shall restrict the development of our theory to real valued functions, relegating to the exercises the extension of some of the results to the complex valued case. Our question is, as usual, whether or not the equation

(1)
$$f(x) = \sum_{n=1}^{\infty} c_n \varphi_n$$

holds, that is, whether or not the Fourier series of f converges to f. The first thing to notice is that it may not if there are not enough functions in the orthogonal system $\{\varphi_n\}$. This was originally pointed out by Fourier himself in a footnote contained in a supplementary short paper that he sent to the *Institut de France* in 1809 and is shown by the following example (which is simpler than Fourier's own): the system $\{\sin x, \ldots, \sin nx, \ldots\}$ is orthogonal on $(-\pi, \pi)$ but (1) does not hold for $f \equiv 1$ because its Fourier coefficients are all zero.

To better understand what is involved we shall establish the following analogy inspired by the work of Erhardt Schmidt (1876–1959), a student of Hilbert. Let c_1, c_2, and c_3 denote the coordinates of a point p in $\mathbb{R}^3$ with respect to the orthogonal vectors $\varphi_1 = (1, 0, 0)$, $\varphi_2 = (0, 1, 0)$, and $\varphi_3 = (0, 0, 1)$, where the duplication in the notation is intentional. Then $p = (c_1, c_2, c_3)$ is the sum of its components with respect to $\{\varphi_1, \varphi_2, \varphi_3\}$:

(2)
$$p = \sum_{n=1}^{3} c_n \varphi_n.$$

However, an arbitrary point p may not be the sum of its components with respect to the smaller orthogonal system $\{\varphi_1, \varphi_2\}$. For instance, $p = (0, 0, 1)$ is not.

In a similar manner, when dealing with integrable functions, an orthogonal system $\{\varphi_n\}$ acts as a reference or coordinate system. Then the Fourier coefficients of a given function f are its coordinates with respect to that system, and the terms $c_n\varphi_n$ of its Fourier series are its components. If (1) holds, then we can say that f is the sum of its components, but we cannot expect (1) to be valid for an arbitrary function f if the orthogonal system $\{\varphi_n\}$ is not large enough. In fact, if there is an f with $\|f\| > 0$ that is orthogonal to all the φ_n, then its Fourier coefficients are all zero, and (1) cannot hold. Based on the preceding observation, we shall call an orthogonal system *complete* if it cannot be enlarged with additional orthogonal functions.

Definition 6.3 *An orthogonal system is called* **complete** *if and only if it is not a proper subset of another orthogonal system, that is, if the situation* $(f, \varphi_n) = 0$ *for all n implies that* $\|f\| = 0$.

We have just seen that the completeness of the orthogonal system is a necessary condition for the convergence of Fourier series in the general case, and this poses the following question: how can we tell if a given orthogonal system is complete? Definition 6.3 is not useful in this respect, and then, in an attempt to answer this question, we shall explore the euclidean space analogy a little further. If $\|p\|$ denotes the distance from a point p in $\mathbb{R}^3$ to the origin, the Pythagorean theorem states that

$$\|p\|^2 = \sum_{n=1}^{3} \|c_n\varphi_n\|^2 = \sum_{n=1}^{3} c_n^2\|\varphi_n\|^2.$$

Clearly, these identities do not hold for all p in $\mathbb{R}^3$ if the orthogonal system $\{\varphi_1, \varphi_2, \varphi_3\}$ is replaced by a smaller one. Conversely, the fact that they hold for all p and the system $\{\varphi_1, \varphi_2, \varphi_3\}$ automatically makes this system complete. Indeed, if a given p is orthogonal to φ_n for each n, then c_n—defined by the dot product $p \cdot \varphi_n$—is zero for each n. Then the Pythagorean property shows that $\|p\| = 0$, and p cannot be used to enlarge $\{\varphi_1, \varphi_2, \varphi_3\}$. Now, if we accept that the norm of a function,

$$\|f\| = \sqrt{\int_a^b rf^2}\,,$$

is a good way to measure how far f is from the zero function, then the result analogous to the Pythagorean theorem should be

(3) $$\|f\|^2 = \sum_{n=1}^{\infty} c_n^2\|\varphi_n\|^2.$$

Equation (3) is usually called *Parseval's identity* because it was first stated in 1799—in a somewhat different form and without proof, for he claimed it was self-evident—by Marc-Antoine Parseval-Deschênes (1755–1836) for power series. Fourier stated at one point that it is easy to evaluate the sum of the series on the right—for the case of trigonometric series—but failed to do so.[1]

---

[1] *The Analytical Theory of Heat*, page 207.

In view of the analogy established with $\mathbb{R}^3$, it is reasonable to conjecture that an orthogonal system is complete if and only if Parseval's identity holds for all integrable functions with respect to that system. We can easily prove now the *if* part of this conjecture. In §6.6, which is an optional section and requires knowledge of the Lebesgue integral, we shall prove the *only if* part in the appropriate context.

Theorem 6.1 *If* (3) *holds for every integrable function* f *on* (a, b) *and some orthogonal system* $\{\varphi_n\}$, *then* $\{\varphi_n\}$ *is complete.*

Proof. If $\{\varphi_n\}$ is not complete, then there is a function f with $\|f\| > 0$ that is orthogonal to all the φ_n and then $c_n = 0$ for this f and all n, leading to the contradiction

$$0 < \|f\|^2 = \sum_{n=1}^{\infty} c_n^2 \|\varphi_n\|^2 = 0,$$

Q.E.D.

In view of this result we can now ask: how can we tell, given an orthogonal system, whether or not Parseval's identity is valid for all integrable f? But this turning of a question into another may start to look like a case of passing the hot potato. And it can be questioned whether the potato is all that hot because the completeness of $\{\varphi_n\}$, which we have just seen to be a necessary condition for the convergence of Fourier series, is certainly not sufficient. In fact, it can be shown—and we shall do it in §6.4—that the usual trigonometric system is complete, and yet we know that a trigonometric Fourier series may not converge at an infinite set of points, as stated at the end of §2.4.

We must ask ourselves whether or not the completeness of an orthogonal system and Parseval's identity are all that relevant. Perhaps we have been barking up the wrong tree in posing these questions. Meanwhile, a Danish actuary, Jørgen Pedersen Gram (1850–1916), had discovered in 1883 an unexpected role of Fourier series in the solution of a certain approximation problem, and this cleared up this point and provided a clue as to what to do next.

§6.2 An Approximation Problem

Gram sought, and found, and answer to the following problem. Given an orthogonal system $\{\varphi_n\}$ and a positive integer N, how can we best approximate a given real valued, integrable function f by a linear combination

$$\sum_{n=1}^{N} k_n \varphi_n$$

of the first N orthogonal functions; that is, how should we choose the constants k_n for the approximation to be as good as possible?[1]

---

[1] Actually, Gram's quest was more ambitious, for he posed this problem for a given set of linearly independent functions, not necessarily orthogonal. To turn this set into an orthogonal system, he developed an orthogonalization process, later rediscovered by Schmidt in 1907 (see Exercise 2.5).

It is necessary, of course, to state what is meant by best approximation. Gram chose to define the *error* in the approximation by

$$E_N = \left\| f - \sum_{n=1}^{N} k_n \varphi_n \right\|^2$$

and the *best approximation* as the one that minimizes E_N. Then he obtained the following result.

Theorem 6.2 (**Gram, 1883**) *If f is an integrable function on (a, b), $\{\varphi_n\}$ is an orthogonal system on (a, b), and N is a positive integer, then the minimum value of E_N is obtained when $k_n = c_n$ for all n, that is, when each k_n is the nth Fourier coefficient of f with respect to $\{\varphi_n\}$, and this minimum value is*

(4)
$$\| f \|^2 - \sum_{n=1}^{N} c_n^2 \|\varphi_n\|^2.$$

Proof. If we put $\sigma_N = \sum_{n=1}^{N} k_n \varphi_n$, where the k_n are constants, then E_N equals

$$
\begin{aligned}
\| f - \sigma_N \|^2 &= (f - \sigma_N, f - \sigma_N) \\
&= (f, f) - (\sigma_N, f) - (f, \sigma_N) + (\sigma_N, \sigma_N) \\
&= \| f \|^2 - 2(f, \sigma_N) + (\sigma_N, \sigma_N) \\
&= \| f \|^2 - 2 \sum_{n=1}^{N} k_n (f, \varphi_n) + \sum_{n=1}^{N} k_n^2 (\varphi_n, \varphi_n)
\end{aligned}
$$

since $(\varphi_m, \varphi_n) = 0$ when $m \neq n$. By Definition 6.2, $(f, \varphi_n) = c_n \|\varphi_n\|^2$, and then the right-hand side above becomes

$$\| f \|^2 + \sum_{n=1}^{N} [k_n^2 \|\varphi_n\|^2 - 2k_n c_n \|\varphi_n\|^2] = \| f \|^2 + \sum_{n=1}^{N} [k_n \|\varphi_n\| - c_n \|\varphi_n\|]^2 - \sum_{n=1}^{N} c_n^2 \|\varphi_n\|^2.$$

It has a minimum when $k_n = c_n$, and then the minimum value of $E_N = \| f - \sigma_N \|^2$ is given by (4), Q.E.D.

It is quite interesting, and no less unexpected, that it should be precisely when the k_n are the Fourier coefficients of f that one obtains the minimum error in the approximation.

Now, it is natural to expect—or at least to wish—that the error in approximating an arbitrary function f by the Nth partial sum of its Fourier series, s_N, approaches zero as $N \to \infty$. To obtain a necessary and sufficient condition for this fact to be true note that Theorem 6.2 shows that

$$\| f - s_N \|^2 = \| f \|^2 - \sum_{n=1}^{N} c_n^2 \|\varphi_n\|^2,$$

and then that the right-hand side approaches zero as $N \to \infty$ if and only if Parseval's identity (3) holds for f. This proves the following corollary.

Corollary 1 *If f is an integrable function on (a, b), $\{\varphi_n\}$ is an orthogonal system on (a, b), and s_N is the Nth partial sum of the Fourier series of f with respect to $\{\varphi_n\}$, then $\| f - s_N \| \to 0$ as $N \to \infty$ if and only if Parseval's identity (3) holds for f.*

We see then that two different problems involving Fourier series highlight the importance of Parseval's identity. Therefore, the need to prove it can no longer be ignored. Furthermore, Parseval's identity is of importance in other applications not related to the current situation. For instance, in Exercises 4.4 and 4.13 we see how it can be of assistance in computing the sum of certain numerical series, in Exercise 4.11 it is shown to hold the key to term by term integrability of Fourier series, and in §6.5 it leads to a surprising conclusion when evaluating the energy of the vibrating string.

We shall return to Parseval's identity in §6.4. For the time being we shall be content with a much weaker result that also follows immediately from Theorem 6.2. For any $N > 0$, the value of (4) is nonnegative since it is the minimum value of $E_N \geq 0$. Since N is arbitrary, we have the following corollary.

Corollary 2 (Bessel's Inequality) *If f is an integrable function on (a, b) and $\{\varphi_n\}$ is an orthogonal system on (a, b), then*

$$(5) \qquad \sum_{n=1}^{\infty} c_n^2 \|\varphi_n\|^2 \leq \|f\|^2.$$

This inequality owes its name to Friedrich Wilhelm Bessel (1784–1846), who proved it in 1828 for the usual trigonometric system on $(-\pi, \pi)$. In its more general form it was first stated by Schmidt in his doctoral dissertation of 1905. Bessel's inequality is the easy half of Parseval's identity, but it is useful by itself. In the next section we shall use it to fill a gap in Chapter 3.

§6.3 The Uniform Convergence of Fourier Series

This section is in the nature of a digression to deal with some unfinished business. We shall use Bessel's inequality to prove Theorem 3.2. It is sufficient to do it for $f : (-\pi, \pi) \to \mathbb{R}$ since a change of variable would take care of the more general situation. Then we are dealing with the orthogonal system

$$(6) \qquad \tfrac{1}{2}, \cos x, \sin x, \ldots, \cos nx, \sin nx, \ldots .$$

Theorem 6.3 (Heine, 1870) *Let $f : [-\pi, \pi] \to \mathbb{R}$ be a continuous function that has a piecewise continuous derivative and is such that $f(-\pi) = f(\pi)$. Then its Fourier series on $(-\pi, \pi)$ with respect to (6) converges uniformly to f on $[-\pi, \pi]$.*

Proof. The uniform convergence of the Fourier series

$$\tfrac{1}{2}a_0 + \sum_{n=1}^{\infty} (a_n \cos nx + b_n \sin nx)$$

of f would follow from the Weierstrass M-test (Theorem A.2) if we show that for each $n \geq 1$ there is a number M_n such that $|a_n \cos nx + b_n \sin nx| \leq M_n$ and that the series $\sum_{n=1}^{\infty} M_n$ converges.

Now, for any real numbers r and s, $0 \leq (r-s)^2 = r^2 + s^2 - 2rs$, so that

(7) $$2rs \leq r^2 + s^2$$

and then

$$(r+s)^2 = r^2 + s^2 + 2rs \leq 2(r^2 + s^2).$$

As a consequence of this inequality we obtain

$$(a_n \cos nx + b_n \sin nx)^2 \leq 2(a_n^2 \cos^2 nx + b_n^2 \sin^2 nx) \leq 2(a_n^2 + b_n^2),$$

and then, for $n \geq 1$, (7) yields

$$|a_n \cos nx + b_n \sin nx| \leq \sqrt{2(a_n^2 + b_n^2)} < \frac{2}{n}\sqrt{n^2(a_n^2 + b_n^2)} \leq \frac{1}{n^2} + n^2(a_n^2 + b_n^2).$$

Since $\sum_{n=1}^{\infty} (1/n^2)$ converges, it remains to be proved that

$$\sum_{n=1}^{\infty} n^2(a_n^2 + b_n^2)$$

converges.

For $n \geq 1$, nb_n and $-na_n$ are precisely the Fourier coefficients of f' on $(-\pi, \pi)$. In fact, for $n \geq 1$, integration by parts shows that

$$\frac{1}{\pi} \int_{-\pi}^{\pi} f'(x) \cos nx \, dx = \frac{1}{\pi}\left[f(x) \cos nx \Big|_{-\pi}^{\pi} + \int_{-\pi}^{\pi} nf(x) \sin nx \, dx \right]$$

$$= \frac{1}{\pi}\left[f(\pi) \cos n\pi - f(-\pi) \cos(-n\pi) + \int_{-\pi}^{\pi} nf(x) \sin nx \, dx \right]$$

$$= \frac{n}{\pi} \int_{-\pi}^{\pi} f(x) \sin nx \, dx$$

$$= nb_n,$$

and, similarly,

$$\frac{1}{\pi} \int_{-\pi}^{\pi} f'(x) \sin nx \, dx = -na_n.$$

Then, since the trigonometric functions in the orthogonal system (6) have norm $\sqrt{\pi}$, Bessel's inequality for f' gives

$$\sum_{n=1}^{\infty} n^2(a_n^2 + b_n^2) \leq \frac{1}{\pi} \int_{-\pi}^{\pi} (f')^2,$$

showing that the series on the left converges, Q.E.D.

This completes the proof of Theorem 3.3.

§6.4 Convergence in the Mean

In view of Gram's result, Ernst Fischer (1875–1959), a professor at the University of Cologne, introduced in 1907 a new concept of convergence.

Definition 6.4 *A sequence* $\{f_n\}$ *of integrable functions on* (a, b) **converges in the mean** *to an integrable function* f *if and only if* $\|f - f_n\| \to 0$ *as* $n \to \infty$. *Then we write*

$$\text{l.i.m. } f_n = f$$

or simply $f_n \to f$ *in the mean as* $n \to \infty$.

Here l.i.m. means *limit in the mean* as $n \to \infty$. Properly speaking we should refer to this type of convergence as *mean-square convergence*, since other kinds of means can be obtained by integrating $|f - f_n|^p$, where p is a positive integer other than 2. However, only the case $p = 2$ will be considered in this book, and it is appropriate to just talk about convergence in the mean in this context. Then Corollary 1 of Gram's theorem can be rephrased as follows.

Theorem 6.4 *Let* f *be an integrable function on* (a, b) *and* $\{\varphi_n\}$ *an orthogonal system on* (a, b). *Then the Fourier series of* f *with respect to* $\{\varphi_n\}$ *converges to* f *in the mean if and only if Parseval's identity* (3) *holds for* f.

Once more, Parseval's identity is shown as holding the key to an important fact—this time convergence in the mean—and we must address the problem of proving it. For the particular case of the usual trigonometric system, this was done independently by Charles Jean de la Vallée Poussin (1866–1962), of Louvain, in 1893, by Aleksandr Mikhailovich Liapunov (1857–1918) in 1896, and by Adolf Hurwitz (1859–1919) in 1903. We shall first present this result in the more general setting in which the orthogonal system consists of the eigenfunctions of a regular Sturm-Liouville problem

$$(pX')' + (\lambda r - q)X = 0,$$
$$a_1 X(a) - b_1 X'(a) = 0$$
$$a_2 X(b) + b_2 X'(b) = 0$$

(for conditions on p, r, q, and the constants see §4.1 and Definition 4.3). Keep in mind, when a norm is written below, that these eigenfunctions are orthogonal with weight function r. The proof of the next theorem is based on Theorem 4.6, which was stated without proof in Chapter 4.

Theorem 6.5 *Let* $f : [a, b] \to \mathbb{R}$ *be an integrable function and let* $\{X_n\}$ *be the sequence of eigenfunctions of a regular Sturm-Liouville problem on* $[a, b]$. *Then Parseval's identity holds for* f *with respect to* $\{X_n\}$

$$(8) \qquad \|f\|^2 = \sum_{n=1}^{\infty} c_n^2 \|X_n\|^2,$$

and the Fourier series of f *with respect to* $\{X_n\}$ *converges to* f *in the mean.*

Proof. We shall show below that

1. f can be approximated as closely as desired by a piecewise constant function g such that $g \equiv 0$ on $[a, a + \eta] \cup (b - \eta, b]$ for some small $\eta > 0$,

2. by rounding the corners, g can be approximated as closely as desired by a function h of class C^1 such that $h(a) = h(b) = h'(a) = h'(b) = 0$,

3. for N large enough, h can be approximated uniformly as closely as desired by the Nth partial sum h_N of its Fourier series with respect to $\{X_n\}$, and

4. as a consequence of the preceding approximations, $\|f - h_N\| \to 0$ as $N \to \infty$.

Accepting these facts, note that h_N is of the form $\sum_{n=1}^{N} k_n X_n$, and then, if s_N is the Nth partial sum of the Fourier series of f with respect to $\{X_n\}$, Theorem 6.2 implies that

$$0 \le \|f - s_N\|^2 \le \|f - h_N\|^2.$$

Therefore, $\|f - s_N\| \to 0$ as $N \to \infty$, that is, $f = \text{l.i.m. } s_N$, and then Parseval's identity holds for f with respect to $\{X_n\}$ by Theorem 6.4. Only statements 1 to 4 remain to be proved.

1. Let $M > 0$ be such that $|f| \le M$ and let $\epsilon > 0$ be arbitrary. Since f is integrable then, by the definition of integral (Definition 2.2), there is a piecewise constant function $g : [a, b] \to \mathbb{R}$ such that $g \le f$, $|g| \le M$ and

$$\int_a^b |f - g| = \int_a^b (f - g) < \epsilon.$$

It is clear that, if η is a small enough positive number, we can modify g so that $g \equiv 0$ on $[a, a + \eta] \cup (b - \eta, b]$ without altering the preceding inequality.

2. If g is continuous define $h \equiv g \equiv 0$. Otherwise, let $x_1 < \cdots < x_n$ be the points in (a, b) at which g is discontinuous. Choose $\delta > 0$ to be so small that the intervals $I_i = [x_i - \delta, x_i + \delta]$ are in (a, b) and pairwise disjoint, and for each i define $h_i : I_i \to \mathbb{R}$ by

$$h_i(x) = \frac{g(x_i + \delta) + g(x_i - \delta)}{2} + \frac{g(x_i + \delta) - g(x_i - \delta)}{2} \left[\frac{x - x_i}{\delta} + \frac{1}{\pi} \sin \frac{\pi}{\delta} (x - x_i) \right].$$

Then h_i is of class C^1, $h_i(x_i \pm \delta) = g(x_i \pm \delta)$, $h_i'(x_i \pm \delta) = 0$, and h_i has values between $g(x_i - \delta)$ and $g(x_i + \delta)$ because its derivative has the same sign as the difference $g(x_i + \delta) - g(x_i - \delta)$ in the interior of I_i. Now define $h : [a, b] \to \mathbb{R}$ so that $h \equiv h_i$ on each I_i and $h \equiv g$ elsewhere. It follows that h is of class C^1, $|h| \le M$, $h(a) = h(b) = h'(a) = h'(b) = 0$, and

$$\int_a^b |g - h| \le \sum_{i=1}^{n} \int_{x_i - \delta}^{x_i + \delta} |g - h_i| = 2\delta \sum_{i=1}^{n} |g(x_i + \delta) - g(x_i - \delta)|,$$

which is smaller than ϵ if δ is small enough.

3. If h_N denotes the Nth partial sum of the Fourier series of h with respect to $\{X_n\}$, we know that $h_N \to h$ uniformly on $[a, b]$ by Theorem 4.6. Then N can be chosen so large that $|h_N| < M + 1$, $|h - h_N| < \epsilon/(b - a)$ on $[a, b]$ and

$$\int_a^b |h - h_N| < \epsilon.$$

4. From

$$\int_a^b |f - h_N| \leq \int_a^b (|f - g| + |g - h| + |h - h_N|) < 3\epsilon$$

and from

$$(f - h_N)^2 = |f - h_N||f - h_N| \leq (|f| + |h_N|)|f - h_N| < (2M + 1)|f - h_N|$$

and if r_M denotes the maximum value of r on $[a, b]$, we conclude that

$$\|f - h_N\|^2 = \int_a^b r(f - h_N)^2 < r_M(2M + 1) \int_a^b |f - h_N| < 3r_M(2M + 1)\epsilon.$$

This shows that $\|f - h_N\|$ is arbitrarily small for N large enough and completes the proof, Q.E.D.

If we compare Theorem 6.5 with Theorem 4.6, we see that convergence in the mean always holds for the Fourier series of an integrable function, while it is necessary to impose some additional smoothness conditions to guarantee its pointwise convergence. It is in this sense that convergence in the mean seems to be natural for work with Fourier series.

We now state the particular version of Theorem 6.5 for the usual trigonometric system. It should be compared with Theorem 2.2.

Theorem 6.6 *Let* $f : (-\pi, \pi) \to \mathbb{R}$ *be an integrable function that has Fourier series*

$$\tfrac{1}{2}a_0 + \sum_{n=1}^{\infty}(a_n \cos nx + b_n \sin nx)$$

on $(-\pi, \pi)$. *Then Parseval's identity*

$$\frac{1}{\pi} \int_{-\pi}^{\pi} f^2 = \tfrac{1}{2}a_0^2 + \sum_{n=1}^{\infty}(a_n^2 + b_n^2)$$

holds for f, *the usual trigonometric system*

(6) $$\tfrac{1}{2}, \cos x, \sin x, \ldots, \cos nx, \sin nx, \ldots$$

is complete, and the stated Fourier series converges to f *in the mean.*

Proof. This proof is identical to that of Theorem 6.5 but replacing $[a, b]$ with $[-\pi, \pi]$, replacing $\{\varphi_n\}$ with (6), using Theorem 6.3 instead of Theorem 4.6, and noticing that $\|1/2\|^2 = \pi/2$ and $\|\sin nx\|^2 = \|\cos nx\|^2 = \pi$, Q.E.D.

★   When a given orthogonal system is not derived from the eigenfunctions of a Sturm-Liouville problem, then it is necessary to provide an independent proof of Parseval's identity. To illustrate this in the case of the Haar functions of Example 6.3 we shall use the following notation and terminology. If a function f has domain in the real line, its *support*, which we shall denote by $S(f)$, is defined to be the set of all x in $\mathbb{R}$ such that $f(x) \neq 0$.

Theorem 6.7 *If f is an integrable function on $[0, 1)$, then Parseval's identity holds for f with respect to the Haar orthogonal system.*

Proof.[1] Let $M > 0$ be such that $|f| \leq M$. Then, according to the definition of integral (Definition 2.2), given any $\epsilon > 0$, there is a positive integer m that is a power of 2, a partition of $[0, 1)$ into $2m$ subintervals of equal length

$$I_{mk} = \left[\frac{k-1}{2m}, \frac{k}{2m}\right),$$

$k = 1, \ldots, 2m$, and a function $g : [0, 1) \to \mathbb{R}$ that is constant on each I_{mk} and such that $g \leq f$, $|g| \leq M$ and

$$\int_0^1 |f - g| = \int_0^1 (f - g) < \frac{\epsilon}{2M}.$$

But then

$$\|f - g\|^2 = \int_0^1 |f - g|^2 \leq \int_0^1 (|f| + |g|)|f - g| \leq 2M \int_0^1 (f - g) < \epsilon.$$

If for each fixed k we define a function u_{mk} to have value 1 on I_{mk} and to be zero elsewhere on $[0, 1)$, then it is clear that g is a linear combination of the u_{mk}. Next we claim that each u_{mk} is a linear combination of the Haar functions $\varphi_0, \varphi_{11}, \ldots, \varphi_{mm}$. Indeed, there are unique positive integers $n_2, n_4, n_8, \ldots, n_m$ such that

$$S(u_{mk}) \subset S(\varphi_{mn_m}) \subset \cdots \subset S(\varphi_{4n_4}) \subset S(\varphi_{2n_2})$$

(refer to Figure 6.1). Now consider the function $\varphi_0 + c_1\varphi_{11}$, where $c_1 = 1$ if $S(u_{mk}) \subset S(\varphi_0 + \varphi_{11})$ or $c_1 = -1$ if $S(u_{mk}) \subset S(\varphi_0 - \varphi_{11})$. Then $\varphi_0 + c_1\varphi_{11}$ has value 2 on its support, which is an interval of length $1/2$. With c_1 already chosen, consider next $\varphi_0 + c_1\varphi_{11} + 2c_2\varphi_{2n_2}$, where $c_2 = 1$ if $S(u_{mk}) \subset S(\varphi_0 + c_1\varphi_{11} + 2\varphi_{2n_2})$ or $c_2 = -1$ if $S(u_{mk}) \subset S(\varphi_0 + c_1\varphi_{11} - 2\varphi_{2n_2})$. Then $\varphi_0 + c_1\varphi_{11} + 2c_2\varphi_{2n_2}$ has value 4 on its support, which is an interval of length $1/4$. Continue in this manner until

---

[1]For a very short proof using linear algebra see, for instance, G. B. Folland, *Fourier Analysis and Its Applications*, Wadsworth & Brooks/Cole, Pacific Grove, California, 1992, page 200. Here we have opted for a constructive proof.

$\varphi_0 + c_1\varphi_{11} + 2c_2\varphi_{2n_2} + 4c_4\varphi_{4n_4} + \cdots + mc_m\varphi_{mn_m}$ has value $2m$ on its support, which has length $1/2m$ and coincides with $S(u_{mk})$. Then

$$u_{mk} = \frac{1}{2m}(\varphi_0 + c_1\varphi_{11} + 2c_2\varphi_{2n_2} + 4c_4\varphi_{4n_4} + \cdots + mc_m\varphi_{mn_m}).$$

It follows that g is a linear combination of $\varphi_0, \varphi_{11}, \ldots, \varphi_{mm}$. If we replace its coefficients by the Fourier coefficients of f with respect to $\varphi_0, \varphi_{11}, \ldots, \varphi_{mm}$, we obtain a partial sum s_m which, by Theorem 6.2, satisfies

$$0 \le \|f - s_m\|^2 \le \|f - g\|^2 < \epsilon.$$

Since ϵ is arbitrary, this shows that $s_m \to f$ in the mean as $m \to \infty$, and, by Theorem 6.4, Parseval's identity holds for f with respect to the Haar functions, Q.E.D.

The Haar functions can be used to point out some interesting developments in the subject of orthogonal systems. If we extend φ_{11} to $\mathbb{R}$ with zero values outside $[0, 1)$, all but the first of the Haar functions—or rather their extensions to $\mathbb{R}$ with zero values outside $[0, 1)$—can be obtained by appropriately translating and dilating φ_{11}. In fact,

$$\varphi_{mn}(x) = \varphi_{11}(mx - n + 1)$$

for each positive integer m of the form 2^k and each $n = 1, \ldots, m$. If, instead, k and n are allowed to asssume any integral values in the preceding formula, we obtain a larger orthogonal system $\{\varphi_{mn}\}$ on $\mathbb{R}$ which can be shown to be complete with respect to to the inner product

$$(f, g) = \int_{-\infty}^{\infty} fg.$$

The function $\varphi_0 \equiv 1$, which is orthogonal to all the φ_{mn}, must be discarded now because its square is not integrable on $\mathbb{R}$ and does not have a norm.

Consider now a function $f : \mathbb{R} \to \mathbb{R}$ that vanishes outside an interval $[-a, a]$ and is integrable on $[-a, a]$. Since $\|\varphi_{mn}\|^2 = 1/m$, f has Fourier series

$$\sum m(f, \varphi_{mn})\varphi_{mn}$$

with respect to $\{\varphi_{mn}\}$. The advantage of this system is that its elements, which we shall still call Haar functions, have a very small support for m large. Given a point x, only those terms of the stated Fourier series for which x is in the support of the corresponding φ_{mn} contribute to the sum and the rest are zero. In a similar vein, to study the behavior of f on a small interval I it is necessary only to look at those terms for which the support of φ_{mn} intersects I. This is in marked contrast with the usual trigonometric Fourier series of f on $[-a, a]$ because any interval intersects the support of any of its terms. In particular, the series expansion in Haar functions is reasonably efficient—because of the local character of its terms—to represent f in a small neigborhood of a point of discontinuity, while we have already seen in §2.6 that the trigonometric Fourier series is totally inadequate for the job.

On the other hand, if f is very smooth, a large number of the steplike Haar functions may be necessary to reconstruct f to a reasonable degree. In other words, the convergence of the expansion of a smooth function in a Fourier series with respect to the Haar functions is necessarily slow. This disadvantage has been overcome by a sequence of developments culminating in the following theorem by Ingrid Daubechies, then at *Vrije Universiteit*, Brussels, and the Belgium National Science Foundation.

Theorem 6.8 (Daubechies, 1988) *For any positive integer p there are functions* $\varphi : \mathbb{R} \to \mathbb{R}$ *of class* C^p *and bounded support such that the functions*

$$\varphi_{mn}(x) = \varphi(mx - n),$$

where $m = 2^k$ *and k and n are arbitrary integers, form a complete orthogonal set with respect to the inner product*

$$(f, g) = \int_{-\infty}^{\infty} fg.$$

The functions φ_{mn} of Theorem 6.8 are called *wavelets*—a term introduced by A. Grossman and J. Morlet in 1984—and their generating function φ is called the *mother wavelet*. It should be said that the mother wavelet cannot be written in closed form, that is, by means of an analytic formula, but its graph can be constructed with arbitrarily high precision via a recursive algorithm. These wavelets represent the best of both worlds in that they are smooth, a property they share with trigonometric sines and cosines, but localized like the Haar functions. In the preceding discussion we have found it convenient, both for simplicity and because they were discovered first, to introduce a complete orthogonal system of wavelets on $(-\infty, \infty)$, but they are not ideal to represent a function defined on a bounded interval because its extension to $\mathbb{R}$ may introduce discontinuities at the endpoints, and then these will affect the series coefficients. There are two ways of dealing with this. The first is to use *periodized* wavelets, which can be obtained from the regular ones by a somewhat involved procedure. But this is more or less equivalent to analyzing the periodic extension of the given function by means of regular wavelets and, once more, discontinuities at the endpoints of the original interval will affect the coefficients in the expansion. There is another solution, proposed by Yves Meyer, of the *Institut Universitaire de France*, in 1992, and it consists of using a complete orthogonal system of wavelets whose support is in the interior of the given bounded interval plus some special functions at the edges. However, the description of this system is not trivial and cannot be included here. The study of all kinds of wavelets—for there are more than just those mentioned above—and their applications is now very vigorously pursued as a distinct field within applied mathematics.[1]

---

[1] For an account of these developments see I. Daubechies, *Ten Lectures on Wavelets*, SIAM, Philadelphia, 1992. Orthogonal wavelets with bounded support are discussed in Chapter 6 and, in particular, Daubechies' algorithm is discussed in §6.5. Periodized wavelets are studied in §9.3 and Meyer's wavelets on an interval, in §10.7. For a historical account of the subject see Y. Meyer, *Bull. Am. Math. Soc.*, **28** (1993) 350–360.

§6.5 Applications to the Vibrating String

We shall use the main results of §6.4 to draw some interesting conclusions about the vibrations of strings.

Example 6.5 As an application of Parseval's identity we shall compute the energy of a vibrating string tied down at its endpoints and arrive at a surprising result. More specifically, assume that its displacement is a solution of the boundary value problem (1)–(5) of Chapter 5. If m is the mass per unit length of the string, its kinetic energy at a fixed instant t is

$$E_k(t) = \frac{m}{2} \int_0^a u_t^2(x, t)\, dx.$$

If we define $y = \pi x/a$ and $f : (-\pi, \pi) \to \mathbb{R}$ to be the odd function such that $f(y) = u_t(ay/\pi, t)$ for $y \geq 0$, then

$$E_k(t) = \frac{am}{2\pi} \int_0^\pi u_t^2\left(\frac{ay}{\pi}, t\right) dy = \frac{am}{4\pi} \int_{-\pi}^\pi f^2(y)\, dy = \frac{am}{4\pi} \| f \|^2.$$

But f is of class C^1 on $(0, \pi)$ because u is of class C^2, and then it equals its Fourier series

(9)
$$f(y) = \sum_{n=1}^\infty b_n \sin ny.$$

By Parseval's identity for f,

$$E_k(t) = \frac{am}{4\pi} \sum_{n=1}^\infty b_n^2 \int_{-\pi}^\pi \sin^2 ny\, dy = \sum_{n=1}^\infty \frac{m}{2} \int_0^a \left[b_n^2 \sin^2 \frac{n\pi}{a} x \right] dx.$$

Since (9) is equivalent to

$$u_t(x, t) = \sum_{n=1}^\infty b_n \sin \frac{n\pi}{a} x$$

on $(0, a)$, we conclude that, at each time t, the kinetic energy of the vibrating string is the sum of the kinetic energies of the component vibrations. This is a surprising fact since we might have expected these vibrations to interfere with one another in such a manner as to invalidate this result. A similar result holds for the potential energy,

$$E_p(t) = \frac{a\tau}{4\pi} \sum_{n=1}^\infty c_n^2 \int_{-\pi}^\pi \sin^2 ny\, dy = \sum_{n=1}^\infty \frac{\tau}{2} \int_0^a \left[c_n^2 \sin^2 \frac{n\pi}{a} x \right] dx,$$

where

$$u_x(x, t) = \sum_{n=1}^\infty c_n \sin \frac{n\pi}{a} x,$$

and, therefore, for the total energy.

Example 6.6 We now return to the boundary value problem of Theorem 5.4 and assume that

$$v(x, t) = \sum_{n=1}^{\infty} \left[A_n \cos \sqrt{\lambda_n}\, t + \frac{B_n}{\sqrt{\lambda_n}} \sin \sqrt{\lambda_n}\, t \right] X_n(x)$$

is not of class C^2, but only a generalized solution in the sense described there. Now let $\tilde{f}$ and $\tilde{g}$ be functions that satisfy the same hypotheses as f and g, respectively, let $\tilde{A}_n$ and $\tilde{B}_n$ be the Fourier coefficients of $\tilde{f} - U$ and $\tilde{g}$ with respect to the eigenfunctions X_n, and assume that for this pair of functions the boundary value problem (18)–(22) of §5.6 has an exact solution

$$\tilde{v}(x, t) = \sum_{n=1}^{\infty} \left[\tilde{A}_n \cos \sqrt{\lambda_n}\, t + \frac{\tilde{B}_n}{\sqrt{\lambda_n}} \sin \sqrt{\lambda_n}\, t \right] X_n(x)$$

of class C^2. For instance, fix N and let $\tilde{f}$ equal U plus the Nth partial sum of the Fourier series of $f - U$ and let $\tilde{g}$ be the Nth partial sum of the Fourier series of g.

Since the Fourier coefficients of $f - \tilde{f}$ and $g - \tilde{g}$ are $A_n - \tilde{A}_n$ and $B_n - \tilde{B}_n$, respectively, and if we assume for convenience that the X_n are normalized, Theorem 6.5 gives

$$\|f - \tilde{f}\|^2 = \sum_{n=1}^{\infty} (A_n - \tilde{A}_n)^2$$

and

$$\|g - \tilde{g}\|^2 = \sum_{n=1}^{\infty} (B_n - \tilde{B}_n)^2.$$

Similarly, since the series expansions for $v(x, t)$ and $\tilde{v}(x, t)$ are, for each fixed t, Fourier series with respect to the X_n,

$$\|v(x, t) - \tilde{v}(x, t)\|^2 = \sum_{n=1}^{\infty} \left[(A_n - \tilde{A}_n) \cos \sqrt{\lambda_n}\, t + \frac{B_n - \tilde{B}_n}{\sqrt{\lambda_n}} \sin \sqrt{\lambda_n}\, t \right]^2.$$

It follows that $\|v(x, t) - \tilde{v}(x, t)\|$ can be made arbitrarily small if $\|f - \tilde{f}\|$ and $\|g - \tilde{g}\|$ are small enough.

Herein lies the value of the generalized solution in applications, since f and g are only mathematical approximations of the actual functions $\tilde{f}$ and $\tilde{g}$ describing the initial conditions. If we know from physical considerations that an exact solution of class C^2 exists for the actual initial conditions, the generalized solution will be a good approximation—in norm, that is, in the mean—of the exact solution if f and g are good approximations of $\tilde{f}$ and $\tilde{g}$.

★   ## §6.6   The Riesz-Fischer Theorem

Theorem 6.5 is formulated in the particular context of a system of eigenfunctions of a regular Sturm-Liouville problem. Ideally, we would like to have a theorem stating that, if f belongs to a sufficiently large class of functions and if $\{\varphi_n\}$ is an *arbitrary* complete orthogonal system in such a class, then the Fourier series of f with respect to $\{\varphi_n\}$ converges to f in the mean. Such a result exists, and it was obtained indirectly in the following manner.[1]

We already know that an orthogonal system has to be large enough—complete—for an arbitrary integrable function to be the sum of its Fourier series. But there is another question that is just as natural to ask: is the space of integrable functions large enough to contain one whose Fourier coefficients are a prescribed set of constants? This is precisely the question that a young Hungarian mathematician, Frigyes Riesz (1880–1956), posed himself, inspired by the work of Hilbert and Schmidt. In view of Bessel's inequality, Riesz posed his question as follows: given a complete orthogonal system $\{\varphi_n\}$ and a sequence of constants $\{c_n\}$ such that $\sum_{n=1}^{\infty} c_n^2 \|\varphi_n\|^2$ converges, is there an integrable function f whose Fourier coefficients are the c_n?

Let us state at the outset that the space of all Riemann integrable functions is not large enough to provide an affirmative answer to the preceding question. This we shall prove later. What Riesz did was to consider the next largest set, that of all measurable functions on $[a, b]$ that are square integrable in the sense of Lebesgue, which we shall denote by $L^2(a, b)$. Here we identify any two functions that differ only on a set of measure zero. It is necessary to specify that the functions be square integrable, a property that is necessary to define their norms (for convenience we shall assume in this section that $r \equiv 1$, but things are easily modified otherwise), because in the unbounded case a function may be Lebesgue integrable while its square is not (Exercise 9.5 of Chapter 2). Riesz was then able to show that $L^2(a, b)$ is large enough to contain a function with prescribed Fourier coefficients.

Before we prove this, note that the sum, product, and product by a constant of measurable functions are measurable.[2] Hence, if f and g are in $L^2(a, b)$, then $\|f\|$ and (f, g) are well defined and the theory developed above for Riemann integrable functions, up to and including Theorem 6.4, remains valid in this context as is. Note also that the *triangle inequality*, $\|f + g\| \le \|f\| + \|g\|$ (see Exercise 6.1 (i)), shows that $\|f + g\|^2 < \infty$, that is, that $f + g$ is in $L^2(a, b)$.

Theorem 6.9 (Riesz, 1907) *If $\{\varphi_n\}$ is an orthogonal system in $L^2(a, b)$ and if $\{c_n\}$ is a sequence of constants such that $\sum_{n=1}^{\infty} c_n^2 \|\varphi_n\|^2$ converges, then there is a function f in $L^2(a, b)$ whose Fourier coefficients with respect to $\{\varphi_n\}$ are the constants c_n, and Parseval's identity*

$$\|f\|^2 = \sum_{n=1}^{\infty} c_n^2 \|\varphi_n\|^2$$

holds for f. Furthermore, if $\{\varphi_n\}$ is complete, then there is a unique function in $L^2(a, b)$ whose Fourier coefficients are the c_n.

---

[1] §2.9 is a necessary prerequisite to understand the rest of this section. In particular, Exercises 9.10, 9.12, and 9.15 of §2.9 will be used in the proof of Theorem 6.9.

[2] See, for instance, H. L. Royden, *Real Analysis*, 2nd ed., Macmillan, New York, 1968, pages 66–67.

Proof. We shall show below that there is a subset E_0 of $[a, b]$ of measure zero such that the series

$$(10) \qquad \sum_{n=1}^{\infty} c_n \varphi_n$$

converges uniformly on $[a, b] - E_0$ and that its sum f, which we complete with arbitrary values on E_0, is in $L^2(a, b)$. Then (10) can be integrated term by term over $[a, b] - E_0$ after multiplication by φ_m (Exercise 9.12 of Chapter 2) and then over $[a, b]$ (Exercise 9.10 of Chapter 2). Doing so we obtain

$$(f, \varphi_m) = \sum_{n=1}^{\infty} c_n(\varphi_n, \varphi_m) = c_m \|\varphi_m\|^2,$$

showing that c_m is the mth Fourier coefficient of f. Similarly, term-by-term integration of (10) after multiplication by f yields

$$\|f\|^2 = (f, f) = \sum_{n=1}^{\infty} c_n(f, \varphi_n) = \sum_{n=1}^{\infty} c_n^2 \|\varphi_n\|^2,$$

proving Parseval's identity for f.

If, in addition, $\{\varphi_n\}$ is complete and if g is a function in $L^2(a, b)$ that has the same Fourier coefficients as f, then all the coefficients of $f - g$ are zero; that is, $f - g$ is orthogonal to all the φ_n. The completeness of $\{\varphi_n\}$ implies that $\|f - g\| = 0$ and then that $f = g$ since we identify functions in $L^2(a, b)$ that differ only on a set of measure zero. This proves uniqueness.

To deal with with the convergence of (10), we rearrange its terms as follows

$$\sum_{n=1}^{\infty} c_n \varphi_n = (c_0 \varphi_0 + \cdots + c_{n_1} \varphi_{n_1}) + (c_{n_1+1} \varphi_{n_1+1} + \cdots + c_{n_2} \varphi_{n_2}) + \cdots,$$

where $n_1, n_2, \ldots$ will be chosen below. If s_{n_k} denotes the partial sum $\sum_{n=1}^{n_k} c_n \varphi_n$ and if K is an arbitrary integer greater than 1, the rearranged series can be written as

$$(11) \qquad \sum_{n=1}^{\infty} c_n \varphi_n = s_{n_1} + \sum_{k=1}^{K-1} (s_{n_{k+1}} - s_{n_k}) + \sum_{k=K}^{\infty} (s_{n_{k+1}} - s_{n_k}).$$

By the Weierstrass M-test and the convergence of the geometric series, we know that the right-hand side converges absolutely and uniformly on any set in which $|s_{n_{k+1}} - s_{n_k}| < 2^{-k}$ for all $k \geq K$. Then what we do is define the set

$$E_k = \{ x \in [a, b] \colon |s_{n_{k+1}} - s_{n_k}| \geq 2^{-k} \}$$

for each $k \geq K$ and remove it from $[a, b]$. Consequently, (11) converges absolutely and uniformly on $[a, b] - \cup_{k=K}^{\infty} E_k$.

Next we show that, if m denotes Lebesgue measure, then $m(\cup_{k=K}^{\infty} E_k)$ is arbitrarily small for K large enough, and, therefore, (11) converges uniformly on $[a, b] - E_0$, where E_0 is a set of measure zero. In fact, since $\sum_{n=1}^{\infty} c_n^2 \|\varphi_n\|^2$ converges, Cauchy's criterion asserts that for any k the integer n_k can be chosen so large that

$$\|s_{n_{k+1}} - s_{n_k}\|^2 = \int_a^b (s_{n_{k+1}} - s_{n_k})^2 = \int_a^b \sum_{n=n_k+1}^{n_{k+1}} c_n^2 \varphi_n^2 = \sum_{n=n_k+1}^{n_{k+1}} c_n^2 \|\varphi_n\|^2 < \frac{1}{8^k}$$

for arbitrary $n_{k+1} > n_k$. Then, $m(E_k) < 2^{-k}$ or else

$$\|s_{n_{k+1}} - s_{n_k}\|^2 = \int_a^b (s_{n_{k+1}} - s_{n_k})^2 \geq \int_{E_k} (s_{n_{k+1}} - s_{n_k})^2 \geq \left(\frac{1}{2^k}\right)^2 m(E_k) \geq \frac{1}{8^k},$$

a contradiction. Accordingly,

$$m\left(\bigcup_{k=K}^{\infty} E_k\right) \leq \sum_{k=K}^{\infty} m(E_k) < \sum_{k=K}^{\infty} \frac{1}{2^k} = \frac{1}{2^K} \sum_{k=0}^{\infty} \left(\frac{1}{2}\right)^k = \frac{1}{2^{K-1}}.$$

It remains to be proved that f is in $L^2(a, b)$. Given $\epsilon > 0$ let k be so large that $8^{-k} < \epsilon/2$ and, then, let n_k be so large that, as above, $\|s_{n_j} - s_{n_k}\|^2 < 8^{-k}$ for $s_{n_j} > s_{n_k}$. Finally, by Fatou's lemma (Exercise 9.15 of Chapter 2), choose $n_j > n_k$ to be so large that

$$\int_a^b (f - s_{n_k})^2 \leq \int_a^b (s_{n_j} - s_{n_k})^2 + \frac{\epsilon}{2} = \|s_{n_j} - s_{n_k}\|^2 + \frac{\epsilon}{2} < \frac{1}{8^k} + \frac{\epsilon}{2} < \epsilon.$$

This shows that $f - s_{n_k}$ is in $L^2(a, b)$, and then $f = f - s_{n_k} + s_{n_k}$ is in $L^2(a, b)$, Q.E.D.

It is now easy to see that this theorem is not valid if $L^2(a, b)$ is replaced by the space of Riemann integrable functions. Let the c_n be the Fourier coefficients of the function g of Exercise 9.2 of Chapter 2, which is in $L^2(a, b)$ but is not Riemann integrable. If we assume that there is a Riemann integrable function f whose Fourier coefficients are the c_n, then the Fourier coefficients of $f - g$ are all zero; that is, $f - g$ is orthogonal to all the φ_n. But then $\|f - g\| = 0$ by the completeness of $\{\varphi_n\}$, and it follows easily from the definition of the Lebesgue integral that $f = g$ except, possibly, on a set of measure zero. This, however, contradicts the choice of g which is discontinuous on a set of positive measure.

As a straightforward consequence of Theorem 6.9 we can now show that the converse of Theorem 6.1 holds in $L^2(a, b)$ and, therefore, that convergence in the mean always holds in $L^2(a, b)$ with respect to a complete orthogonal system.

Theorem 6.10 *Let $\{\varphi_n\}$ be a complete orthogonal system in $L^2(a, b)$, and let f be arbitrary in $L^2(a, b)$. Then Parseval's identity holds for f, and its Fourier series with respect to $\{\varphi_n\}$ converges to f in the mean; that is, if c_n is the nth Fourier coefficient of f with respect to $\{\varphi_n\}$,*

$$f = \text{l.i.m.} \sum_{n=1}^{\infty} c_n \varphi_n.$$

Proof. The last assertion is a consequence of Theorem 6.4. Recall now that the results of §6.2 apply to functions in $L^2(a, b)$. Then Bessel's inequality

$$\sum_{n=1}^{\infty} c_n^2 \|\varphi_n\|^2 \leq \|f\|^2$$

shows that the series on its left-hand side converges. By Theorem 6.9, there is a function g in $L^2(a, b)$ whose nth Fourier coefficient is c_n and such that

$$\|g\|^2 = \sum_{n=1}^{\infty} c_n^2 \|\varphi_n\|^2.$$

Since the Fourier coefficients of $f - g$ are all zero, the completeness of $\{\varphi_n\}$ implies that $\|f - g\| = 0$ and then that $f = g$ and $\|f\|^2 = \|g\|^2$, showing that Parseval's identity holds for f because it holds for g, Q.E.D.

A few months after Riesz proved his theorem, Fischer showed that a sequence of functions $\{f_n\}$ in $L^2(a, b)$ converges in the mean to a function in $L^2(a, b)$ if and only if $\|f_m - f_n\| \to 0$ as $m, n \to \infty$ and then deduced Theorem 6.9 as a corollary. For this reason, Theorem 6.9 is usually known as the *Riesz-Fischer Theorem*. Fischer also showed that the use of $L^2(a, b)$ is necessary in that no smaller set of functions will suffice. Thus, the results of Riesz and Fischer served to highlight the importance of the Lebesgue integral, which has since become an indispensable tool in mathematical analysis.

We have also seen a notion of distance from a function f to the zero function, namely $\|f\|$, and an analogy that regards $L^2(a, b)$ as an infinite-dimensional Euclidean space in which every function is the sum of its components—in the mean—and with Parseval's identity playing the role of the Pythagorean theorem. At about this time, in his doctoral dissertation of 1906, Maurice Fréchet (1878–1973) introduced a general concept of distance in spaces of functions, which then become *metric spaces*. A *distance* is a nonnegative function d that associates to each pair of functions f and g a real number $d(f, g)$ with the following properties:

1. $d(f, g) = 0$ if and only if $f = g$,

2. $d(f, g) = d(g, f)$, and

3. $d(f, g) \leq d(f, h) + d(h, g)$ for any other function h.

Then $L^2(a, b)$ becomes a metric space if we define $d(f, g) = \|f - g\|$, as Fréchet himself did. Another metric space, now denoted by l^2, is the set of all sequences $\{k_n\}$ of constants such that $\sum_{n=1}^{\infty} k_n^2$ converges. This space was formally introduced by Schmidt in 1908, building on the work of Hilbert who had used these sequences in solving certain integral equations. Both Schmidt and Fréchet had noticed that $L^2(a, b)$ and l^2 are geometrically analogous, a fact that became completely clear with the proof of the Riesz-Fischer theorem. In fact, given a complete orthogonal system $\{\varphi_n\}$ in $L^2(a, b)$, there is a one-to-one correspondence between this space and l^2 established as follows. To each function f in $L^2(a, b)$ we associate the sequence $\{c_n\|\varphi_n\|\}$ in l^2, where the c_n

DAVID HILBERT *circa* 1900
Courtesy of Springer-Verlag Archives.

are the Fourier coefficients of f with respect to $\{\varphi_n\}$; to each sequence in l^2 we associate the function whose existence is asserted by the Riesz-Fischer theorem.

Since the work of Hilbert had fallen very short of defining l^2, both this space and $L^2(a, b)$ were denominated *Hilbert space* by Riesz in a book published in 1913. They are the prototypes of a larger class of space, an infinite-dimensional equivalent of Euclidean space for which John von Neumann (1903–1957), born János Neumann in Budapest, would later borrow the denomination *Hilbert space* and develop a complete axiomatic theory in the late 1920s at Berlin. But this was not the end of the story on function spaces, which continued with the introduction by Riesz of the $L^p(a, b)$ spaces, $p > 1$. The inner product (f, g) was abandoned, and only the concept of norm was retained, since the distance is defined from the norm. The most influential work in this direction was that of Stefan Banach (1892–1945), in particular through the publication in 1932 of his book *Théorie des opérations linéaires*.

It was then at this very fruitful point in time that the work of Hilbert and Schmidt, the discovery of Lebesgue integration, the introduction of metric spaces by Fréchet, and the connection among these ideas established by the Riesz-Fischer theorem blended together to give birth to a new mathematical discipline: *functional analysis*, as it was named by Paul P. Lévy (1886–1971).[1]

---

[1] For a detailed account of these developments see J. Dieudonné, *History of Functional Analysis*, North-Holland, Amsterdam, 1981.

EXERCISES

Unless otherwise specified, the weight function in any of the inner products below is $r \equiv 1$.

1.1 Let $f: \mathbb{R} \to \mathbb{R}$ be an odd, periodic function of period 2 such that $f \equiv 1$ on $(0, 1)$, and, for each positive integer n, define $\varphi_n(x) = f(2^n x)$. Show that $\{\varphi_n\}$ is an orthonormal system on $(0, 1)$. It is called the *Rademacher system*, as it was introduced by H. Rademacher in 1922.

1.2 A finite number of functions $f_1, \ldots, f_n$ defined on an interval I is said to be *linearly independent* on I if and only if the identity

$$c_1 f_1 + \cdots + c_n f_n = 0,$$

where $c_1, \ldots, c_n$ are constants, implies that $c_1 = \cdots = c_n = 0$. An infinite set of functions defined on I is said to be linearly independent on I if and only if every finite subset is linearly independent on I. Prove that
  (*i*) every orthogonal system on (a, b) is linearly independent on (a, b) and
  (*ii*) the sequence of functions $1, x, x^2, \ldots, x^n, \ldots$ is linearly independent on any interval.

1.3 Use De Moivre's formula $\cos n\theta + i \sin n\theta = (\cos \theta + i \sin \theta)^n$ to verify that the function $C_n(x) = \cos(n \arccos x)$ is a polynomial.

1.4 Show that the Fourier series of $f(x) = x(\pi - x)$ with respect to the orthogonal systems

$$S_1 = \{\sin x, \ldots, \sin nx, \ldots\} \quad \text{and} \quad S_2 = \{1, \cos x, \ldots, \cos nx, \ldots\}$$

on $(0, \pi)$ are

$$\frac{8}{\pi} \sum_{n=1}^{\infty} \frac{1}{(2n-1)^3} \sin(2n-1)x \quad \text{and} \quad \frac{\pi^2}{6} - \sum_{n=1}^{\infty} \frac{1}{n^2} \cos 2nx,$$

respectively.

1.5 Find the complex Fourier series of $f(x) = e^{ax}$ on $(-\pi, \pi)$ if
  (*i*) a is a real, positive number and
  (*ii*) $a = i\alpha$, where α is a real number but not an integer.

1.6 If X_n are the eigenfunctions of the Sturm-Liouville problem (1)–(3) of Chapter 4 and if r is the function of §4.1, the system $\{X_n\}$ is orthogonal with weight function r by Theorem 4.1, and then, if $\varphi_n = \sqrt{r}\, X_n$, the system $\{\varphi_n\}$ is orthogonal with weight function $r \equiv 1$. Note that the Fourier series of f with respect to $\{X_n\}$ (that is, as in Definition 4.5) is *not* the same as the Fourier series of f with respect to $\{\varphi_n\}$ and verify that, instead, the correct identification is, $\sqrt{r}$ times the Fourier series of f with respect to $\{X_n\}$ is the same as the Fourier series of $\sqrt{r}\, f$ with respect to $\{\varphi_n\}$.

1.7 Prove or disprove that the Rademacher system of Exercise 1.1 is complete.

1.8 If the function $f \equiv 1$ is added to the Rademacher system of Exercise 1.1, is the new enlarged system complete? (**Hint:** consider a function $g: (0, 1) \to \mathbb{R}$ that is orthogonal to f and such that $g(x) = g(1 - x)$.)

1.9 Let $\{\varphi_n\}$ be a complete orthogonal system on $[a, b]$. Prove that, if $f, g: [a, b] \to \mathbb{R}$ are continuous functions that have the same Fourier coefficients with respect to $\{\varphi_n\}$, then $f \equiv g$ on $[a, b]$. Is the result also true if f and g are just integrable functions?

1.10 Let $\{\varphi_n\}$ be a complete orthogonal system of continuous functions on $[a, b]$ and let $f : [a, b] \to \mathbb{R}$ be a continuous function. Use the result of Exercise 1.9 to prove that, if the Fourier series of f converges uniformly on $[a, b]$, then it converges to $f(x)$ at every point x of $[a, b]$. Is the result also true if f is just an integrable function?

2.1 Use Theorem 6.2 and Theorem 2.7 of Chapter 2 to give a short proof of the fact that, if $f : [-\pi, \pi] \to \mathbb{R}$ is continuous on $[-\pi, \pi]$ and periodic of period 2π, then Parseval's identity holds for f with respect to the usual trigonometric system

$$\tfrac{1}{2}, \; \cos x, \; \sin x, \; \ldots, \; \cos nx, \; \sin nx, \; \ldots$$

and that this identity is

$$\frac{1}{\pi} \int_{-\pi}^{\pi} f^2 = \tfrac{1}{2} a_0^2 + \sum_{n=1}^{\infty} (a_n^2 + b_n^2).$$

2.2 Let $f : (a, b) \to \mathbb{R}$ be an integrable function, let $\{\varphi_n\}$ be an orthogonal system on (a, b), and, for a given positive integer N and given constants $k_1, \ldots, k_N$, define

$$M_N = \max_{x \in [a,b]} \left| f(x) - \sum_{n=1}^{N} k_n \varphi_n(x) \right|.$$

Prove or disprove that the constants that give the smallest value of M_N are $k_n = c_n$, the first N Fourier coefficients of f with respect to $\{\varphi_n\}$, and that this smallest value of M_N is the same as that of the error E_N (**Hint:** consider an orthogonal system on $[-1, 1]$ starting with $\varphi_1 \equiv 0$ and $\varphi_2(x) = x$).

2.3 Modify the statement and proof of Theorem 6.2 and Corollary 2 so that they apply to complex valued functions.

2.4 Prove or disprove that, if $\{\varphi_n\}$ is a complete orthogonal system on (a, b), then there is an integrable function $f : (a, b) \to \mathbb{R}$ with positive norm that is orthogonal to $\varphi_1 + \varphi_n$ for each n.

2.5 Use induction to prove the so-called *Gram-Schmidt orthogonalization process*: if $\{f_n\}$ is a sequence of linearly independent continuous functions on an interval I, as defined in Exercise 1.2, then the functions defined by

$$\varphi_1 = f_1$$

$$\varphi_2 = f_2 - \frac{(f_2, \varphi_1)}{\|\varphi_1\|^2} \varphi_1$$

$$\cdot \quad \cdot \quad \cdot \quad \cdot \quad \cdot \quad \cdot \quad \cdot$$

$$\varphi_n = f_n - \left[\frac{(f_n, \varphi_1)}{\|\varphi_1\|^2} \varphi_1 + \cdots + \frac{(f_n, \varphi_{n-1})}{\|\varphi_{n-1}\|^2} \varphi_{n-1} \right],$$

and so on, form an orthogonal system on I.

2.6 Does the conclusion in Exercise 2.5 still hold if the functions f_n are just integrable? (**Hint:** show that the sequence $\{f_n\}$, where $f_n : [0, 1] \to \mathbb{R}$ is defined by

$$f_n(x) = \begin{cases} 1, & x \neq 1/n \\ 0, & x = 1/n, \end{cases}$$

is linearly independent on $[0, 1]$.)

2.7   Apply the Gram-Schmidt orthogonalization process of Exercise 2.5 with $f_n(x) = x^n$ on $[-1, 1]$ for each $n \geq 0$ to obtain the first five elements of an orthogonal set of polynomials $P_n(x)$ with $P_n(1) = 1$ for all $n \geq 0$. These are called the *Legendre polynomials* and will play a fundamental role in the solution of boundary value problems in spherical coordinates in Chapter 11.

2.8   The Gram-Schmidt orthogonalization process of Exercise 2.5 applied to the functions defined by $f_n(x) = (2x)^n e^{-x^2/2}$ on $(-\infty, \infty)$ for each $n \geq 0$ generates an orthogonal system $\{\varphi_n\}$ whose elements are called the *Hermite functions*, and the products $H_n(x) = e^{x^2/2}\varphi_n(x)$ are called the *Hermite polynomials*. Show that the first four Hermite polynomials are $H_0 \equiv 1$,

$$H_1(x) = 2x, \qquad H_2(x) = 4x^2 - 2, \qquad \text{and} \qquad H_3(x) = 8x^3 - 12x$$

and verify that they satisfy the *Hermite differential equation* $X'' - 2xX' + 2nX = 0$ that arises in some boundary value problems in parabolic coordinates.

2.9   The Gram-Schmidt orthogonalization process of Exercise 2.5 applied to the functions defined by $f_n(x) = (-x)^n e^{-x/2}$ on $[0, \infty)$ for each $n \geq 0$ generates an orthogonal system $\{\varphi_n\}$ whose elements are called the *Laguerre functions*, and the products

$$L_n(x) = \frac{e^{x^2/2}}{n!}\varphi_n(x)$$

are called the *Laguerre polynomials*. Show that the first four Laguerre polynomials are $L_0 \equiv 1$,

$$L_1(x) = -x+1, \quad L_2(x) = \tfrac{1}{2}(x^2-4x+2), \quad \text{and} \quad L_3(x) = \tfrac{1}{6}(-x^3+9x^2-18x+6)$$

and verify that they satisfy the *Laguerre differential equation* $xX'' + (1-x)X' + nX = 0$ that arises in some boundary value problems in spherical coordinates.

2.10   Find the best approximation, in the sense that E_3 is a minimum, of
 (*i*) $f(x) = e^x$ on $[-1, 1]$,
 (*ii*) $f(x) = e^{-x^2}$ on $(-\infty, \infty)$, and
 (*iii*) $f(x) = e^{-3x/2}$ on $[0, \infty)$
of the form $a\varphi_1(x) + b\varphi_2(x) + c\varphi_3(x)$, where a, b, and c are constants and φ_1, φ_2, and φ_3 are the first three (*i*) Legendre polynomials, (*ii*) Hermite functions, and (*iii*) Laguerre functions, of Exercises 2.7, 2.8 and 2.9, respectively.

3.1   Let $f : [-\pi, \pi] \to \mathbb{R}$ be a continuous function with a piecewise continuous derivative such that $f(-\pi) = f(\pi)$. Prove that its usual Fourier series on $(-\pi, \pi)$ can be differentiated term by term and that the differentiated series converges to $f'(x)$ at each point x in $(-\pi, \pi)$, where f' is continuous and has one-sided derivatives.

3.2   If f is as in Exercise 3.1, show that Parseval's identity holds for f with respect to the usual trigonometric system (6).

3.3   State and prove a result similar to that of Exercise 3.1 for the Fourier sine series and the Fourier cosine series of a function $f : (0, \pi) \to \mathbb{R}$.

3.4   Assume that $f : [0, \pi] \to \mathbb{R}$ is a function of class C^1 such that $f(0) = f(\pi) = 0$. Prove that

$$\int_0^\pi f^2 \leq \int_0^\pi (f')^2.$$

For what functions does equality hold? (**Hint:** use the results of Exercises 3.3 and 2.1.)

3.5 Show that, if a trigonometric series of the form

$$\tfrac{1}{2}a_0 + \sum_{n=1}^{\infty}(a_n \cos nx + b_n \sin nx)$$

is such that

$$a_n^2 \le \frac{1}{n^4} \qquad \text{and} \qquad b_n^2 \le \frac{1}{n^4},$$

then it is a uniformly convergent Fourier series on $[-\pi, \pi]$.

4.1 Consider a sequence of integrable functions $\{f_n\}$ defined on $[0, 1]$ by

$$f_n(x) = \begin{cases} n^{p/4}, & 0 < x < 1/n \\ 0, & \text{otherwise,} \end{cases}$$

where p is a positive integer. Show that

(*i*) $\lim_{n \to \infty} f_n(x) = 0$,

(*ii*) $f_n \to 0$ in the mean if $p = 1$, and

(*iii*) $f_n \not\to 0$ in the mean if $p > 1$.

4.2 Give an example of a sequence of integrable functions $\{f_n\}$ on $[0, 1]$ such that $f_n \to 0$ in the mean but $f_n(x) \not\to 0$ for all x in $[0, 1]$.

4.3 Consider the orthogonal systems S_1 and S_2 on $(0, \pi)$ of Exercise 1.4.

(*i*) Prove that Parseval's identity holds for any function integrable on $(0, \pi)$ with respect to S_1 and S_2.

(*ii*) Prove that S_1 and S_2 are complete on $(0, \pi)$.

(*iii*) Are S_1 and S_2 complete on $(-\pi, \pi)$?

(*iv*) Show that, if an integrable function f has Fourier series

$$\sum_{n=1}^{\infty} b_n \sin nx \qquad \text{and} \qquad a_0 + \sum_{n=1}^{\infty} a_n \cos nx$$

on $(0, \pi)$ with respect to S_1 and S_2, respectively, then

$$\sum_{n=1}^{\infty} b_n^2 = 2a_0^2 + \sum_{n=1}^{\infty} a_n^2.$$

(**Hint:** for (*i*) consider the usual Fourier series of the odd extension of a function $f : (0, \pi) \to \mathbb{R}$ to $(-\pi, \pi)$).

4.4 Use the function $f(x) = x(\pi - x)$ and the result of Exercise 4.3 (*i*) to prove the identities

$$\sum_{n=1}^{\infty} \frac{1}{(2n-1)^6} = \frac{\pi^6}{960} \qquad \text{and} \qquad \sum_{n=1}^{\infty} \frac{1}{n^4} = \frac{\pi^4}{90}.$$

4.5 Let s_N denote the sum of all the terms in any of the Fourier series for $f(x) = x(\pi - x)$ found in Exercise 1.4 up to and including the term for $n = N$.

(*i*) Use Parseval's identity, proved in Exercise 4.3 (*i*), to show that the error in approximating f by s_N is given by

$$E_N = \| f - s_N \|^2 = \sum_{n=N+1}^{\infty} c_n^2 \| \varphi_n \|^2.$$

(*ii*) Use the stated expression for E_N to decide which of the two series gives the best approximation, that is, the smallest value of E_N.

(*iii*) Use the result of Exercise 4.4 to find the smallest value of N so that, in each case, $E_N < 0.006$.

4.6   It is shown in the proof of the integral test in calculus that, if $F : [N, \infty) \to \mathbb{R}$, where N is a positive integer, is continuous and decreasing, then

$$\sum_{n=N+1}^{\infty} F(n) \le \int_N^{\infty} F.$$

Use this inequality, the function $f(x) = x(\pi - x)$, and the results of Exercise 4.5 to show that

$$E_N \le \frac{32}{10\pi(2N-1)^5} \quad \text{for } S_1 \qquad \text{and} \qquad E_N \le \frac{\pi}{12N^3} \quad \text{for } S_2.$$

4.7   Prove the following extension—discovered by Pierre Fatou in 1906—of Parseval's identity for trigonometric Fourier series: if $f, g : (-\pi, \pi) \to \mathbb{R}$ are integrable functions that have Fourier series

$$\tfrac{1}{2}a_0 + \sum_{n=1}^{\infty}(a_n \cos nx + b_n \sin nx) \qquad \text{and} \qquad \tfrac{1}{2}\alpha_0 + \sum_{n=1}^{\infty}(\alpha_n \cos nx + \beta_n \sin nx)$$

on $(-\pi, \pi)$, then

$$\frac{1}{\pi}\int_{-\pi}^{\pi} fg = \tfrac{1}{2}a_0\alpha_0 + \sum_{n=1}^{\infty}(a_n\alpha_n + b_n\beta_n).$$

Use this formula to evaluate the integral

$$\int_{-\pi}^{\pi} \frac{e^x}{1 + \tan^2 x}\, dx.$$

(**Hint:** apply Theorem 6.6 to $f + g$ and to $f - g$.)

4.8   If f and g are real valued and integrable on (a, b), prove the *Schwarz inequality*,

$$|(f, g)| \le \|f\|\,\|g\|,$$

and find a necessary and sufficient condition for equality to hold if, in addition, f and g are continuous. Then extend the result to complex valued functions (**Hint:** start with the identity

$$\int_a^b \left(\int_a^b |f(x)g(y) - f(y)g(x)|^2\, dy\right) dx \ge 0$$

and then expand the integrand).[1]

---

[1] Although this inequality was frequently used by Hermann Amandus Schwarz (1843–1921) of Berlin, it was actually discovered in 1859 by Viktor Yakovlevich Bunyakovskii (1804–1889). A similar inequality for finite sums had already been proved by Cauchy in Note *II* to his *Cours d'analyse*.

4.9   Use the Schwarz inequality of Exercise 4.8 to prove the following modified version of the Gram-Schmidt orthogonalization process: if $\{f_n\}$ is a sequence of linearly independent integrable functions with positive norms defined on an interval I, then the functions defined by

$$\varphi_1 = f_1$$

$$\cdot \quad \cdot \quad \cdot \quad \cdot \quad \cdot \quad \cdot \quad \cdot$$

$$\varphi_n = f_n - [c_1\varphi_1 + \cdots + c_{n-1}\varphi_{n-1}],$$

where $c_m = 0$ if $\|\varphi_m\| = 0$ and

$$c_m = \frac{(f_n, \varphi_m)}{\|\varphi_m\|^2}$$

if $\|\varphi_m\| \neq 0$ for $1 \leq m \leq n - 1$, and so on, are orthogonal to each other on I.

4.10   Use the *Schwarz inequality* of Exercise 4.8 to show that, if a sequence $\{f_n\}$ of integrable functions on (a, b) converges in the mean to an integrable function f, then

$$\int_a^x f_n \to \int_a^x f$$

uniformly on $[a, b]$.

4.11   Use the result of Exercise 4.10 to prove that, if Parseval's identity holds for an integrable function f on (a, b) with respect to an orthogonal system $\{\varphi_n\}$ on (a, b), then the Fourier series of f with respect to $\{\varphi_n\}$ can be integrated term by term from a to x for any x in $[a, b]$.

4.12   Show that, if f is complex valued in Theorem 6.6, then its conclusion must be modified as follows:

$$\frac{1}{\pi} \int_{-\pi}^{\pi} |f|^2 = \frac{1}{2}|a_0|^2 + \sum_{n=1}^{\infty} (|a_n|^2 + |b_n|^2).$$

Then use this formula to prove Parseval's identity for the complex Fourier series of f on $(-\pi, \pi)$:

$$\frac{1}{2\pi} \int_{-\pi}^{\pi} |f|^2 = \sum_{n=-\infty}^{\infty} |c_n|^2.$$

This, in turn, proves the completeness of the orthogonal system $\{e^{inx}\}$ on $(-\pi, \pi)$ (**Hint:** show first that

$$c_n = \tfrac{1}{2}(a_n - ib_n) \quad \text{and} \quad c_{-n} = \tfrac{1}{2}(a_n + ib_n)$$

for all $n \geq 0$, where we define $b_0 = 0$).

4.13   Use the results of Exercises 1.5 and 4.12 to show that,
(*i*) if $a > 0$, then

$$\sum_{n=1}^{\infty} \frac{1}{n^2 + a^2} = \frac{\pi}{2a} \coth a\pi - \frac{1}{2a^2},$$

and
(*ii*) if α is not an integer, then

$$\sum_{n=-\infty}^{\infty} \frac{1}{(n - \alpha)^2} = \frac{\pi^2}{\sin^2 \alpha\pi}.$$

4.14 Prove that the orthogonal system of Chebyshev polynomials of Example 6.1 is complete (**Hint:** use the result of Exercise 4.3 (*ii*)).

6.1 Show that the Schwarz inequality is valid for functions f and g in $L^2(a, b)$ and then that

(*i*) $\|f + g\| \le \|f\| + \|g\|$ (the *triangle inequality*),

(*ii*) $\|f - g\| \ge |\|f\| - \|g\||$, and

(*iii*) $\|f + g\|^2 = 2(\|f\|^2 + \|g\|^2)$ (the *parallelogram law*).

6.2 Let $\{f_n\}$ be a sequence of functions in $L^2(a, b)$. We say, with Riesz, that f_n *converges weakly* to a function f in $L^2(a, b)$ as $n \to \infty$ if and only if $(f_n, g) \to (f, g)$ as $n \to \infty$ for all g in $L^2(a, b)$. Show that, if $f_n \to f$ in the mean as $n \to \infty$, then

(*i*) f_n converges weakly to f as $n \to \infty$ and

(*ii*) $\|f_n\| \to \|f\|$ as $n \to \infty$

(**Hint:** for (*ii*) use the triangle inequality of Exercise 6.1).

6.3 Using $f_n(x) = \sin nx$ on $[0, \pi]$ and the Riemann-Lebesgue theorem—also valid in $L^2(a, b)$—show that the converse of the statements made in Exercise 6.2 are not true. However, prove that, if both f_n converges weakly to f and $\|f_n\| \to \|f\|$ as $n \to \infty$, then $f_n \to f$ in the mean.

6.4 Let $\{f_n\}$ be a sequence of functions in $L^2(a, b)$ such that $f_n(x) \to f(x)$ as $n \to \infty$ for all x in $[a, b]$, where f is in $L^2(a, b)$. Prove that, if there is a function g in $L^2(a, b)$ such that $|f_n| \le |g|$ on $[a, b]$, then $f_n \to f$ in the mean (**Hint:** use the Lebesgue dominated convergence theorem of Exercise 9.13 of Chapter 2).

7

FOURIER TRANSFORMS

§7.1 The Laplace Equation

In his 1811 memoir Fourier considered the problem of heat propagation in an ideal bar of infinite length whose initial temperature is a known function f. He soon discovered that a series solution was not possible in this case and proposed, instead, an integral solution. To satisfy the initial condition, this integral must equal $f(x)$ when $t = 0$, and then a certain unknown function in the integrand—rather than a sequence of coefficients as in the case of a finite bar—has to be determined. This function is known today as the *Fourier transform* of f, and it will be properly defined below.

The Fourier transform was also discovered a little later by Cauchy and Poisson—in papers read at the *Institut de France* in 1815—working independently on fluid flow. They arrived at the transform from their study of the following equation

$$(1) \qquad\qquad u_{xx} + u_{yy} = 0.$$

This is usually known as the *Laplace equation* or the *potential equation* because its three-dimensional form appears prominently in Laplace's 1782 study of Newtonian potential, specifically of the gravitational attraction of spheroids and the shape of the planets. However, it was originally discovered by Euler, also in its three-dimensional form, in his *Principia motus fluidorum* of 1752. Since then it has been of use in many disciplines, including fluid flow, gravitational potential, propagation of heat, electricity, and magnetism. In 1778 Lagrange used the two-dimensional Laplace equation to study the propagation of waves in a thin layer of water, but died before he could incorporate any results in a revised edition of his *Mécanique analytique*, which he started in 1810. Perhaps for this reason, the *Institut de France* decided that this should be the subject for the 1815 prize competition, and Cauchy was the winner with the paper *Théorie de la propagation des ondes à la surface d'un fluide pesant d'une profondeur indéfinie*, in which he introduced the Fourier transform. At that time Cauchy was unaware of Fourier's earlier discovery, but in 1818 he acknowledged Fourier's priority. Poisson,

PIERRE SIMON LAPLACE
From H. S. Williams, *Modern Development of the
Physical Sciences, Book III*, The Goodhue Company, 1912.

who could not compete for the prize because he was a member of the *Institut*, presented
a paper, later incorporated into his *Mémoire sur la théorie des ondes* of 1818, containing
similar results and also using the Fourier transform.

Next we shall derive the two-dimensional Laplace equation and then pose a boundary
value problem that leads to the Fourier transform. Consider a thin layer of water, and
assume that its molecules move in planes parallel to the xy-plane, so that, essentially,
we have a two-dimensional situation. We shall work under the following hypotheses,
which approximately describe the behavior of liquids such as water.

1. The liquid is *homogeneous*; that is, at each time it has the same density at all
   points.

2. The liquid is *incompressible*; that is, at each point it has the same density at all
   times, which is therefore constant and we denote it by ρ.

3. The amount of work done in moving a particle from one point in the liquid to
   another is independent of the path followed in the motion, a fact that is usually
   expressed by saying that the velocity field of the liquid is *conservative*.

Let $v(x, y, t)$ denote the velocity of the fluid at a point (x, y) at time t, and let v_1 and v_2

be the components of v in the x and y directions, respectively. We assume that these components are of class C^1 and consider a small rectangle $[x, x + h] \times [y, y + k]$. If v_1 were independent of y, the instantaneous rate of mass flow through the side at x at time t would be $\rho k v_1(x, y, t)$, but since v_1 depends on y this rate is given by the integral

$$\int_y^{y+k} \rho v_1(x, \eta, t)\, d\eta.$$

Similar expressions hold for the other three sides, and then, since the flow is incompressible, the principle of conservation of mass demands that the total instantaneous rate of mass flow into the rectangle be zero. Hence,

$$\int_y^{y+k} \rho[v_1(x, \eta, t) - v_1(x+h, \eta, t)]\, d\eta + \int_x^{x+h} \rho[v_2(\xi, y, t) - v_2(\xi, y+k, t)]\, d\xi = 0.$$

Dividing by $-\rho hk$ and using the mean value theorem of the integral calculus, we obtain

$$\frac{v_1(x + h, \eta^*, t) - v_1(x, \eta^*, t)}{h} + \frac{v_2(\xi^*, y + k, t) - v_2(\xi^*, y, t)}{k} = 0,$$

where $y \leq \eta^* \leq y + k$ and $x \leq \xi^* \leq x + h$. Then, letting h and k approach zero, it is easily seen, as in Exercise 2.1 of Chapter 1, that

$$v_{1_x}(x, y, t) + v_{2_y}(x, y, t) = 0,$$

which is called the *equation of continuity*.

The force acting on a particle of mass m located at a point (x, y) at time t has components $m v_{1_t}(x, y, t)$ and $m v_{2_t}(x, y, t)$. Now, if γ is a smooth path from a point P to a point Q, the work done in the motion of that particle along γ is defined to be the line integral along γ of the scalar product of the force and the tangent vector to the path. More precisely, if γ is given in parametric form by $\gamma(s) = (x(s), y(s))$, its tangent vector at s has components $x'(s)$ and $y'(s)$, and then the work done in the motion of that particle from $P = \gamma(s_0)$ to $Q = \gamma(s)$ is

$$m \int_{s_0}^s [v_{1_t}(\gamma(\sigma), t)x'(\sigma) + v_{2_t}(\gamma(\sigma), t)y'(\sigma)]\, d\sigma.$$

The assumption that the velocity field is conservative means that, if P is fixed, this work is a function of Q and t only, that is, of $\gamma(s)$ and t only. If we denote this integral by $u(\gamma(s), t)$, using first the chain rule and then the fundamental theorem of calculus, we obtain

$$\frac{d}{ds} u(\gamma(s), t) = u_x(\gamma(s), t)x'(s) + u_y(\gamma(s), t)y'(s)$$
$$= v_{1_t}(\gamma(s), t)x'(s) + v_{2_t}(\gamma(s), t)y'(s).$$

Since these equations hold for any s and t, it follows that

$$u_x = v_{1_t} \qquad \text{and} \qquad u_y = v_{2_t}$$

and then that

$$u_{xx} + u_{yy} = v_{1_{tx}} + v_{2_{ty}} = (v_{1_x} + v_{2_y})_t.$$

In view of the equation of continuity, the right-hand side is zero, and this establishes the two-dimensional Laplace equation.

In 1848, William Thomson introduced the term *harmonic function*, which has become standard, for a solution of the Laplace equation. Finding a harmonic function corresponding to given boundary conditions in a bounded rectangular domain is a problem that can be solved using the techniques of the previous chapters (see Exercises 1.1 to 1.6). We shall consider here the following boundary value problem, known as the *Dirichlet problem* for the half-plane.[1] If

$$D = \{ (x, y) \in \mathbb{R}^2 : -\infty < x < \infty, \ y > 0 \}$$

and

$$\overline{D} = \{ (x, y) \in \mathbb{R}^2 : -\infty < x < \infty, \ y \geq 0 \},$$

find a continuous function $u : \overline{D} \to \mathbb{R}$ that is of class C^2, harmonic in D, and such that

$$(2) \qquad u(x, 0) = f(x), \qquad -\infty < x < \infty,$$

where f is a given bounded, continuous function. We may relax this condition and require only that f be bounded and piecewise continuous on every bounded interval, but then u can be required only to be continuous in D and (2) is replaced by

$$\lim_{y \to 0} u(x, y) = \tfrac{1}{2}[f(x+) + f(x-)], \qquad -\infty < x < \infty.$$

We shall solve this problem by separation of variables. If there is a solution of the form $u(x, y) = X(x)Y(y)$, substitution into (1) leads to $X''Y + XY'' = 0$ and then to

$$\frac{X''}{X} = -\frac{Y''}{Y}.$$

Since the left-hand side is a function of x only and the right-hand side is a a a function of y only, each of them must equal the same constant, which we denote by $-\lambda$, and this yields the pair of ordinary differential equations

$$X'' + \lambda X = 0 \qquad \text{and} \qquad Y'' - \lambda Y = 0.$$

---

[1]In his Winter lectures of 1856–1857, published in 1876, Dirichlet asserted that, for any bounded domain, there is a unique solution of the Laplace equation with given boundary values and that this solution is the function that minimizes a certain integral. The existence of this unique minimizing function was called the *Dirichlet principle* by Riemann because he learned it from Dirichlet, although it was first explicitly stated by Thomson in 1847 in a paper that he published in French in Liouville's journal. When Dirichlet's assumption of the existence of the minimizing function was challenged by Weierstrass in 1870 for lack of proof (a correct proof was first given by Hilbert in 1900), it could no longer be a principle. However, the modified term *Dirichlet problem* has remained to refer to the problem of finding a harmonic function with given boundary values.

For $\lambda < 0$ the solutions for X are unbounded because they contain the exponentials $e^{\pm\sqrt{-\lambda}\,x}$ that are unbounded on $(-\infty, \infty)$. For $\lambda = 0$ the only bounded solutions for both X and Y are constant, and for $\lambda > 0$ the general solutions of the preceding equations are

$$X(x) = A \cos \sqrt{\lambda}\,x + B \sin \sqrt{\lambda}\,x$$
$$Y(y) = C e^{\sqrt{\lambda}\,y} + D e^{-\sqrt{\lambda}\,y}.$$

To obtain a solution that is bounded for all $y > 0$ we must choose $C = 0$, and then, incorporating the constant D into A and B, we obtain

$$u(x, y) = X(x)Y(y) = e^{-\sqrt{\lambda}\,y}\left(A \cos \sqrt{\lambda}\,x + B \sin \sqrt{\lambda}\,x\right).$$

If we put $\omega = \sqrt{\lambda}$ and note that for $\omega = 0$ this expression becomes a constant, then all the bounded solutions found above are of the form

$$u(x, y) = e^{-\omega y}\left[A(\omega) \cos \omega x + B(\omega) \sin \omega x\right]$$

for $\omega \geq 0$. The notation $A(\omega)$, $B(\omega)$ is meant to indicate that there is a pair of constants for each ω.

In order to satisfy the initial condition (2) we shall use superposition of solutions, but, since every nonnegative value of ω is permissible, we shall form an infinite integral

$$\int_0^\infty e^{-\omega y}\left[A(\omega) \cos \omega x + B(\omega) \sin \omega x\right] d\omega$$

rather than an infinite series. Note that $B(0)$ is completely arbitrary, and we choose $B(0) = 0$. Then, if we define a new function $\hat{f}$ by

$$\hat{f}(\omega) = \begin{cases} \frac{1}{2}[A(\omega) - i B(\omega)] & \text{if} \quad \omega \geq 0 \\ \frac{1}{2}[A(-\omega) + i B(-\omega)] & \text{if} \quad \omega < 0 \end{cases}$$

so that

$$A(\omega) = \hat{f}(\omega) + \hat{f}(-\omega) \qquad \text{and} \qquad B(\omega) = i[\hat{f}(\omega) - \hat{f}(-\omega)]$$

and if we recall that

$$\sin \omega x = \frac{e^{i\omega x} - e^{-i\omega x}}{2i} \qquad \text{and} \qquad \cos \omega x = \frac{e^{i\omega x} + e^{-i\omega x}}{2},$$

the preceding integral becomes, after simplification,

$$\int_0^\infty e^{-\omega y}\left[\hat{f}(\omega)e^{i\omega x} + \hat{f}(-\omega)e^{-i\omega x}\right] d\omega$$

$$= \int_0^\infty \hat{f}(\omega)e^{-\omega y}e^{i\omega x}\, d\omega + \int_{-\infty}^0 \hat{f}(\alpha)e^{\alpha y}e^{i\alpha x}\, d\alpha$$

$$= \int_{-\infty}^\infty \hat{f}(\omega)e^{-|\omega|y}e^{i\omega x}\, d\omega.$$

Therefore, the original problem poses the following: find a function $\hat{f}: \mathbb{R} \to \mathbb{C}$ such that

$$\int_{-\infty}^{\infty} \hat{f}(\omega) e^{-|\omega|y} e^{i\omega x} \, d\omega$$

is a harmonic function in D and, for $y = 0$,

(3) $$\int_{-\infty}^{\infty} \hat{f}(\omega) e^{i\omega x} \, d\omega = f(x).$$

§7.2 Fourier Transforms

We shall start with the second part of our task: find a function $\hat{f}$ such that (3) is satisfied. As we did in Chapter 2 for the Fourier coefficients, we shall first find $\hat{f}$ by an informal procedure and save the justification for later (Theorem 7.1).

Let a and α be arbitrary nonzero real numbers. If we assume that (3) holds, if we multiply both sides by $e^{-i\alpha x}$ and integrate from $-a$ to a, and if the change of order of integration below is valid, we have

$$\int_{-a}^{a} f(x) e^{-i\alpha x} \, dx = \int_{-a}^{a} \left(\int_{-\infty}^{\infty} \hat{f}(\omega) e^{i\omega x} \, d\omega \right) e^{-i\alpha x} \, dx$$

$$= \int_{-a}^{a} \left(\int_{-\infty}^{\infty} \hat{f}(\omega) e^{i(\omega - \alpha)x} \, d\omega \right) dx$$

$$= \int_{-\infty}^{\infty} \left(\int_{-a}^{a} \hat{f}(\omega) e^{i(\omega - \alpha)x} \, dx \right) d\omega$$

$$= \int_{-\infty}^{\infty} \hat{f}(\omega) \frac{e^{i(\omega - \alpha)a} - e^{-i(\omega - \alpha)a}}{i(\omega - \alpha)} \, d\omega$$

$$= \int_{-\infty}^{\infty} \hat{f}(\omega) \frac{2 \sin(\omega - \alpha)a}{(\omega - \alpha)} \, d\omega$$

$$= 2 \int_{-\infty}^{\infty} \hat{f}(\alpha + t) \frac{\sin at}{t} \, dt.$$

Since

(4) $$\int_{-\infty}^{\infty} \frac{\sin at}{t} \, dt = \pi$$

(either from a table of integrals or from Exercise B.3),

$$\int_{-a}^{a} f(x) e^{-i\alpha x} \, dx = 2 \int_{-\infty}^{\infty} \frac{\hat{f}(\alpha + t) - \hat{f}(\alpha)}{t} \sin at \, dt + 2\pi \hat{f}(\alpha).$$

Now, if $\hat{f}$ were such that the quotient on the right is integrable and if the Riemann-Lebesgue Theorem were valid for integrals on $(-\infty, \infty)$, we would be able to conclude that

$$\lim_{a \to \infty} \int_{-a}^{a} f(x)e^{-i\alpha x}\, dx = 2\pi\, \hat{f}(\alpha).$$

This leads to the reasonable conjecture that

(5)
$$\hat{f}(\omega) = \frac{1}{2\pi} \int_{-\infty}^{\infty} f(x)e^{-i\omega x}\, dx.$$

Equations (3) and (5) are, in the present context, the counterparts of the concepts of Fourier series and Fourier coefficients. In order to give precise definitions and to justify manipulations like the ones above, we shall, from now on, rely on the terminology and the development of improper integrals contained in Appendix B.

Definition 7.1 *Let* $f : \mathbb{R} \to \mathbb{R}$ *be locally integrable. If the integral below converges, then the function* $\hat{f} : \mathbb{R} \to \mathbb{C}$ *defined by*

(5)
$$\hat{f}(\omega) = \frac{1}{2\pi} \int_{-\infty}^{\infty} f(x)e^{-i\omega x}\, dx.$$

is called the **Fourier transform** *of* f. *The integral*

$$\int_{-\infty}^{\infty} \hat{f}(\omega)e^{i\omega x}\, d\omega$$

is called the **Fourier integral** *of* f.[1]

Example 7.1 Find the Fourier transform of

$$f(x) = \begin{cases} 1 & \text{if} \quad |x| < \delta \\ 0 & \text{if} \quad |x| \ge \delta, \end{cases}$$

---

[1]The Fourier transform and integral are not defined in a unique way in the literature on the subject. However, all such definitions are of the form

$$\hat{f}(\omega) = \frac{C}{2\pi} \int_{-\infty}^{\infty} f(x)e^{-ik\omega x}\, dx$$

for the Fourier transform and

$$\frac{|k|}{C} \int_{-\infty}^{\infty} \hat{f}(\omega)e^{i\omega x}\, d\omega$$

for the Fourier integral, where C and k are constants. We have chosen $C = k = 1$. Other definitions use combinations of $C = 1$, $C = 2\pi$, $C = \sqrt{2\pi}$, $k = \pm 1$, and $k = \pm 2\pi$. The value of C is arbitrary, and that of k is then selected so that the Fourier integral of f converges to f under appropriate conditions.

where δ is a fixed positive number. We have

$$\hat{f}(\omega) = \frac{1}{2\pi} \int_{-\delta}^{\delta} e^{-i\omega x} dx$$

$$= \frac{1}{2\pi} \int_{-\delta}^{\delta} (\cos \omega x - i \sin \omega x) dx$$

$$= \frac{1}{2\pi} \int_{-\delta}^{\delta} \cos \omega x \, dx$$

$$= \frac{\sin \omega \delta}{\pi \omega}.$$

Example 7.2 If $y > 0$ is a constant, find the Fourier transform of $f(x) = e^{-|x|y}$.

$$\hat{f}(\omega) = \frac{1}{2\pi} \int_{-\infty}^{\infty} e^{-|x|y} e^{-i\omega x} dx$$

$$= \frac{1}{2\pi} \left[\int_{0}^{\infty} e^{-(i\omega+y)x} dx + \int_{-\infty}^{0} e^{-(i\omega-y)x} dx \right]$$

$$= \frac{1}{2\pi} \left[\frac{1}{i\omega + y} - \frac{1}{i\omega - y} \right]$$

$$= \frac{y}{\pi(\omega^2 + y^2)}.$$

In these examples the existence of $\hat{f}$ has been shown by direct computation. Next we give sufficient conditions for the existence of the Fourier transform and the convergence of the Fourier integral. Theorem 7.1 does for Fourier integrals what Theorem 2.2 did for Fourier series. The terminology is explained in Appendix B.

Theorem 7.1 *Let $f : \mathbb{R} \to \mathbb{R}$ be integrable on $\mathbb{R}$. Then its Fourier transform $\hat{f}$ is well defined, bounded, and continuous, and*

$$\text{PV} \int_{-\infty}^{\infty} \hat{f}(\omega) e^{i\omega x} d\omega = \tfrac{1}{2}[f(x+) + f(x-)]$$

at every point x in $\mathbb{R}$ such that f has right-hand and left-hand derivatives at x and such that f is piecewise continuous on an arbitrarily small interval centered at x. If, in addition, $\hat{f}$ is integrable on $\mathbb{R}$, then f is continuous and

$$\int_{-\infty}^{\infty} \hat{f}(\omega) e^{i\omega x} d\omega = f(x).$$

Proof. First note that $|f(x)e^{-i\omega x}| = |f(x)|$, and then, since f is integrable on $\mathbb{R}$,

$$\int_{-\infty}^{\infty} f(x)e^{-i\omega x} dx$$

converges uniformly with respect to ω on $\mathbb{R}$ by Theorem B.5. This shows that $\hat{f}$ is well defined, and it is bounded by

$$\int_{-\infty}^{\infty} |f|$$

by Theorem B.3. To show that $\hat{f}$ is continuous, let a and b be real numbers with $a < b$ and use the triangle inequality to show that

$$2\pi |\hat{f}(\omega) - \hat{f}(\upsilon)| \le \left| \int_{-\infty}^{\infty} f(x)e^{-i\omega x}\, dx - \int_{a}^{b} f(x)e^{-i\omega x}\, dx \right|$$

$$+ \left| \int_{-\infty}^{\infty} f(x)e^{-i\upsilon x}\, dx - \int_{a}^{b} f(x)e^{-i\upsilon x}\, dx \right|$$

$$+ \left| \int_{a}^{b} f(x)e^{-i\omega x}\, dx - \int_{a}^{b} f(x)e^{-i\upsilon x}\, dx \right|.$$

By the uniform convergence of the integral defining $\hat{f}$ and Definition B.3, the numbers a and b can be chosen so that the first two terms on the right-hand side above are as small as desired. With a and b already chosen, the third term, which is no larger than

$$\max_{x\in[a,b]} |e^{-i\omega x} - e^{-i\upsilon x}| \int_{-\infty}^{\infty} |f|,$$

can also be made as small as desired by the continuity of the exponential if $\omega - \upsilon$ is small enough. This shows that $\hat{f}$ is continuous.

Now, for any real number a,

$$\int_{-a}^{a} \hat{f}(\omega)e^{i\omega x}\, d\omega = \int_{-a}^{a} \left(\frac{1}{2\pi} \int_{-\infty}^{\infty} f(s)e^{-i\omega s}\, ds \right) e^{i\omega x}\, d\omega$$

$$= \frac{1}{2\pi} \int_{-a}^{a} \left(\int_{-\infty}^{\infty} f(s)e^{i\omega(x-s)}\, ds \right) d\omega.$$

Here f is locally integrable, $e^{i\omega(x-s)}$ is continuous, and the inner integral converges uniformly with respect to ω on $[-a, a]$ by Theorem B.5 with $|f|$ in the role of g. Then Theorem B.8 applies to reverse the order of integration, and we obtain

(6) $$\int_{-a}^{a} \hat{f}(\omega)e^{i\omega x}\, d\omega = \frac{1}{2\pi} \int_{-\infty}^{\infty} \left(\int_{-a}^{a} f(s)e^{i\omega(x-s)}\, d\omega \right) ds$$

$$= \frac{1}{2\pi} \int_{-\infty}^{\infty} f(s) \frac{2\sin a(x-s)}{x-s}\, ds$$

$$= \frac{1}{\pi} \int_{-\infty}^{\infty} f(x+t) \frac{\sin at}{t}\, dt.$$

Now,

$$\int_0^\infty f(x+t)\frac{\sin at}{t}\,dt = \int_0^\infty \frac{f(x+t) - f(x+)}{t}\sin at\,dt + f(x+)\int_0^\infty \frac{\sin at}{t}\,dt.$$

If f has a right-hand derivative at x, the quotient in the first integral on the right-hand side is a piecewise continuous function of t on $(0, \delta)$ for δ small enough and integrable on (δ, ∞). It is then integrable on $(-\infty, \infty)$, and this integral approaches zero as $a \to \infty$ by the extended Riemann-Lebesgue theorem (Theorem B.6). It follows from (3) that the last integral on the right, whose integrand is an even function of t, has value $\pi/2$ and this proves that

$$\lim_{a \to \infty} \int_0^\infty f(x+t)\frac{\sin at}{t}\,dt = \frac{\pi}{2}f(x+).$$

Similarly,

$$\lim_{a \to \infty} \int_{-\infty}^0 f(x+t)\frac{\sin at}{t}\,dt = \frac{\pi}{2}f(x-).$$

Then it follows from (6) that

$$\text{PV} \int_{-\infty}^\infty \hat{f}(\omega)e^{i\omega x}\,d\omega = \tfrac{1}{2}[f(x+) + f(x-)].$$

If, in addition, $\hat{f}$ is integrable, then the Fourier integral of f converges because of the identity $|\hat{f}(\omega)e^{i\omega x}| = |\hat{f}(\omega)|$, and it equals its principal value by Theorem B.1. Therefore,

$$\int_{-\infty}^\infty \hat{f}(\omega)e^{i\omega x}\,d\omega = \tfrac{1}{2}[f(x+) + f(x-)],$$

and since the left-hand side is 2π times the Fourier transform of $\hat{f}$ evaluated at $-x$ it is continuous as shown above. Thus, f is continuous and $\tfrac{1}{2}[f(x+) + f(x-)] = f(x)$, Q.E.D.

The second equation in Theorem 7.1 is usually referred to as the *inversion formula* for $\hat{f}$. Sometimes, it can be used indirectly to compute a Fourier transform.

Example 7.3 If $y > 0$ is a constant, find the Fourier transform of

$$f(x) = \frac{2y}{x^2 + y^2}.$$

A direct attempt to compute $\hat{f}$ involves an integral that is not easy to evaluate without sophisticated techniques. However, we can exploit the similarity between this integral and the Fourier integral and also the similarity between the given function and the Fourier transform of Example 7.2. Indeed, it is easy to see that $e^{-|x|y}$ is integrable on $\mathbb{R}$, and so is its transform since

$$\int_0^\infty \frac{y}{\pi(\omega^2 + y^2)}\,d\omega = \frac{1}{\pi}\lim_{a \to \infty}\int_0^a \frac{y}{\omega^2 + y^2}\,d\omega = \frac{1}{\pi}\lim_{a \to \infty}\left(\arctan\frac{a}{y}\right) = \frac{1}{2}$$

and similarly on $(-\infty, 0)$. Then, according to Theorem 7.1, its Fourier integral converges and the inversion formula is

$$\int_{-\infty}^{\infty} \frac{y}{\pi(\omega^2 + y^2)} e^{i\omega x} \, d\omega = e^{-|x|y},$$

that is,

$$e^{-|\omega|y} = \int_{-\infty}^{\infty} \frac{y}{\pi(s^2 + y^2)} e^{is\omega} \, ds = \frac{1}{2\pi} \int_{-\infty}^{\infty} \frac{2y}{x^2 + y^2} e^{-ix\omega} \, dx = \hat{f}(\omega).$$

This evaluates the Fourier transform we were seeking.

§7.3 Properties of the Fourier Transform

First we learn how to find the Fourier transforms of some new functions from those of known ones.

Theorem 7.2 *Let f and g be locally integrable and have Fourier transforms $\hat{f}$ and $\hat{g}$, respectively. If α is a real number and a is a real or complex constant, then the Fourier transform of*

(i) $f + g$ *is* $\hat{f} + \hat{g}$,

(ii) af *is* $a\hat{f}$,

(iii) $f(x)\cos\alpha x$ *is* $\dfrac{\hat{f}(\omega - \alpha) + \hat{f}(\omega + \alpha)}{2}$,

(iv) $f(x)\sin\alpha x$ *is* $\dfrac{\hat{f}(\omega - \alpha) - \hat{f}(\omega + \alpha)}{2i}$,

(v) $f(x - \alpha)$ *is* $e^{-i\omega\alpha}\hat{f}(\omega)$, *and*

(vi) $f(\alpha x)$ *is* $\dfrac{1}{|\alpha|}\hat{f}\left(\dfrac{\omega}{\alpha}\right)$ *if* $\alpha \neq 0$.

The proof involves only straightforward computations and is left as an exercise.

Theorem 7.3 *Let $f: \mathbb{R} \to \mathbb{R}$ be continuous and integrable on $\mathbb{R}$ and such that $f(x) \to 0$ as $x \to \pm\infty$. If f' is piecewise continuous and integrable on $\mathbb{R}$, then*

$$\widehat{f'}(\omega) = i\omega\hat{f}(\omega).$$

Proof. Since f and f' are integrable, their transforms are well defined as in the proof of Theorem 7.1. Now choose $a > 0$ such that f' is continuous at $\pm a$, let $x_1, x_2, \ldots, x_n$ be the points at which f' is discontinuous in the interval $(-a, a)$, and define $x_0 = -a$ and $x_{n+1} = a$. Then integration by parts shows that

$$\int_{x_k}^{x_{k+1}} f'(x)e^{-i\omega x} \, dx = f(x_{k+1}-)e^{-i\omega x_{k+1}} - f(x_k+)e^{-i\omega x_k} + i\omega \int_{x_k}^{x_{k+1}} f(x)e^{-i\omega x} \, dx.$$

Using Theorem B.1, adding the equations that result from substituting $k = 0, \ldots, n$, and noting that the continuity of f implies that $f(x_k-) = f(x_k+) = f(x_k)$ for $0 < k < n + 1$, we obtain

$$\int_{-\infty}^{\infty} f'(x)e^{-i\omega x}\, dx = \lim_{a \to \infty} \int_{-a}^{a} f'(x)e^{-i\omega x}\, dx$$

$$= \lim_{a \to \infty} \left[f(a)e^{-i\omega a} - f(-a)e^{i\omega a} + i\omega \int_{-a}^{a} f(x)e^{-i\omega x}\, dx \right].$$

Since $f(\pm a) \to 0$ as $a \to \infty$, this shows that $2\pi \widehat{f'}(\omega) = i\omega(2\pi \hat{f}(\omega))$, Q.E.D.

Corollary *If f, f', f'', $\ldots$, $f^{(n-1)}$ satisfy the same hypotheses that f satisfies in the theorem, then*

$$\widehat{f^{(n)}}(\omega) = (i\omega)^n \hat{f}(\omega).$$

The preceding theorem and corollary provide us with a new method to solve ordinary and partial differential equations. It will be fully explained in §7.6. At this point, we shall consider a simple example to see what is involved.

Example 7.4 There are well-known methods to find a particular solution of the differential equation $-y'' + y = f$ if f is in a certain class of elementary functions. If it is not, but if it is integrable, then it has a Fourier transform. If we assume that y also has a Fourier transform and if Theorem 7.2(i) and the corollary above are applicable, the differential equation is transformed into the algebraic equation $\omega^2 \hat{y}(\omega) + \hat{y}(\omega) = \hat{f}(\omega)$. This one is easy to solve:

$$\hat{y}(\omega) = \hat{f}(\omega)\frac{1}{\omega^2 + 1}.$$

From Example 7.2 we know that the quotient on the right is the Fourier transform of

$$g(x) = \pi e^{-|x|}.$$

Then $\hat{y} = \hat{f}\hat{g}$, where $\hat{f}$ and $\hat{g}$ are known, and all this poses the following question: can we find y in terms of the known functions f and g and, if so, what is it? We shall investigate this matter in the next section.

§7.4 Convolution

By making certain assumptions and through a few informal manipulations, we shall gain an idea of what the answer to the question posed in the preceding example should be. Once in possesion of this knowledge, it will be feasible to prove that our conjecture is correct. If $\hat{y} = \hat{f}\hat{g}$, if y equals its Fourier integral, if the integrals below converge, and if the reversal of order of integration is valid, then

$$y(x) = \int_{-\infty}^{\infty} \hat{f}(\omega)\hat{g}(\omega)e^{i\omega x}\, d\omega$$

$$= \int_{-\infty}^{\infty} \hat{f}(\omega)\left(\frac{1}{2\pi}\int_{-\infty}^{\infty} g(t)e^{i\omega(x-t)}\, dt\right) d\omega$$

$$= \int_{-\infty}^{\infty} \hat{f}(\omega)\left(-\frac{1}{2\pi}\int_{\infty}^{-\infty} g(x-s)e^{i\omega s}\, ds\right) d\omega$$

$$= \frac{1}{2\pi}\int_{-\infty}^{\infty}\left(\int_{-\infty}^{\infty} \hat{f}(\omega)g(x-s)e^{i\omega s}\, ds\right) d\omega$$

$$= \frac{1}{2\pi}\int_{-\infty}^{\infty}\left(\int_{-\infty}^{\infty} \hat{f}(\omega)g(x-s)e^{i\omega s}\, d\omega\right) ds$$

$$= \frac{1}{2\pi}\int_{-\infty}^{\infty}\left(\int_{-\infty}^{\infty} \hat{f}(\omega)e^{i\omega s}\, d\omega\right)g(x-s)\, ds$$

$$= \frac{1}{2\pi}\int_{-\infty}^{\infty} f(s)g(x-s)\, ds.$$

Although some of the preceding manipulations are unjustified, this is enough to conjecture that the function of x defined by the last right-hand side has Fourier transform $\hat{f}\hat{g}$. The rest of this section is devoted to proving this result (Theorem 7.6), stating sufficient conditions for its validity. We start with formal definition.

Definition 7.2 *If f, $g: \mathbb{R} \to \mathbb{R}$ are locally integrable, the* **convolution** *of f and g is the function denoted by $f*g$ and defined by*

(7)
$$f*g(x) = \int_{-\infty}^{\infty} f(s)g(x-s)\, ds$$

for all x in $\mathbb{R}$ if this integral converges.

The proof of the next theorem is left as an exercise.

Theorem 7.4 *If the following convolutions are defined, then*

(i) $(f+g)*h = f*h + g*h$,

(ii) $f*(g+h) = f*g + f*h$, *and*

(iii) $f*g = g*f$.

Next we give sufficient conditions for the existence of the convolution.

Theorem 7.5 *If* $f : \mathbb{R} \to \mathbb{R}$ *is integrable and* $g : \mathbb{R} \to \mathbb{R}$ *is bounded and piecewise continuous, then the integral*

(8)
$$\int_{-\infty}^{\infty} |f(s)g(x-s)|\, ds$$

and the integral in (7) *converge uniformly with respect to* x *on* $\mathbb{R}$ *and* $f * g$ *is continuous.*

Proof. If $|g| < M$, then

$$\int_0^a |f(s)g(x-s)|\, ds < M \int_0^a |f| < M \int_0^\infty |f|.$$

Then the left-hand side is an increasing function of a bounded by the right-hand side and has a limit as $a \to \infty$. A similar result holds for the integral on $(-a, 0)$, which shows that (8) converges. Then the integral in (7) also converges by Theorem B.3. Their convergence is uniform by Theorem B.5 with $M|f|$ in the role of g. Since g is piecewise continuous $g(x-s)$, as a function of (s, x), is patchwise continuous on $\mathbb{R} \times I$ in the sense of Definition B.6. Indeed, it may be discontinuous only on lines of the form $s = x - x_0$, where x_0 is a point of discontinuity of g, and its one-sided limits in a horizontal or vertical direction exist because $g(x - s)$ has the same values on lines parallel to $s = x$. Then $f * g$ is continuous by Theorem B.7, Q.E.D.

Theorem 7.6 *If* f, $g : \mathbb{R} \to \mathbb{R}$ *are integrable and* $g : \mathbb{R} \to \mathbb{R}$ *is piecewise continuous, then* $f * g$ *is integrable on* $\mathbb{R}$ *and*

$$\widehat{f * g} = 2\pi\, \hat{f} \hat{g}.$$

Proof. By Theorem 7.5, $|f * g|$ is continuous and, therefore, integrable on $[0, a]$ for any $a > 0$. Since f is integrable, $|g(x - s)|$ was seen to be patchwise continuous in the proof of Theorem 7.5, and (8) converges uniformly on $\mathbb{R}$, we can use Theorem B.8 to obtain

$$\int_0^a |f * g| = \int_0^a \left| \int_{-\infty}^{\infty} f(s)g(x-s)\, ds \right| dx$$

$$\leq \int_0^a \left(\int_{-\infty}^{\infty} |f(s)g(x-s)|\, ds \right) dx$$

$$= \int_{-\infty}^{\infty} \left(\int_0^a |f(s)g(x-s)|\, dx \right) ds$$

$$= \int_{-\infty}^{\infty} |f(s)| \left(\int_0^a |g(x-s)|\, dx \right) ds$$

$$\leq \int_{-\infty}^{\infty} |f(s)| \left(\int_{-\infty}^{\infty} |g(x-s)|\, dx \right) ds$$

$$= \int_{-\infty}^{\infty} |f(s)| \left(\int_{-\infty}^{\infty} |g(t)| \, dt \right) ds$$

$$= \left(\int_{-\infty}^{\infty} |f| \right) \left(\int_{-\infty}^{\infty} |g| \right).$$

The right-hand side is well defined because f and g are integrable, and then the left-hand side has a limit as $a \to \infty$ because it is increasing and bounded by the right-hand side. This is also the case for the integral of $|f*g|$ on $(-\infty, 0)$, showing that $f*g$ is integrable, and then its Fourier transform is well defined. We have

$$4\pi^2 \hat{f}(\omega)\hat{g}(\omega) = \left(\int_{-\infty}^{\infty} f(s)e^{-i\omega s} \, ds \right) \left(\int_{-\infty}^{\infty} g(t)e^{-i\omega t} \, dt \right)$$

$$= \int_{-\infty}^{\infty} f(s)e^{-i\omega s} \left(\int_{-\infty}^{\infty} g(x-s)e^{-i\omega(x-s)} \, dx \right) ds$$

$$= \int_{-\infty}^{\infty} \left(\int_{-\infty}^{\infty} f(s)g(x-s)e^{-i\omega x} \, dx \right) ds.$$

To reverse the order of integration, note that the hypotheses of Theorem B.9 are satisfied since f is integrable, $g(x-s)e^{-i\omega x}$ is piecewise continuous, the integral

$$\int_{-\infty}^{\infty} f(s)g(x-s)e^{-i\omega x} \, ds$$

converges uniformly with respect to x on $\mathbb{R}$ by Theorem 7.5, and the integral

$$\int_{-\infty}^{\infty} g(x-s)e^{-i\omega x} \, dx$$

converges uniformly with respect to s on $\mathbb{R}$ because $|g(x-s)e^{-i\omega x}| = |g(x-s)|$ and g is integrable. Therefore, we can reverse the order of integration, and

$$4\pi^2 \hat{f}(\omega)\hat{g}(\omega) = \int_{-\infty}^{\infty} \left(\int_{-\infty}^{\infty} f(s)g(x-s) \, ds \right) e^{-i\omega x} \, dx$$

$$= \int_{-\infty}^{\infty} f*g \, e^{-i\omega x} \, dx$$

$$= 2\pi \widehat{f*g}(\omega),$$

Q.E.D.

Example 7.5 Returning to the question posed in Example 7.4, we can now reasonably conjecture that, if $\hat{y} = \hat{f}\hat{g}$, then

$$y(x) = \frac{1}{2\pi} f*g(x) = \frac{1}{2\pi} \int_{-\infty}^{\infty} f(s)g(x-s) \, ds = \frac{1}{2} \int_{-\infty}^{\infty} f(s)e^{-|x-s|} \, ds.$$

Since many of the assumptions leading to this conclusion were unjustified, this solution must be verified. The details are left to Exercise B.12.

Example 7.6 For each $y > 0$ define a function $P^y : \mathbb{R} \to \mathbb{R}$ by

$$P^y(x) = \frac{2y}{x^2 + y^2}.$$

Then P^y is bounded, continuous, and integrable on $\mathbb{R}$, and its Fourier transform was found in Example 7.3 to be $\widehat{P^y}(\omega) = e^{-|\omega|y}$. Then, if $f : \mathbb{R} \to \mathbb{R}$ is integrable,

$$\widehat{f * P^y} = 2\pi \hat{f}(\omega) e^{-|\omega|y}.$$

§7.5 Solution of the Dirichlet Problem for the Half-Plane

In §7.1 we were looking for a function $\hat{f} : \mathbb{R} \to \mathbb{C}$ such that

$$\int_{-\infty}^{\infty} \hat{f}(\omega) e^{-|\omega|y} e^{i\omega x}\, d\omega$$

is a harmonic function in D and, for $y = 0$,

$$\int_{-\infty}^{\infty} \hat{f}(\omega) e^{i\omega x}\, d\omega = f(x).$$

At least this should be so for the principal values of these integrals. Theorem 7.1 suggests that $\hat{f}$ should be the Fourier transform of f, and then the result of Example 7.6 suggests that we should consider the function

$$\frac{1}{2\pi} f * P^y$$

as the possible solution of our problem. We can now prove that this is indeed the solution.

Theorem 7.7 *Let $f : \mathbb{R} \to \mathbb{R}$ be a bounded, piecewise continuous, integrable function. Then the function $u : D \to \mathbb{R}$ defined by*

(9) $$u(x, y) = \frac{1}{2\pi} f * P^y,$$

where

$$P^y(x) = \frac{2y}{x^2 + y^2},$$

is a bounded harmonic function in D such that

$$\lim_{y \to 0} u(x, y) = \tfrac{1}{2}[f(x+) + f(x-)]$$

for every x in $\mathbb{R}$.[1]

SIMÉON DENIS POISSON
Courtesy of *Archives de l'Académie des Sciences, Paris.*

Proof. To prove that u is harmonic, note that

$$\frac{\partial^2}{\partial x^2} P^y(x - s) + \frac{\partial^2}{\partial y^2} P^y(x - s) = 0,$$

so that it suffices to show that we can differentiate

$$f * P^y(x) = \int_{-\infty}^{\infty} f(s) P^y(x - s)\, ds$$

under the integral sign. If $|P^y| < K$, this integral converges uniformly with respect to x in $\mathbb{R}$, by Theorem B.5, with $K|f|$ in the role of g. Similarly, the convergence is uniform with respect to y in $\mathbb{R}$.

---

[1] This solution appears on page 86 of Poisson's 1818 memoir on the motion of waves. When the right-hand side of (9) is replaced with its integral form (9) is known as *Poisson's integral formula for the half-plane* and is probably the first recorded use of convolution. To avoid the constant $1/2\pi$, the function P^y is often replaced with the quotient $P^y/2\pi$, called the *Poisson kernel.*

Now,

$$\frac{\partial}{\partial x} P^y(x - s) = -\frac{4(x - s)y}{(x - s)^2 + y^2}$$

is continuous and bounded for each $y > 0$, and then the integral

(10)
$$\int_{-\infty}^{\infty} f(s) \frac{\partial}{\partial x} P^y(x - s) \, ds$$

also converges uniformly with respect to x in $\mathbb{R}$. Then, by Theorem B.10, $f * P^y$ is differentiable with respect to x and its derivative is given by (10) for each $y > 0$.

Similarly,

$$\frac{\partial}{\partial y} P^y(x - s) = \frac{2[(x - s)^2 - y^2]}{[(x - s)^2 + y^2]^2}$$

is continuous and bounded by

$$\frac{2}{(x - s)^2 + y^2} = \frac{2}{y^2} \frac{1}{1 + \left(\dfrac{s - x}{y}\right)^2}.$$

Then, if c and M are positive constants, if $y \geq c$, and if $|f| \leq M$, we have

$$\left| \int_b^{\infty} f(s) \frac{\partial}{\partial y} P^y(x-s) \, ds \right| \leq \frac{2M}{c} \int_b^{\infty} \frac{1}{1 + \left(\dfrac{s - x}{y}\right)^2} \frac{1}{y} \, ds = \frac{2M}{c} \left[\frac{\pi}{2} - \arctan \frac{b - x}{y} \right]$$

for $b > x$. Since the right-hand side approaches zero as $b \to \infty$, the integral

$$\int_0^{\infty} f(s) \frac{\partial}{\partial y} P^y(x - s) \, ds$$

converges uniformly with respect to y on $[c, \infty)$. This is also the case for the integral of the same function from $-\infty$ to 0, and then, by Theorem B.10,

$$\frac{\partial}{\partial y} f * P^y(x) = \int_{-\infty}^{\infty} f(s) \frac{\partial}{\partial y} P^y(x - s) \, ds$$

for $y \geq c$ and, since $c > 0$ is arbitrary, for all $y > 0$ and each x in $\mathbb{R}$. The same type of argument shows that the second derivatives can be carried out inside the integral sign, and this completes the proof that u satisfies the Laplace equation in D.

It is easy to see that u is bounded because, if $|f| \leq M$ on $\mathbb{R}$, we have

$$|f * P^y(x)| \leq M \int_{-\infty}^{\infty} P^y(x - s) \, ds = M \int_{-\infty}^{\infty} P^y(u) \, du,$$

and, by direct computation,

(11)
$$\int_0^{\infty} P^y = \int_{-\infty}^{0} P^y = \pi.$$

Finally, to prove that

$$\lim_{y \to 0} u(x, y) = \frac{1}{2\pi} \lim_{y \to 0} f * P^y(x) = \tfrac{1}{2}[f(x+) + f(x-)]$$

it suffices to show that

$$\lim_{y \to 0} \int_0^\infty P^y(s) f(x - s) \, ds - \pi f(x-) = \lim_{y \to 0} \int_{-\infty}^0 P^y(s) f(x - s) \, ds - \pi f(x+) = 0$$

for each fixed x in $\mathbb{R}$. Now, using (11) we see that

$$\int_0^\infty P^y(s) f(x - s) \, ds - \pi f(x-) = \int_0^\infty P^y(s)[f(x - s) - f(x-)] \, ds$$

$$= \left(\int_0^\delta + \int_\delta^\infty \right) P^y(s)[f(x - s) - f(x-)] \, ds$$

for any $\delta > 0$. If $\epsilon > 0$ is given, choose δ to be so small that

$$|f(x - s) - f(x-)| < \frac{\epsilon}{2\pi}$$

if $0 < s < \delta$, which is possible because $f(x-)$ exists. Then

$$\left| \int_0^\delta P^y(s)[f(x - s) - f(x-)] \, ds \right| < \frac{\epsilon}{2\pi} \int_0^\delta P^y < \frac{\epsilon}{2\pi} \int_0^\infty P^y < \frac{\epsilon}{2}.$$

Also, if $|f| \le M$ on $\mathbb{R}$,

$$\left| \int_\delta^\infty P^y(s)[f(x - s) - f(x-)] \, ds \right| < 4M \int_\delta^\infty \frac{y}{s^2 + y^2} \, ds = 4M \left[\frac{\pi}{2} - \arctan \frac{\delta}{y} \right],$$

which is less that $\epsilon/2$ if $y > 0$ is small enough. This proves the first limit, and the second is analogous, Q.E.D.

When f is continuous we may expect the solution of our boundary value problem to be continuos on $\overline{D}$. This is, in fact, the case.

Theorem 7.8 *Let $f : \mathbb{R} \to \mathbb{R}$ be a bounded, continuous, integrable function. Then the function $u : \overline{D} \to \mathbb{R}$ defined by*

$$u(x, y) = \begin{cases} \dfrac{1}{2\pi} f * P^y(x) & \text{if } y > 0 \\[2mm] f(x) & \text{if } y = 0, \end{cases}$$

where P^y is as above, is bounded and continuous on $\overline{D}$ and harmonic in D.

Proof. We know from Theorem 7.7 that u is bounded, continuous, and harmonic in D. It is also continuous for $y = 0$ because it equals f there and f is continuous. Then it remains to be shown that, for any x_0 in $\mathbb{R}$ and $y > 0$, $u(x, y) \to f(x_0)$ as $(x, y) \to (x_0, 0)$. To start with, the triangle inequality gives

$$|u(x, y) - f(x_0)| \leq |u(x, y) - f(x)| + |f(x) - f(x_0)|.$$

By the continuity of f at x_0, given $\epsilon > 0$, there is an $\eta > 0$ such that

$$|f(x) - f(x_0)| < \frac{\epsilon}{2}$$

if $|x - x_0| < \eta$. Also, as in the proof of Theorem 7.7, $u(x, y) \to f(x)$ as $y \to 0$, but now this convergence is uniform with respect to x in $[x_0 - \eta, x_0 + \eta]$ because f is uniformly continuous on every closed bounded interval. In fact, for any $\epsilon > 0$ there is a $\delta > 0$ such that

$$|f(x - s) - f(x)| < \frac{\epsilon}{2\pi}$$

for all x in $[x_0 - \eta, x_0 + \eta]$ if $0 < s < \delta$. That is, the δ in the proof of Theorem 7.7 can be chosen independent of x in $[x_0 - \eta, x_0 + \eta]$, proving that

$$|u(x, y) - f(x)| < \frac{\epsilon}{2}$$

for all x in $[x_0 - \eta, x_0 + \eta]$ if $y > 0$ is small enough. Then, for such x and y,

$$|u(x, y) - f(x_0)| < \frac{\epsilon}{2} + \frac{\epsilon}{2} = \epsilon,$$

Q.E.D.

The preceding theorems show only the existence of a solution to the boundary value problem in our hands. It remains to be shown that it is well posed in the sense of Hadamard. This would require the development of a maximum principle for the Laplace equation and then a certain amount of work. Since time, space, and energy are also bounded, this proof will be omitted.[1]

§7.6   The Fourier Transform Method

We have solved the Dirichlet problem for the half-plane going all the way back to separation of variables and discovering the Fourier transform in the process. Tedious as this may have been, it has provided us with an invaluable tool. For once this discovery has been made it is no longer necessary to start the solution of this type of boundary value problem with separation of variables. Instead, we can proceed as follows:

---

[1]The maximum principle for the Laplace equation will be presented in its three-dimensional version as Theorem 9.6. For the rest of the proof see, for instance, R. T. Seeley, *An Introduction to Fourier Series and Integrals*, Benjamin, New York, 1966, pages 72–77.

1. take the Fourier transforms of all equations involved, assuming that they exist and that all necessary theorems are applicable,

2. solve the equations resulting from this procedure to obtain the Fourier transforms of all unknown functions,

3. recover the unknown functions from their Fourier transforms, and

4. verify the solutions so obtained.

This is called the *Fourier transform method*, which we partially applied in Example 7.4. Usually, the first two steps are the easiest. Recovering a function from its Fourier transform is theoretically possible via the Fourier integral, but this may not be easy to evaluate. In this respect it is helpful to have as complete a table of Fourier transforms as possible (see Appendix C for a small one). Note that, if f is continuous and integrable on $\mathbb{R}$, Theorem 7.1 shows that the correspondence $f \leftrightarrow \hat{f}$ is one to one. That is, only one continuous function f corresponds to a given transform $\hat{f}$, and this makes it possible to read f from $\hat{f}$ from the table. The verification of the solution is usually hard, as shown by the proofs of the theorems in §7.5, and different problems require different techniques. To illustrate the Fourier transform method we consider the original problem that led Fourier to the discovery of the transform.

Example 7.7 Let D and $\bar{D}$ be as in §7.1, but with the variable y replaced by t. Find a bounded, continuous function $u : \bar{D} \to \mathbb{R}$ that is of class C^2 in D and such that

$$u_t = k u_{xx} \qquad \text{in } D$$
$$u(x, 0) = f(x) \qquad -\infty < x < \infty,$$

where $f : \mathbb{R} \to \mathbb{R}$ is integrable, bounded, and continuous.

Assume that, for each fixed $t \geq 0$, u, considered a function of x only, has a Fourier transform $\hat{u}(\omega, t)$. Transforming the given equations for u gives

$$\widehat{u_t}(\omega, t) = k\widehat{u_{xx}}(\omega, t) = k(i\omega)^2 \hat{u}(\omega, t)$$
$$\hat{u}(\omega, 0) = \hat{f}(\omega).$$

Now, if

$$\hat{u}(\omega, t) = \frac{1}{2\pi} \int_{-\infty}^{\infty} u(x, t) e^{-i\omega x} \, dx$$

can be differentiated with respect to t under the integral sign, we obtain

$$\hat{u}_t(\omega, t) = \frac{1}{2\pi} \int_{-\infty}^{\infty} u_t(x, t) e^{-i\omega x} \, dx = \widehat{u_t}(\omega, t).$$

Then the transformed equations become

$$\hat{u}_t(\omega, t) = -k\omega^2 \hat{u}(\omega, t)$$
$$\hat{u}(\omega, 0) = \hat{f}(\omega).$$

For each fixed but arbitrary ω, this is an initial value problem for a first order ordinary differential equation in the variable t, which is, in any case, a much simpler problem than a boundary value problem for a second-order partial differential equation. In this particular case the ordinary differential equation is linear, and the initial value problem has the well-known solution

$$\hat{u}(\omega, t) = \hat{f}(\omega)e^{-k\omega^2 t}.$$

From a table of Fourier transforms (in this particular instance refer to Exercise 3.5) we learn that, if a is a positive constant, then $e^{-a\omega^2}$ is the Fourier transform of $\sqrt{\pi/a}\, e^{-x^2/4a}$. That is, for $t > 0$,

$$\hat{u}(\omega, t) = \hat{f}(\omega)\hat{g}(\omega),$$

where $\hat{g}(\omega)$ is the Fourier transform of

$$g(x) = \sqrt{\frac{\pi}{kt}} e^{-x^2/4kt}.$$

Since g is integrable, bounded, and continuous, then f and g satisfy the conditions of Theorem 7.6 and we have

$$\hat{u}(\omega, t) = \frac{1}{2\pi} \widehat{f*g}.$$

This suggests that

(12) $$u(x, t) = \frac{1}{2\pi} f*g(x) = \frac{1}{\sqrt{4\pi kt}} \int_{-\infty}^{\infty} f(s)e^{-(x-s)^2/4kt}\, ds.$$

The verification of this solution is very similar to that of Theorems 7.7 and 7.8. Since f is integrable and the exponential in the last integral is bounded by 1, the integral converges uniformly with respect to x and t. Then differentiation under the integral sign shows that u is a solution of the heat equation. Using the substitution $s = x + \sqrt{4kt}\, z$ and the fact that the integral of e^{-z^2} from $-\infty$ to ∞ is $\sqrt{\pi}$, we obtain

$$u(x, t) - f(x) = \frac{1}{\sqrt{\pi}} \int_{-\infty}^{\infty} \left[f\left(x + \sqrt{4kt}\, z \right) - f(x) \right] e^{-z^2}\, dz$$

for any x in $\mathbb{R}$. Arguments analogous to those in the proofs of Theorems 7.7 and 7.8 show that $u(x, t) - f(x) \to 0$ as $t \to 0$ and that $u(x, t) - f(x_0) \to 0$ as $(x, t) \to (x_0, 0)$. This shows that u is a solution of our boundary value problem.

To show uniqueness, let u_1 and u_2 be solutions and define $v = u_1 - u_2$. Then v is a bounded, continuous solution of the heat equation in D such that $v(x, 0) = 0$. If $|v| \le M$ in D and $a > 0$ is fixed, the function

$$v(x, t) - \frac{Mx^2}{a^2} - \frac{2Mkt}{a^2}$$

satisfies the heat equation in the domain $\{ (x, t) \in \mathbb{R}^2 : -a < x < a, t > 0 \}$ and is never positive on the boundary. By the maximum principle for the heat equation

(Theorem 3.4, whose statement and proof remain valid if the interval $[0, a]$ is replaced by $[-a, a]$), $v \le 0$ in the stated domain. Letting $a \to \infty$, we see that $v \le 0$ in the upper half-plane. Similarly, $-v \le 0$ there, and then $v \equiv u_1 - u_2 \equiv 0$. This shows that (12) is the unique solution of our boundary value problem.

Now, if u and v are the solutions corresponding to initial conditions f and g and if M_{f-g} denotes the maximum value of $|f - g|$, we have

$$
\begin{aligned}
|u(x, t) - v(x, t)| &= \frac{1}{\sqrt{4\pi kt}} \left| \int_{-\infty}^{\infty} [f(s) - g(s)] e^{-(x-s)^2/4kt} \, ds \right| \\
&\le \frac{M_{f-g}}{\sqrt{4\pi kt}} \int_{-\infty}^{\infty} e^{-(x-s)^2/4kt} \, ds \\
&= \frac{M_{f-g}}{\sqrt{\pi}} \int_{-\infty}^{\infty} e^{-u^2} \, du \\
&= M_{f-g},
\end{aligned}
$$

where we have used the substitution

$$
u = \frac{x - s}{\sqrt{4kt}}
$$

and the last integral has been taken from a table of integrals. This shows continuous dependence on initial conditions and completes the proof that the boundary value problem is well posed in the sense of Hadamard.

EXERCISES

Only formal solutions are sought for the boundary value problems in any of the following exercises, and no verification is required.

1.1 Solve the boundary value problem

$$
\begin{aligned}
u_{xx} + u_{yy} &= 0 & 0 < x < a, \quad 0 < y < b \\
u(x, 0) &= 0 & 0 \le x \le a \\
u(x, b) &= \begin{cases} Cx, & 0 \le x < a/2 \\ C(a - x), & a/2 \le x \le a \end{cases} \\
u(0, y) &= 0 & 0 \le y \le b \\
u(a, y) &= 0 & 0 \le y \le b,
\end{aligned}
$$

where C is a constant.

1.2 Repeat Exercise 1.1 if the boundary conditions $u(0, y) = u(a, y) = 0$ are replaced by $u_x(0, y) = u_x(a, y) = 0$.

1.3 Show that the method of separation of variables applied to the boundary value problem

$$u_{xx} + u_{yy} = 0 \qquad 0 < x < a, \quad 0 < y < b$$
$$u(x, 0) = f(x) \qquad 0 \le x \le a$$
$$u(x, b) = g(x) \qquad 0 \le x \le a$$
$$u(0, y) = 0 \qquad 0 \le y \le b$$
$$u(a, y) = 0 \qquad 0 \le y \le b$$

leads to a solution of the form

$$u(x, y) = \sum_{n=1}^{\infty} \left[a_n \frac{\sinh \frac{n\pi}{a} y}{\sinh \frac{n\pi}{a} b} + b_n \frac{\sinh \frac{n\pi}{a}(b - y)}{\sinh \frac{n\pi}{a} b} \right] \sin \frac{n\pi}{a} x,$$

where a_n and b_n are the coefficients of the Fourier sine series of g and f, respectively, on $(0, a)$. Obtain the particular form of this solution for the case $b = a$ and

$$f(x) = g(x) = \begin{cases} Cx, & 0 \le x < a/2 \\ C(a - x), & a/2 \le x \le a. \end{cases}$$

In this case, find the approximate value of u in the center of the square.

1.4 Solve the boundary value problem

$$u_{xx} + u_{yy} = 0 \qquad 0 < x < a, \quad 0 < y < b$$
$$u(x, 0) = f_1(x) \qquad 0 \le x \le a$$
$$u(x, b) = g_1(x) \qquad 0 \le x \le a$$
$$u(0, y) = f_2(y) \qquad 0 \le y \le b$$
$$u(a, y) = g_2(y) \qquad 0 \le y \le b$$

(**Hint:** solve first the boundary value problem

$$u_{xx} + u_{yy} = 0 \qquad 0 < x < a, \quad 0 < y < b$$
$$u(x, 0) = 0 \qquad 0 \le x \le a$$
$$u(x, b) = 0 \qquad 0 \le x \le a$$
$$u(0, y) = f_2(y) \qquad 0 \le y \le b$$
$$u(a, y) = g_2(y) \qquad 0 \le y \le b$$

and then combine this solution and that of Exercise 1.3).

1.5 Solve the boundary value problem

$$u_{xx} + u_{yy} = 0 \qquad 0 < x < a, \quad 0 < y < b$$
$$u(x, 0) = 0 \qquad 0 \le x \le a$$
$$u(x, b) = f(x) \qquad 0 \le x \le a$$
$$u(0, y) = 0 \qquad 0 \le y \le b$$
$$u_x(a, y) = -hu(a, y) \qquad 0 \le y \le b,$$

where h is a positive constant (**Hint:** use the eigenvalues of §3.6).

1.6 Repeat Exercise 1.5 if the boundary condition $u(0, y) = 0$ is replaced by $u_x(0, y) = 0$.

The next two exercises are intended as motivation for the definitions in Exercises 2.5 and 2.6.

1.7 Show that, if the method of separation of variables is applied to find a bounded solution of the boundary value problem

$$
\begin{array}{ll}
u_{xx} + u_{yy} = 0 & 0 < x < a, \quad y > 0 \\
u(x, 0) = 0 & 0 \le x \le a \\
u(0, y) = f(y) & y \ge 0 \\
u(a, y) = g(y) & y \ge 0,
\end{array}
$$

it leads to an integral solution involving functions $A(\omega)$ and $B(\omega)$ such that

$$
\int_0^\infty A(\omega) \sin \omega y \, d\omega = f(y)
$$

and

$$
\int_0^\infty B(\omega) \sin \omega y \, d\omega = g(y)
$$

(**Hint:** write the general solution of $X'' - \lambda X = 0$, $\lambda > 0$, in the form

$$
X(x) = A \cosh \sqrt{\lambda}\, x + \frac{B - A \cosh \sqrt{\lambda}\, a}{\sinh \sqrt{\lambda}\, a} \sinh \sqrt{\lambda}\, x,
$$

where A and B are arbitrary constants, and write $\omega = \sqrt{\lambda}$).

1.8 Show that, if the method of separation of variables is applied to find a bounded solution of the boundary value problem

$$
\begin{array}{ll}
u_{xx} + u_{yy} = 0 & x > 0, \quad y > 0 \\
u(x, 0) = f(x) & x \ge 0 \\
u_x(0, y) = 0 & y \ge 0,
\end{array}
$$

it leads to an integral solution involving a function $A(\omega)$ such that

$$
\int_0^\infty A(\omega) \cos \omega x \, d\omega = f(x).
$$

2.1 Find the Fourier transform of

(i) $f(x) = \begin{cases} 1 - \frac{1}{2}|x|, & |x| \le 2 \\ 0, & |x| > 2 \end{cases}$ (ii) $f(x) = e^{-|x|} \sin x$

2.2 If a is a positive constant and n is a positive integer, find the Fourier transform of

$$
f(x) = \begin{cases} 0, & x < 0 \\ x^n e^{-ax}, & x \ge 0. \end{cases}
$$

Use the particular case for $n = 0$ and $a = 1$ to show that $\hat{f}$ need not be integrable even if f is.

2.3 Asume that the even extension of $f : [0, \infty) \to \mathbb{R}$ is integrable. Show that, if we define

$$
\hat{f}_C(\omega) = \frac{2}{\pi} \int_0^\infty f(x) \cos \omega x \, dx
$$

for $\omega \geq 0$, called the *Fourier cosine transform* of f, then

$$\int_0^\infty \hat{f}_C(\omega) \cos \omega x \, d\omega = \tfrac{1}{2}[f(x+) + f(x-)]$$

at every $x \geq 0$ such that the extension of f has one-sided derivatives at x and is piecewise continuous on an arbitrarily small interval centered at x (**Hint:** the function

$$\int_{-\infty}^\infty f_E(u) \sin \omega(x - u) \, du,$$

where f_E is the even extension of f to $\mathbb{R}$, is odd in ω, but it is even if the sine is replaced by the cosine).

2.4 Assume that the odd extension of $f : [0, \infty) \to \mathbb{R}$ (it may be necessary to redefine f at the origin) is integrable. Show that, if we define

$$\hat{f}_S(\omega) = \frac{2}{\pi} \int_0^\infty f(x) \sin \omega x \, dx,$$

for $\omega \geq 0$, called the *Fourier sine transform* of f, then

$$\int_0^\infty \hat{f}_S(\omega) \sin \omega x \, d\omega = \tfrac{1}{2}[f(x+) + f(x-)]$$

at every $x \geq 0$ such that the extension of f has one-sided derivatives at x and is piecewise continuous on an arbitrarily small interval centered at x.

2.5 Show that the Fourier sine and cosine transforms of an integrable function are bounded and continuous. Then use this result to sharpen the conclusions in Exercises 2.3 and 2.4 under the additional assumptions that $\hat{f}_C$ and $\hat{f}_E$ are integrable on $(0, \infty)$.

2.6 If a is a positive constant, find the Fourier sine and cosine transforms of $f(x) = e^{-ax}$. Use the corresponding inversion formulas to show that the Fourier cosine transform of

$$\frac{a}{a^2 + x^2}$$

and the Fourier sine transform of

$$\frac{x}{a^2 + x^2}$$

are both equal to $e^{-\omega a}$.

2.7 If a is a positive constant, find the Fourier sine and cosine transforms of

$$(i) \ f(x) = \begin{cases} a - x, & 0 \leq x \leq a \\ 0, & x > a \end{cases} \qquad (ii) \ f(x) = \begin{cases} \cos x, & 0 \leq x \leq \pi/2 \\ 0, & x > \pi/2. \end{cases}$$

2.8 Let f_E and f_O be the even and odd extensions, respectively, of $f : (0, \infty) \to \mathbb{R}$ to the entire real line, and assume that these extensions are integrable. Show that for $\omega \geq 0$ the Fourier cosine and sine transforms of f are given by

$$\hat{f}_C(\omega) = 2\hat{f}_E(\omega) \qquad \text{and} \qquad \hat{f}_S(\omega) = 2i\,\hat{f}_O(\omega).$$

3.1 Prove Theorem 7.2.

3.2   If $\delta > 0$ and $\omega_0 > \delta$, find the Fourier transform of

$$f(x) = \begin{cases} \cos \omega_0 x, & |x| < \delta \\ 0, & |x| > \delta \end{cases}$$

and sketch the graph of $\hat{f}$ for $\delta \ll \omega_0$.

3.3   Show that the hypothesis $f(x) \to 0$ as $|x| \to \infty$ in Theorem 7.3 is superfluous.

3.4   Show that, if f is piecewise continuous and integrable on $\mathbb{R}$ and if $xf(x)$ is integrable on $\mathbb{R}$, then the Fourier transform of $xf(x)$ is $i\hat{f}'(\omega)$.

3.5   Show that the Fourier transform of $f(x) = e^{-ax^2}$, where a is a positive constant, is

$$\hat{f}(\omega) = \frac{1}{2\sqrt{a\pi}} e^{-\omega^2/4a}$$

(**Hint:** obtain a differential equation for $\hat{f}$ by differentiating under the integral sign).

3.6   Use the results of Exercises 3.4 and 3.5 to find the Fourier transforms of

(i) $f(x) = xe^{-ax^2}$ (ii) $f(x) = x^2 e^{-ax^2}$,

where a is a positive constant. Show that, if $f(x) = xe^{-x^2/2}$, then $f = \sqrt{2\pi}\, i\hat{f}$.

3.7   Use the function

$$f(x) = \begin{cases} \pi, & |x| < \delta \\ 0, & |x| \geq \delta, \end{cases}$$

where $\delta > 0$, to show that for all very large $a > 0$ there is a point $x > -\delta$ near $-\delta$ such that

$$\int_{-a}^{a} \hat{f}(\omega) e^{i\omega x}\, d\omega$$

exceeds by about 9% the value π of the total jump at $-\delta$ and similarly near δ. Draw the graph of the previous integral as a function of x for very large a. This is called the *Gibbs phenomenon* for Fourier integrals.

3.8   Let g be piecewise continuous and integrable on $\mathbb{R}$ and define $f(x) = \int_0^x g$. Show that, if $f(x) \to 0$ as $|x| \to \infty$, then $\hat{f}(0) = \hat{f}_C(0) = \hat{f}_S(0) = 0$ and, for $\omega \neq 0$,

(i) $\hat{f}(\omega) = \dfrac{1}{i\omega}\hat{g}(\omega)$ (ii) $\hat{f}_C(\omega) = -\dfrac{1}{\omega}\hat{g}_S(\omega)$ ·   (iii) $\hat{f}_S(\omega) = \dfrac{1}{\omega}\hat{g}_C(\omega)$.

3.9   Using the same hypotheses as those in Theorem 7.3, prove the following formulas for the Fourier cosine and sine transforms of derivatives

(i) $\widehat{f'}_C(\omega) = -\dfrac{2}{\pi}f(0) + \omega\hat{f}_S(\omega)$ (ii) $\widehat{f'}_S(\omega) = -\omega\hat{f}_C(\omega)$

(iii) $\widehat{f''}_C(\omega) = -\dfrac{2}{\pi}f'(0) - \omega^2\hat{f}_C(\omega)$ (iv) $\widehat{f''}_S(\omega) = \dfrac{2}{\pi}\omega f(0) - \omega^2\hat{f}_S(\omega)$.

3.10   Let $f, g : \mathbb{R} \to \mathbb{R}$ be integrable functions, and let f be even and g be odd. Use Theorem 7.2(iv) and Exercise 2.9 to show that, if α is a real constant, then
(i) the Fourier cosine transform of $f(x + \alpha) + f(x - \alpha)$ is $2\hat{f}_C(\omega)\cos\omega\alpha$,
(ii) the Fourier sine transform of $f(x + \alpha) - f(x - \alpha)$ is $-2\hat{f}_C(\omega)\sin\omega\alpha$,

(*iii*) the Fourier sine transform of $g(x + \alpha) + g(x - \alpha)$ is $2\hat{g}_S(\omega)\cos \omega\alpha$, and
(*iv*) the Fourier cosine transform of $g(x + \alpha) - g(x - \alpha)$ is $2\hat{g}_S(\omega)\sin \omega\alpha$.

4.1 Show that, if f and g are as in Theorem 7.6, if g has a finite number of discontinuities, and if g' is bounded and piecewise continuous, then

$$\text{PV} \int_{-\infty}^{\infty} \hat{f}(\omega)\hat{g}(\omega)e^{i\omega x}\, d\omega = \frac{1}{2\pi} f*g(x)$$

at every x in $\mathbb{R}$ (**Hint:** use Theorem B.10 to show that $f*g$ has a continuous derivative).

4.2 Let $f : \mathbb{R} \to \mathbb{R}$ be an integrable function such that f and f' are bounded and piecewise continuous and f has a finite number of discontinuities. Use Exercise 4.1 to prove the *Plancherel formula*

$$\int_{-\infty}^{\infty} |\hat{f}|^2 = \frac{1}{2\pi} \int_{-\infty}^{\infty} f^2,$$

which is the Fourier integral counterpart of Parseval's identity for Fourier series.

4.3 In the same way as Parseval's identity can be used to evaluate the sum of certain infinite series, the Plancherel formula can be used to evaluate certain improper integrals. State the equation that results by applying the Plancherel formula to the function of
    (*i*) Example 7.1 and
    (*ii*) Exercise 2.1 (*i*).

4.4 State and prove equations analogous to the Plancherel formula for the Fourier cosine and sine transforms.

4.5 If $f(x) = e^{-|x|}$ for all x in $\mathbb{R}$, find $\hat{f}*\hat{f}$ (**Hint:** use the equation in Exercise 4.1 and the transform from Example 7.2).

4.6 Assume that the even extensions of $f, g : [0, \infty) \to \mathbb{R}$ to $\mathbb{R}$ satisfy the hypotheses of Theorem 7.6. Use the result of Exercise 2.9 to show that $\hat{f}_C\hat{g}_C$ is the Fourier cosine transform of

$$\frac{1}{\pi} \int_0^{\infty} f(s)[g_E(x - s) + g_E(x + s)]\, ds.$$

4.7 Assume that the even extension of f and the odd extension of g satisfy the hypotheses of Theorem 7.6. Use the result of Exercise 2.9 to show that $\hat{f}_C\hat{g}_S$ is the Fourier sine transform of

$$\frac{1}{\pi} \int_0^{\infty} f(s)[g_O(x - s) + g_O(x + s)]\, ds = \frac{1}{\pi} \int_0^{\infty} g(s)[f_E(x - s) - f_E(x + s)]\, ds.$$

6.1 Use the Fourier transform method to solve the boundary value problem of §7.5:

$$\begin{aligned} u_{xx} + u_{yy} &= 0 & -\infty < x < \infty, \quad y > 0 \\ u(x, 0) &= f(x) & -\infty < x < \infty \end{aligned}$$

(**Hint:** use the results of Exercise 2.3 and Example 7.6).

6.2 Use the Fourier transform method to solve the boundary value problem

$$\begin{aligned} u_t &= ku_{xx} & -\infty < x < \infty, \quad t > 0 \\ u(x, 0) &= e^{-x^2} & -\infty < x < \infty \end{aligned}$$

(**Hint:** use the substitution

$$y = \sqrt{1 + \alpha}\, s - \frac{\alpha x}{\sqrt{1 + \alpha}},$$

where $\alpha = 1/4kt$).

6.3   Use the Fourier transform method to solve the boundary value problem

$$u_t = ku_{xx} + hu_x \qquad -\infty < x < \infty, \quad t > 0$$
$$u(x, 0) = f(x) \qquad -\infty < x < \infty,$$

where h is a positive constant.

6.4   Use the Fourier transform method to solve the boundary value problem

$$u_{tt} = c^2 u_{xx} \qquad -\infty < x < \infty, \quad t > 0$$
$$u(x, 0) = f(x) \qquad -\infty < x < \infty$$
$$u_t(x, 0) = 0 \qquad -\infty < x < \infty.$$

There is also a Fourier cosine transform method and a Fourier sine transform method. The general procedure is as in the usual Fourier transform method, but uses only cosine or only sine transforms and the properties listed in Exercise 3.9. The method will not work if a first-order partial derivative is transformed because this introduces both the cosine and the sine transforms of the function, as shown by (*i*) and (*ii*) of Exercise 3.9.

6.5   Use the Fourier sine transform method to obtain the integral solution of Exercise 1.7. How is the solution to be modified if the boundary condition $u(x, 0) = 0$ is replaced by $u(x, 0) = h(x)$? (**Hint:** for the last question, find the solution of Exercise 1.3 with $g \equiv 0$.)

6.6   Use the Fourier cosine transform method to solve the boundary value problem of Exercise 1.8 (**Hint:** use the results of Exercises 2.6, 2.7 and 4.6).

6.7   Use the Fourier sine transform method to solve the boundary value problem

$$u_t = ku_{xx} \qquad 0 < x < \infty, \quad t > 0$$
$$u(x, 0) = f(x) \qquad 0 < x < \infty$$
$$u(0, t) = 0 \qquad t \geq 0$$

(**Hint:** use the results of Exercises 3.5, 2.9 and 4.7).

6.8   Repeat Exercise 6.7 if the boundary condition $u(0, t) = 0$ is replaced by $u_x(0, t) = 0$.

6.9   Use the Fourier sine transform method to solve the boundary value problem

$$u_{tt} = c^2 u_{xx} \qquad x > 0, \quad t > 0$$
$$u(x, 0) = f(x) \qquad x \geq 0$$
$$u_t(x, 0) = g(x) \qquad x \geq 0$$
$$u(0, t) = 0 \qquad t \geq 0.$$

How is the solution to be modified if the boundary condition $u(0, t) = 0$ is replaced by $u_x(0, t) = 0$? (**Hint:** for the first part, use the results of Exercises 3.8 and 3.10.)

6.10   Solve the boundary value problem

$$u_{xx} + u_{yy} = 0 \qquad x > 0, \quad y > 0$$
$$u(x, 0) = f(x) \qquad x \geq 0$$
$$u_x(0, y) = g(y) \qquad y \geq 0$$

(**Hint:** solve this problem for $f \equiv 0$ and then add the solution to that of Exercise 6.6).

8

LAPLACE TRANSFORMS

§8.1 The Laplace Transform and the Inversion Theorem

Some elementary functions, such as a constant, the exponential, the sine, and the cosine, do not have Fourier transforms because they are not integrable on $\mathbb{R}$. The straightforward remedy to this is very simple: just multiply these functions, before transforming, by a suitable factor that makes the product integrable, and then adjust the corresponding inversion formula in a suitable way.

The first of these vague steps—resting on whatever reasonable meaning can be assigned to the word *suitable*—is the easiest, in particular when dealing with problems in which the variable is time. It is usual, as we did in Chapters 3 and 5, to state these problems for values of time starting at some initial instant t_0, which we shall take, without loss of generality, to be $t_0 = 0$. Then, given a function f that is not integrable on $\mathbb{R}$, such as those already mentioned or $f(t) = t^n$, where n is a positive integer, define or redefine f to be identically zero for $t < 0$, and then $f(t)e^{-\sigma t}$ may be integrable for some real value of σ. If it is, this modified function has the Fourier transform

$$\frac{1}{2\pi} \int_0^\infty f(t)e^{-(\sigma+i\omega)t}\, dt.$$

Omitting the factor $1/2\pi$, adopting the letter t for the variable, and denoting $\sigma + i\omega$ by s, lead to the following modified transform.

Definition 8.1 *Let* $f : [0, \infty) \to \mathbb{C}$ *be such that* $f(t)e^{-\sigma t}$ *is integrable on* $(0, \infty)$ *for some* σ *in* $\mathbb{R}$. *Then the function* $F : \mathbb{C} \to \mathbb{C}$ *defined by*

$$F(s) = \int_0^\infty f(t)e^{-st}\, dt$$

with domain

$$D_F = \{\, s \in \mathbb{C} : f(t)e^{-st} \text{ is integrable}\,\}$$

is called the **Laplace transform** *of* f. *It is also denoted by* $\mathcal{L}[f]$.

It is quite easy to compute the Laplace transforms of some of the elementary functions.

Example 8.1 If $f \equiv a$ on $[0, \infty)$, where a is a real or complex constant, then

$$F(s) = \int_0^\infty a e^{-st} \, dt = \lim_{b \to \infty} \left[\frac{a e^{-st}}{-s} \right]_0^b = \frac{a}{s}$$

if the real part of s, which we denote by $\mathrm{Re}(s)$, is greater than zero.

Example 8.2 If $f(t) = e^{at}$ on $[0, \infty)$, where a is a real or complex constant, then

$$F(s) = \int_0^\infty e^{(a-s)t} \, dt = \lim_{b \to \infty} \left[\frac{e^{(a-s)t}}{a - s} \right]_0^b = \frac{1}{s - a}$$

for $\mathrm{Re}(s) > \mathrm{Re}(a)$.

We have allowed f to be complex-valued in Definition 8.1, which may seem unnecessarily general for physical applications. However, this does not introduce any additional difficulty and will be found to be very useful in §8.2, through the use of these and similar examples.

The reason for the name of the transform is that the integral in Definition 8.1 appears in Laplace's *Théorie analytique des probabilités* of 1812. However, its re-introduction and importance in solving differential equations and boundary value problems only go back to the 1920s when it replaced Oliver Heaviside's operational calculus in applications to electrical engineering. Heaviside (1850–1925), who had no university degree, had developed a strong interest in long distance telegraphy when he worked as an operator at the Great Northern Telegraph Company, in Newcastle, from 1868 until his resignation in 1874 (this was the only employment he ever held in his life, which he largely spent in dire poverty and seclusion). To solve many of the problems posed by telegraphy, he perfected a method to solve ordinary and partial differential equations that regards differentiation as an operator. For instance, to solve the simple differential equation $y'' - y = 1$ for $t > 0$ subject to initial conditions $y(0) = y'(0) = 0$, Heaviside would denote the differential operator d/dt by p—he had used the letter D until 1886—and obtain the equation $p^2 y - y = \mathbf{1}$, where $\mathbf{1}$ denotes the function that vanishes for $t < 0$ and has value 1 for $t \geq 0$. Hence, if we treat p as an algebraic quantity,

$$(1) \qquad\qquad y = \frac{1}{p^2 - 1} \mathbf{1}.$$

This cannot be the end of the line, of course. To obtain the actual solution from this operational solution let us accept that the geometric series expansion is valid for this operator, and then

$$\frac{1}{p^2 - 1} \mathbf{1} = \frac{1}{p^2} \frac{1}{1 - \dfrac{1}{p^2}} \mathbf{1} = \frac{1}{p^2} \left[1 + \frac{1}{p^2} + \frac{1}{p^4} + \cdots \right] \mathbf{1} = \left[\frac{1}{p^2} + \frac{1}{p^4} + \frac{1}{p^6} + \cdots \right] \mathbf{1}.$$

Since p represents differentiation, Heaviside regarded $1/p$ as integration from zero to t. In this manner,

$$\frac{1}{p}\mathbf{1} = t \qquad \text{and} \qquad \frac{1}{p^n}\mathbf{1} = \frac{t^n}{n!}$$

for any positive integer n, leading to the actual solution

$$y(t) = \frac{t^2}{2!} + \frac{t^4}{4!} + \frac{t^6}{6!} + \cdots = \frac{e^t + e^{-t}}{2} - 1,$$

which, perhaps surprisingly, is the correct solution.

Heaviside's methods were, to say the least, unrigorous and even more confusing when applied to partial differential equations. For instance, when he applied them to the equation for a semi-infinite cable he needed to define $p^{1/2}$ (see Exercise 4.1) and other fractional powers of p.

Heaviside's pioneering work, both practical and theoretical, was the object of frequent and serious opposition, but his genius was eventually recognized by Lord Kelvin and Sir Oliver Lodge, and he was elected a Fellow of the Royal Society in 1891. Heaviside published two papers in the *Proceedings of the Royal Society*, but when no more of them were accepted because his methods were unrigorous—although he always obtained correct results—he resumed publishing in the *Electrician*, his natural scientific habitat at the time, and held the referees of the Royal Society and other *scienticulists* guilty of looking at *the gift horse in the mouth*. His numerous journal articles—which he started publishing at the age of 22—were eventually gathered in two collections: *Electrical Papers*, in two volumes, in 1892 and *Electromagnetic Theory*, in three volumes, in 1893, 1899, and 1912.[1] From now on we shall refer to these collections as *EP* and *EMT*, indicating the volume by a roman numeral. *EMT*II starts with an assertion of Heaviside's belief that mathematics is an experimental science—the only mathematical work that he truly admired was Fourier's—and then he put his view of mathematical rigor in a nutshell by asking: *shall I refuse my dinner because I do not fully understand the process of digestion?* (*EMT*II, page 9).

Attempts were made by other researchers to justify Heaviside's operational calculus in solving a differential equation with **1** on the right-hand side. As early as 1915, Thomas John I'A. Bromwich (1875–1929), an English mathematician, did it by resorting to the theory of complex integration, but in 1927 he admitted that Heaviside's method was easier to use. However, at that time he developed a second method, soon favored by many engineers, in which he followed Heaviside to obtain the operational solution, but then used complex integration to recover the actual solution from it (essentially using the method of §8.5 below). Meanwhile, John R. Carson, of the American Telephone and Telegraph Company, had come up with some ideas of his own in 1917, which he perfected—perhaps inspired by a private communication from Harry Bateman (1882–

---

[1] The latest editions are by Chelsea, New York, in 1970 and 1971, respectively. The latter also contains Heaviside's works after 1912.

1946)—by 1922.[1] He also obtained the operational solution first *à la* Heaviside, call it $G(p)$, and then showed that the actual solution $y(t)$ is a solution of the integral equation

$$G(p) = p \int_0^\infty y(t)e^{-pt}\, dt.$$

The right-hand side is the Laplace transform of y except for the notation and the extra p in front of the integral, whose presence will be explained below. The first one to regard this equation as a transformation of $y(t)$ into $G(p)$, and then Bromwich's complex integration as the inverse transformation that obtains $y(t)$ from $G(p)$, was Balthasar van der Pol (1889–1959), of the *Philips Gloeilampenfabriken* in Holland, in 1929.

We have seen how Bromwich and Carson, as well as several other workers in the field, followed Heaviside's method to find the operational solution. For this they were to be taken to task by the German mathematician Gustav Doetsch (1892–)—who had already been using the Laplace transform for some time in solving integral and differential equations—in his 1930 review of Carson's 1926 book *Electric Circuit Theory and Operational Calculus*. Doetsch's view was that Carson should have transformed the equations from the start instead of using Heaviside's unrigorous method to obtain the operational solution. Doetsch also stressed the importance of the corresponding inversion formula—already anticipated by Poisson in 1823—and went on to develop the properties and applications of the transform in his 1937 book *Theorie und Anwengdun der Laplace-Transformation*, the first of several he wrote on the subject. He also switched to the letter s as the variable for the transform, stating as a reason that p looks like a positive constant while s looks like a variable and is just next to t, the variable for f, in the alphabet.

Here is how Doetsch's method replaces Heaviside's operators in solving the initial value problem $y'' - y = 1$, $y(0) = y'(0) = 0$. First multiply the equation by e^{-st}, where s is real and positive, and then assume that $y(t)e^{-st}$ and $y'(t)e^{-st}$ approach zero as $t \to \infty$ and integrate from 0 to ∞. Assuming also that the integrals below converge, integrating by parts, and denoting the transform of y by Y, we obtain

$$\int_0^\infty y'(t)e^{-st}\, dt = \left[y(t)e^{-st} \right]_0^\infty + s \int_0^\infty y(t)e^{-st}\, dt = -y(0) + sY(s) = sY(s)$$

$$\int_0^\infty y''(t)e^{-st}\, dt = \left[y'(t)e^{-st} \right]_0^\infty + s \int_0^\infty y'(t)e^{-st}\, dt = -y'(0) + s^2 Y(s) = s^2 Y(s).$$

Then, the given initial value problem is transformed into $s^2 Y(s) - Y(s) = \mathcal{L}[1](s)$, that is,

$$(2) \qquad\qquad Y(s) = \frac{1}{s^2 - 1}\mathcal{L}[1](s),$$

---

[1] Bateman may have been the first to use the Laplace transform method in its present form in a 1910 paper on the equations of radioactive disintegration, but it seems that this publication did not make an impact at the time.

which is a more palatable version of (1). Using the result of Example 8.1, $\mathcal{L}[\mathbf{1}](s) = 1/s$, and then (2) becomes

$$Y(s) = \frac{1}{(s^2 - 1)s} = \frac{\frac{1}{2}}{s - 1} + \frac{\frac{1}{2}}{s + 1} - \frac{1}{s}.$$

Then the transforms in Examples 8.1 and 8.2 suggest that the actual solution of the given initial value problem is

$$y(t) = \tfrac{1}{2}(e^t + e^{-t}) - 1 = \cosh t - 1,$$

as indeed it is.

Note that, if we rewrite (2) as

$$\frac{1}{s^2 - 1} = s \int_0^\infty y(t)e^{-st}\, dt,$$

then it is just Carson's integral equation for this example. The factor s on the right arises from the transform of **1**.

It should not go without saying that not every function has a Laplace transform.

Example 8.3 If $f(t) = e^{t^2}$ on $[0, \infty)$, then f has no Laplace transform because, for any σ in $\mathbb{R}$, $f(t)e^{-\sigma t} \to \infty$ as $t \to \infty$, which shows that $f(t)e^{-\sigma t}$ is not integrable.

It is not accidental that in the first two examples the domain of the Laplace transform is a half-plane in $\mathbb{C}$. D_F can be either empty, as in the preceding example, or the entire complex plane, as for $f \equiv 0$. But it can be shown that, otherwise, there is a real number r such that D_F is the open half-plane $\{\, s \in \mathbb{C} : \operatorname{Re}(s) > r \,\}$ or the closed half-plane $\{\, s \in \mathbb{C} : \operatorname{Re}(s) \geq r \,\}$ (see Exercise 1.6). D_F is an open half-plane in Examples 8.1 and 8.2, and in the next example it is a closed half-plane.

Example 8.4 If $f(t) = 1/(1 + t^2)$ on $[0, \infty)$ and if $\operatorname{Re}(s) \geq 0$, then

$$\int_0^x \left| \frac{1}{1+t^2} e^{-st} \right| dt \leq \int_0^x \frac{1}{1+t^2}\, dt < \int_0^\infty \frac{1}{1+t^2}\, dt = \frac{\pi}{2},$$

which shows, by letting $x \to \infty$, that $f(t)e^{-st}$ is integrable for $\operatorname{Re}(s) \geq 0$. Clearly, $f(t)e^{-st}$ is not integrable for $\operatorname{Re}(s) < 0$, as it approaches ∞ as $t \to \infty$. Hence, $D_F = \{s \in \mathbb{C} : \operatorname{Re}(s) \geq 0\}$. While it is a relatively easy matter to determine D_F for the given function, computing its Laplace transform is rather involved.[1]

The next question is how to recover a function from its Laplace transform. Theoretically this is an easy task because, according to Definition 8.1, $F(s)$ is actually a Fourier transform—that of the function that vanishes for $t < 0$ and equals $2\pi f(t)e^{-\sigma t}$ for $t \geq 0$, where $\sigma = \operatorname{Re}(s)$—and then we can use Theorem 7.1. Therefore, from now on we restrict ourselves to the following class of functions.

---

[1] See, for instance, R. V. Churchill, *Operational Mathematics*, 3rd ed., McGraw-Hill, New York, 1972, pages 110–111.

Definition 8.2 *Let* $\mathcal{L}$ *be the set of all locally integrable functions* $f : \mathbb{R} \to \mathbb{C}$ *such that* f *vanishes for* $t < 0$ *and* $f(t)e^{-\sigma t}$ *is integrable for some* σ *in* $\mathbb{R}$.

Theorem 8.1 *Let* $f \in \mathcal{L}$ *have Laplace transform* F *and let* σ *be a real number such that* $f(t)e^{-\sigma t}$ *is integrable on* $(0, \infty)$. *Then*

(3)
$$\frac{1}{2\pi} \, \text{PV} \int_{-\infty}^{\infty} F(\sigma + i\omega)e^{(\sigma+i\omega)t} \, d\omega = \tfrac{1}{2}[f(t+) + f(t-)]$$

at every point t *in* $[0, \infty)$ *such that* f *has right-hand and left-hand derivatives at* t *and* f *is piecewise continuous on an arbitrarily small interval centered at* t.

Proof. Note that $f(t)e^{-(\sigma+i\omega)t}$ is integrable on $(0, \infty)$ because $f(t)e^{-\sigma t}$ is, and then F is defined at $\sigma + i\omega$. Hence,

$$F(\sigma + i\omega) = \int_0^{\infty} f(t)e^{-(\sigma+i\omega)t} \, dt = \int_{-\infty}^{\infty} f(t)e^{-\sigma t} e^{-i\omega t} \, dt = 2\pi \hat{g}(\omega),$$

if g is the function defined by $g(t) = 0$ for $t < 0$ and $g(t) = f(t)e^{-\sigma t}$ for $t \geq 0$. Since g is integrable, Theorem 7.1 implies that

$$\text{PV} \int_{-\infty}^{\infty} \frac{1}{2\pi} F(\sigma + i\omega)e^{i\omega t} \, d\omega = \text{PV} \int_{-\infty}^{\infty} \hat{g}(\omega)e^{i\omega t} \, d\omega = \tfrac{1}{2}[g(t+) + g(t-)]$$

at every t at which g has one-sided limits and one-sided derivatives. Replacing $g(t)$ with $f(t)e^{-\sigma t}$ proves (3), Q.E.D.

In particular, this result shows that if $f \in \mathcal{L}$ is continuous and has Laplace transform F, then f is the unique function in $\mathcal{L}$ with this transform, and we can denote it by $\mathcal{L}^{-1}[F]$. As in the case of Fourier transforms, if we have an extensive table of precomputed Laplace transforms it may be possible to use it to find $\mathcal{L}^{-1}[F]$ for a given transform F. This is usually simpler than evaluating the principal value of the integral in (3).

§8.2 Properties of the Laplace Transform

The results in this section will allow us to compute the Laplace transforms of new functions from those of old ones. The first theorem contains a partial list of several properties proved by Abel in a posthumous memoir on what he called a *generating function*, which he applied to obtain power series expansions and to evaluate certain definite integrals. Although the context and motivation are entirely different from ours, Abel's generating function is an indefinite integral whose integrand differs from that of the Laplace transform only in that $-s$ is replaced by x.

Theorem 8.2 *Let* $f, g \in \mathcal{L}$ *have Laplace transforms* F *and* G, *respectively. If* τ *is a positive real number and* a *is an arbitrary real or complex constant, then the Laplace transform of*

(i) $f + g$ is $F + G$,

(ii) af is aF,

(iii) $f(t)e^{at}$ is $F(s - a)$,

(iv) $f(t - \tau)$ is $e^{-\tau s}F(s)$,   and

(v) $f(\tau t)$ is $\dfrac{1}{\tau}F\left(\dfrac{s}{\tau}\right)$.

The proof is left as an exercise.

Example 8.5 If $f(t) = \sinh at$ on $[0, \infty)$, where a is a real or complex constant, then $f(t) = (e^{at} - e^{-at})/2$ and Theorem 8.2(i)–(ii) together with the result of Example 8.2 show that

$$F(s) = \frac{1}{2}\left(\frac{1}{s - a} - \frac{1}{s + a}\right) = \frac{a}{s^2 - a^2}.$$

Example 8.6 If τ is a positive real number and if H^τ is defined to be zero for $t < \tau$ and one for $t \geq \tau$, then $H^\tau(t) = \mathbf{1}(t - \tau)$, where $\mathbf{1}$ is the *Heaviside function*, which we shall henceforth denote by H in Heaviside's honor. According to Theorem 8.2(iv) and using the result of Example 8.1,

$$\mathcal{L}[H^\tau](s) = \frac{e^{-\tau s}}{s}.$$

Example 8.7 If τ and T are positive real numbers and if $f(t) = 1$ for $\tau \leq t < \tau + T$ and $f(t) = 0$ otherwise, then $f = H^\tau - H^{\tau + T}$, and

$$F(s) = \frac{e^{-\tau s}}{s} - \frac{e^{-(\tau + T)s}}{s} = \frac{e^{-\tau s}}{s}(1 - e^{-Ts}).$$

To emphasize the fact that the function of time in Theorem 8.2(iv) vanishes for $t < \tau$, this property can be restated as follows: the Laplace transform of $f(t - \tau)H(t - \tau)$ is $e^{-\tau s}F(s)$, where H is the Heaviside function.

Example 8.8 If τ is a positive real number and if for any positive integer n we define $f(t) = n$ for $(n - 1)\tau \leq t < n\tau$ and $f(t) = 0$ otherwise, then it follows from an examination of the graph of f, shown in Figure 8.1, that $f(t) = f(t - \tau)H(t - \tau) + H(t)$ for all $t \geq 0$. The Laplace transform turns this equation into

$$F(s) = e^{-\tau s}F(s) + \frac{1}{s},$$

and then

$$F(s) = \frac{1}{s(1 - e^{-\tau s})}.$$

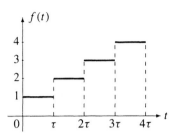

Figure 8.1

Theorem 8.3 *Let $f \in \mathcal{L}$ be continuous for $t > 0$ and assume that it has a piecewise continuous derivative and that there is a real number σ such that $f(t)e^{-\sigma t} \to 0$ as $t \to \infty$. Then $f' \in \mathcal{L}$ and, if F is the Laplace transform of f,*

$$\mathcal{L}[f'](s) = sF(s) - f(0+)$$

for $\operatorname{Re}(s) \geq \sigma$.

Proof. Choose $a > 0$ such that f' is continuous at a, let $t_1, t_2, \ldots, t_n$ be the points at which f' is discontinuous in the interval $(0, a)$, and define $t_0 = 0$ and $t_{n+1} = a$. Then integration by parts shows that

$$\int_{t_k}^{t_{k+1}} f'(t)e^{-st} \, dt = f(t_{k+1}-)e^{-st_{k+1}} - f(t_k+)e^{-st_k} + s\int_{t_k}^{t_{k+1}} f(t)e^{-st} \, dt.$$

Adding the equations that result from substituting $k = 0, \ldots, n$ and noticing that the continuity of f implies that $f(t_k-) = f(t_k+) = f(t_k)$ for $0 < k < n + 1$, we obtain

$$\int_0^a f'(t)e^{-st} \, dt = f(a)e^{-sa} - f(0+) + s\int_0^a f(t)e^{-st} \, dt.$$

Since $f(t)e^{-\sigma t} \to 0$ as $t \to \infty$, we conclude that, for $\operatorname{Re}(s) \geq \sigma$,

$$\int_0^\infty f'(t)e^{-st} \, dt = \lim_{a \to \infty}\left[f(a)e^{-sa} - f(0+) + s\int_0^a f(t)e^{-st} \, dt \right] = sF(s) - f(0+),$$

Q.E.D.

Corollary *Let $f, f', f'', \ldots, f^{(n-1)}$ satisfy the same hypotheses that f satisfies in the theorem. Then $f^{(n)} \in \mathcal{L}$ and, if F is the Laplace transform of f,*

$$\mathcal{L}[f^{(n)}](s) = s^n F(s) - s^{n-1} f(0+) - s^{n-2} f'(0+) - \cdots - f^{(n-1)}(0+)$$

for $\operatorname{Re}(s) \geq \sigma$.

Proof. Applying Theorem 8.3 n times we obtain

$$\mathcal{L}[f'](s) = sF(s) - f(0+)$$

$$\mathcal{L}[f''](s) = s\mathcal{L}[f'](s) - f'(0+)$$

$$\cdot \quad \cdot \quad \cdot \quad \cdot \quad \cdot \quad \cdot \quad \cdot \quad \cdot \quad \cdot$$

$$\mathcal{L}[f^{(n)}](s) = s\mathcal{L}[f^{(n-1)}](s) - f^{(n-1)}(0+).$$

The result follows by eliminating $\mathcal{L}[f']$, ..., $\mathcal{L}[f^{(n-1)}]$, Q.E.D.

Example 8.9 Find the Laplace transforms of the functions defined by $f(t) = \sin at$ and $g(t) = \cos at$ on $[0, \infty)$, where a is a constant. Since $f'(t) = ag(t)$ and $g'(t) = -af(t)$, transforming this system and using Theorem 8.3 gives

$$sF(s) - f(0+) = aG(s) \ .$$
$$sG(s) - g(0+) = -aF(s).$$

Putting $f(0+) = 0$ and $g(0+) = 1$ and solving, we obtain

$$F(s) = \frac{a}{s^2 + a^2} \qquad \text{and} \qquad G(s) = \frac{s}{s^2 + a^2}.$$

The following technical lemma will be useful in establishing the next theorem.

Lemma 8.1 *If $f(t)e^{-st}$ is integrable on $(0, \infty)$, then $t^n f(t)e^{-(s+\epsilon)t}$ is integrable on $(0, \infty)$ for any positive integer n and any $\epsilon > 0$.*

Proof. Since the product $t^n e^{-\epsilon t}$ is bounded on $[0, \infty)$, we have

$$|t^n f(t)e^{-(s+\epsilon)t}| \leq t^n e^{-\epsilon t}|f(t)e^{-st}| \leq M|f(t)e^{-st}|$$

for some $M > 0$. Then the left-hand side is integrable on $(0, \infty)$ by Theorem B.2(iii) because $f(t)e^{-st}$ is integrable, Q.E.D.

Theorem 8.4 *If $f \in \mathcal{L}$, then its Laplace transform F is differentiable in the interior of D_F and*

$$F'(s) = -\int_0^\infty tf(t)e^{-st} \, dt.$$

Proof. Note first that, if s is in the interior of D_F, so is $s - \epsilon$ for $\epsilon > 0$ small enough. Then, $f(t)e^{-(s-\epsilon)t}$ is integrable on $(0, \infty)$, and, by Lemma 8.1, so is $tf(t)e^{-st}$. To compute $F'(s)$ we shall use the definition of derivative. If $h \neq 0$, we have

$$\frac{F(s+h) - F(s)}{h} + \int_0^\infty tf(t)e^{-st} \, dt = \int_0^\infty \left(\frac{e^{-ht} - 1}{h} + t\right) f(t)e^{-st} \, dt,$$

and then, to see that the integral on the right converges, we use the MacLaurin series for e^{-ht}. This gives

$$\frac{e^{-ht} - 1}{h} = \frac{1}{h}\left(-ht + \frac{h^2 t^2}{2!} - \frac{h^3 t^3}{3!} + \frac{h^4 t^4}{4!} - \cdots\right) = -t + \frac{ht^2}{2!} - \frac{h^2 t^3}{3!} + \frac{h^3 t^4}{4!} - \cdots$$

and then, since $n! > (n-1)!$ for $n > 1$,

$$\left|\frac{e^{-ht} - 1}{h} + t\right| = |h|t^2\left|\frac{1}{2!} - \frac{ht}{3!} + \frac{h^2 t^2}{4!} - \cdots\right|$$

$$< |h|t^2\left(1 + \frac{|h|t}{2!} + \frac{|h|^2 t^2}{3!} + \cdots\right)$$

$$= |h|t^2 e^{|h|t}.$$

If s is in the interior of D_F, so is $s - 2|h|$ for $|h|$ small enough. This means that $f(t)e^{-(s-2|h|)t}$ is integrable on $(0, \infty)$, and then so is $t^2 f(t)e^{-(s-|h|)t}$ by Lemma 8.1 with $\epsilon = |h|$. Therefore, the integral in question converges by Theorems B.2(iii) and B.3, and we have

$$\left|\frac{F(s+h) - F(s)}{h} + \int_0^\infty tf(t)e^{-st}\,dt\right| \le |h|\int_0^\infty \left|t^2 f(t)e^{-(s-|h|)t}\right|\,dt.$$

The theorem follows by letting $h \to 0$, Q.E.D.

Corollary *If $f \in \mathcal{L}$ and n is a positive integer, then the Laplace transform of $t^n f(t)$ is $(-1)^n F^{(n)}(s)$ for s in the interior of D_F.*

Proof. For $n = 1$ the result is true by the theorem, and then it follows by induction for arbitrary $n > 1$, Q.E.D.

Example 8.10 If $f(t) = t^n$ on $[0, \infty)$, where n is a positive integer, and if H is the Heaviside function, then $f(t) = t^n H(t)$ and the Corollary gives

$$F(s) = (-1)^n H^{(n)}(s) = (-1)^n \frac{d^n}{ds^n}\left(\frac{1}{s}\right) = (-1)^n \frac{(-1)^n n!}{s^{n+1}} = \frac{n!}{s^{n+1}}.$$

In applications to ordinary differential equations (see Exercises 2.7 and 3.1) one frequently encounters a Laplace transform $F(s)$ that is a quotient of polynomials in s with real coefficients and such that the degree of the numerator is smaller than that of the denominator. In this case it is quite easy to find $\mathcal{L}^{-1}[F]$ without performing the integration in (3) of Theorem 8.1. Indeed, if the denominator has zeros at points a_1, $a_2, \ldots, a_n$ in $\mathbb{C}$ with multiplicities $m_1, m_2, \ldots, m_n$, respectively, then it can be written as the product of a constant times $(s - a_1)^{m_1}(s - a_2)^{m_2} \cdots (s - a_n)^{m_n}$. This leads us to consider a rational function

$$F(s) = \frac{P(s)}{(s - a_1)^{m_1}(s - a_2)^{m_2} \cdots (s - a_n)^{m_n}},$$

where $a_1, a_2, \ldots, a_n$ are real or complex numbers and $P(s)$ is a polynomial of degree smaller than $m_1 + \cdots + m_n$. Then $F(s)$ can be written as a sum of partial fractions,

$$F(s) = \sum_{j=1}^{n} \left[\frac{A_{j1}}{s - a_j} + \frac{A_{j2}}{(s - a_j)^2} + \cdots + \frac{A_{jm_j}}{(s - a_j)^{m_j}} \right].$$

According to Theorem 8.2 (ii)–(iii) and the result of Example 8.10, each partial fraction

$$\frac{A_{jk}}{(s - a_j)^k}$$

is the Laplace transform of

$$A_{jk} \frac{t^{k-1}}{(k - 1)!} e^{a_j t},$$

and then

$$\mathcal{L}^{-1}[F](t) = \sum_{j=1}^{n} \left[A_{j1} + A_{j2}t + \cdots + A_{jm_j} \frac{t^{m_j - 1}}{(m_j - 1)!} \right] e^{a_j t}.$$

The simplicity of this procedure is due to the fact that the denominator of each partial fraction is a power of a linear term of the form $s - a_j$, which is possible because complex values are allowed for the a_j. If only real constants were allowed in factoring the denominator of $F(s)$, then it may contain irreducible quadratic factors, and the corresponding partial fractions would be less simple. The price to pay for the simplicity of this method is that each term in the sum above may be a complex-valued function of t, even if the sum itself is real. This is a reason why we have allowed f to be complex-valued in Definition 8.1.

Example 8.11 If $F(s) = 1/(s^2 + a^2)^2$, where a is a real constant, find $\mathcal{L}^{-1}[F]$. Since $(s^2 + a^2)^2 = (s + ia)^2(s - ia)^2$, expanding $F(s)$ into partial franctions yields

$$\frac{1}{(s^2 + a^2)^2} = \frac{1}{4a^2} \left[\frac{i/a}{s + ia} - \frac{1}{(s + ia)^2} - \frac{i/a}{s - ia} - \frac{1}{(s - ia)^2} \right].$$

This is the Laplace transform of

$$f(t) = \frac{1}{4a^2} \left[\left(\frac{i}{a} - t \right) e^{-iat} - \left(\frac{i}{a} + t \right) e^{iat} \right] = \frac{1}{2a^3} (\sin at - at \cos at),$$

and, since this is a continuous function, $\mathcal{L}^{-1}[F] = f$.

It is not usual to find rational functions of s in applications to partial differential equations. We start with a very simple example.

Example 8.12 A semi-infinite string stretched along the positive x-axis is initially at rest, and its end at $x = 0$ is kept fixed at the origin. If g is the acceleration of gravity, the motion of the string is governed by the equation $u_{tt} = c^2 u_{xx} - g$, as shown in §5.2. This poses the boundary value problem

$$
\begin{aligned}
u_{tt} &= c^2 u_{xx} - g &\quad x &> 0, \quad t > 0 \\
u(0, t) &= 0 &\quad t &\geq 0 \\
u(x, 0) &= 0 &\quad x &\geq 0 \\
u_t(x, 0) &= 0 &\quad x &\geq 0.
\end{aligned}
$$

We shall use the Laplace transform method. That is, assuming that for each fixed $x > 0$ the Laplace transform of $u(x, t)$ with respect to t, which we denote by $U(x, s)$, exists, that the Corollary of Theorem 8.3 is applicable, and that $\mathcal{L}[u_{xx}] = U_{xx}$, the first two equations are transformed into

$$
s^2 U(x, s) - su(x, 0) - u_t(x, 0) = c^2 U_{xx}(x, s) - \frac{g}{s}
$$

$$
U(0, s) = 0.
$$

The initial conditions reduce the differential equation to $s^2 U(x, s) = c^2 U_{xx}(x, s) - g/s$, whose general solution is

$$
U(x, s) = A(s)e^{-sx/c} + B(s)e^{sx/c} - \frac{g}{s^3}
$$

for each fixed s. Since $e^{sx/c}$ is not a Laplace transform (see Exercise 1.7), we choose $B \equiv 0$, and then $U(0, s) = 0$ implies that $A(s) = g/s^3$. Hence

$$
U(x, s) = \frac{g}{s^3}(e^{-sx/c} - 1).
$$

Using Theorem 8.2(iv) and Example 8.10, we recognize this as the transform of

$$
u(x, t) = \frac{g}{2}\left(t - \frac{x}{c}\right)^2 H\left(t - \frac{x}{c}\right) - \frac{gt^2}{2} H(t).
$$

The study of this solution will be completed in Exercise 2.9.

Usually, more sophisticated techniques are needed in applications to partial differential equations. The next example provides a Laplace transform that is of interest in both heat conduction and some telegraphy problems.

Example 8.13 Let a be a positive constant and define

$$
f(t) = \frac{a}{t\sqrt{\pi t}}e^{-a^2/t}
$$

for $t > 0$. To compute the Laplace transform of f, let s be a positive real number and make the substitution $z = \sqrt{st} - a/\sqrt{t}$. Then

$$
st + \frac{a^2}{t} = z^2 + 2a\sqrt{s} \quad\text{and}\quad \frac{a}{t} + \frac{z}{\sqrt{t}} - \sqrt{s} = 0.
$$

Solving the second of these equations, which is quadratic in $1/\sqrt{t}$, gives

$$\frac{1}{\sqrt{t}} = \frac{-z + \sqrt{z^2 + 4as^{1/2}}}{2a},$$

where the choice of sign is determined by the fact that $t > 0$. Differentiating with respect to t and simplifying, we obtain

$$\frac{1}{t\sqrt{t}} = \frac{1}{a}\left(1 - \frac{z}{\sqrt{z^2 + 4as^{1/2}}}\right)\frac{dz}{dt},$$

and then

$$\int_0^\infty \frac{a}{t\sqrt{\pi t}}e^{-st-a^2/t}\,dt = \frac{1}{\sqrt{\pi}}\int_{-\infty}^\infty \left(1 - \frac{z}{\sqrt{z^2 + 4as^{1/2}}}\right)e^{-z^2-2a\sqrt{s}}\,dz.$$

Since the quotient in parentheses is an odd function of z and the integral of e^{-z^2} from $-\infty$ to ∞ has value $\sqrt{\pi}$, as found in the tables, the right-hand side becomes $e^{-2a\sqrt{s}}$. Finally, it is a standard result of complex analysis that, if two functions of the complex variable s are differentiable on the half-plane $\mathrm{Re}(s) > 0$ and are identical on the positive real axis, then they are identical for $\mathrm{Re}(s) > 0$.[1] Since F must be differentiable by Theorem 8.4 and equals $e^{-2a\sqrt{s}}$ for all real $s > 0$, we conclude that it is defined for $\mathrm{Re}(s) > 0$ and that

$$F(s) = e^{-2a\sqrt{s}}.$$

§8.3 Convolution

Convolution is important for Laplace transforms for the same reason that it is for Fourier transforms. It arises naturally when solving differential equations, ordinary or partial, by the transform method. The definition of convolution is as before, namely, if f, $g \in \mathcal{L}$,

$$f*g(t) = \int_{-\infty}^\infty f(\tau)g(t-\tau)\,d\tau,$$

but now this assumes a particularly simple form because f and g vanish outside $[0, \infty)$, so that $f(\tau)g(t - \tau) = 0$ for τ outside the interval $[0, t]$, $t > 0$, and the convolution of f and g becomes

$$(4) \qquad f*g(t) = \begin{cases} 0 & \text{if } t \le 0 \\ \displaystyle\int_0^t f(\tau)g(t-\tau)\,d\tau & \text{if } t > 0. \end{cases}$$

In this way $f*g$ is always well defined because f and g are locally integrable. If, in addition, g is piecewise continuous, then $g(t - \tau)$, considered a function of (τ, t), is patchwise continuous on $(0, \infty) \times (0, \infty)$ in the sense of Definition B.6, and $f*g$ is continuous by Theorem B.7.

---

[1] See, for instance, L. Ahlfors, *Complex Analysis*, 3rd ed., McGraw-Hill, New York, 1979, page 127.

Theorem 8.5 *If* f, $g \in \mathcal{L}$ *have Laplace transforms* F *and* G, *respectively, and if* f *is integrable and* g *is piecewise continuous, then* $f*g \in \mathcal{L}$, *and*

$$\mathcal{L}[f*g] = FG.$$

Proof. Let σ be a nonnegative real number such that $g(t)e^{-\sigma t}$ is integrable on $(0, \infty)$, and let $\operatorname{Re}(s) \geq \sigma$. Then

$$F(s)G(s) = \left(\int_0^\infty f(\tau)e^{-s\tau}\, d\tau \right)\left(\int_0^\infty g(u)e^{-su}\, du \right)$$

$$= \int_0^\infty f(\tau)e^{-s\tau}\left(\int_0^\infty g(u)e^{-su}\, du \right) d\tau.$$

Making the substitution $u = t - \tau$ and then noticing that $g(t - \tau) = 0$ for $0 \leq t < \tau$, the right-hand side becomes

$$\int_0^\infty f(\tau)e^{-s\tau}\left(\int_\tau^\infty g(t-\tau)e^{-s(t-\tau)}\, dt \right) d\tau = \int_0^\infty \left(\int_0^\infty f(\tau)g(t-\tau)e^{-st}\, dt \right) d\tau.$$

Next we shall show that the order of integration can be reversed. First note that f is integrable and also that

$$\int_0^\infty f(\tau)g(t - \tau)e^{-st}\, d\tau$$

converges uniformly with respect to t on an arbitrary bounded interval $[0, a]$ by Definition B.3 because, if $b > a$, then $g(t - \tau) = 0$ if $\tau > b$ and the integral of $|f(\tau)g(t - \tau)e^{-st}|$ on (b, ∞) is zero. Also,

$$\int_0^\infty g(t - \tau)e^{-st}\, dt$$

converges uniformly with respect to τ on $[0, \infty)$ because the substitution $u = t - \tau$, the identity $g(u) = 0$ for $u < 0$, and the inequalities $\operatorname{Re}(s) \geq \sigma \geq 0$ yield

$$\int_0^\infty |g(t - \tau)e^{-st}|\, dt = \int_{-\tau}^\infty |e^{-s\tau}g(u)e^{-su}|\, du \leq \int_0^\infty |g(u)e^{-\sigma u}|\, du.$$

Since $g(t)e^{-\sigma u}$ is integrable, Theorem B.9 is applicable and

$$F(s)G(s) = \int_0^\infty e^{-st}\left(\int_0^\infty f(\tau)g(t-\tau)\, d\tau \right) dt = \int_0^\infty f*g(t)e^{-st}\, dt = \mathcal{L}[f*g](s),$$

Q.E.D.

Example 8.14 Solve the boundary value problem for a semi-infinite bar, $x \geq 0$, whose initial temperature is zero and whose temperature at $x = 0$ is a given function of time. More precisely, if g is a bounded function with a bounded, continuous derivative on $(0, \infty)$, such that $g(0) = 0$, we want to find a bounded solution of the boundary value problem

$$
\begin{aligned}
u_t &= k u_{xx} & x > 0, \quad t > 0 \\
u(0, t) &= g(t) & t > 0 \\
u(x, 0) &= 0 & x > 0.
\end{aligned}
$$

Assuming that for each fixed $x > 0$ the Laplace transform of $u(x, t)$ with respect to t, which we denote by $U(x, s)$, exists, that Theorem 8.3 is applicable, and that $\mathcal{L}[u_{xx}] = U_{xx}$, the first two equations are transformed into

$$
\begin{aligned}
sU(x, s) - u(x, 0) &= kU_{xx}(x, s) \\
U(0, s) &= G(s).
\end{aligned}
$$

The condition $u(x, 0) = 0$ reduces the differential equation to $sU(x, s) = kU_{xx}(x, s)$, whose general solution, for each fixed s, is

$$
U(x, s) = A(s)e^{-x\sqrt{s/k}} + B(s)e^{x\sqrt{s/k}}.
$$

Since the second exponential is not a Laplace transform (see Exercise 1.7), we choose $B \equiv 0$, and then the endpoint condition $U(0, s) = G(s)$ implies that

$$
U(x, s) = G(s)e^{-x\sqrt{s/k}}.
$$

This is a product of Laplace transforms, that of g and that of the function of Example 8.13 for the choice $2a = x/\sqrt{k}$, which we can consider as having been placed in a table of transforms, and then Theorem 8.5 shows that

$$
u(x, t) = \frac{1}{2\sqrt{\pi k}} \int_0^t \frac{x}{\tau\sqrt{\tau}} e^{-x^2/4k\tau} g(t - \tau) \, d\tau.
$$

for $x > 0$ and $t > 0$.

It can be verified that u is bounded and has the required boundary values as limiting values.[1] It can also be shown that this problem is well posed in the sense of Hadamard (see Exercise 3.2).

§8.4 The Telegraph Equation

It is appropriate, in honor of Heaviside, to present some applications to telegraphy. We consider the voltage $u(x, t)$ and intensity of current $i(x, t)$ at a point x on a horizontal electric transmission line at time t. The first partial differential equation describing

---

[1] For the most involved parts of the verification see, for instance, A. Broman, *Introduction to Partial Differential Equations from Fourier Series to Boundary Value Problems*, Dover, New York, 1989, pages 155–156.

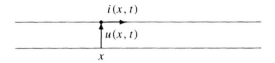

Figure 8.2

the phenomenon was found by William Thomson (1824–1907), of the University of Glasgow, in 1855. Using the laws of circuit theory he deduced that

$$-u_x = ri$$
$$-i_x = cu_t$$

where r and c are the resistance and capacitance of the line per unit length, and then, eliminating the intensity of current i, we get the following equation for the voltage

$$u_{xx} = cru_t.$$

This is known as *Thomson's cable equation*, and the theory built on it made possible the success of the Atlantic submarine cables of 1865 and 1866, much as its neglect made the earlier cable of 1858 a failure.[1] Note that Thomson's equation is just the well-known heat equation with $k = 1/rc$, and this is both good and bad.[2] It is good because the equation is solvable—we have seen that repeatedly—so that a theory for the transmission of electric signals on such a cable can be developed. It is bad because the propagation of heat, or the propagation of electric signals governed by such equation, is essentially a process of diffusion rather than transmission. For instance, if we apply a Morse code pulse at the start of the line, that is, if we assume that the voltage at $x = 0$ is a positive constant E (for electromotive force) for $0 \le t < T$ and zero for any other t, then it can be shown that the voltage at any point $x > 0$ on the line rises slowly for a while and then declines slowly (see Exercise 4.1). The pulse has been slurred rather than transmitted at whatever rate, and only slow signaling is possible on such a cable.

Now, Thomson's equation is simple because it disregards the leakage conductance g and the inductance l of the line per unit length. These were considered by Heaviside, the conductance in 1874 (*EPI*, page 48) and the inductance—but no conductance now—in 1876 (*EPI*, page 54). If we consider both at once, as Heaviside did in 1887, the laws of circuit theory lead to the system

$$-u_x = li_t + ri$$
$$-i_x = cu_t + gu$$

---

[1]Thomson's success was rewarded with a knighthood in 1866. He would become Lord Kelvin in 1892.

[2]Thomson read Fourier's *Théorie analytique de la chaleur* during his summer vacation of 1840, at the age of 16, and published his first research paper, on Fourier series, the year after. Eight of his next ten papers were on the theory of heat, a favorite of his throughout his life, and it is natural that he should use a familiar theory to treat a new problem in electricity.

(*EP*II, page 123). These are the well-known equations that Heaviside used repeatedly from then on, and eliminating i leads to

$$(5) \qquad u_{xx} = clu_{tt} + (cr + gl)u_t + gru,$$

an equation that Heaviside never wrote. It was originally obtained by Gustav Robert Kirchhoff (1824–1887) in 1857—a fact that Heaviside ignored—and is now called the *telegraph equation*. The same equation is obtained for i by eliminating u.

Let us assume first, as Heaviside did in 1876, that $g = 0$. Then the telegraph equation can be written in the form

$$\frac{1}{cl}u_{xx} = u_{tt} + \frac{r}{l}u_t,$$

and it is readily seen that, if the quotient r/l is so small that the term on the right can be neglected, this becomes the wave equation, but with $1/cl$ in place of the constant c^2 used in Chapter 5. If we use this equation, rather than Thomson's, as a model for telegraphic transmission things are radically different. Recalling the general solution of the wave equation in terms of traveling waves, we can expect that a Morse code pulse applied at $x = 0$ would start traveling to the right with speed $1/\sqrt{cl}$. If the line is very long—ideally, infinitely long—there will be no reflection of the pulse at the other end, and it has been transmitted undistorted (we shall actually show this in detail as part of Example 8.15).

Undistorted transmission on a line or cable became imperative precisely in 1876 because this was the year in which a new device, the telephone, was patented by Alexander Graham Bell.[1] Thomson's cable, in which r and c dominate, may have been useful for slow telegraphic signaling, but rapid speech—translated into electrical signals—would be distorted beyond recognition on the shortest cable. Unless, of course, one paid heed to Heaviside's advice; to wit, that the way to make the quotient r/l very small is to *increase the inductance greatly* by artificial means (*EP*II, page 123). There is certainly a limit to how small the resitance can be made in practice. In fact, all four constants are positive on a real line, so Heaviside sought a condition for such a line to be distortionless and in 1887 he found it to be

$$\frac{g}{c} = \frac{r}{l}$$

(*EP*II, page 123). We shall show this in Example 8.15, but it is preferable to start in a more general setting and to use the mathematical tools developed in this chapter. Consider, then, the problem of finding the response of the line to a Morse code pulse, that is, if $u(0, t) = E[H(t) - H^T(t)]$, we want to find $u(x, t)$ for all $x > 0$ and all $t > 0$. In other words, using the function of Example 8.7, we pose the boundary value problem consisting of solving the telegraph equation subject to the conditions

$$
\begin{aligned}
u(0, t) &= E[H(t) - H^T(t)] & t &\geq 0 \\
u(x, 0) &= 0 & x &> 0 \\
u_t(x, 0) &= 0 & x &> 0.
\end{aligned}
$$

---

[1] Bell's representative arrived at the patent office only a few hours ahead of Elisha Grey, of Chicago, the inventor of an almost identical device.

Assuming that the Laplace transform of $u(x, t)$ with respect to t, which we denote by $U(x, s)$, exists for each fixed $x > 0$, that the Corollary of Theorem 8.3 is applicable, and that $\mathcal{L}[u_{xx}] = U_{xx}$, the telegraph equation is transformed into

$$U_{xx}(x, s) = cl[s^2 U(x, s) - su(x, 0) - u_t(x, 0)]$$

$$+(cr + gl)[sU(x, s) - u(x, 0)] + grU(x, s).$$

Using now the initial conditions $u(x, 0) = u_t(x, 0) = 0$ and then Example 8.7 to transform the endpoint condition at $x = 0$ gives the system

$$U_{xx}(x, s) = [cls^2 + (cr + gl)s + gr]U(x, s) = [(cs + g)(ls + r)]U(x, s)$$

$$U(0, s) = \frac{E}{s}(1 - e^{-Ts}),$$

which is, for each fixed s, an initial value problem for an ordinary differential equation in x. If we define

$$\lambda^2 = (cs + g)(ls + r),$$

the general solution of $U_{xx} = \lambda^2 U$ is

$$U(x, s) = Ae^{\lambda x} + Be^{-\lambda x},$$

where A and B are constants (for each fixed s) that, in view of the initial condition, must satisfy

$$A + B = \frac{E}{s}(1 - e^{-Ts}).$$

Solving for B in terms of A and taking this value to the general solution,

(6) $$U(x, s) = Ae^{\lambda x} + \left[\frac{E}{s}(1 - e^{-Ts}) - A\right]e^{-\lambda x}.$$

The constant A cannot be determined at this stage. There is insufficient information in the absence of any knowledge about the length of the line and conditions at that end. Before we specify any, let us remark on the fact that, in general, recovering $u(x, t)$ from $U(x, s)$ is no simple matter because λ is a rather involved function of s. In some cases it requires knowledge of functions that will not be introduced until Chapter 10.[1] Some reasonable simplification must be made at this point, and we shall study what Heaviside labeled a *distortionless circuit*, that is, a cable in which the constants have been adjusted so that $g/c = r/l$.

---

[1] The reader who eventually acquires knowledge of the methods of §8.5 and the functions of §10.10 will find a number of interesting applications to transmission lines, for any value of λ, in H. S. Carslaw and J. C. Jaeger, *Operational Methods in Applied Mathematics*, 2nd ed., Dover, New York, 1963, pages 197–209.

Example 8.15 Consider a semi-infinite line for $x > 0$ with $g/c = r/l$. This is an idealization of a very long line that stretches over several miles of desert or for many miles under the ocean. Since $gl = cr = \sqrt{glcr}$, λ becomes a linear function of s. Indeed,

$$\lambda = \sqrt{(cs + g)(ls + r)} = \sqrt{cls^2 + 2gls + gr} = \sqrt{cl}\, s + \sqrt{gr}.$$

In this case $e^{\lambda x}$ cannot be a Laplace tranform because it is not bounded for $\mathrm{Re}(\lambda) > 0$ (see Exercise 1.7), and we choose $A = 0$ in (6), which then becomes

$$U(x, s) = \frac{E e^{-x\sqrt{gr}}}{s} (1 - e^{-Ts}) e^{-x\sqrt{cls}}.$$

Since the numerator of the fraction on the right-hand side is independent of s, $U(x, s)$ is the product of a constant, the transform of Example 8.7, and an exponential of the form $e^{-\tau s}$. Using Theorem 8.2 (iv) with $\tau = x\sqrt{cl}$,

$$u(x, t) = E e^{-x\sqrt{gr}} [H(t - x\sqrt{cl}) - H^T(t - x\sqrt{cl})].$$

This means that there is no voltage at a point x on the line until $t = x\sqrt{cl}$. At this time the pulse arrives attenuated by the losses along the line—represented by the exponential $e^{-x\sqrt{gr}}$—and then stays at this constant value until $t = x\sqrt{cl} + T$, at which time the voltage returns to zero. That is, except for the damping, the line does not distort the pulse. Hence, Heaviside's use of the term *distortionless*. This solution is also valid in the ideal case $g = r = 0$, but then there is no attenuation of the pulse.

The proposed solution is easily verified except for the fact that it is not differentiable with respect to t at $t = x\sqrt{cl}$ and $t = x\sqrt{cl} + T$. Neither is the input voltage at the jump discontinuities, which unavoidably propagate along the line. In this respect, the situation is similar to that already encountered for the vibrating string at the end of §5.3.

This propagation of the attenuated pulse along the line immediateley poses a natural question: if the line is finite, what happens to the pulse on reaching the end? This situation will be examined in two particular cases, one in Example 8.16 and one in Exercise 4.3.

Example 8.16 Consider a finite, distortionless line stretching from $x = 0$ to $x = a$ and short-circuited at $x = a$, that is, $u(a, t) = 0$ for $t > 0$. There is, however, no reason for alarm. This is not a short circuit of the fuse-blowing type because there is enough resistance along the line to make this a perfectly safe situation.

For the sake of simplicity, let us denote the input voltage by $f(t)$, even if we still choose to think of it as a pulse, and its Laplace tranform by $F(s)$. Then (6) is replaced by

$$U(x, s) = A e^{\lambda x} + [F(s) - A] e^{-\lambda x},$$

and then the condition $u(a, t) = 0$ implies that $U(a, s) = 0$. This allows us to evaluate A, and then

$$U(x, s) = F(s) \frac{e^{\lambda(a-x)} - e^{-\lambda(a-x)}}{e^{\lambda a} - e^{-\lambda a}} = F(s) \frac{e^{-\lambda x} - e^{-\lambda(2a-x)}}{1 - e^{-2\lambda a}}.$$

We shall find the inverse transform by expanding this function into what Heaviside called a *wave series* in 1895 (*EMT*II, page 70). This will show the propagation of the pulse along the line as t increases. Since $|e^{-2\lambda a}| < 1$ for $\text{Re}(s) > 0$, we can use the geometric series expansion and then substitute $\lambda = \sqrt{cl}\, s + \sqrt{gr}$ to obtain

$$U(x, s) = F(s)\left[e^{-\lambda x} - e^{-\lambda(2a-x)}\right] \sum_{n=0}^{\infty} e^{-2\lambda a n}$$

$$= \sum_{n=0}^{\infty} F(s)e^{-\lambda(2an+x)} - \sum_{n=0}^{\infty} F(s)e^{-\lambda[2a(n+1)-x]}$$

$$= \sum_{n=0}^{\infty} e^{-\sqrt{gr}(2an+x)} F(s)e^{-\sqrt{cl}(2an+x)s} - \sum_{n=1}^{\infty} e^{-\sqrt{gr}(2an-x)} F(s)e^{-\sqrt{cl}(2an-x)s}.$$

The terms of each of these series are Laplace transforms of known functions, and it would be easy to find $u(x, t)$ if the inversion could be performed term by term. We shall assume that this is the case,[1] and then

$$u(x, t) = \sum_{n=0}^{\infty} e^{-\sqrt{gr}(2an+x)} f[t - \sqrt{cl}(2an+x)] - \sum_{n=1}^{\infty} e^{-\sqrt{gr}(2an-x)} f[t - \sqrt{cl}(2an-x)],$$

which gives the solution as a series of attenuated traveling waves. This solution is easily verified for the case of $f(t) = E[H(t) - H^T(t)]$ except, of course, at the points of discontinuity (see Exercise 4.2).

To give a physical interpretation of what goes on along the line, think of $f(t)$ as a narrow pulse of value $E > 0$ for $0 \le t < T$ and assume that the line is so long that $T \ll a\sqrt{cl}$. For $t < a\sqrt{cl}$ every term in the series solution for $u(x, t)$ is zero except the one in the first series for $n = 0$. The front of this original pulse travels to the right with speed $1/\sqrt{cl}$ and reaches the end of the line at $t = a\sqrt{cl}$. In the xt-plane this wave front moves along the line $t = \sqrt{cl}x$, as shown in Figure 8.3, with the end of the pulse following T seconds later along the line $t = \sqrt{cl}x + T$. The voltage at an arbitrary point x_0 becomes positive during a time interval of length T, as shown by the dark vertical band in Figure 8.3, and then zero again. At $x = a$ the first term of the first series becomes zero for $t > a\sqrt{cl} + T$, but the term for $n = 1$ in the second series has become active T seconds before. This means that the front of the pulse has been reflected, has changed sign—it is upside down—and is now traveling to the left with speed $1/\sqrt{cl}$ along the line $t = \sqrt{cl}(2a - x)$, with its end trailing T seconds behind along the line $t = \sqrt{cl}(2a - x) + T$. Eventually the front of this upside down pulse is reflected again at the origin of the line at $t = 2a\sqrt{cl}$, changes sign again, and, as the second term of the first series becomes active, starts moving to the right along the line $t = \sqrt{cl}(x + 2a)$. This phenomenon goes on indefinitely: the pulse is reflected every $a\sqrt{cl}$ seconds at each end of the line, changes sign, and starts traveling in the opposite direction. As

---

[1] For a justification of this fact see, for instance, G. Doetsch, *Introduction to the Theory and Application of the Laplace Transformation*, Springer-Verlag, Berlin/Heidelberg, 1974, pages 192–200.

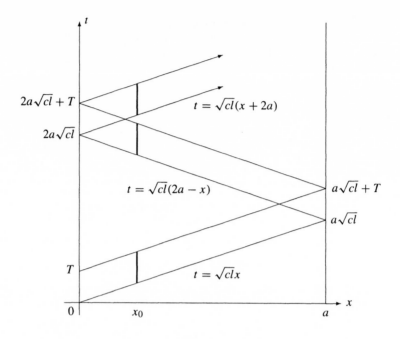

Figure 8.3

each term of one series becomes zero the next term from the other series takes over T seconds before. Except for some overlap at the ends of the line, each point x_0 sees the pulse come and then go after T seconds infinitely many times, as shown by the dark vertical bands in Figure 8.3. The situation is very similar to that for the vibrating string discussed in §5.3, but in this case the pulse is subject to the attenuation represented by the exponentials. In spite of the attenuation, these endless reflections may be considered an undesirable effect since they will interfere with each other. Exercise 4.6 shows how to avoid them by replacing the short circuit at the end of the line with the appropriate kind of circuit.

In all the preceding examples we have been able to recognize $U(x, s)$ as the Laplace transform of some known function. But this is not always possible, and then we must evaluate the integral in the left-hand side of (3). This usually requires knowledge of complex analysis and is the object of §8.5.

§8.5 The Method of Residues

When the integral in the left-hand side of (3) must be evaluated, as is typical in applications to partial differential equations, the easiest way is to resort to Cauchy's residue theorem of complex analysis, as Bromwich did. Indeed, with $s = \sigma + i\omega$,

$$\text{PV} \int_{-\infty}^{\infty} F(\sigma + i\omega)e^{(\sigma+i\omega)t}\, d\omega = \lim_{b\to\infty} \int_{-b}^{b} F(\sigma + i\omega)e^{(\sigma+i\omega)t}\, d\omega$$

$$= \frac{1}{i} \lim_{b\to\infty} \int_{\gamma_{\sigma b}} F(s)e^{st}\, ds,$$

where $\gamma_{\sigma b}$ is the segment in $\mathbb{C}$ from $\sigma - ib$ to $\sigma + ib$. It suffices to compute the limit as $N \to \infty$ of the integral along $\gamma_{\sigma b_N}$, where $\{b_N\}$ is a sequence of positive numbers such that $b_N \to \infty$ as $N \to \infty$. To evaluate the latter integral, let γ_{b_N} be an arc of curve from $\sigma + ib_N$ to $\sigma - ib_N$ such that $\gamma_{\sigma b_N} + \gamma_{b_N}$ is a simple closed path in the half-plane $\{s \in \mathbb{C} : \text{Re}(s) \leq \sigma\}$, as shown in Figure 8.4.

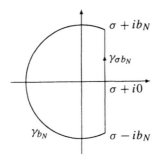

Figure 8.4

Theorem 8.6 *Let $f \in \mathcal{L}$ have Laplace transform F, let σ be a real number such that $f(t)e^{-\sigma t}$ is integrable on $(0, \infty)$, and assume that the continuation of F is analytic for $\text{Re}(s) < \sigma$ except, possibly, at a finite or countable set S of isolated points, none of which are on the γ_{b_N}. If*

$$\lim_{N\to\infty} \int_{\gamma_{b_N}} F(s)e^{st}\, ds = C,$$

where C is a constant, and if $R_{s_n}[F(s)e^{st}]$ denotes the residue of $F(s)e^{st}$ at s_n for s_n in S,

$$\tfrac{1}{2}[f(t+) + f(t-)] = \text{PV} \int_{-\infty}^{\infty} \frac{1}{2\pi} F(\sigma + i\omega)e^{(\sigma+i\omega)t}\, d\omega = \sum_{s_n \in S} R_{s_n}[F(s)e^{st}] - \frac{C}{2\pi i}$$

at every point t in $[0, \infty)$ at which f has one-sided derivatives.

Proof. For each positive integer N, let S_N be the subset of S contained in the open set bounded by $\gamma_{\sigma b_N} + \gamma_{b_N}$. Cauchy's residue theorem gives

$$\int_{\gamma_{\sigma b_N}} F(s)e^{st}\,ds + \int_{\gamma_{b_N}} F(s)e^{st}\,ds = 2\pi i \sum_{s_n \in S_N} R_{s_n}[F(s)e^{st}],$$

and, letting $N \to \infty$,

$$\lim_{N \to \infty} \int_{\gamma_{\sigma b_N}} F(s)e^{st}\,ds = 2\pi i \sum_{s_n \in S} R_{s_n}[F(s)e^{st}] - C.$$

That is,

$$\lim_{N \to \infty} \int_{-b_N}^{b_N} F(\sigma + i\omega)e^{(\sigma+i\omega)t}\,d\omega = \frac{1}{i}\lim_{N \to \infty} \int_{\gamma_{\sigma b_N}} F(s)e^{st}\,ds$$

$$= 2\pi \sum_{s_n \in S} R_{s_n}[F(s)e^{st}] - \frac{C}{i},$$

and, if the right-hand side is a series, its convergence is guaranteed by that of the left-hand side. The remaining equality is from Theorem 8.1, Q.E.D.

To compute the integral along γ_{b_N} this path should have some simple geometric shape. It is usually chosen to consist, in part or entirely, of circular arcs with center at the origin, straight line segments, or even parabolic arcs. Furthermore, if the integral is to approach a finite limit, we shall require that, as $N \to \infty$, $|F(s)e^{st}| \to 0$ on γ_{b_N} more rapidly than the length of γ_{b_N} increases. For the case of circular arcs the following result is frequently useful.[1]

Lemma 8.2 *Let R_N be the distance from the origin to $\sigma + ib_N$, and assume that γ_{b_N} contains an arc γ of the circle $s(\theta) = R_N e^{i\theta}$. If there are positive constants M and m such that*

$$|F(s)| < \frac{M}{|s|^m}$$

for $|s|$ large enough, then

$$\lim_{N \to \infty} \int_{\gamma} F(s)e^{st}\,ds = 0$$

for $t > 0$.

Proof. If γ is in the right half-plane, which is possible only if $\sigma > 0$, then its length approaches a limit $l \le \sigma$ as $N \to \infty$. Since

$$|F(s)e^{st}| < \frac{M}{R_N^m}e^{\sigma t}$$

---

[1] For the other types of arcs see, for instance, R. V. Churchill, *op. cit.*, pages 211–213.

if N is large enough and $\text{Re}(s) \leq \sigma$, it follows that the integral of $F(s)e^{st}$ along γ approaches zero as $N \to \infty$. If, instead, γ is in the left half-plane, use the fact that $|e^{st}| = e^{R_N t \cos \theta}$, and then for N large enough,

$$\left| \int_\gamma F(s)e^{st}\, ds \right| < \frac{M}{R_N^m} \left| \int_\gamma e^{st}\, ds \right|$$

$$\leq \frac{M}{R_N^m} \int_{\pi/2}^{3\pi/2} e^{R_N t \cos \theta} R_N\, d\theta$$

$$= \frac{2M}{R_N^{m-1}} \int_{\pi/2}^{\pi} e^{R_N t \cos \theta}\, d\theta.$$

Making the substitution $\varphi = \theta - \pi/2$ and since the graph of $\sin \varphi$ shows that $\sin \varphi > 2\varphi/\pi$ for $0 < \varphi < \pi/2$, it is seen that the right-hand side above equals

$$\frac{2M}{R_N^{m-1}} \int_0^{\pi/2} e^{-R_N t \sin \varphi}\, d\varphi < \frac{2M}{R_N^{m-1}} \int_0^{\pi/2} e^{-2R_N \varphi t/\pi}\, d\varphi = \frac{\pi M}{R_N^m t}(1 - e^{-R_N t}),$$

which also approaches zero as $N \to \infty$, Q.E.D.

Example 8.17 If h and T are positive constants, solve the boundary value problem

$$\begin{array}{ll} u_t = k u_{xx} & x > 0, \quad t > 0 \\ u_x(0, t) = h u(0, t) & t > 0 \\ u(x, 0) = T & x > 0 \end{array}$$

corresponding to a semi-infinite bar, $x \geq 0$, whose initial temperature is T and whose end at $x = 0$ loses heat by convection to the surrounding medium, assumed to be at zero temperature.

Assuming that for each $x > 0$ the Laplace transform of $u(x, t)$ with respect to t, which we denote by $U(x, s)$, exists, that Theorem 8.3 is applicable, that $\mathcal{L}[u_x] = U_x$, and that $\mathcal{L}[u_{xx}] = U_{xx}$ leads to the transformed system

$$sU(x, s) - T = kU_{xx}(x, s)$$

$$U_x(0, s) = hU(0, s).$$

For each fixed s, the general solution of the differential equation is

$$U(x, s) = A(s)e^{-x\sqrt{s/k}} + B(s)e^{x\sqrt{s/k}} + \frac{T}{s}.$$

Since the second exponential is not a Laplace transform (see Exercise 1.7), we choose $B \equiv 0$, and then the endpoint condition $U_x(0, s) = hU(0, s)$ implies that

$$U(x, s) = \frac{T}{s}\left[1 - \frac{h}{(\sqrt{s/k} + h)} e^{-x\sqrt{s/k}} \right] = \frac{T}{s} - \frac{T}{s} e^{-x\sqrt{s/k}} + \frac{T}{\sqrt{sk}(\sqrt{s/k} + h)} e^{-x\sqrt{s/k}}.$$

We already know the inverse transforms of the first two terms on the right-hand side from Examples 8.1 and 8.14 with $G(s) = T/s$, but the third transform is unfamiliar to us. Denote this third term by $F(x, s)$ and assume that, for each $x > 0$, it is the transform of a function $f(x, t)$ in $\mathcal{L}$ that is continuous in t. Observe that, for each fixed x, $F(x, s)e^{st}$ is an analytic function of s in the domain $-\pi < \arg s < \pi$ and has no singularities—and therefore no residues—there. Then, if $\sigma > 0$ is as in Theorem 8.6 and γ_{b_N} is a path in this domain as described above,

$$f(x, t) = -\frac{1}{2\pi i} \lim_{N \to \infty} \int_{\gamma_{b_N}} F(x, s)e^{st} \, ds = -\frac{1}{2\pi i} \lim_{N \to \infty} \int_{\gamma_{b_N}} \frac{T e^{st - x\sqrt{s/k}}}{\sqrt{sk}(\sqrt{s/k} + h)} \, ds.$$

We shall use the path in Figure 8.5, consisting of the arcs AB and EF of the circle

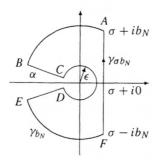

Figure 8.5

$s(\theta) = R_N e^{i\theta}$, where R_N is the distance from the origin to $\sigma + ib_N$, the segments BC and DE—on lines through the origin—and the arc of circle CD centered at the origin, where the integrand has a branch point. Note that $\mathrm{Re}\sqrt{s/k} > 0$ and then, since $h > 0$,

$$\left| \sqrt{s/k} + h \right| > \left| \sqrt{s/k} \right| \qquad \text{and} \qquad \left| e^{-x\sqrt{s/k}} \right| < 1.$$

It follows that $|F(x, s)| < T/|s|$ along the arcs AB and EF, and then, for $t > 0$, Lemma 8.2 guarantees that the integrals along these arcs approach zero as $N \to \infty$. Therefore, as $N \to \infty$, the integral along γ_{b_N} approaches the sum of those along BC, CD and DE. On CD we have $s(\theta) = \epsilon e^{i\theta}$, and then the inequality $\left| \sqrt{s/k} + h \right| > h$ implies that

$$\left| F(x, s)e^{st} \right| < \frac{T|e^{st}|}{h|\sqrt{sk}|} = \frac{T e^{\epsilon t \cos\theta}}{h\sqrt{\epsilon k} e^{\cos\theta/2}}.$$

Since the length of CD is smaller than $2\pi\epsilon$, the absolute value of the integral along CD approaches zero as $\epsilon \to 0$ for any N. On the segments BC and DE, we have $s = re^{i(\pi - \alpha)}$ and $s = re^{i(-\pi + \alpha)}$, $r = |s|$, respectively. But the integrand is independent of α, and then these integrals can be replaced by their limits as $\alpha \to 0$ and $\epsilon \to 0$. Now, as $\alpha \to 0$, we see that $s \to -r$ and

$$\sqrt{s} = \sqrt{r}e^{i(\pm\pi \mp \alpha)/2} \to \sqrt{r}e^{\pm i\pi/2} = \pm i\sqrt{r}.$$

Hence, as $\alpha \to 0$ and $\epsilon \to 0$, the sum of the integrals along BC and DE has the limit

$$-\int_{R_N}^0 \frac{e^{-rt}e^{-ix\sqrt{r/k}}}{i\sqrt{r}(i\sqrt{r/k}+h)}\,dr - \int_0^{R_N} \frac{e^{-rt}e^{ix\sqrt{r/k}}}{-i\sqrt{r}(h-i\sqrt{r/k})}\,dr$$

$$= \frac{1}{i}\int_0^{R_N} \frac{e^{-rt}}{\sqrt{r}}\left[\frac{e^{ix\sqrt{r/k}}}{(h-i\sqrt{r/k})} + \frac{e^{-ix\sqrt{r/k}}}{(h+i\sqrt{r/k})}\right]dr$$

$$= \frac{2}{i}\int_0^{R_N} \frac{e^{-rt}}{\sqrt{r}}\frac{h\cos x\sqrt{r/k} - \sqrt{r/k}\sin x\sqrt{r/k}}{h^2 + r/k}\,dr$$

$$= \frac{4}{i}\sqrt{k}\int_0^{\sqrt{R_N/k}} e^{-kz^2 t}\frac{h\cos xz - z\sin xz}{h^2 + z^2}\,dz,$$

where we have made the substitution $r = kz^2$ in the last step. It is well-known that

$$\int_0^\infty e^{-h\tau}\cos\tau z\,d\tau = \frac{h}{h^2+z^2} \qquad \text{and} \qquad \int_0^\infty e^{-h\tau}\sin\tau z\,d\tau = \frac{z}{h^2+z^2},$$

and then, as $N \to \infty$,

$$f(x,t) = \frac{2T}{\pi}\int_0^\infty e^{-kz^2 t}\left(\int_0^\infty e^{-h\tau}\cos(x+\tau)z\,d\tau\right)dz$$

$$= \frac{2T}{\pi}\int_0^\infty e^{-h\tau}\left(\int_0^\infty e^{-kz^2 t}\cos(x+\tau)z\,dz\right)d\tau$$

$$= \frac{T}{\sqrt{\pi kt}}\int_0^\infty e^{-h\tau - (x+\tau)^2/4kt}\,d\tau,$$

where the last inner integral has been obtained from the tables. For $x > 0$ and $t > 0$ the complete solution is then assembled from Examples 8.1 and 8.14, with $G(s) = T/s$, and this expression. We obtain

$$u(x,t) = T - \frac{T}{2\sqrt{\pi k}}\int_0^t \frac{x}{\tau\sqrt{\tau}}e^{-x^2/4k\tau}\,d\tau + \frac{T}{\sqrt{\pi kt}}\int_0^\infty e^{-h\tau - (x+\tau)^2/4kt}\,d\tau.$$

In this example there were no residues to compute. When there are, the following result from complex analysis is frequently useful.

Proposition 8.1 *If g is analytic in a neighborhood of a point s_0 except at s_0, where it has a pole of order m, then its residue at s_0 is*

$$R_{s_0}[g(s)] = \lim_{s \to s_0}\frac{1}{(m-1)!}\frac{d^{m-1}}{ds^{m-1}}[(s-s_0)^m g(s)].$$

Example 8.18 Sometimes the method of residues will lead to a form of the solution that is different from that obtained by another method. For instance, consider again the problem of Example 8.16, that is, that of a distortionless electric line of finite length a whose end at $x = a$ is short-circuited. We shall assume that none of the line constants is zero, but the result is easily modified if $gr = 0$. The transform $U(x, s)$ obtained there can be rewritten as

$$U(x, s) = F(s) \frac{\sinh \lambda(a - x)}{\sinh \lambda a}.$$

In view of the identity $\sinh \lambda a = -i \sin i \lambda a$, it has poles, all simple, for $\lambda a = \pm n\pi i$, where n is any positive integer. Substituting the value of $\lambda = \sqrt{cl}\, s + \sqrt{gr}$ and recalling that $g/c = r/l$, we see that these poles are at

$$s_n = -\sqrt{\frac{gr}{cl}} + \frac{n\pi}{a\sqrt{cl}}i = -\frac{r}{l} + \frac{n\pi}{a\sqrt{cl}}i$$

and at $\bar{s}_n$, the complex conjugate of s_n. We shall apply Theorem 8.6, choosing γ_{b_N} to be an arc of circle centered at the origin with radius

$$R_N = \frac{(2N + 1)}{2a\sqrt{cl}}\pi$$

and N so large that

$$\frac{r}{l} < \frac{\sqrt{4N + 1}}{4a\sqrt{cl}}\pi.$$

This guarantees that $|s_N| < R_N < |s_{N+1}|$, so that γ_{b_N} does not pass through any of the s_n or $\bar{s}_n$. The residue of $U(x, s)e^{st}$ at each of these points can be computed using Proposition 8.1. Since the poles are simple, $m = 1$, and we get

$$R_{s_n}[U(x, s)e^{st}] = \lim_{s \to s_n} (s - s_n)F(s)e^{st} \frac{\sinh(\sqrt{cl}\, s + \sqrt{gr}\,)(a - x)}{\sinh(\sqrt{cl}\, s + \sqrt{gr}\,)a}.$$

If we note that $\sqrt{cl}\, s_n + \sqrt{gr} = n\pi i/a$ and that, according to l'Hôpital's rule,

$$\lim_{s \to s_n} \frac{\sinh(\sqrt{cl}\, s + \sqrt{gr}\,)a}{(s - s_n)} = a\sqrt{cl} \cosh(\sqrt{cl}\, s_n + \sqrt{gr}\,)a$$

$$= a\sqrt{cl} \cosh n\pi i$$

$$= a\sqrt{cl} \cos n\pi,$$

we obtain

$$R_{s_n}[U(x, s)e^{st}] = F(s_n)e^{s_n t} \frac{\sinh n\pi i(a - x)/a}{a\sqrt{cl} \cos n\pi} = \frac{(-1)^n}{a\sqrt{cl}} i F(s_n)e^{s_n t} \sin \frac{n\pi}{a}(a - x).$$

Similarly, one obtains the residue at $\bar{s}_n$, and then the sum of the two residues is

$$(7) \qquad \frac{(-1)^n e^{-rt/l}}{a\sqrt{cl}} \left[i F(s_n)e^{in\pi t/a\sqrt{cl}} - i F(\bar{s}_n)e^{-in\pi t/a\sqrt{cl}} \right] \sin \frac{n\pi}{a}(a - x).$$

To continue building the solution we must specify the input voltage. If $f(t) = EH(t)$, then $F(s) = E/s$, which has an additional simple pole at the origin. The corresponding residue is

$$R_0[U(x, s)e^{st}] = \lim_{s \to 0} s \frac{E \sinh \lambda(a - x)}{s \sinh \lambda a} e^{st} = E \frac{\sinh \sqrt{gr} \,(a - x)}{\sinh \sqrt{gr} \, a}.$$

If we now write $s_n = |s_n|e^{i(\pi - \varphi_n)}$, where $\varphi_n = \arctan(n\pi l/ar\sqrt{cl})$, the difference in the brackets in (7) becomes

$$2\mathrm{Re}\left[i F(s_n) e^{in\pi t/a\sqrt{cl}} \right] = \frac{2E}{|s_n|} \mathrm{Re}\left[i e^{-i(\pi - \varphi_n) + in\pi t/a\sqrt{cl}} \right] = \frac{2E}{|s_n|} \sin \left(\frac{n\pi t}{a\sqrt{cl}} + \varphi_n \right).$$

By Lemma 8.2, which is applicable because $|F(s)| = E/|s|$ and because it is possible to show that the quotient of hyperbolic sines in the expression for $U(x, s)$ is bounded (see Exercise 5.2 for an outline of proof), the integral of along γ_{b_N} is zero, and then Theorem 8.6 yields

$$u(x, t) = E \frac{\sinh \sqrt{gr} \,(a - x)}{\sinh \sqrt{gr} \, a} + \frac{2Ee^{-rt/l}}{a\sqrt{cl}} \sum_{n=1}^{\infty} \frac{(-1)^n}{|s_n|} \sin \left(\frac{n\pi t}{a\sqrt{cl}} + \varphi_n \right) \sin \frac{n\pi}{a}(a - x).$$

The response to $f(t) = E[H(t) - H^T(t)]$ is readily obtained from $u(x, t)$.

This solution in terms of trigonometric functions resembles Bernoulli's solution for the vibrating string, while the solution of the same problem, found in Example 8.16 by Heaviside's wave series method, is similar to d'Alembert's. As in the case of the vibrating string, the trigonometric solution cannot be verified by term by term differentiation. But now there is a significant difference. While d'Alembert's solution was criticized as uninformative from a physical point of view, exactly the opposite is true now. To show the transmission of the pulse along the line in this case, the solution obtained by the wave series method is far superior to the trigonometric solution.

§8.6 Historical Epilogue

Oliver Heaviside's legacy to mathematics and electromagnetism is impressive. In addition to perfecting the operational calculus that later inspired the Laplace transform method, he developed vector calculus in 1885, starting with the definitions of scalar and vector products as used today (*EP*II, pages 4 and 5).[1] In the same year he formulated what has become the cornerstone of electromagnetic theory. Heaviside refers to his discovery as follows:

> I here introduce a new method of treating the subject [Maxwell's theory of electromagnetism], which may perhaps be appropriately named the Duplex method, since its main characteristic is the exhibition of the electric, magnetic, and electromagnetic equations in a duplex form.

---

[1] The vector calculus was simultaneously discovered on the other side of the ocean—without the benefit of Atlantic telegraphy—by Josiah W. Gibbs.

This was the first appearance in print of the famous *Maxwell's equations* of electromagnetic theory (*EPI*, pages 447, 448, 452 and 475), which are not included in Maxwell's treatise. Maxwell was not a very clear writer—his treatise is almost unreadable from a certain point on—so that many future expositors of the subject preferred to follow Heaviside's interpretation of Maxwell and did not realize that the duplex equations were Heaviside's own.

OLIVER HEAVISIDE
From *Electrical Papers*, Vol. 1,
The Copley Publishers, Boston, 1925.

Heaviside's contribution to telegraphy and telephony was invaluable, but for the longest time they fell on deaf ears in his own country. He found a formidable obstacle in William Henry Preece, *Electrician to the Post Office*. Preece's opposition was based on a two-pillar foundation. The first was his ignorance of what really took place in the transmission of electric signals, as shown in a paper published in 1887. The second was Heaviside's reply containing the following assessment:

> Either, firstly, the accepted theory of electromagnetism must be most profoundly modified; or, secondly, the views expressed by Mr. Preece in his paper are profoundly erroneous ... Mr. Preece is wrong, not merely in some points of detail, but radically wrong, generally speaking, in methods, reasoning, results, and conclusions.

The fact is that Mr. Preece, later Sir William, never accepted the validity of Heaviside's advice to increase both the inductance and, to a lesser extent to avoid excessive attenuation, the leakage conductance, so as to approximate the distortionless condition $g/c = r/l$. "More capacitance," seems to have been Preece's *motto*. The outcome of all this is that Heaviside started having trouble publishing his papers, which were in opposition to the official view, that the British government sank a fortune building the wrong kind of lines, and that England lost its leadership in this field to America. Mihajlo Idvorsky Pupin, a Serbian immigrant from the Austrian village of Idvor who became professor of mathematics at Columbia University, was the first to build a line using increased inductance, but in his paper of May 1900 acknowledged the source of his theoretical background, stating that

> Mr. Oliver Heaviside of England, to whose profound researches most of the existing mathematical theory of electrical wave propagation is due, was the originator and most ardent advocate of wave conductors of high inductance.

Soon afterwords, the American Telephone and Telegraph Company succeeded in establishing coast to coast telephone communication by using increased inductance.

Heaviside was good with words in many ways. To him we owe, for instance, the terms inductance, attenuation, and magnetic reluctance (*EP*II, page 28) and the use of voltage for electromotive force (*EMT*I, page 26). He was a colorful, entertaining and opinionated writer, as shown by the following additional quotations:

> Self-induction is salvation. [*EMT*II, page 354]; As critics cannot always find time to read more than the preface, the following remarks may serve to direct their attention to some of the leading points in this volume. [*EMT*I, Preface]; And that there is a natural tendency for both the human body and understanding to move in circles is proved by the accounts of the doings of belated travellers in the wild, and by the contents of a great mass of books. [*EP*I, page 353]; Electric and magnetic force. May they live for ever and never be forgot. [*EMT*III, page 1]; When Prof. Hughes speaks of the resistance of a wire, he does not *always* mean what common men, men of ohms, volts and farads, mean by the resistance of a wire—only sometimes. [*EP*II, page 28]; Different men have different opinions—some like apples, some like inions [*EMT*I, page 352].

Mathematician at large, electrician, philosopher, acid humorist, iconoclast *extraordinaire*, he was awarded—but declined—the Hughes Medal of the Royal Society in 1904, received a Ph.D *honoris causa* from the University of Göttingen in 1906, was made an honorary member of the Institution of Electrical Engineers of Great Britain in 1908 and of the American Institute of Electrical Engineers in 1918, and was awarded the first Faraday Medal of the Institution of Electrical Engineers in 1923.

At his country home in Torquay, where he spent the last seventeen years of his life mostly alone and in great financial trouble—despite a small government pension that he accepted only on the condition that it be in recognition of his scientific work—things were less rosy. For lack of payment the bank was after his home, and the gas company cut off his gas. A victim of lumbago and rheumatic gout, he had to eat cold food and

live in a cold house. On arriving at his door in the winter of 1921, a distinguished
visitor found a note stating that Heaviside had gone to bed to keep warm. Stuffed in the
cracks of the door, to prevent any cold drafts, there was an assortment of papers: some
advertisements, an invitation by the President of the Royal Society, threats from the gas
company about cutting off the gas . . . . The following Spring Heaviside wrote:

> Could not wear boots at all. Could not get proper bed socks to walk about in.
> Buried under all the blankets I have. Now and then I scribbled a sort of diary
> about my persecution by the Poor and the Gas and others.[1]

Irrepressible in his writing, he continued working on his scientific papers, many of which
were found posthumously. He died in a nursing home on February 3, 1925.

EXERCISES

1.1 Find the Laplace transforms of the functions defined on $[0, \infty)$ by

$$(i) \ f(t) = \begin{cases} t, & 0 \le t < 1 \\ 1, & 1 \le t < \infty \end{cases} \qquad (ii) \ f(t) = \begin{cases} e^t, & 0 \le t < 1 \\ e, & 1 \le t < \infty \end{cases}$$

and state their domains

1.2 Use repeated integration by parts to find the Laplace transform of the function defined
by $f(t) = t^n e^{at}$ on $[0, \infty)$, where n is a positive integer and a is an arbitrary real or
complex constant.

1.3 If a function defined on $(0, \infty)$ is unbounded at the origin, its integral from zero to ∞
is defined as the limit of its integral from $a > 0$ to ∞ as $a \to 0$, provided that the
latter integral and limit exist. In this manner some unbounded functions have Laplace
transforms. Show that the Laplace transform of the function defined by $f(t) = 1/\sqrt{t}$
on $[0, \infty)$ exists for $\text{Re}(s) > 0$ and then find its value at each real $s > 0$ (**Hint:** make
the substitution $x = \sqrt{st}$ and then use a table of integrals).

1.4 Does the function defined by $f(t) = \cos(e^{t^2})$ on $[0, \infty)$ have a Laplace transform?

1.5 Show that a Laplace transform is a bounded function of s.

1.6 Let $f : \mathbb{R} \to \mathbb{C}$ be such that $f(t)e^{-\sigma t}$ is integrable for some σ in $\mathbb{R}$. Prove that either
$D_F = \mathbb{C}$ or there is an $r \le \sigma$ in $\mathbb{R}$ such that D_F is the open half-plane $\{s \in \mathbb{C} : \text{Re}(s) > r\}$
or the closed half-plane $\{s \in \mathbb{C} : \text{Re}(s) \ge r\}$.

1.7 Is $F(s) = e^{a\sqrt{s}}$, where $a > 0$, the Laplace transform of a function in $\mathcal{L}$?

---

[1]Quoted from Sir Edmund T. Whittaker, "Oliver Heaviside," *Bull. Calcutta Math. Soc.*, **20** (1928–
1929) 199–220.

2.1 If a and b are nonzero constants, find the Laplace transforms of the functions defined on $[0, \infty)$ by

(i) $f(t) = e^{-3(t+1)}$ (ii) $f(t) = t^5 - te^t$ (iii) $f(t) = e^{at} \cos bt$

(iv) $f(t) = t \sin at$ (v) $f(t) = \sin at \sin bt$ (vi) $f(t) = \dfrac{1}{a} \sin at - \dfrac{1}{b} \sin bt$.

2.2 If $r_1 < r_2 < r_3 < r_4$ are given positive numbers, sketch the function

$$f(t) = (t - r_1)H^{r_1}(t) - (t - r_2)H^{r_2}(t) - (t - r_3)H^{r_3}(t) + (t - r_4)H^{r_4}(t)]$$

and choose r_4 so that $f \equiv 0$ for $t > r_4$. Then find its Laplace transform.

2.3 Let T be a positive constant and let $f \in \mathcal{L}$ be such that $f(t + T) = f(t)$ for all $t \geq 0$. Show that

$$F(s) = \frac{1}{1 - e^{-Ts}} \int_0^T f(t)e^{-st}\, dt.$$

2.4 Use the result of the previous exercise to find the Laplace transforms of

(i) $f(t) = |\sin at|$ on $[0, \infty)$ for $a > 0$, that is, a full-wave rectified sine oscillation,

(ii) a series of rectangular pulses of width T, that is,

$$f(t) = E \sum_{n=0}^{\infty} [H^{2nT}(t) - H^{(2n+1)T}(t)],$$

where E is a positive constant, and

(iii) the periodic extension to $[0, \infty)$ of period $r_1 + r_4$ of the function of Exercise 2.2.

2.5 It is shown in complex analysis that, if F and G are differentiable functions of s on the half-plane $\mathrm{Re}(s) > 0$ and if $F \equiv G$ on the positive real axis, then $F \equiv G$ on the whole half-plane. Use this result to find the Laplace transforms of the functions defined on $[0, \infty)$ by

(i) $f(t) = \dfrac{1}{\sqrt{t}}$ (ii) $f(t) = \sqrt{t}$ (iii) $f(t) = \dfrac{a}{\sqrt{\pi t}}e^{-a^2/t}$,

where a is a positive constant (**Hint:** for (i) refer to Exercise 1.3 and for (iii), to Example 8.13).

2.6 Find the inverse Laplace transforms of

(i) $F(s) = \dfrac{\pi e^{-2s}}{s^2 + \pi^2}$ (ii) $F(s) = \dfrac{\sqrt{\pi} e^{-s}}{s^{3/2}}$

(iii) $F(s) = \dfrac{s^6 - 3s^5 - 7s^4 - 27s^3 - 162}{s^3(s^4 - 81)}$ (iv) $F(s) = \dfrac{10s}{(s + 3)(s^2 + 4s + 8)}$.

(**Hint:** for (ii) refer to Exercise 2.5).

2.7 Use the Laplace transform method to solve the initial value problems:

(i) $y'' + ty' - 2y = 4$, $y(0) = y'(0) = 0$ and

(ii) $y'' + ty' - 2y = 0$, $y(0) = 2, y'(0) = 0$.

2.8 A uniform elastic bar of length a is located along the x-axis between $x = 0$ and $x = a$. If the bar is initially at rest, the end at $x = 0$ is kept fixed, and a constant force F acts parallel to the bar at the end $x = a$, the longitudinal displacement of the bar for $t > 0$ is a solution of the boundary value problem

$$
\begin{aligned}
u_{tt} &= c^2 u_{xx} & 0 < x < a, \quad t > 0 \\
u(0, t) &= 0 & t \geq 0 \\
Eu_x(a, t) &= F & t \geq 0 \\
u(x, 0) &= 0 & 0 \leq x \leq a \\
u_t(x, 0) &= 0 & 0 \leq x \leq a,
\end{aligned}
$$

where c and E are positive constants that depend on the material. Use the Laplace transform method to find a solution of this problem (**Hint:** show that $U(x, s)$ is a constant times the Laplace transform of the function in Exercise 2.4(*iii*) for $r_1 = (a - x)/c$, $r_2 = (a + x)/c$, $r_3 = (3a - x)/c$, and $r_4 = (3a + x)/c$).

2.9 Sketch the solution found in Example 8.12. What happens to the portion of the string for $x \geq ct$ as $t \to \infty$?

2.10 Use the Laplace transform method to solve the boundary value problem for a semi-infinite string for $x > 0$ that is initially at rest if its endpoint at $x = 0$ moves according to the equation $u(0, t) = f(t)$, where $f \in \mathcal{L}$ is such that $f(0) = 0$. Describe the motion of the string if $f(t) = \sin t$ for $0 \leq t \leq \pi$ and $f(t) = 0$ for $t > \pi$.

2.11 The motion of a semi-infinite string for $x > 0$ subject to an external force $f(t)$ is described by the equation $u_{tt} = c^2 u_{xx} + f$. If its end at $x = 0$ is kept fixed at the origin and the string is initially at rest, state the corresponding boundary value problem and then use the Laplace transform method to find a bounded solution for $x > 0$ and $t > 0$ if $f(t) = A \sin \omega t$, where A and ω are positive constants.

3.1 Use the Laplace transform method to find the general solutions of the ordinary differential equations

   (*i*) $y'' + y = 2 \sec^3 t$ and

   (*ii*) $y''' + y' = \tan t$

   in the interval $-\pi/2 < t < \pi/2$,

3.2 Accepting that the solution of the boundary value problem of Example 8.14 has been verified, show that the problem is well posed in the sense of Hadamard (**Hint:** follow the method of Example 7.7 and, at the appropriate moment, make the substitution $z = x/2\sqrt{k\tau}$).

3.3 Use the Laplace transform method to solve the boundary value problem

$$
\begin{aligned}
u_t &= k u_{xx} & x > 0, \quad t > 0 \\
u(0, t) &= f(t) & t > 0 \\
u(x, 0) &= T & x > 0,
\end{aligned}
$$

where T is a positive constant, $f \in \mathcal{L}$, and $f(0) = T$.

3.4 Use the Laplace transform method to solve the boundary value problem

$$
\begin{aligned}
u_t &= k u_{xx} & x > 0, \quad t > 0 \\
u_x(0, t) &= -f(t) & t > 0 \\
u(x, 0) &= 0 & x > 0,
\end{aligned}
$$

where $f \in \mathcal{L}$ is never negative, so that the condition at $x = 0$ represents an incoming flow of heat (**Hint:** use the solution of Exercise 2.5(*iii*)).

3.5 Use the Laplace transform method to show that, if a semi-infinite string for $x > 0$ is initially at rest, if its end at $x = 0$ is kept fixed at the origin, and if its initial displacement is given by $u(x, 0) = f(x)$, where $f \in \mathcal{L}$ is of class C^2 and such that $f(0) = f''(0) = 0$, then its displacement is given by

$$u(x, t) = \tfrac{1}{2}[\tilde{f}(x + ct) + \tilde{f}(x - ct)],$$

where $\tilde{f}$ is the odd extension of f to $\mathbb{R}$. Verify the solution and show that the hypotheses on f cannot be weakened (**Hint:** Solve the ordinary differential equation in x arising from the Laplace transform method with respect to t by the Laplace transform method with respect to x; then, in the expression for $U(x, s)$, the coefficient of $e^{sx/c}$ must approach zero as $x \to \infty$ by Exercise 1.7).

4.1 Solve the boundary value problem of Example 8.15 for Thomson's cable with $g = l = 0$, rather than for a distortionless line.

4.2 Verify the solution of the boundary value problem of Example 8.16.

4.3 Solve the boundary value problem corresponding to a distortionless line stretching from $x = 0$ to $x = a$ and open at $x = a$, that is, $i(a, t) = 0$ for $t > 0$.

4.4 Use Heaviside's wave series method to solve the following boundary value problem for a string of length a,

$$
\begin{aligned}
u_{tt} &= c^2 u_{xx} & 0 &< x < a, \quad t > 0 \\
u(0, t) &= f(t) & t &> 0 \\
u(a, t) &= 0 & t &> 0 \\
u(x, 0) &= 0 & 0 &\le x \le a \\
u_t(x, 0) &= 0 & 0 &\le x \le a,
\end{aligned}
$$

where $f \in \mathcal{L}$ is such that $f(0) = 0$. Give a physical interpretation of the motion of the string if f is positive for $0 < t < T \ll a/c$ and zero for $t \ge T$.

The next two exercises require some familiarity with the concept of impedance from electric circuit theory.

4.5 Consider a finite line stretching from $x = 0$ to $x = a$ that is initially at rest; that is, the input voltage and current are zero at $t = 0$. At this time an electromotive force $f(t)$ is applied at the input by attaching a generator with internal impedance Z_i, and the line is loaded at $x = a$ with an impedance Z_l, as shown in Figure 8.6. Use the general

Figure 8.6

solution for $U(x, s)$, the transform of the first Heaviside equation $-u_x = l i_t + r i$, and the transformed endpoint conditions

$$U(0, s) = F(s) - Z_i(s)I(0, s)$$
$$U(a, s) = Z_l(s)I(a, s)$$

to show that

$$U(x, s) = F(s)Z \frac{(Z + Z_l)e^{-\lambda x} - (Z - Z_l)e^{-\lambda(2a-x)}}{(Z + Z_i)(Z + Z_l) - (Z - Z_i)(Z - Z_l)e^{-2\lambda a}},$$

where

$$Z(s) = \frac{ls + r}{\lambda} = \sqrt{\frac{ls + r}{cs + g}}$$

is called the *characteristic impedance* of the line.[1]

4.6   Find $U(x, s)$ if the line of Exercise 4.5 is distortionless, if $f(t)$ is a rectangular pulse, and if $Z_l = Z$ (this is called matching impedances). Show that under these conditions there are no reflections at the end of the line, which absorbs the pulse just as an infinite line would. For instance, solve the boundary value problem in the particular case $Z_i = 0$ and compare the solution with the solution of Example 8.15.

5.1   How are the problem of Example 8.17 and its solution to be modified if the surrounding medium is at a constant temperature T_m?

5.2   Complete the details in the following sketch of proof that the quotient of hyperbolic sines in Example 8.18 is bounded for all s of the form $R_N e^{i\theta}$, where $0 \le \theta \le 2\pi$,

$$R_N = \frac{(2N + 1)}{2a\sqrt{cl}}\pi, \quad \text{and} \quad \frac{r}{l} < \frac{\sqrt{4N + 1}}{4a\sqrt{cl}}\pi.$$

Write $A = \sqrt{cl}$, $B = \sqrt{gr}$, and let s be as above. We have

$$R_N = \frac{(2N + 1)}{2aA}\pi, \quad Ba < \frac{\sqrt{4N + 1}}{4}\pi,$$

and

$$2|\sinh(As + B)|^2 = \cosh[(2N + 1)\pi \cos\theta + 2Ba] - \cos[(2N + 1)\pi \sin\theta].$$

Let θ_0 be the largest value of $\theta < \pi/2$ for which $\cos[(2N + 1)\pi \sin\theta] = 0$, that is,

$$\sin\theta_0 = \frac{4N + 1}{4N + 2} \quad \text{and} \quad \cos\theta_0 = \frac{\sqrt{8N + 3}}{4N + 2}.$$

Then, for $\theta_0 \le \theta \le \pi - \theta_0$, we have $\cos[(2N + 1)\pi \sin\theta] \le 0$ and

$$2|\sinh(As + B)|^2 \ge \cosh[(2N + 1)\pi \cos\theta + 2Ba].$$

For any other θ, $\cosh[(2N + 1)\pi \cos\theta + 2Ba]$ has a minimum value $m > 1$, and then

$$2|\sinh(As + B)|^2 \ge C \cosh[(2N + 1)\pi \cos\theta + 2Ba],$$

where $C < 1 - 1/m < 1$. The inequality $(a - x)/a < 1$ implies that the quotient in question approaches zero as $N \to \infty$.

---

[1]This equation for $U(x, s)$ was derived by F. Di Pasquantonio in "Applicazione della teoria distribuzioni all'analisi dei transitori delle linee electriche. I. Problema della linea inizialmente a riposo," *Alta Frequenza*, **34** (1965) 707–738.

5.3   Solve the boundary value problem of Example 8.18 if the end of the line at $x = a$ is open instead of short-circuited, that is, if $i(a, t) = 0$ for $t > 0$.

5.4   How is the solution of Exercise 5.3 to be modified if the line is not distortionelss but, instead,

$$\frac{\pi}{a\sqrt{cl}} > \left| \frac{r}{l} - \frac{g}{c} \right|?$$

5.5   Solve the boundary value problem of Example 8.18 for Thomson's cable with $g = l = 0$, rather than for a distortionless line, and verify the solution.

9

BOUNDARY VALUE PROBLEMS
IN HIGHER DIMENSIONS

§9.1 Electrostatic Potential in a Charged Box

We shall pose a simple boundary value problem in electricity, that of finding the electrostatic potential at each point inside a conducting rectangular box, open at the top, that is kept at zero voltage and filled with some nonconducting material, assuming that there is a known electric charge distribution in this material and that the potential on the top surface is known. It would be impossible, in the space available, to do full justice to the theory of electrostatics in explaining the concepts that are necessary to properly pose this problem and to derive the relevant partial differential equation. The reader who is not interested in its derivation from physical principles may proceed directly to equation (3). For the rest, the following informal outline may be more illuminating than none at all. We shall assume that the reader is familiar with the concepts of surface and volume integrals, at least at an elementary level.

It has been known since antiquity that there are two kinds of electricity, *positive* and *negative* in the terminology introduced by Benjamin Franklin (1706–1790), and that there is an attractive force between charges of opposite signs and a repulsive force between those of the same sign. About 1777, Charles Augustin de Coulomb (1736–1806) set up some experiments to measure this force, concluding as a result that the force exerted by a charge q located at a point P on a unit charge of the same sign located at another point Q that is at a distance d from P is proportional to the quotient q/d^2 and directed from P to Q. To write an equation describing this observation, denote the electric force by E and the vector from P to Q by r, and write the constant of proportionality in the form $k = 1/4\pi\epsilon$, where ϵ is is another constant that depends on the medium containing the charges and the 4π is introduced for convenience so that, as we shall soon see, it disappears in subsequent calculations. If we define the inner product of two vectors $v_1 = (x_1, x_2, x_3)$ and $v_2 = (y_1, y_2, y_3)$ to be $\langle v_1, v_2 \rangle = x_1 y_1 + x_2 y_2 + x_3 y_3$ and then $\|r\|^2$ as the inner product $\langle r, r \rangle$, we see that $r/\|r\|$ is a unit vector in the same direction as r and $d = \|r\|$. This leads to the following equation

$$E = \frac{q}{4\pi\epsilon\|r\|^3}r,$$

which is known today as *Coulomb's law*, while the force E itself is usually called the *electric field* or *electric field intensity* at the point Q.[1]

Now consider a sphere S of radius R centered at P, where the charge q is located. If Q is an arbitrary point of S, let r denote the vector from P to Q; n, the outward unit vector perpendicular to S at Q; and E, the electric field at Q. If $\langle E, n \rangle$ denotes the inner product of E and n, then, since $\|r\| = R$ and $\langle r, n \rangle = \langle r, r/\|r\| \rangle = \|r\| = R$, a simple calculation gives

(1) $$\epsilon \int_S \langle E, n \rangle = \frac{q}{4\pi} \int_S \frac{\langle r, n \rangle}{\|r\|^3} = \frac{q}{4\pi R^2} \int_S 1 = \frac{q}{4\pi R^2} 4\pi R^2 = q.$$

The calculation of this surface integral was simple because S is a sphere, there is only one charge at its center, and r has constant norm R. This simple fact, arising directly from Coulomb's experimental law, has been generalized into one of the fundamental postulates of electrostatics, but to show this we shall need a bit of mathematics. Note first that the electric field is just an example of a *vector field*, that is, a rule that assigns a vector to each point (x, y, z) in a certain region of space. In 1828 Mikhail Vasilevich Ostrogradski (1801–1861) discovered the following result, which he published three years later in a paper on heat propagation in the memoirs of the Academy of Sciences of Saint Petersburg. If F is a vector field in three-space with components F_1, F_2, and F_3 that are of class C^1 and if S is a smooth closed surface bounding a region V in which this field is defined, then

$$\int_V (F_{1x} + F_{2y} + F_{3z}) = \int_S \langle F, n \rangle,$$

where the letter subscripts indicate, as usual, partial derivatives. An earlier version of this type of theorem, establishing an important identity between a volume integral and a surface integral, had already been published by Poisson in 1826, but Ostrogradski was unaware of it. As was Carl Friedrich Gauss (1777–1855)—called the *prince of mathematicians*—of Ostrogradski's result, which he rediscovered later independently. We shall refer to it as the *Ostrogradski-Gauss theorem*.[2] In addition, Gauss deduced the following corollary, whose proof is left to the reader (Exercise 1.1). If P is any point in the interior of S and if r denotes the vector from P to a point of S, so that r is a function

---

[1] Coulomb was not aware of the fact that similar experiments and an identical result had already been obtained by Henry Cavendish (1731–1810), also the discoverer of the fact that water is made up of hydrogen and oxygen. Dominated by a retiring and secretive personality to a rather bizarre extent, Cavendish never published any results, but they were found after his death by Maxwell in manuscript form.

[2] For a proof, at an advanced level, see, for instance, M. Spivak, *Calculus on Manifolds*, Benjamin, Menlo Park, California, 1965, page 135.

CARL FRIEDRICH GAUSS IN 1828
Portrait by Bendixen.
Reprinted with permission from
W. Kaufmann-Bühler, *Gauss. A Biographical Study*,
Springer-Verlag, Berlin/Heidelberg/New York, 1981.

of this point, then

$$\int_S \frac{\langle r, n \rangle}{\|r\|^3} = 4\pi.$$

This means that (1) is also valid if S is an arbitrary surface surrounding the charge q at P. Since P is arbitrary inside S, it follows by additivity that, if V, the interior of S, contains several charges, $q_1, \ldots, q_K$, then

$$\epsilon \int_S \langle E, n \rangle = \sum_{k=1}^{K} q_K.$$

Finally, if there is a continuous charge density distribution ρ inside S, then the total charge in V is the volume integral of ρ over V, and the previous identity must be replaced by

$$\epsilon \int_S \langle E, n \rangle = \int_V \rho.$$

This is known in electrostatics as *Gauss' law*. Combining it with the Ostrogradski-Gauss theorem applied to the electric field, we obtain

$$\epsilon \int_V (E_{1x} + E_{2y} + E_{3z}) = \int_V \rho,$$

and it is a reasonable expectation that, as the surface S and its interior V shrink to a point, we shall have

(2) $$E_{1x} + E_{2y} + E_{3z} = \frac{\rho}{\epsilon}$$

at that point. Indeed, this is the first of *Maxwell's equations*, the starting point of electromagnetic theory. The opposite of the sum $E_{1x} + E_{2y} + E_{3z}$ was called the *convergence* of E by Maxwell—for reasons that would take us too far afield to explain— but it did not turn out to be as useful as the sum itself, which was later called the *divergence* of E by M. Abraham and P. Langevin and written as div E. It is in divergence form that Heaviside stated this equation (*EPI*, page 475), and that is how it became popular. The Ostrogradski-Gauss theorem can then be written in the form

$$\int_V \operatorname{div} F = \int_S \langle F, n \rangle$$

and is usually known today as the *divergence theorem*.[1] Needless to say, this theorem has an obvious two-dimensional version in which the volume V is replaced by an area and the bounding surface or surfaces—because there may be more than one surface bounding V, as is the case for the region between two concentric spheres—by a curve or curves.

Another experimental observation made by James Clerk Maxwell (1831–1879), of Cambridge University, is that an electric field is *conservative*, that is, that the amount of work done in moving a charge from one point in the vector field to another is independent of the path followed in the motion. If γ is a smooth path from a point P to a point Q, the work done in the motion of that charge along γ is defined to be the line integral over γ of the inner product of the force and the tangent vector to the path. More precisely, if γ is given in parametric form by $\gamma(s) = (x(s), y(s), z(s))$, its tangent vector at s has components $x'(s)$ $y'(s)$, and $z'(s)$, and then the work done in the motion of a unit charge from $P = \gamma(s_0)$ to $Q = \gamma(s)$ is

$$\int_{s_0}^{s} \langle E(\gamma(\sigma)), \gamma'(\sigma) \rangle \, d\sigma.$$

Then, since the field is conservative and if P is fixed, this work is a function of $Q = \gamma(s)$ only. We shall denote the opposite of this integral by $u(\gamma(s))$ and refer to it as the

---

[1] The reader who is interested in the reason for the term divergence, and familiar with the calculus of differential forms, may wish to see the author's "On the Concept of Divergence" *Int. J. Math. Educ. Sci. Technol.*, **12** (1981) 399–406.

electrostatic potential at Q. It is usually measured in volts, and there must always be a point of zero potential or zero voltage, chosen arbitrarily such as P above, to which the potential of other points is compared. Differentiating with respect to s,

$$\frac{d}{ds}u(\gamma(s)) = u_x(\gamma(s))x'(s) + u_y(\gamma(s))y'(s) + u_z(\gamma(s))z'(s)$$
$$= -[E_1(\gamma(s))x'(s) + E_2(\gamma(s))y'(s) + E_3(\gamma(s))z'(s)].$$

Since the last equation equation holds for any s, it follows that

$$u_x = -E_1, \qquad u_y = -E_2, \qquad \text{and} \qquad u_x = -E_3,$$

and then, combining this with (2),

$$(3) \qquad\qquad u_{xx} + u_{yy} + u_{zz} = -\frac{\rho}{\epsilon}.$$

This is the partial differential equation for the electrostatic potential u in a medium containing an electric charge distribution of density ρ, where ϵ is a constant of the medium.

To pose our boundary value problem in precise terms, let a, b, and c be positive numbers, and consider the sets

$$D = (0, a) \times (0, b) \times (0, c)$$
$$\overline{D} = [0, a] \times [0, b] \times [0, c]$$

and

$$D_c = \{(x, y, z) \in \overline{D} : z < c\}.$$

Then, if $\rho: (0, a) \times (0, b) \times (0, c) \to \mathbb{R}$ and $f: [0, a] \times [0, b] \to \mathbb{R}$ are known bounded and continuous functions, we pose the problem of finding a bounded function $u: \overline{D} \to \mathbb{R}$ that is continuous on D_c and satisfies (3) in D and the following boundary conditions

$$
\begin{aligned}
u(0, y, z) &= u(a, y, z) = 0 & 0 \le y \le b, \quad 0 \le z \le c \\
u(x, 0, z) &= u(x, b, z) = 0 & 0 \le x \le a, \quad 0 \le z \le c \\
u(x, y, 0) &= 0 & 0 \le x \le a, \quad 0 \le y \le b \\
u(x, y, c) &= f(x, y) & 0 \le x \le a, \quad 0 \le y \le b.
\end{aligned}
$$

Equation (3) is called the *Poisson equation* because it was introduced by Poisson in 1813 for the gravitational potential inside a body. Poisson knew that his proof of this equation was incomplete, but a complete proof was provided in 1839 by Gauss. It was first used to describe electric phenomena by George Green (1793–1841), the son of a prosperous miller and baker from Sneinton, near Nottingham, who had taught himself continental mathematics, and appears as equation (1) in *An Essay on the Application of Mathematical Analysis to the Theories of Electricity and Magnetism*, privately printed by subscription in 1828 in Nottingham.[1] This essay went largely ignored until it was rediscovered and its value recognized by William Thomson, then of St Peter's College, who had it reprinted later—between 1850 and 1854—in a regular mathematics journal in three parts. Green, who also anticipated the Dirichlet principle in 1833, the year that he took residence at Gonville and Caius College in Cambridge, would earn his Bachelor of Arts degree four years later as Fourth Wrangler—a low position that "most inadequately represented his mathematical power," in the words of Norman Ferrers, future editor of Green's collected papers—and then became a Fellow of the College in 1839, only two years before his premature death. It was on page 22 of his 1828 *Essay* that Green introduced the term *potential function* for a solution of the Poisson equation, which Gauss would later shorten to just *potential*. Then, on pages 23 and 25 he stated two useful theorems (see Exercise 1.6) that are equivalent to the divergence theorem of Ostrogradski and Gauss and that we shall have frequent occasion to use later on.[2] Furthermore, on pages 32 and 33 he introduced what Riemann later called *Green functions*, the basis of a new method of solving partial differential equations that we shall explore in greater detail in Chapter 12.

It was through Thomson that *le Mémoire de Green* became known in the continent. Thomson had read about the Essay in a memoir on definite integrals by Robert Murphy, a contemporary of Green and a Fellow at Caius, but was unable to find a copy in any of Cambridge's book shops. The day before departing for a visit to Paris after taking his examinations, Thomson mentioned his lack of success in this search to his tutor, William Hopkins, who said "I have some copies of it," and gave him three. Thomson was able to read some of the Essay on the top of a diligence on his trip to Paris and immediately after his arrival, on January 31, 1845, he visited Liouville and gave him one of the copies. The established mathematician took an interest in the novice's ideas, produced pen and paper, spent a pleasant time working on various subjects of common interest, and invited the young man to visit again.

By contrast, Thomson was informed by the *concierge* on his first attempted visit to Cauchy that his master—who seems to have ruled his personal life as well as mathematics

---

[1] For almost one hundred years, since the death of Newton in 1727 until a few years before the publication of Green's Essay, English mathematics had become stagnant, basking in Newton's glory. The teaching at Cambridge Univerity was firmly rooted in Newtonian mathematics and the recent advances by the French and other continental mathematicians were ignored. The great French secular schools, where only merit was a requirement for admission, had no counterpart in England, where Cambridge was the center of mathematical learning. Here a requirement of celibacy was imposed on the faculty, a Jew like James Joseph Sylvester—Second Wrangler in 1837 (that is, he finished in second place in the Senate House examinations)—could not be awarded a B.A., and a Unitarian like Augustus de Morgan—Fourth Wrangler in 1827—was debarred from a Fellowship. It was at about this time that a few works by the French masters started to be circulated in Cambridge, but it is not entirely clear how Green got hold of them out at the Sneinton mill, where he was required by his father to work full-time as a baker. The fact is that he became acquainted with the works of Fourier, Laplace, Poisson, and others. With the publication of his Essay he became the first great English mathematician since Newton.

[2] These, as well as an 1850 theorem by Thomson, which is usually known as *Stokes' theorem* because it was used by Stokes as one of the questions in the annual Smith's Prize Examination at Cambridge, are particular cases of a more general result—an n-dimensional analogue of the fundamental theorem of the integral calculus—that is also known as Stokes' theorem. For a proof of this theorem, as well as a derivation of the other results from it, see M. Spivak, *op. cit.*, pages 124–136.

WILLIAM THOMSON IN 1846
From A. Gray, *Lord Kelvin: an Account of
his Scientific Life and Work*, Dent, 1908.

by *epsilonics*—received visitors only on Fridays from half past three to five. On a second successful attempt, Cauchy made sure to ascertain whether or not Thomson knew the differential calculus before proceeding to tell him all about his own discoveries. In a letter to his father, Thomson later said that Cauchy—who had invited Thomson to attend his *belmère*'s Tuesday evening soirées—tried to convert him to Roman Catholicism. Meanwhile, Liouville had been talking about Green's memoir, which prompted Sturm to pay Thomson a visit when he was absent. He left his card with a message that he would call again next morning, but impatience won over good manners and at ten o'clock that evening an agitated Sturm stormed into Thomson's lodgings and, in Thomson's own words,

> did not lose much time in asking about Green's memoir, which he looked over with great avidity. When I pointed out to him one thing which he had himself about a year ago in Liouville's Journal, he exclaimed, "Ah! mon Dieu, oui."[1]

Returning to the problem at hand, after indulging in a natural tendency to digress, we shall consider first the simpler situation in which there are no charges inside the box;

---

[1] Quoted from S. P. Thompson, *The Life of William Thomson; Baron Kelvin of Largs*, Vol. 1, Macmillan, London, 1910, page 121.

that is, the Poisson equation is replaced by the Laplace equation

(4)
$$u_{xx} + u_{yy} + u_{zz} = 0.$$

We will then return to the charged box in §9.4.

Using the method of separation of variables, we assume that there is a solution of the form

$$u(x, y, z) = X(x)Y(y)Z(z).$$

Then (4) implies that

$$X''YZ + XY''Z + XYZ'' = 0,$$

and dividing by XYZ and rearranging terms, we obtain

$$\frac{X''}{X} + \frac{Y''}{Y} = -\frac{Z''}{Z}.$$

Since the left-hand side is a function of x and y only, while the right-hand side is a function of z only, each side must equal the same constant that we shall denote by $-\lambda$, that is,

$$\frac{X''}{X} + \frac{Y''}{Y} = -\frac{Z''}{Z} = -\lambda$$

and, similarly,

$$\frac{X''}{X} = -\lambda - \frac{Y''}{Y} = -\mu,$$

where μ is a constant.

These equations and the boundary conditions lead to the Sturm-Liouville problems

$$
\begin{aligned}
X'' + \mu X &= 0, & X(0) = X(a) &= 0 \\
Y'' + (\lambda - \mu)Y &= 0, & Y(0) = Y(b) &= 0
\end{aligned}
$$

and to the initial value problem $Z'' - \lambda Z = 0$, $Z(0) = 0$. The eigenvalues and eigenfunctions of the Sturm-Liouville problems are

$$
\begin{aligned}
\mu &= \frac{m^2\pi^2}{a^2}, & X(x) &= \sin\frac{m\pi}{a}x \\
\lambda - \mu &= \frac{n^2\pi^2}{b^2}, & Y(y) &= \sin\frac{n\pi}{b}y,
\end{aligned}
$$

where m and n are arbitrary positive integers, and then

$$\lambda = \frac{n^2\pi^2}{b^2} + \frac{m^2\pi^2}{a^2} = \frac{\pi^2}{a^2b^2}(a^2n^2 + b^2m^2).$$

The solution of the initial value problem for Z is

$$Z(z) = C \sinh\sqrt{\lambda}\,z,$$

where C is an arbitrary constant, and all this leads to the product solution

(5) $\qquad u_{mn}(x, y, z) = C_{mn} \sin \dfrac{m\pi}{a} x \, \sin \dfrac{n\pi}{b} y \, \sinh \dfrac{\pi}{ab} \sqrt{a^2 n^2 + b^2 m^2} \, z,$

where C_{mn} is an arbitrary constant. It satisfies all the boundary conditions except for the one at $z = c$.

In order to obtain a solution such that $u(x, y, c) = f(x, y)$, we shall proceed as in the one-dimensional case and propose as a solution an infinite series whose terms are the u_{mn}. Then we must find values of the constants C_{mn} such that, for $z = c$,

(6) $\qquad \displaystyle\sum_{m,n=1}^{\infty} C_{mn} \sinh \frac{\pi c}{ab} \sqrt{a^2 n^2 + b^2 m^2} \, \sin \frac{m\pi}{a} x \, \sin \frac{n\pi}{b} y = f(x, y).$

This poses the question of the convergence of such a series.

§9.2 Double Fourier Series

To see how we should define a double Fourier series we shall first proceed informally. Let $R_\pi = [-\pi, \pi] \times [-\pi, \pi]$ and assume for now that $f : R_\pi \to \mathbb{R}$ is as nice as necesssary to make the following steps valid. First fix y and expand the resulting function of x in a convergent Fourier series

(7) $\qquad f(x, y) = \tfrac{1}{2} a_0(y) + \displaystyle\sum_{m=1}^{\infty} [a_m(y) \cos mx + b_m(y) \sin mx],$

where

$$a_m(y) = \frac{1}{\pi} \int_{-\pi}^{\pi} f(x, y) \cos mx \, dx, \qquad m \geq 0,$$

and

$$b_m(y) = \frac{1}{\pi} \int_{-\pi}^{\pi} f(x, y) \sin mx \, dx, \qquad m \geq 1.$$

Next expand a_m and b_m in Fourier series of their own

(8) $\qquad a_m(y) = \tfrac{1}{2} a_{m0} + \displaystyle\sum_{n=1}^{\infty} (a_{mn} \cos ny + b_{mn} \sin ny)$

and

(9) $\qquad b_m(y) = \tfrac{1}{2} c_{m0} + \displaystyle\sum_{n=1}^{\infty} (c_{mn} \cos ny + d_{mn} \sin ny),$

where

$$a_{mn} = \frac{1}{\pi^2} \int_{-\pi}^{\pi} \int_{-\pi}^{\pi} f(x, y) \cos mx \, \cos ny \, dx \, dy, \qquad m, n \geq 0$$

$$b_{mn} = \frac{1}{\pi^2} \int_{-\pi}^{\pi} \int_{-\pi}^{\pi} f(x, y) \cos mx \, \sin ny \, dx \, dy, \qquad m \geq 0, \ n \geq 1$$

$$c_{mn} = \frac{1}{\pi^2} \int_{-\pi}^{\pi} \int_{-\pi}^{\pi} f(x, y) \sin mx \cos ny \, dx \, dy, \qquad m \geq 1, \ n \geq 0$$

$$d_{mn} = \frac{1}{\pi^2} \int_{-\pi}^{\pi} \int_{-\pi}^{\pi} f(x, y) \sin mx \sin ny \, dx \, dy, \qquad m, n \geq 1.$$

Substituting (8) and (9) into (7) yields

$$f(x, y) = \tfrac{1}{4} a_{00} + \tfrac{1}{2} \sum_{m=1}^{\infty} (a_{m0} \cos mx + c_{m0} \sin mx) + \tfrac{1}{2} \sum_{n=1}^{\infty} (a_{0n} \cos ny + b_{0n} \sin ny)$$

$$+ \sum_{m=1}^{\infty} \left[\sum_{n=1}^{\infty} (a_{mn} \cos mx \cos ny + b_{mn} \cos mx \sin ny \right.$$

$$\left. + c_{mn} \sin mx \cos ny + d_{mn} \sin mx \sin ny) \right].$$

Of course, this does not prove convergence, and it leads to nested single series rather than to a double series. However, it suffices to motivate the following definition.

Definition 9.1 *Let* $f : R_\pi \to \mathbb{R}$ *be an integrable function. Then the constants* $a_{mn}, b_{mn}, c_{mn},$ *and* d_{mn} *defined above are called the* **Fourier coefficients** *of* f *on* R_π, *and we write*

$$f(x, y) \sim \tfrac{1}{4} a_{00} + \tfrac{1}{2} \sum_{m=1}^{\infty} (a_{m0} \cos mx + c_{m0} \sin mx) + \tfrac{1}{2} \sum_{n=1}^{\infty} (a_{0n} \cos ny + b_{0n} \sin ny)$$

$$\text{(10)}$$

$$+ \sum_{m,n=1}^{\infty} (a_{mn} \cos mx \cos ny + b_{mn} \cos mx \sin ny$$

$$+ c_{mn} \sin mx \cos ny + d_{mn} \sin mx \sin ny).$$

The series on the right-hand side is called the **Fourier series** *of* f *on* R_π.

Example 9.1 We shall find the double Fourier series of the function defined on R_π by $f(x, y) = x^2 y^2$. Note first that two integrations by parts yield

$$\int_{-\pi}^{\pi} x^2 \cos mx \, dx = \frac{2}{m^2} [\pi \cos m\pi + \pi \cos(-m\pi)] = (-1)^m \frac{4\pi}{m^2},$$

and then

$$a_{00} = \frac{1}{\pi^2} \int_{-\pi}^{\pi} \int_{-\pi}^{\pi} x^2 y^2 \, dx \, dy = \frac{1}{\pi^2} \left(\int_{-\pi}^{\pi} x^2 \, dx \right)^2 = \frac{1}{\pi^2} \left(\frac{2\pi^3}{3} \right)^2 = \frac{4\pi^4}{9}$$

$$a_{m0} = \frac{1}{\pi^2} \int_{-\pi}^{\pi} \int_{-\pi}^{\pi} x^2 y^2 \cos mx \, dx \, dy$$

$$= \frac{1}{\pi^2} \left(\int_{-\pi}^{\pi} x^2 \cos mx \, dx \right) \left(\int_{-\pi}^{\pi} y^2 \, dy \right) = (-1)^m \frac{8\pi^2}{3m^2}$$

$$a_{0n} = \frac{1}{\pi^2} \int_{-\pi}^{\pi} \int_{-\pi}^{\pi} x^2 y^2 \cos ny \, dx \, dy = (-1)^n \frac{8\pi^2}{3n^2}$$

$$a_{mn} = \frac{1}{\pi^2} \int_{-\pi}^{\pi} \int_{-\pi}^{\pi} x^2 y^2 \cos mx \, \cos ny \, dx \, dy$$

$$= \frac{1}{\pi^2} \left(\int_{-\pi}^{\pi} x^2 \cos mx \, dx \right) \left(\int_{-\pi}^{\pi} y^2 \cos ny \, dy \right) = (-1)^{m+n} \frac{16}{m^2 n^2}.$$

Also,

$$\int_{-\pi}^{\pi} x^2 \sin mx \, dx = 0$$

because the integrand is odd, which implies that $b_{mn} = c_{mn} = d_{mn} = 0$ for all permissible values of the subscripts, and then

$$f(x, y) \sim \frac{\pi^4}{9} + \frac{4\pi^2}{3} \sum_{m=1}^{\infty} \frac{(-1)^m}{m^2} \cos mx + \frac{4\pi^2}{3} \sum_{n=1}^{\infty} \frac{(-1)^n}{n^2} \cos ny$$

$$+ 16 \sum_{m,n=1}^{\infty} \frac{(-1)^{m+n}}{m^2 n^2} \cos mx \, \cos ny.$$

Next we define convergence in the following way. Let s_{MN} be the sum of all terms on the right-hand side of (10) with $m \leq M$ and $n \leq N$.

Definition 9.2 *The Fourier series of an integrable function $f : R_\pi \to \mathbb{R}$ converges to a sum S at a point (x, y) in R_π if and only if for every $\epsilon > 0$*

$$|S - s_{MN}(x, y)| < \epsilon$$

for all sufficiently large M and N.

The theory of double Fourier series is considerably more complex than the one-dimensional theory. For one thing, there are alternate definitions of convergence for double series. The type just defined is called *rectangular convergence*, but there are also *restricted rectangular convergence* if the values of M and N in the last line of Definition 9.2 are restricted so that $[0, M] \times [0, N]$ is nearly a square, *square convergence* if s_{MN} is replaced with s_{NN}, and *circular convergence* if it is replaced with the sum of those terms such that $m^2 + n^2 \leq N^2$ and $N \to \infty$. Also, a convergence theorem for double Fourier series requires stronger hypotheses than those in the one-variable case and localization (refer to the remark after Theorem 2.2) does not hold. These assertions are validated by the fact that there is a nontrivial example of a function that has partial derivatives at every point and is identically zero in a neighborhood of $(0, 0)$ but whose Fourier series diverges at $(0, 0)$ regardless of what definition is used. Furthermore, it is not necessarily true that the coefficients of a convergent double Fourier series approach zero as $m, n \to \infty$.[1] All this notwithstanding, it is possible to prove a useful convergence theorem rather simply. First, we shall find it convenient, as in the one-dimensional case, to obtain an integral representation for s_{MN}. From the equations for the Fourier coefficients we obtain

$$a_{m0} \cos mx + c_{m0} \sin mx = \frac{1}{\pi^2} \int_{-\pi}^{\pi} \int_{-\pi}^{\pi} f(s, t) \cos m(s - x) \, ds \, dt$$

$$a_{0n} \cos ny + b_{0n} \sin ny = \frac{1}{\pi^2} \int_{-\pi}^{\pi} \int_{-\pi}^{\pi} f(s, t) \cos n(t - y) \, ds \, dt$$

and

$$a_{mn} \cos mx \cos ny + b_{mn} \cos mx \sin ny + c_{mn} \sin mx \cos ny + d_{mn} \sin mx \sin ny$$

$$= \frac{1}{\pi^2} \int_{-\pi}^{\pi} \int_{-\pi}^{\pi} f(s, t) \cos m(s - x) \cos n(t - y) \, ds \, dt.$$

Therefore,

$$
\begin{aligned}
s_{MN}(x, y) &= \frac{1}{\pi^2} \int_{-\pi}^{\pi} \int_{-\pi}^{\pi} f(s, t) \left[\frac{1}{2} + \sum_{m=1}^{M} \cos m(s - x) \right] \left[\frac{1}{2} + \sum_{n=1}^{N} \cos n(t - y) \right] ds \, dt \\
&= \frac{1}{\pi^2} \int_{-\pi}^{\pi} \int_{-\pi}^{\pi} f(s, t) \frac{\sin \left(M + \frac{1}{2} \right)(s - x)}{2 \sin \frac{1}{2}(s - x)} \frac{\sin \left(N + \frac{1}{2} \right)(t - y)}{2 \sin \frac{1}{2}(t - y)} ds \, dt \\
&= \frac{1}{4\pi^2} \int_{-\pi}^{\pi} \int_{-\pi}^{\pi} f(s, t) D_M(s - x) D_N(t - y) \, ds \, dt,
\end{aligned}
$$

(11)

where D_M and D_N represent the Dirichlet kernel of §2.4.

Next we must keep in mind, in order to give a useful convergence theorem, that functions found in applications frequently have discontinuities, but they seldom have wild kinds of discontinuities. For a function of two variables it is usually sufficient to allow jump discontinuities on a finite number of nice arcs of curve, such as those in the following definition.

Definition 9.3 *If $R = [a, b] \times [c, d]$ is a rectangle in $\mathbb{R}^2$, a function $f : R \to \mathbb{R}$ is defined to be* **piecewise continuous** *if and only if*

(i) *R can be divided into a finite number of subregions by horizontal segments, vertical segments, and arcs of the form $g : I \to [c, d]$, where I is a closed subinterval of $[a, b]$ and g is continuously differentiable and has a continuously differentiable inverse, and*

---

[1] An excellent review of these and other issues for double Fourier series, accessible to readers superficially familiar with the concept of an L^p function (see §6.6), can be found in J. M. Ash, "Multiple Trigonometric Series," in *Studies in Harmonic Analysis*, Mathematical Association of America, Washington, DC., 1976, pages 76–96.

(ii) f *is continuous inside each of the subregions so determined and can be extended to its boundary as a continuous function.*

If, in addition, the partial derivatives f_x *and* f_y *are piecewise continuous, then* f *is said to be* **piecewise continuously differentiable**.

The condition on the inverse of g guarantees that this definition is symmetric in x and y. It should be noted that, if f is piecewise continuous or piecewise continuously differentiable on R as defined above, then it is also piecewise continuous or piecewise continuously differentiable in both x and y in the sense of Definitions B.5 and B.8. This will allow us to use some of the results in Appendix B.

The proof of convergence will then be based on the following two-dimensional extensions of the Riemann-Lebesgue theorem. The proof of the first is a straightforward modification of the one-dimensional proof and it will be left as an exercise.

Lemma 9.1 *Let R be a bounded rectangle in the plane and let $f : R \to \mathbb{R}$ be an integrable function. Then,*

$$\lim_{r,s \to \infty} \int_R f(x, y) \sin rx \, \sin sy \, dx \, dy = \lim_{r,s \to \infty} \int_R f(x, y) \cos rx \, \cos sy \, dx \, dy = 0.$$

Lemma 9.2 *Let $R = [a, b] \times [c, d]$ be a rectangle in the plane such that at least one of the intervals $[a, b]$ and $[c, d]$ contains the origin. Let $f : R \to \mathbb{R}$ be a piecewise continuous function, and assume that there is an $\epsilon > 0$ such that f is piecewise continuously differentiable on the rectangle*
  (i) $\{ (x, y) \in R : |y| \le \epsilon \}$ *if $[c, d]$ contains the origin and*
  (ii) $\{ (x, y) \in R : |x| \le \epsilon \}$ *if $[a, b]$ contains the origin.*
Then

$$\lim_{r,s \to \infty} \int_R f(x, y) \sin rx \, \frac{\sin sy}{y} \, dx \, dy = \lim_{r,s \to \infty} \int_R f(x, y) \frac{\sin rx}{x} \, \sin sy \, dx \, dy = 0.$$

Proof. If $[c, d]$ does not contain the origin, then the quotient $f(x, y)/y$ is piecewise continuous on R, and therefore integrable, and the first limit follows from Lemma 9.1. If, instead, $[c, d]$ contains the origin, define

$$F(y, r) = \int_a^b f(x, y) \sin rx \, dx.$$

Then the first integral in the statement of the lemma becomes

$$(12) \quad \int_c^d F(y, r) \frac{\sin sy}{y} \, dy = \int_c^d \frac{F(y, r) - F(0, r)}{y} \sin sy \, dy + F(0, r) \int_c^d \frac{\sin sy}{y} \, dy.$$

The integral in the last term on the right is bounded independently of s since the substitution $t = sy$ and Example B.3 show that it is a continuous function of s that has a limit as $s \to \infty$. Then this term can be made as small as desired by choosing r large enough because $F(0, r) \to 0$ as $r \to \infty$ by the Riemann-Lebesgue theorem. Fixing the choice of r, note that F is a piecewise continuous function of y on $[c, d]$ by Lemma B.1 and that it has one-sided derivatives with respect to y at $y = 0$ by Lemmas B.4 and 2.2. Then the quotient in the first integral on the right-hand side of (12) is piecewise continuous and this term approaches zero as $s \to \infty$ by the Riemann-Lebesgue theorem. This proves the first limit, and the proof of the second is analogous, Q.E.D.

Finally, we can proceed to the convergence proof. As in the one-dimensional case, we shall find it convenient, without loss of generality, to extend f to the entire plane as a periodic function of period 2π in each variable.

Theorem 9.1 *Let* $f : \mathbb{R}^2 \to \mathbb{R}$ *be periodic of period* 2π *in each variable and piecewise continuous on every bounded rectangle. Then its Fourier series on* R_π *converges to* $f(x, y)$ *at each point* (x, y) *in* $\mathbb{R}^2$ *such that, for some positive number* $\epsilon < \pi$,

(i) f *is piecewise continuously differentiable on* $[x - \epsilon, x + \epsilon] \times [y - \pi, y + \pi]$ *and on* $[x - \pi, x + \pi] \times [y - \epsilon, y + \epsilon]$, *and*

(ii) f *is of class* C^1 *on the square* $[x - \epsilon, x + \epsilon] \times [y - \epsilon, y + \epsilon]$ *and* f_{xy} *exists and is bounded there.*

Proof. Putting $u = s - x$ and $v = t - y$ in (11) and using the periodicity of f, we obtain

$$s_{MN}(x, y) = \frac{1}{4\pi^2} \int_{-\pi}^{\pi} \int_{-\pi}^{\pi} f(x + u, y + v) D_M(u) D_N(v) \, du \, dv.$$

Then, since the integral of D_N from $-\pi$ to π has value 2π for any N,

$$(13) \quad s_{MN}(x, y) - f(x, y) = \frac{1}{4\pi^2} \int_{-\pi}^{\pi} \int_{-\pi}^{\pi} [f(x + u, y + v)$$

$$- f(x, y)] D_M(u) D_N(v) \, du \, dv.$$

If $\delta \leq \epsilon$ and if $-\delta \leq u, v \leq \delta$, then, by the mean value theorem, there is a σ between x and $x + u$ and then a τ between y and $y + v$ such that

$$f(x + u, y + v) - f(x, y)$$
$$= f(x + u, y + v) - f(x, y + v) + f(x, y + v) - f(x, y)$$
$$= u f_x(\sigma, y + v) + f(x, y + v) - f(x, y)$$
$$= u[f_x(\sigma, y + v) - f_x(\sigma, y)] + u f_x(\sigma, y) + f(x, y + v) - f(x, y)$$
$$= uv f_{xy}(\sigma, \tau) + u f_x(\sigma, y) + f(x, y + v) - f(x, y).$$

Then the contribution to the integral in (14) over the square $S_\delta = [-\delta, \delta] \times [-\delta, \delta]$ can be split as follows

$$\int_{S_\delta} f_{xy}(\sigma, \tau) uv D_M(u) D_N(v) \, du \, dv + \int_{-\delta}^{\delta} \left(\int_{-\delta}^{\delta} f_x(\sigma, y) u D_M(u) \, du \right) D_N(v) \, dv$$

$$(14) \qquad + \int_{-\delta}^{\delta} \left(\int_{-\delta}^{\delta} \frac{f(x, y + v) - f(x, y)}{v} \frac{v}{\sin \frac{1}{2} v} \sin \left(N + \tfrac{1}{2} \right) v \, dv \right) D_M(u) \, du.$$

Since the integrands in the first term and in the inner integral in the second term of (14) are bounded with bounds independent of M and N, these integrals can be made as small as desired by making δ small enough. Also, the inner integral in the last term approaches zero as $N \to \infty$ by the Riemann-Lebesgue theorem since the fractions in the integrand are continuous for $v \neq 0$ and have the limits $f_y(x, y)$ and 2, respectively, as $v \to 0$. It remains to deal with the integral of D_N from $-\delta$ to δ, which equals

$$\int_{-\delta}^{\delta} D_N(v)\, dv = 2 \int_0^{\delta} D_N(v)\, dv$$

$$= 4 \int_0^{\delta} \frac{\sin\left(N + \frac{1}{2}\right)v}{v}\, dv + 2 \int_0^{\delta} \frac{v - 2\sin\frac{1}{2}v}{v\sin\frac{1}{2}v} \sin\left(N + \frac{1}{2}\right)v.$$

The fraction in the last integrand is piecewise continuous because it is continuous on $(0, \delta]$ and, by l'Hôpital's rule, approaches zero as $v \to 0$. Therefore, that integral approaches zero as $N \to \infty$ by the Riemann-Lebesgue theorem, while

$$\int_0^{\delta} \frac{\sin\left(N + \frac{1}{2}\right)v}{v}\, dv = \int_0^{\left(N + \frac{1}{2}\right)\delta} \frac{\sin t}{t}\, dt$$

approaches $\pi/2$ as $N \to \infty$. This shows that the integral of D_N from $-\delta$ to δ is bounded independently of N, proving that (14) can be made as small as desired by choosing δ small and N large.

Having chosen δ, the contribution to the integral in (14) over $[-\pi, \pi] \times [-\pi, \pi] - S_\delta$ can be written as the sum of those over the rectangles $R_1 = [-\pi, \pi] \times [\delta, \pi]$, $R_2 = [-\pi, -\delta] \times [-\delta, \delta]$, $R_3 = [-\pi, \pi] \times [-\pi, -\delta]$, and $R_4 = [\delta, \pi] \times [-\delta, \delta]$. The integral over R_1 is

$$\int_{R_1} [f(x + u, y + v) - f(x, y)] \frac{u}{\sin\frac{1}{2}u \sin\frac{1}{2}v} \frac{\sin\left(M + \frac{1}{2}\right)u}{u} \sin\left(N + \frac{1}{2}\right)v \, du\, dv.$$

Since $f(x + u, y + v)$ is a piecewise continuous function of (u, v) on R_1 and piecewise continuously differentiable in some vertical strip of R_1 centered at $u = 0$, and since the first fraction in the integrand has continuous partial derivatives on R_1 (this is easily seen using the definition of derivative and l'Hôpital's rule), this integral approaches zero as $M, N \to \infty$ by Lemma 9.2, and an analogous argument applies to the integrals over R_2 to R_4, Q.E.D.

The uniform convergence of one-dimensional Fourier series was proved as a consequence of Bessel's inequality, and the same method can be used for double Fourier series.

If f and g are integrable on a rectangle R in the plane and if r is a fixed function that is continuous and positive on R (usually $r \equiv 1$), we shall denote the integral of the product rfg on R by (f, g) and refer to this expression as the *inner product* of f and g. As in §6.1,

$$\|f\| = \sqrt{(f, f)}$$

will be called the *norm* of f. The concept of *orthogonal system* on R is also as in §6.1, as is the definition of the *Fourier coefficients* of $f : R \rightarrow \mathbb{R}$ with respect to a given orthogonal system $\{\varphi_k\}$,

$$c_k = \frac{1}{\|\varphi_k\|^2} (f, \varphi_k),$$

and of its *Fourier series*

$$\sum_{k=1}^{\infty} c_k \varphi_k.$$

Theorem 9.2 (Bessel's Inequality) *If f and $\{\varphi_n\}$ are as above, then*

$$\sum_{k=1}^{\infty} c_k^2 \|\varphi_k\|^2 \leq \|f\|^2.$$

The proof is identical to that of Corollary 2 of Theorem 6.2.

Example 9.2 The functions

$$1, \quad \cos mx, \quad \sin mx, \quad \cos ny, \quad \sin ny, \quad \ldots$$
$$\cos mx \cos ny, \quad \sin mx \sin ny, \quad \cos mx \sin ny, \quad \sin mx \cos ny, \quad \ldots,$$

where m and n are arbitrary positive integers, form an orthogonal system on R_π. The norm of the first is 2π, the norm of the remaining functions in the first row is $\sqrt{2}\,\pi$, and the norm of each of the others is π. Bessel's inequality for $f : R_\pi \rightarrow \mathbb{R}$ with respect to this system is

$$\tfrac{1}{4}a_{00}^2 + \tfrac{1}{2}\sum_{m=1}^{\infty}(a_{m0}^2 + c_{m0}^2) + \tfrac{1}{2}\sum_{n=1}^{\infty}(a_{0n}^2 + b_{0n}^2) + \sum_{m,n=1}^{\infty}(a_{mn}^2 + b_{mn}^2 + c_{mn}^2 + d_{mn}^2) \leq \frac{1}{\pi^2}\|f\|^2.$$

Theorem 9.3 *Let $f : \mathbb{R}^2 \rightarrow \mathbb{R}$ be a function of class C^1 and periodic of period 2π with respect to each variable and assume that it has a mixed partial derivative f_{xy} at every point. Then its Fourier series on R_π converges uniformly to f on R_π.*

Proof. The series converges to $f(x, y)$ at every (x, y) in the plane by Theorem 9.1. To show that the convergence is uniform and referring to (10), it is enough to show that for any $m, n \geq 1$ there is a number M_{mn} such that the mnth term in (10) is no larger than M_{mn} in absolute value and such that

$$\sum_{m,n=1}^{\infty} M_{mn}$$

converges. As in the proof of Theorem 6.3, it is easily shown that

$$|a_{mn} \cos mx \cos ny + b_{mn} \cos mx \sin ny + c_{mn} \sin mx \cos ny + d_{mn} \sin mx \sin ny|$$

$$\leq |a_{mn} \cos mx + c_{mn} \sin mx| + |b_{mn} \cos mx + d_{mn} \sin mx|$$

$$< \frac{1}{m^2} + m^2(a_{mn}^2 + c_{mn}^2) + \frac{1}{m^2} + m^2(b_{mn}^2 + b_{mn}^2)$$

$$= \frac{2}{m^2} + m^2(a_{mn}^2 + b_{mn}^2 + c_{mn}^2 + d_{mn}^2).$$

Using now the identity $f(-\pi, y) = f(\pi, y)$, which is a consequence of the continuity and the periodicity of f, it is easily seen that the Fourier coefficients of f_x are

$$A_{mn} = mc_{mn}, \quad B_{mn} = md_{mn}, \quad C_{mn} = -ma_{mn}, \quad \text{and} \quad D_{mn} = -mb_{mn},$$

and then Bessel's inequality for f_x shows that

$$\sum_{m,n=1}^{\infty} (A_{mn}^2 + B_{mn}^2 + C_{mn}^2 + D_{mn}^2)$$

converges. Since $\sum_{m=1}^{\infty} (2/m^2)$ also converges, it is enough to take

$$M_{mn} = \frac{2}{m^2} + m^2(a_{mn}^2 + b_{mn}^2 + c_{mn}^2 + d_{mn}^2),$$

Q.E.D.

Assume now that f is integrable on R_π. As in the proof of Theorem 6.5, f can be approximated by a piecewise constant function which, in turn, can be approximated by a function of class C^2 that is zero in a small neighborhood of the boundary. By Theorem 9.3, this function can be approximated uniformly by its double Fourier series. These facts can be used as in the proof of Theorem 6.5 to establish the following result.

Theorem 9.4 *Let* $f: \mathbb{R}^2 \to \mathbb{R}$ *be an integrable function. Then Parseval's identity*

$$\tfrac{1}{4}a_{00}^2 + \tfrac{1}{2}\sum_{m=1}^{\infty}(a_{m0}^2 + c_{m0}^2) + \tfrac{1}{2}\sum_{n=1}^{\infty}(a_{0n}^2 + b_{0n}^2) + \sum_{m,n=1}^{\infty}(a_{mn}^2 + b_{mn}^2 + c_{mn}^2 + d_{mn}^2) = \frac{1}{\pi^2}\|f\|^2$$

holds for f *and its double Fourier series on* R_π *converges to* f *in the mean, that is,* $\|f - s_{MN}\| \to 0$ *as* $n \to \infty$.

The problem of defining the Fourier series of a function f on an arbitrary rectangle $R_{ab} = (\alpha - a, \alpha + a) \times (\beta - b, \beta + b)$, where a, α, b, and β are real numbers and a and b are positive, is reduced to the previous case by making the substitution

$$u = \frac{\pi}{a}(x - \alpha) \quad \text{and} \quad v = \frac{\pi}{b}(y - \beta),$$

defining $h(u, v) = f(x, y)$, expanding h as in (10), and then undoing the substitution. We obtain

$$(15) \quad f(x, y) \sim \tfrac{1}{4} a_{00} + \tfrac{1}{2} \sum_{m=1}^{\infty} \left[a_{m0} \cos \frac{m\pi}{a} (x - \alpha) + c_{m0} \sin \frac{m\pi}{a} (x - \alpha) \right]$$

$$+ \tfrac{1}{2} \sum_{n=1}^{\infty} \left[a_{0n} \cos \frac{n\pi}{b} (y - \beta) + b_{0n} \sin \frac{n\pi}{b} (y - \beta) \right]$$

$$+ \sum_{m,n=1}^{\infty} \left[a_{mn} \cos \frac{m\pi}{a} (x - \alpha) \cos \frac{n\pi}{b} (y - \beta) \right.$$

$$+ b_{mn} \cos \frac{m\pi}{a} (x - \alpha) \sin \frac{n\pi}{b} (y - \beta)$$

$$+ c_{mn} \sin \frac{m\pi}{a} (x - \alpha) \cos \frac{n\pi}{b} (y - \beta)$$

$$\left. + d_{mn} \sin \frac{m\pi}{a} (x - \alpha) \sin \frac{n\pi}{b} (y - \beta) \right],$$

where

$$d_{mn} = \frac{1}{ab} \int_{\alpha-a}^{\alpha+a} \int_{\beta-b}^{\beta+b} f(x, y) \sin \frac{m\pi}{a} (x - \alpha) \sin \frac{n\pi}{b} (y - \beta) \, dx \, dy,$$

and similarly for the other coefficients.

The series on the right-hand side of (15) is called the **Fourier series** of f on R_{ab}. Results analogous to Theorems 9.1 and 9.3 hold in this case.

§9.3　The Dirichlet Problem in a Box

The method of separation of variables applied to the boundary value problem of §9.1 led to the problem of finding constants C_{mn} such that (6) holds. To do this we first extend f to the entire plane as an odd, periodic function of x of period $2a$ and as an odd, periodic function of y of period $2b$, and we assume that it is piecewise continuously differentiable and that f_{xy} exists on any region in which f is of class C^1. Then its Fourier coefficients a_{mn}, b_{mn}, and c_{mn} vanish, and for (6) to be the Fourier series of f on $(-a, a) \times (-b, b)$ we must choose C_{mn} so that

$$C_{mn} \sinh \frac{\pi c}{ab} \sqrt{a^2 n^2 + b^2 m^2} = d_{mn},$$

where d_{mn} is as stated at the end of §9.2 with $\alpha = \beta = 0$. Taking the values of C_{mn} obtained from this equation to (5) and forming an infinite series with the resulting functions we obtain the formal solution

$$(16) \qquad u(x, y, z) = \sum_{m,n=1}^{\infty} d_{mn} \frac{\sinh \dfrac{\pi}{ab} \sqrt{a^2 n^2 + b^2 m^2}\, z}{\sinh \dfrac{\pi c}{ab} \sqrt{a^2 n^2 + b^2 m^2}} \sin \frac{m\pi}{a} x \, \sin \frac{n\pi}{b} y,$$

which must be verified. We know that (16) converges for $z = c$ and that its sum is $f(x, y)$ at any point (x, y) in $(0, a) \times (0, b)$ in whose neighborhood f is of class C^1. Clearly, u vanishes for $x = 0, a$; $y = 0, b$; and $z = 0$. Now fix $z_0 < c$ and consider the parallelpiped $D_{z_0} = \{(x, y, z) \in D_c : z \le z_0\}$. It follows from the definition of d_{mn} that these coefficients are bounded by some constant M for all m and n, and then, if we define

$$F(m, n) = \frac{\pi}{ab} \sqrt{a^2 n^2 + b^2 m^2}$$

and recall that $\sinh z$ is a positive, increasing function of z for $z > 0$, the mnth term in (16) is bounded by

$$M \frac{e^{F(m,n)z} - e^{-F(m,n)z}}{e^{F(m,n)c} - e^{-F(m,n)c}} = M e^{-F(m,n)(c-z)} \frac{1 - e^{-2F(m,n)z}}{1 - e^{-2F(m,n)c}}$$

$$< e^{-F(m,n)(c-z_0)} \frac{1}{1 - e^{-2F(1,1)c}}.$$

Therefore, (16) converges uniformly on D_{z_0} by the Weierstrass M-test (Theorem A.2) if we show that

$$(17) \qquad \sum_{m,n=1}^{\infty} e^{-F(m,n)(c-z_0)}$$

converges. To see this, assume, without loss of generality, that $b \ge a$, so that

$$F(m, n) \ge \frac{\pi}{b} \sqrt{m^2 + n^2},$$

define $K = \pi(c - z_0)/b$, and note that, using the Maclaurin series for the exponential,

$$e^{-K\sqrt{m^2+n^2}} = \frac{1}{e^{K\sqrt{m^2+n^2}}} < \frac{4!}{K^4 (m^2 + n^2)^2}.$$

Then (17) converges by comparison with

$$\frac{4!}{K^4} \sum_{m,n=1}^{\infty} \frac{1}{(m^2 + n^2)^2},$$

which is known to converge (see Exercise 3.1).

Since the terms of (16) are continuous and since the convergence is uniform on D_{z_0}, u is continuous on D_{z_0} by Theorem A.4, and since $z_0 < c$ is arbitrary, u is continuous on D_c.

Now, for each fixed (y, z) with $0 < y < b$ and $0 < z < c$, differentiation of (16) term by term twice with respect to x introduces in the general term the factor $-m^2\pi^2/a^2$, which increases with m less rapidly than the exponential in (17) decreases. Then it is seen, as above, that the derived series converges uniformly with respect to x on $[0, a]$, and then, by Theorem A.6, its sum is $u_{xx}(x, y, z)$ if $0 < x < a$. Similarly, the sum of the series obtained by differentiation of (16) term by term twice with respect to y is $u_{yy}(x, y, z)$. For each fixed (x, y) with $0 < x < a$ and $0 < y < b$, two differentiations with respect to z introduces in the general term the factor

$$\left(\frac{\pi}{ab}\sqrt{a^2n^2 + b^2m^2}\right)^2 = \frac{m^2\pi^2}{a^2} + \frac{n^2\pi^2}{b^2},$$

and, as before, the derived series converges to $u_{zz}(x, y, z)$. Comparison of these three derived series shows that u satisfies the Laplace equation in D.

Assume now that the odd, periodic extension of f to $\mathbb{R}^2$ is of class C^1 and that f_{xy} exists everywhere. This implies, in particular, that f vanishes on the boundary of the rectangle $[0, a] \times [0, b]$, and then the electrostatic potential varies continuously over the four edges of the box at $z = c$. Now define

$$f_{mn}(x, y, z) = d_{mn} \sin\frac{m\pi}{a}x \, \sin\frac{n\pi}{b}y$$

and

$$g_{mn}(x, y, z) = \frac{\sinh\dfrac{\pi}{ab}\sqrt{a^2n^2 + b^2m^2}\, z}{\sinh\dfrac{\pi c}{ab}\sqrt{a^2n^2 + b^2m^2}}.$$

By Theorem 9.3,

$$\sum_{m,n=1}^{\infty} f_{mn}(x, y, z)$$

converges uniformly, and then (16) also converges uniformly on $\overline{D}$ by Theorem A.3 because $\{g_{mn}\}$ is a nonincreasing sequence if its terms are ordered so that $a^2n^2 + b^2m^2$ is nondecrasing and because $|g_{mn}(x, y, z)| \leq 1$ if $0 \leq z \leq c$. Under these conditions the sum of (16) is continuous on $\overline{D}$ by Theorem A.4. We have proved the following theorem.

Theorem 9.5 *Let $f : [0, a] \times [0, b] \to \mathbb{R}$ be a function whose extension to the entire plane as an odd, periodic function of x of period $2a$ and as an odd, periodic function of y of period $2b$ is piecewise continuously differentiable and has a mixed second-order partial derivative on any region in which it is of class C^1. Then the function $u : \overline{D} \to \mathbb{R}$ defined by*

$$(15) \qquad u(x, y, z) = \sum_{m,n=1}^{\infty} d_{mn} \frac{\sinh\dfrac{\pi}{ab}\sqrt{a^2n^2 + b^2m^2}\, z}{\sinh\dfrac{\pi c}{ab}\sqrt{a^2n^2 + b^2m^2}} \sin\frac{m\pi}{a}x \, \sin\frac{n\pi}{b}y,$$

where

$$d_{mn} = \frac{4}{ab} \int_0^a \int_0^b f(x, y) \sin \frac{m\pi}{a} x \sin \frac{n\pi}{b} y \, dx \, dy \, dz,$$

is continuous on the set $D_c = \{(x, y, z) \in \bar{D}: z < c\}$ and satisfies the boundary value problem

$$
\begin{array}{lll}
u_{xx} + u_{yy} + u_{zz} = 0 & \text{in } D & \\
u(0, y, z) = u(a, y, z) = 0 & 0 \le y \le b, & 0 \le z \le c \\
u(x, 0, z) = u(x, b, z) = 0 & 0 \le x \le a, & 0 \le z \le c \\
u(x, y, 0) = 0 & 0 \le x \le a, & 0 \le y \le b \\
u(x, y, c) = f(x, y) & 0 \le x \le a, & 0 \le y \le b.
\end{array}
$$

The last equation is valid on any region in which f is of class C^1. If, in addition, the odd extension of f is of class C^1 everywhere, then u is continuous on $\bar{D}$.

It is also possible to show that our boundary value problem is well posed in the sense of Hadamard using the following result.

Theorem 9.6 (The maximum principle for the Laplace equation) *Let D be an open, bounded set in $\mathbb{R}^3$, let ∂D denote its boundary, and let $\bar{D}$ be the union of D and ∂D. If $u: \bar{D} \to \mathbb{R}$ is a continuous function that satisfies the Laplace equation in D, then u attains its maximum and minimum values on ∂D.*

Proof. Assume that u has its maximum value M at a point (x_0, y_0, z_0) in D, and let $m < M$ be its maximum value on ∂D. Then, if d is the diameter of a sphere containing D, define $v: \bar{D} \to \mathbb{R}$ by

$$v(x, y, z) = u(x, y, z) + \frac{M - m}{2d^2}[(x - x_0)^2 + (y - y_0)^2 + (z - z_0)^2].$$

If (x, y, z) is in ∂D, we have

$$v(x, y, z) < u(x, y, z) + \frac{M - m}{2} \le m + \frac{M - m}{2} < M.$$

Since $v(x_0, y_0, z_0) = u(x_0, y_0, z_0) = M$, v attains its maximum value at some point in D, and this implies that v_{xx}, v_{yy}, and v_{zz} are less than or equal to zero at that point. But this is impossible because

$$v_{xx} + v_{yy} + v_{zz} = u_{xx} + u_{yy} + u_{zz} + \frac{3}{d^2}(M - m) = \frac{3}{d^2}(M - m) > 0,$$

and this contradiction shows that u attains its maximum value on ∂D. The same type of argument shows that $-u$ attains its maximum value on ∂D and then u attains its minimum value on ∂D, Q.E.D.

Theorem 9.7 *The boundary value problem of Theorem 9.5 is well posed in the sense of Hadamard.*

Proof. Let u_1 and u_2 be solutions corresponding to electrostatic potentials given by functions f and g at $z = c$. By the maximum principle for the Laplace equation, $|u_1 - u_2| \le M_{f-g}$, where M_{f-g} denotes the maximum value of $|f - g|$. This shows that the solution depends continuously on boundary values, and putting $f \equiv g$ shows uniqueness, Q.E.D.

§9.4 Return to the Charged Box

We can use the knowledge acquired in solving the simpler problem for the Laplace equation to solve the following problem for the Poisson equation in a box, which is even more general than the one posed in §9.1,

(17)	$u_{xx} + u_{yy} + u_{zz} = -\dfrac{\rho}{\epsilon}$	in D	
(18)	$u(0, y, z) = f_1(y, z)$	$0 \le y \le b,$	$0 \le z \le c$
(19)	$u(a, y, z) = f_2(y, z)$	$0 \le y \le b,$	$0 \le z \le c$
(20)	$u(x, 0, z) = f_3(x, z)$	$0 \le x \le a,$	$0 \le z \le c$
(21)	$u(x, b, z) = f_4(x, z)$	$0 \le x \le a,$	$0 \le z \le c$
(22)	$u(x, y, 0) = f_5(x, y)$	$0 \le x \le a,$	$0 \le y \le b$
(23)	$u(x, y, c) = f_6(x, y)$	$0 \le x \le a,$	$0 \le y \le b,$

where ρ and the f_i, $1 \le i \le 6$, are functions of class C^1.

Although this problem may seem rather involved, the solution is quite straightforward. Assume first that $\rho \equiv 0$, and then, as in §9.3, solve six separate problems, in each of which only one of the f_i is not identically zero. The sum of these six solutions of the Laplace equation is also a solution by the linearity of the equation and cleary satisfies (18)–(23). Then, if we find a solution of the Poisson equation corresponding to zero boundary values—this will be outlined below—it can be added to the sum of the previous six solutions of the Laplace equation to yield a function that satisfies (17)–(23). Finally, if u and v are solutions of (17)–(23) corresponding to sets of boundary value functions f_i and g_i, then $u - v$ satisfies the Laplace equation with boundary value functions $f_i - g_i$. If the electrostatic potential is continuous all around the box, the maximum principle can be used as above to show that the problem is well posed in the sense of Hadamard.

It remains to find a solution of the Poisson equation corresponding to zero boundary values. The fastest way to do this can be sketched as follows. Each term of the triple series

$$\sum_{m,n,p=1}^{\infty} C_{mnp} \sin \frac{m\pi}{a} x \, \sin \frac{n\pi}{b} y \, \sin \frac{p\pi}{c} z$$

satisfies the zero boundary conditions. If we find constants C_{mnp} such that its sum defines a function $u: D \to \mathbb{R}$ of class C^2 that can be differentiated term by term as necessary, then substitution in the Poisson equation gives

$$\sum_{m,n,p=1}^{\infty} C_{mnp}\left(\frac{m^2}{a^2} + \frac{n^2}{b^2} + \frac{p^2}{c^2}\right)\pi^2 \sin\frac{m\pi}{a}x \sin\frac{n\pi}{b}y \sin\frac{p\pi}{c}z = -\frac{\rho(x,y,z)}{\epsilon}.$$

Now, it can be shown that, if the extension of $-\rho/\epsilon$ to $\mathbb{R}^3$ that is odd and periodic of period $2a$ in x, odd and periodic of period $2b$ in y, and odd and periodic of period $2c$ in z is sufficiently smooth, then it is the sum of a uniformly convergent triple Fourier series

$$\sum_{m,n,p=1}^{\infty} d_{mnp} \sin\frac{m\pi}{a}x \sin\frac{n\pi}{b}y \sin\frac{p\pi}{c}z = -\frac{\rho(x,y,z)}{\epsilon},$$

where the definition of d_{mnp} is left to the reader. Comparison of the last two equations gives

$$C_{mnp} = \frac{d_{mnp}}{\left(\dfrac{m^2}{a^2} + \dfrac{n^2}{b^2} + \dfrac{p^2}{c^2}\right)\pi^2},$$

and this leads to the formal solution

$$u(x,y,z) = \sum_{m,n,p=1}^{\infty} \frac{d_{mnp}}{\left(\dfrac{m^2}{a^2} + \dfrac{n^2}{b^2} + \dfrac{p^2}{c^2}\right)\pi^2} \sin\frac{m\pi}{a}x \sin\frac{n\pi}{b}y \sin\frac{p\pi}{c}z$$

of the Poisson equation with zero boundary conditions. It should be verified for each particular choice of the function ρ.

§9.5   The Multiple Fourier Transform Method

It is much simpler to present, or at least to outline without proofs, the multiple Fourier transform method than multiple Fourier series, and we shall consider the case of arbitrary dimension n. If $f : \mathbb{R}^n \to \mathbb{R}$ is an integrable function and if $x = (x_1, \ldots, x_n)$ and $\omega = (\omega_1, \ldots, \omega_n)$ are in $\mathbb{R}^n$, we define the *Fourier transform* $\hat{f}$ of f by

$$\hat{f}(\omega) = \frac{1}{(2\pi)^n} \int_{\mathbb{R}^n} f(x) e^{-i\langle \omega, x \rangle} \, dx,$$

where

$$\langle \omega, x \rangle = \sum_{j=1}^{n} \omega_j x_j.$$

Example 9.3 If $x = (x_1, \ldots, x_n)$ is in $\mathbb{R}^n$ and if $\|x\|^2 = \langle x, x \rangle$, find the Fourier transform of the function $g : \mathbb{R}^n \to \mathbb{R}$ defined, for each fixed value of $t > 0$, by

$$g(x) = \left(\frac{\pi}{kt}\right)^{n/2} e^{-\|x\|^2/4kt}.$$

Using the Fourier transform of

$$\sqrt{\frac{\pi}{kt}}\, e^{-x_j^2/4kt},$$

obtained in Example 7.7, we have

$$
\begin{aligned}
\hat{g}(\omega) &= \frac{1}{(2\pi)^n} \int_{\mathbb{R}^n} \left(\frac{\pi}{kt}\right)^{n/2} e^{-\|x\|^2/4kt}\, e^{-i\langle \omega, x\rangle}\, dx \\
&= \frac{1}{(2\pi)^n} \int_{\mathbb{R}^n} \left(\prod_{j=1}^{n} \sqrt{\frac{\pi}{kt}}\, e^{-x_j^2/4kt}\, e^{-i\omega_j x_j}\right) dx \\
&= \prod_{j=1}^{n} \left(\frac{1}{2\pi} \int_{-\infty}^{\infty} \sqrt{\frac{\pi}{kt}}\, e^{-x_j^2/4kt}\, e^{-i\omega_j x_j}\, dx_j\right) \\
&= \prod_{j=1}^{n} e^{-k\omega_j^2 t} \\
&= e^{-k\|\omega\|^2 t}.
\end{aligned}
$$

It should be clear that the Fourier transform is a linear operation, that is, that, if f and g have transforms $\hat{f}$ and $\hat{g}$ and if c is a constant, then the transform of $cf + g$ is $c\hat{f} + \hat{g}$. If f and $\hat{f}$ are integrable and continuous, then it can be shown that the inversion formula

$$(24) \qquad\qquad f(x) = \int_{\mathbb{R}^n} \hat{f}(\omega) e^{i\langle \omega, x\rangle}\, d\omega$$

holds. Furthermore, if the partial derivative f_{x_j} is continuous and integrable,

$$(25) \qquad\qquad \widehat{f_{x_j}} = i\omega_j \hat{f},$$

and if the convolution $f * g$ is defined by

$$f * g(x) = \int_{\mathbb{R}^n} f(s) g(x - s)\, ds$$

its Fourier transform is

$$(26) \qquad\qquad \widehat{f * g} = (2\pi)^n \hat{f}\hat{g}.$$

As we recall from the one-dimensional case, it is not essential in applications of the Fourier transform method to boundary value problems to provide proofs of the preceding facts or to state precise conditions for their validity, since the ultimate success of the method is determined by verification of the solution *a posteriori*. For the same reason, any specific integrals over $\mathbb{R}^n$ that appear in applying the Fourier transform method can be evaluated as iterated integrals in any order that seems convenient.

Example 9.4 If the temperature in $\mathbb{R}^3$ satisfies the three-dimensional heat equation

$$u_t = k(u_{xx} + u_{yy} + u_{zz})$$

for $t > 0$ and if the initial temperature distribution is given by

$$u(x, y, z, 0) = f(x, y, z),$$

find the temperature for $t > 0$.

We start by transforming the the heat equation using (25),

$$\hat{u}_t(\omega, t) = k[(i\omega_1)^2 \hat{u}(\omega, t) + (i\omega_2)^2 \hat{u}(\omega, t) + (i\omega_3)^2 \hat{u}(\omega, t)] = -k\|\omega\|^2 \hat{u}(\omega, t).$$

Now, if we assume that

$$\hat{u}(\omega, t) = \frac{1}{(2\pi)^3} \int_{-\infty}^{\infty} \int_{-\infty}^{\infty} \int_{-\infty}^{\infty} u(x, t) e^{-i(\omega_1 x + \omega_2 y + \omega_3 z)} \, dx \, dy \, dz$$

can be differentiated with respect to t under the integral sign, we obtain

$$\hat{u}_t(\omega, t) = \frac{1}{(2\pi)^3} \int_{-\infty}^{\infty} \int_{-\infty}^{\infty} \int_{-\infty}^{\infty} u_t(x, t) e^{-i(\omega_1 x + \omega_2 y + \omega_3 z)} \, dx \, dy \, dz = \widehat{u_t}(\omega, t).$$

This leads to the transformed equation

$$\hat{u}_t(\omega, t) = -k\|\omega\|^2 \hat{u}(\omega, t),$$

an ordinary differential equation in t for each fixed ω, whose solution corresponding to the initial condition $\hat{u}(\omega, 0) = \hat{f}(\omega)$ is

$$\hat{u}(\omega, t) = \hat{u}(\omega, 0) e^{-k\|\omega\|^2 t} = \hat{f}(\omega)\hat{g}(\omega),$$

where g is the function of Example 9.3. From (26) we conjecture that

$$\hat{u}(\omega, t) = \frac{1}{(2\pi)^3} \widehat{f * g}(\omega)$$

and then, if we define $r = (x, y, z)$, that

$$u(r, t) = \frac{1}{(2\pi)^3} f * g(r)$$

$$= \frac{1}{(2\pi)^3} \int_{\mathbb{R}^3} f(s) \left(\frac{\pi}{kt}\right)^{3/2} e^{-\|r-s\|^2/4kt} \, ds$$

$$= \frac{1}{(4\pi kt)^{3/2}} \int_{-\infty}^{\infty} \int_{-\infty}^{\infty} \int_{-\infty}^{\infty} f(\xi, \eta, \zeta) e^{-[(x-\xi)^2 + (y-\eta)^2 + (z-\zeta)^2]/4kt} \, d\xi \, d\eta \, d\zeta.$$

This solution was originally found by Laplace. Its verification is similar to that in Example 7.7.

The result of Example 9.4 can also be used to solve problems involving an equation of heat propagation that includes a time-dependent term of internal heat generation. The solution is based on the following result of Jean Marie Constant Duhamel (1797–1883), then a *répétiteur* at the *École Polytechnique*, who presented it to the *Académie des Sciences* in 1830 and published it in 1833. For convenience, we state it for $n = 3$ but it should be clear that both the statement and its proof are valid for arbitrary dimension.

Theorem 9.8 (Duhamel's Principle for the Heat Equation) *Let D be a region in $\mathbb{R}^3$ (by this we mean an open, connected set with no holes), and for each function $q : D \times [0, \infty) \to \mathbb{R}$ and each fixed value of τ define a function $q_\tau : D \to \mathbb{R}$ by $q_\tau(x, y, z) = q(x, y, z, \tau)$. If u_{q_τ} is a solution of the boundary value problem*

$$
\begin{aligned}
u_t &= k\,(u_{xx} + u_{yy} + u_{zz}) & (x, y, z) \text{ in } D, &\quad t > 0 \\
u(x, y, z, t) &= 0 & (x, y, z) \text{ in } \partial D, &\quad t \geq 0 \\
u(x, y, z, 0) &= q_\tau(x, y, z) & (x, y, z) \text{ in } D,
\end{aligned}
$$

where the last condition is present only if D has a boundary ∂D, then the boundary value problem

$$
\begin{aligned}
u_t &= k\,(u_{xx} + u_{yy} + u_{zz}) + q(x, y, z, t) & (x, y, z) \text{ in } D, &\quad t > 0 \\
u(x, y, z, t) &= 0 & (x, y, z) \text{ in } \partial D, &\quad t \geq 0 \\
u(x, y, z, 0) &= 0 & (x, y, z) \text{ in } D
\end{aligned}
$$

has the solution

$$
u(x, y, z, t) = \int_0^t u_{q_\tau}(x, y, z, t - \tau)\, d\tau.
$$

Proof. It is clear that $u(x, y, z, 0) = 0$ and also that, if $u_{q_\tau} \equiv 0$ on ∂D, then its integral $u \equiv 0$ on ∂D. Also,

$$
\begin{aligned}
u_t(x, y, z, t) &= u_{q_t}(x, y, z, 0) + \int_0^t \frac{\partial}{\partial t} u_{q_\tau}(x, y, z, t - \tau)\, d\tau \\[2mm]
&= q_t(x, y, z) + \int_0^t k\left(\frac{\partial^2}{\partial x^2} + \frac{\partial^2}{\partial y^2} + \frac{\partial^2}{\partial z^2}\right) u_{q_\tau}(x, y, z, t - \tau)\, d\tau \\[2mm]
&= q(x, y, z, t) + k\left(\frac{\partial^2}{\partial x^2} + \frac{\partial^2}{\partial y^2} + \frac{\partial^2}{\partial z^2}\right) \int_0^t u_{q_\tau}(x, y, z, t - \tau)\, d\tau \\[2mm]
&= q(x, y, z, t) + k\,(u_{xx} + u_{yy} + u_{zz}),
\end{aligned}
$$

Q.E.D.

Example 9.5 Using Duhamel's principle and the solution of Example 9.4 we see that, if $r = (x, y, z)$, the boundary value problem

$$
\begin{aligned}
u_t &= k\,(u_{xx} + u_{yy} + u_{zz}) + q(r, t) \\
u(r, 0) &= 0
\end{aligned}
$$

has the solution

$$u(r, t) = \int_0^t \int_{\mathbb{R}^3} \frac{1}{[4\pi k(t - \tau)]^{3/2}} \, e^{-\|r-s\|^2/4k(t-\tau)} q(s, \tau) \, ds \, d\tau.$$

It can also be obtained by direct application of the Fourier transform method.

While Duhamel's principle is applicable only if $u(x, y, z, 0) = 0$, the solution of a problem involving the more general initial condition $u(x, y, z, 0) = f(x, y, z)$ can be obtained as the superposition of the solutions of two problems, one for heat equation and one with zero initial condition (see Exercise 5.3).

Example 9.6 From 1808 until 1818 Poisson carried out a certain amount of work on the propagation of waves, and we have already mentioned that this led him to rediscover the Fourier transform and to discover convolution and the Poisson integral formula for the half-plane. His main result in wave propagation was the solution of the boundary value problem

$$u_{tt} = c^2(u_{xx} + u_{yy} + u_{zz})$$
$$u(x, y, z, 0) = f(x, y, z)$$
$$u_t(x, y, z, 0) = g(x, y, z)$$

in the entire space $\mathbb{R}^3$. The three-dimensional wave equation describes first of all the propagation of acoustic waves in space but is also satisfied, as we know today, by the electric and magnetic fields of electromagnetic theory, where c is then the speed of light.

Assuming that $\hat{u}_{tt}(\omega, t) = \widehat{u_{tt}}(\omega, t)$, which is true if differentiation under the integral sign is permissible, the given problem is transformed into

$$\hat{u}_{tt} = c^2[(i\omega_1)^2\hat{u} + (i\omega_2)^2\hat{u} + (i\omega_3)^2\hat{u}] = -\|\omega\|^2 c^2 \hat{u}$$
$$\hat{u}(\omega, 0) = \hat{f}(\omega)$$
$$\hat{u}_t(\omega, 0) = \hat{g}(\omega).$$

The first of these equations is an ordinary differential equation in t for each fixed ω. Its general solution is

$$\hat{u}(\omega, t) = A \cos \|\omega\| ct + B \sin \|\omega\| ct,$$

where A and B are constants, and the initial conditions give

$$A = \hat{f}(\omega) \qquad \text{and} \qquad B\|\omega\|c = \hat{g}(\omega),$$

so that

$$\hat{u}(\omega, t) = \hat{f}(\omega) \cos \|\omega\| ct + \hat{g}(\omega) \frac{\sin \|\omega\| ct}{\|\omega\| c}.$$

Then, under appropriate conditions and if we define $r = (x, y, z)$, the inversion formula (24) provides the solution

$$u(r, t) = \int_{\mathbb{R}^3} \left[\hat{f}(\omega) \cos \|\omega\| ct + \hat{g}(\omega) \frac{\sin \|\omega\| ct}{\|\omega\| c} \right] e^{i \langle \omega, r \rangle} \, d\omega.$$

In order to write this solution in terms of f and g rather than their transforms, let s and ω be vectors in $\mathbb{R}^3$, let ϕ be the angle between them, choose a coordinate system as shown in Figure 9.1, and let S_R be a sphere of radius $R = \|s\|$ centered at the origin. Then,

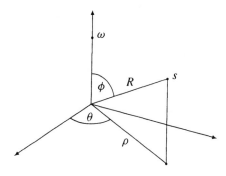

Figure 9.1

using spherical coordinates R, θ, and ϕ, we evaluate the surface integral

$$\int_{S_R} e^{i \langle \omega, s \rangle} \, dS_s = \int_{S_R} e^{i \|\omega\| \|s\| \cos \phi} \, dS_s$$

$$= \int_0^\pi \left(\int_0^{2\pi} e^{i \|\omega\| R \cos \phi} R^2 \sin \phi \, d\theta \right) d\phi$$

$$= 2\pi R^2 \int_0^\pi e^{i \|\omega\| R \cos \phi} \sin \phi \, d\phi$$

$$= 4\pi R^2 \frac{\sin \|\omega\| R}{\|\omega\| R},$$

from which, putting $R = ct$, we obtain

$$\frac{\sin \|\omega\| ct}{\|\omega\| c} = \frac{1}{4\pi c^2 t} \int_{S_{ct}} e^{i \langle \omega, s \rangle} \, dS_s.$$

The notation dS_s indicates that this is a surface integral and that the variable is s. Then, if the change of order of integration performed below is permissible, we have

$$
\int_{\mathbb{R}^3} \hat{g}(\omega) \frac{\sin \|\omega\| ct}{\|\omega\| c} e^{i\langle \omega, r \rangle} d\omega = \int_{\mathbb{R}^3} \hat{g}(\omega) \left(\frac{1}{4\pi c^2 t} \int_{S_{ct}} e^{i\langle \omega, s \rangle} dS_s \right) e^{i\langle \omega, r \rangle} d\omega
$$

$$
= \frac{1}{4\pi c^2 t} \int_{S_{ct}} \left(\int_{\mathbb{R}^3} \hat{g}(\omega) e^{i\langle \omega, r+s \rangle} d\omega \right) dS_s
$$

$$
= \frac{1}{4\pi c^2 t} \int_{S_{ct}} g(r+s) \, dS_s.
$$

Similarly,

$$
\int_{\mathbb{R}^3} \hat{f}(\omega) \cos \|\omega\| ct \, e^{i\langle \omega, r \rangle} d\omega = \int_{\mathbb{R}^3} \hat{f}(\omega) \frac{d}{dt} \frac{\sin \|\omega\| ct}{\|\omega\| c} e^{i\langle \omega, r \rangle} d\omega
$$

$$
= \frac{d}{dt} \left[\frac{1}{4\pi c^2 t} \int_{S_{ct}} f(r+s) \, dS_s \right],
$$

and then

$$
(27) \qquad u(r, t) = \frac{d}{dt} \left[\frac{1}{4\pi c^2 t} \int_{S_{ct}} f(r+s) \, dS_s \right] + \frac{1}{4\pi c^2 t} \int_{S_{ct}} g(r+s) \, dS_s.
$$

This solution can be rewritten in several different ways. The first, using spherical coordinates, that is, putting $s = (R \sin \phi \cos \theta, R \sin \phi \sin \theta, R \cos \phi)$ with $R = ct$, is the form originally given by Poisson in 1818:[1]

$$
u(x, y, z, t) = \frac{1}{4\pi} \frac{d}{dt} \left[t \int_0^\pi \left(\int_0^{2\pi} f(x + ct \sin \phi \cos \theta, y + ct \sin \phi \sin \theta, \right. \right.
$$

$$
\left. \left. z + ct \cos \phi) \sin \phi \, d\theta \right) d\phi \right]
$$

$$
+ \frac{t}{4\pi} \int_0^\pi \left(\int_0^{2\pi} g(x + ct \sin \phi \cos \theta, y + ct \sin \phi \sin \theta, z + ct \cos \phi) \sin \phi \, d\theta \right) d\phi.
$$

This formula can now be rewritten in a more compact form if we note that the last two integrals are surface integrals over the sphere S_1 of radius one centered at the origin. Then, using the outer normal unit vector $n = (R \sin \phi \cos \theta, R \sin \phi \sin \theta, R \cos \phi)$ as a new variable on S_1 we obtain

$$
u(r, t) = \frac{d}{dt} \left[\frac{t}{4\pi} \int_{S_1} f(r + ctn) \, dS_n \right] + \frac{t}{4\pi} \int_{S_1} g(r + ctn) \, dS_n.
$$

---

[1] It is, however, known as *Kirchhoff's formula* because in 1882 Kirchhoff proved a similar but more general formula for an arbitrary region.

These last two forms are more convenient than (27) for the purpose of verifying that any of them is a solution of our problem because they depend on t only through the integrands, and this makes it easier to differentiate them with respect to t. It is enough to assume that f and g are of class C^2, so that differentiation under the integral sign is possible, while $\hat{f}$ and $\hat{g}$ do not even have to exist (Exercise 5.4).

To obtain the last form define $\sigma = r + s$, so that $\|s\| = ct$ becomes $\|\sigma - r\| = ct$. Then, if $S_{ct}(r)$ denotes the sphere with center r and radius ct, the integral

$$\frac{1}{4\pi c^2 t^2} \int_{S_{ct}} f(r+s)\, dS_s = \frac{1}{4\pi (ct)^2} \int_{S_{ct}(r)} f(\sigma)\, dS_\sigma$$

is the average value of f over $S_{ct}(r)$, which we shall denote by $A_f(ct, r)$, and similarly for g. Therefore, (27) becomes

$$u(r, t) = \frac{d}{dt}[t A_f(ct, r)] + t A_g(ct, r).$$

This is known as *Poisson's formula*, and it emphasizes the important fact that $u(r, t)$ depends only on the values of f and g on the sphere $S_{ct}(r)$, which is called the *domain of dependence* of the point (r, t). Assume now that a sound is created at a point r_0 of $\mathbb{R}^3$ at $t = 0$, say, a gong is struck at this location. Then f and g are concentrated at r_0 and vanish elsewhere. If r is any other point in $\mathbb{R}^3$, $S_{ct}(r)$ contains r_0 only if t is such that $ct = \|r - r_0\|$, and then $A_f(ct, r) = A_g(ct, r) = 0$ except for that value of t. That is, an observer at r will hear the sound exactly at a time $t > 0$ such that $ct = \|r - r_0\|$. In fact, the sound reaches all the points on the sphere $S_{ct}(r_0)$ with center r_0 and radius ct at this same time, and the union of these spheres for all values of $t > 0$ is called the *domain of influence* of the point r_0. This is the same type of situation already described in Chapter 5 for the plucked string. We saw there that a disturbance originating at a point x_0 was felt at two other points, $x = x_0 \pm ct$ at time t. These points form a zero-dimensional sphere of center x_0 and radius ct, and the disturbance reaches them along the characteristics, that is, the lines $x = x_0 \pm ct$. This is also true in the present case, but now the *characteristics* are the surfaces $\|r - r_0\| = ct$, four-dimensional cones formed by the lines $r = r_0 + vt$ with $\|v\| = ct$.

We have seen that the sound produced at a certain instant at one point of $\mathbb{R}^3$ will be heard a little later, but also at a very definite instant, at any other point in space. In the same manner, a flash of light originating at a given point at a given time will reach any other point in $\mathbb{R}^3$ at a specific instant, no sooner and no later. The fact that waves in three-dimensional space propagate exactly with speed c is known in physics as *Huygens' Principle*, after Christiaan Huygens (1629–1695) of The Hague, the originator of the wave theory of light, who first stated it in a more detailed form in his *Traité de la lumière* of 1690. Basically, it means that in three-dimensional space we hear sharp sounds and see sharp pictures. But then, of course, this is our everyday experience and hardly seems worth mentioning. However, such daily experience would never take place in a two-dimensional universe. Our own senses can confirm this statement at once. A pebble dropped in a pond generates two-dimensional circular waves that move away from the point of the original disturbance. These waves reach any other point at a certain specific instant but they do not disappear after that moment. Rather, they linger, theoretically forever in the absence of friction, in a wake of ripples that do not go away. The same would be true of sound or light, and this can be demonstrated mathematically in a rather simple manner: use the solution found above but assume that the functions u, f, and g are independent of z. This is called *the method of descent*, and the details are provided in Exercise 5.9. An examination of the solution so obtained shows that the domain of dependence of a point (r, t) is a whole disc. Sounds or light emanating from all points of this disc would arrive at the point r at the same time. The conclusion is that a two-dimensional universe would be a universe of blurred sounds and sights. Life in *Flatland* would lose its edge.

Although our derivation of Poisson's formula using spherical coordinates is three dimensional, it can be shown using more advanced methods[1] that the same solution is valid in $\mathbb{R}^n$ if $n > 3$ is odd. Thus, Huygens' principle holds in any odd-dimensional universe of dimension three or higher, but the method of descent would show that it is not valid in even-dimensional universes or in a one-dimensional universe (see Exercise 5.10 for the last assertion). We are privileged to live in the lowest-dimensional universe in which sharp sounds and sights are possible.

There is a version of Duhamel's principle for the wave equation that would allow us to deal with problems involving an equation of wave propagation that includes a term representing time-dependent external forces (see Exercises 5.11 to 5.13).

Example 9.7 We shall now solve the Dirichlet problem for the upper half space of $\mathbb{R}^3$, that is,

$$u_{xx} + u_{yy} + u_{zz} = 0 \qquad z > 0$$
$$u(x, y, 0) = f(x, y),$$

where we assume that f is integrable, bounded, and continuous on $\mathbb{R}^2$. For each fixed z we shall denote by $\hat{u}$ the double Fourier transform of u with respect to (x, y). Assuming that $\hat{u}_{zz}(\omega, z) = \widehat{u_{zz}}(\omega, z)$, which is true if differentiation under the integral sign is permissible, the transformed problem is

$$(i\omega_1)^2 \hat{u} + (i\omega_2)^2 \hat{u} + \hat{u}_{zz} = -\|\omega\|^2 \hat{u} + \hat{u}_{zz} = 0$$
$$\hat{u}(\omega, 0) = \hat{f}(\omega)$$

The first of these equations is an ordinary differential equation in z for each fixed ω. Its general solution is

$$\hat{u}(\omega, z) = A e^{\|\omega\|z} + B e^{-\|\omega\|z},$$

where A and B are constants. Since the Fourier transform of an integrable function is necessarily bounded according to the definition, we choose $A = 0$, and then the boundary condition gives $B(\omega) = \hat{f}(\omega)$, so that

$$\hat{u}(\omega, z) = \hat{f}(\omega) e^{-\|\omega\|z}.$$

---

[1]See, for instance, P. R. Garabedian, *Partial Differential Equations*, John Wiley & Sons, New York, 1964, pages 191–202.

If $P^z(x, y)$ is the inverse Fourier transform of $e^{-\|\omega\|z}$ and if we define $r = (x, y)$, then we conjecture from (26) that

$$u(r, z) = \frac{1}{(2\pi)^2} f * P^z(r).$$

The evaluation of the inverse Fourier transform of $e^{-\|\omega\|z}$ is not straightforward. It will be computed indirectly from an integral representation of $e^{-\|\omega\|z}$, which, in turn, will be obtained from an unlikely source. Example 8.13 shows that for $s > 0$

$$\frac{1}{a} e^{-2a\sqrt{s}} = \int_0^\infty \frac{1}{t\sqrt{\pi t}} e^{-a^2/t} e^{-st} \, dt,$$

and then this equation can be differentiated with respect to s using Theorem 8.4, yielding

$$\frac{e^{-2a\sqrt{s}}}{\sqrt{s}} = \int_0^\infty \frac{1}{\sqrt{\pi t}} e^{-a^2/t} e^{-st} \, dt.$$

Putting $s = 1$ and $a = \|\omega\|z/2$ we obtain the desired integral representation

$$e^{-\|\omega\|z} = \int_0^\infty \frac{1}{\sqrt{\pi t}} e^{-\|\omega\|^2 z^2/4t} e^{-t} \, dt.$$

Taking this to the inversion formula (24) and reversing the order of integration we have

$$(28) \quad P^z(r) = \int_{\mathbb{R}^2} e^{-\|\omega\|z} e^{i\langle \omega, r\rangle} \, d\omega = \int_0^\infty e^{-t} \frac{1}{\sqrt{\pi t}} \left(\int_{\mathbb{R}^2} e^{-\|\omega\|^2 z^2/4t} e^{i\langle \omega, r\rangle} \, d\omega \right) dt.$$

The inner integral is now evaluated using the result of Example 9.3 for $n = 2$ but replacing x with ω, ω with $-r$, and t with t/kz^2, that is,

$$e^{-\|r\|^2 t/z^2} = \frac{1}{4\pi^2} \int_{\mathbb{R}^2} \frac{\pi z^2}{t} e^{-\|\omega\|^2 z^2/4t} e^{i\langle r, \omega\rangle} \, d\omega = \frac{z^2}{4\pi t} \int_{\mathbb{R}^2} e^{-\|\omega\|^2 z^2/4t} e^{i\langle \omega, r\rangle} \, d\omega.$$

Using this equation to replace the inner integral in (28) and then making the substitution

$$\tau = \frac{\|r\|^2 + z^2}{z^2} t$$

we obtain

$$P^z(r) = \int_0^\infty e^{-t} \frac{1}{\sqrt{\pi t}} \frac{4\pi t}{z^2} e^{-\|r\|^2 t/z^2} \, dt$$

$$= \frac{4\sqrt{\pi}}{z^2} \int_0^\infty \sqrt{t} \, e^{-(\|r\|^2 + z^2)t/z^2} \, dt$$

$$= \frac{4\sqrt{\pi}}{z^2} \int_0^\infty \frac{z\sqrt{\tau}}{\sqrt{\|r\|^2 + z^2}} e^{-\tau} \frac{z^2}{\|r\|^2 + z^2} \, d\tau$$

$$= \frac{4\sqrt{\pi}\, z}{(\|r\|^2 + z^2)^{3/2}} \int_0^\infty \sqrt{\tau}\, e^{-\tau} \, d\tau.$$

Finally, integration by parts reduces the last integral to

$$\int_0^\infty \frac{1}{2\sqrt{\tau}} e^{-\tau}\, d\tau = \int_0^\infty e^{-v^2}\, dv = \frac{\sqrt{\pi}}{2}.$$

Therefore,

$$P^z(r) = \frac{2\pi z}{(\|r\|^2 + z^2)^{3/2}},$$

and the solution to our problem is

$$u(r, z) = \frac{1}{(2\pi)^2}\, f * P^z(r) = \frac{1}{2\pi}\int_{\mathbb{R}^2} f(s)\frac{z}{(\|r - s\|^2 + z^2)^{3/2}}\, ds$$

or, returning to cartesian coordinates,

$$u(x, y, z) = \frac{z}{2\pi}\int_{-\infty}^\infty \int_{-\infty}^\infty \frac{f(\xi, \eta)}{[(x - \xi)^2 + (y - \eta)^2 + z^2]^{3/2}}\, d\xi\, d\eta.$$

This is called *Poisson's integral formula for the upper half-space*. It is verified just as its two-dimensional counterpart in Theorems 7.7 and 7.8.

§9.6 The Double Laplace Transform Method

It is easy to define the multiple Laplace transform. However, its theoretical treatment is complicated by the presence of several complex variables, especially in those aspects involving the theory of residues, while its practical use is limited because it requires knowledge of many initial and boundary conditions. For these reasons, we shall restrict ourselves to the two-dimensional case, omitting a theoretical discussion.[1]

If f is a real-valued function defined on the first quadrant of the xy-plane and if p and s are complex variables, we define the double Laplace transform of f to be

$$\mathcal{L}[f](p, s) = F(p, s) = \int_0^\infty \int_0^\infty f(x, y)e^{-(px+sy)}\, dx\, dy.$$

We shall need the double Laplace transforms of certains types of function.

Example 9.8 Find the double Laplace transform of $f(x, y) = g(x)h(y)$, where g and h are in $\mathcal{L}$ in the sense of Definition 8.2. We have

$$(29) \qquad F(p, s) = \int_0^\infty \int_0^\infty g(x)h(y)e^{-(px+sy)}\, dx\, dy$$

$$= \left(\int_0^\infty g(x)e^{-px}\, dx\right)\left(\int_0^\infty h(y)e^{-sy}\, dy\right)$$

$$= G(p)H(s).$$

---

[1]For a more extensive treatment see, for instance, V. A. Ditkin and A. P. Prudnikov, *Operational Calculus in Two Variables and Its Applications*, Pergamon Press, New York, 1962.

Example 9.9 Find the double Laplace transform of $f(x, y) = g(x + y)$, where g is in $\mathcal{L}$ in the sense of Definition 8.2; that is, f depends on x and y through their sum $x + y$. To evaluate the integral giving the transform we shall use the change of variables $x = u - v$ and $y = v$. Then the inequalities $x > 0$ and $y > 0$ are equivalent to $u > v$ and $v > 0$, and the function $h(u, v) = (u - v, v)$ maps the set $S = \{(u, v) \in \mathbb{R}^2 : u > v, v > 0\}$ in a one-to-one manner onto the first quadrant. Since the Jacobian of h is

$$J(h) = \begin{vmatrix} 1 & -1 \\ 0 & 1 \end{vmatrix} = 1,$$

the change of variables formula

$$\int_{h(S)} f = \int_S (f \circ h)|J(h)|,$$

where the circle denotes composition of functions, gives

(30)
$$F(p, s) = \int_0^\infty \int_0^\infty g(x + y)e^{-(px+sy)}\, dx\, dy$$

$$= \int_0^\infty \left(\int_0^u g(u)e^{-[p(u-v)+sv]}\, dv \right) du$$

$$= \int_0^\infty g(u) \left(\int_0^u e^{(p-s)v}\, dv \right) e^{-pu}\, du$$

$$= \int_0^\infty g(u) \frac{e^{(p-s)u} - 1}{p - s} e^{-pu}\, du$$

$$= \frac{1}{p - s} \int_0^\infty g(u)(e^{-su} - e^{-pu})\, du$$

$$= \frac{G(s) - G(p)}{p - s},$$

where G the Laplace transform of g.

Example 9.10 Find the double Laplace transform of $f(x, y) = g(x - y)$, where g is either even or odd and its restriction to $(0, \infty)$ is in $\mathcal{L}$ in the sense of Definition 8.2. This example may look like the previous one, but, as we shall see, there is a very relevant difference in the solution. Now we shall use the change of variables $x = u + v$ and $y = v$. Then the inequalities $x > 0$ and $y > 0$ are equivalent to $u > -v$ and $v > 0$, and the function $h(u, v) = (u + v, v)$ maps the set $S = \{(u, v) \in \mathbb{R}^2 : u > -v, v > 0\}$ in a one-to-one manner onto the first quadrant. The reader should construct this set graphically to better understand the limits of integration used below. Since the Jacobian of h is

$$J(h) = \begin{vmatrix} 1 & 1 \\ 0 & 1 \end{vmatrix} = 1,$$

then, for $\text{Re}(p + s) > 0$, the change of variables formula gives

$$F(p, s) = \int_0^\infty \int_0^\infty g(x - y)e^{-(px+sy)} \, dx \, dy$$

$$= \int_{-\infty}^0 \left(\int_{-u}^\infty g(u)e^{-[p(u+v)+sv]} \, dv \right) du + \int_0^\infty \left(\int_0^\infty g(u)e^{-[p(u+v)+sv]} \, dv \right) du$$

$$= \int_{-\infty}^0 g(u) \frac{e^{(p+s)u}}{p+s} e^{-pu} \, du + \int_0^\infty g(u) \frac{1}{p+s} e^{-pu} \, du$$

$$= \frac{1}{p+s} \left(\int_{-\infty}^0 g(u)e^{su} \, du + \int_0^\infty g(u)e^{-pu} \, du \right)$$

$$= \frac{1}{p+s} \left(\int_0^\infty g(-\tau)e^{-s\tau} \, d\tau + G(p) \right).$$

If g is odd, then $g(-\tau) = -g(\tau)$ and the first integral equals $-G(s)$. However, if g is even, then $g(-\tau) = g(\tau)$ and the first integral equals $G(s)$. In conclusion,

$$(31) \qquad F(p, s) = \begin{cases} \dfrac{G(p) - G(s)}{p+s} & \text{if } g \text{ is odd} \\[3mm] \dfrac{G(p) + G(s)}{p+s} & \text{if } g \text{ is even.} \end{cases}$$

It is easy to see that, if f and g have double Laplace transforms F and G, respectively, and if c is an arbitrary real or complex constant, then the double Laplace transform of $cf + g$ is $cF + G$. Furthermore, if a and b are positive real numbers, we have the following shifting property:

$$(32) \qquad \mathcal{L}[f(ax, by)](p, s) = \int_0^\infty \int_0^\infty f(ax, by)e^{-(px+sy)} \, dx \, dy$$

$$= \int_0^\infty \int_0^\infty f(u, v)e^{-\left(\frac{p}{a}u + \frac{s}{b}v\right)} \frac{1}{ab} \, du \, dv$$

$$= \frac{1}{ab} F\left(\frac{p}{a}, \frac{s}{b} \right).$$

If F_{10}, F_{01}, F_{01}^x, and F_{10}^y denote the one-dimensional Laplace transforms of the functions $f(\cdot, 0)$, $f(0, \cdot)$, $f_x(0, \cdot)$, and $f_y(\cdot, 0)$, respectively, then we can formally establish the following equations for the transforms of derivatives. If $\text{Re}(p) > 0$,

$$(33) \quad \mathcal{L}[f_x](p, s) = \int_0^\infty \int_0^\infty f_x(x, y)e^{-(px+sy)}\, dx\, dy$$

$$= \int_0^\infty \left(f(x, y)e^{-(px+sy)}\Big|_{x=0}^{x=\infty} + \int_0^\infty f(x, y)pe^{-(px+sy)}\, dx \right) dy$$

$$= p \int_0^\infty \int_0^\infty f(x, y)e^{-(px+sy)}\, dx\, dy - \int_0^\infty f(0, y)e^{-sy}\, dy$$

$$= pF(p, s) - F_{01}(s)$$

$$(34) \quad \mathcal{L}[f_{xx}](p, s) = \int_0^\infty \int_0^\infty f_{xx}(x, y)e^{-(px+sy)}\, dx\, dy$$

$$= \int_0^\infty \left(f_x(x, y)e^{-(px+sy)}\Big|_{x=0}^{x=\infty} + \int_0^\infty f_x(x, y)pe^{-(px+sy)}\, dx \right) dy$$

$$= p \int_0^\infty \int_0^\infty f_x(x, y)e^{-(px+sy)}\, dx\, dy - \int_0^\infty f_x(0, y)e^{-sy}\, dy$$

$$= p\mathcal{L}[f_x](p, s) - F_{01}^x(s),$$

$$= p^2 F(p, s) - pF_{01}(s) - F_{01}^x(s)$$

and, similarly,

$$(35) \qquad \mathcal{L}[f_y](p, s) = sF(p, s) - F_{10}(p)$$

$$(36) \qquad \mathcal{L}[f_{yy}](p, s) = s^2 F(p, s) - sF_{10}(p) - F_{10}^y(p)$$

$$(37) \qquad \mathcal{L}[f_{xy}](p, s) = psF(p, s) - pF_{10}(p) - sF_{01}(s) + f(0, 0).$$

Equations (33) and (35) allow us to compute, at least formally, the double Laplace transforms of certain convolution integrals. Indeed,

$$(38) \qquad \mathcal{L}\left[\int_0^x f(\sigma, y)g(x - \sigma)\, d\sigma \right](p, s) = F(p, s)G(p)$$

because (33) implies that

$$F(p, s)G(p) = \left(\int_0^\infty \int_0^\infty f(x, y)e^{-(px+sy)}\, dx\, dy \right)\left(\int_0^\infty g(u)e^{-pu}\, du \right)$$

$$= \int_0^\infty \left(\int_0^\infty f(x, y)e^{-px}\, dx \right) \left(\int_0^\infty g(u)e^{-pu}\, du \right) e^{-sy}\, dy$$

$$= \int_0^\infty F_{10}(p)G(p)e^{-sy}\, dy.$$

But, according to Theorem 8.5 on the transform of a one-dimensional convolution,

$$F_{10}(p)G(p) = \int_0^\infty \left(\int_0^x f(\sigma, y)g(x - \sigma)\, d\sigma \right) e^{-px}\, dx,$$

and then

$$F(p, s)G(p) = \int_0^\infty \int_0^\infty \left(\int_0^x f(\sigma, y)g(x - \sigma)\, d\sigma \right) e^{-(px+sy)}\, dx\, dy$$

$$= \mathcal{L}\left[\int_0^x f(\sigma, y)g(x - \sigma)\, d\sigma \right](p, s).$$

Similarly,

(39) $$\mathcal{L}\left[\int_0^y f(x, \eta)g(y - \eta)\, d\eta \right](p, s) = F(p, s)G(s).$$

After this development of the double Laplace transform we are ready for some applications.

Example 9.11 Solve the boundary value problem

$$
\begin{array}{ll}
u_{tt} = c^2 u_{xx} & x > 0, \quad t > 0 \\
u(0, t) = 0 & t \geq 0 \\
u(x, 0) = f(x) & x \geq 0 \\
u_t(x, 0) = g(x) & x \geq 0
\end{array}
$$

for a semi-infinite string that is tied down at $x = 0$ and whose initial position and velocity are known. Since the string is fixed at its endpoint, we assume that $f(0) = g(0) = 0$. Using (36) and (34) the wave equation is transformed into

$$s^2 U(p, s) - sU_{10}(p) - U_{10}'(p) = c^2[p^2 U(p, s) - pU_{01}(s) - U_{01}^x(s)],$$

and the initial and boundary conditions give $U_{10}(p) = F(p)$, $U_{10}'(p) = G(p)$, and $U_{01}(s) = 0$, but $U_{01}^x(s)$ is undetermined. From these values we obtain

$$U(p, s) = \frac{c^2 U_{01}^x(s) - sF(p) - G(p)}{c^2 p^2 - s^2},$$

and we see that, while the problem is well posed with the given initial and boundary conditions, the double Laplace transform method requires additional knowledge. In this particular case, it can be obtained indirectly as follows. It can be shown that, since

$U(p, s)$ is a Laplace transform, it must be defined for all values of p and s with sufficiently large real parts and in particular for $p = s/c$. Since the denominator vanishes for such values of p, the numerator must vanish too, and then

$$c^2 U_{01}^x(s) - s F(s/c) - G(s/c) = 0,$$

which gives

$$U(p, s) = \frac{s F(s/c) - s F(p) + G(s/c) - G(p)}{c^2 p^2 - s^2}$$

$$= \frac{1}{2} \left[\frac{1}{c} \frac{F(s/c) - F(p)}{p - s/c} + \frac{1}{c} \frac{F(p) - F(s/c)}{p + s/c} \right]$$

$$+ \frac{1}{2s} \left[\frac{1}{c} \frac{G(s/c) - G(p)}{p - s/c} + \frac{1}{c} \frac{G(p) - G(s/c)}{p + s/c} \right].$$

Now, since f and g are defined only for $x \geq 0$ and since $f(0) = g(0) = 0$, we can extend them to the entire real line as odd functions. Then, if these extensions are denoted by $\tilde{f}$ and $\tilde{g}$, (30), (31), and (32) show that the terms in the brackets are the double Laplace transforms of $\tilde{f}(x + ct) + \tilde{f}(x - ct)$ and $\tilde{g}(x + ct) + \tilde{g}(x - ct)$, respectively, and then, if we recall that the one-dimensional Laplace transform of the Heaviside function H is $1/s$ and that $H(t - \eta)$ has value 1 for $0 < \eta < t$, (39) shows that $U(p, s)$ is the double Laplace transform of

$$u(x, t) = \frac{1}{2} [\tilde{f}(x + ct) + \tilde{f}(x - ct)] + \frac{1}{2} \left(\int_0^t \tilde{g}(x + c\eta)\, d\eta + \int_0^t \tilde{g}(x - c\eta)\, d\eta \right)$$

$$= \frac{1}{2} [\tilde{f}(x + ct) + \tilde{f}(x - ct)] + \frac{1}{2c} \left(\int_x^{x+ct} \tilde{g}(\sigma)\, d\sigma - \int_x^{x-ct} \tilde{g}(\sigma)\, d\sigma \right)$$

$$= \frac{1}{2} [\tilde{f}(x + ct) + \tilde{f}(x - ct)] + \frac{1}{2c} \int_{x-ct}^{x+ct} \tilde{g}.$$

This solution has the same form as the one obtained in Theorem 5.3 for a finite string.

In solving the preceding problem the fact that $U_{01}^x(s)$ was undetermined represented only a temporary difficulty. However, in other cases the demand for a large number of initial and boundary conditions may render the double Laplace transform method inapplicable. This is certainly true in higher dimensions and even in two-dimensional problems involving the Laplace or the Poisson equations. For example, the Dirichlet problem for the first quadrant cannot be solved by this method (Exercise 6.6). On the other hand, it is precisely the presence of these additional initial and boundary conditions that allows us to solve new types of problems by the double Laplace transform method, as in the following simple case.

Example 9.12 Solve the boundary value problem

$$
\begin{aligned}
u_{xx} + u_{yy} &= 0 & x > 0, \quad y > 0 \\
u(x, 0) &= 0 & x \ge 0 \\
u(0, y) &= 0 & y \ge 0 \\
u_y(x, 0) &= x & x \ge 0 \\
u_x(0, y) &= y & x \ge 0.
\end{aligned}
$$

Using (34) and (36) the Laplace equation is transformed into

$$
p^2 U(p, s) - p U_{01}(s) - U_{01}^x(s) + s^2 U(p, s) - s U_{10}(p) - U_{10}^t(p) = 0,
$$

while the one-dimensional transform of Example 8.10 with $n = 1$ shows that the initial and boundary conditions become $U_{10}(p) = 0$, $U_{01}(s) = 0$, $U_{10}^t(p) = 1/p^2$, and $U_{01}^x(s) = 1/s^2$. We obtain

$$
U(p, s) = \frac{U_{01}^x(s) + U_{10}^t(p)}{p^2 + s^2} = \frac{\dfrac{1}{s^2} + \dfrac{1}{p^2}}{p^2 + s^2} = \frac{1}{p^2}\frac{1}{s^2},
$$

which, according to Example 9.8, is the double Laplace transform of $u(x, y) = xy$.

In order to solve more sophisticated boundary value problems by this method it would be necessary to evaluate a larger collection of double Laplace transforms.[1] Exercises 6.2 to 6.5 deal with some simple examples.

EXERCISES

1.1 Prove Gauss' corollary of the Ostrogradski-Gauss theorem (**Hint:** in the three-dimensional case remove a small sphere centered at P from the interior of S and apply the Ostrogradski-Gauss theorem to the remaining volume).

1.2 Assume that a certain material has heat capacity c and density ρ and that heat is generated internally at a rate q per unit volume. Use the divergence theorem of Ostrogradski-Gauss to derive the equation of heat propagation

$$
c\rho u_t = \kappa \operatorname{div}(\operatorname{grad} u) + q,
$$

where κ is a constant (**Hint:** in the three-dimensional case apply Fourier's law as follows: if S is a sphere interior to the solid body, the heat that enters S at any of its points P is proportional to the directional derivative of u in the direction of the outer normal to S at P).

---

[1] For an extensive table see V. A. Ditkin and A. P. Prudnikov, *op. cit.*, pages 100–162, but note that the entries in this table are not double Laplace transforms as defined above but their products by ps.

1.3 An elastic, flexible, homogeneous membrane of mass per unit area m is stretched over a frame in the xy-plane. Let $u(x, y, t)$ denote its vertical displacement and assume that there is no horizontal displacement. As in the case of the string, the equilibrium of horizontal forces gives a constant tension τ. If D is an arbitrary disk in this membrane bounded by a circumference C, if n is the outward unit normal vector on C, and if $D_n u$ is the directional derivative of u in the direction of n, the equilibrium of vertical forces on this disk gives

$$\int_D m u_{tt} = \int_C \tau D_n u.$$

Use the two-dimensional divergence theorem of Ostrogradski-Gauss to show that the equation of motion of the membrane is

$$m u_{tt} = \tau \operatorname{div} (\operatorname{grad} u).$$

1.4 Show that, if we define $k = \kappa/c\rho$ and if $q \equiv 0$, then the equation of heat propagation in Exercise 1.2 becomes

$$u_t = k (u_{xx} + u_{yy} + u_{zz})$$

in rectangular coordinates. This is called the three-dimensional *heat equation*.[1] Similarly, if we define $c^2 = \tau/m$, the equation of motion of the membrane in Exercise 1.3 becomes

$$u_{tt} = c^2 (u_{xx} + u_{yy})$$

in rectangular coordinates. This is called the two-dimensional *wave equation*.

1.5 Let I be a closed interval, let $\gamma : I \to \mathbb{R}$ be a one-to-one, differentiable, closed curve bounding a plane region D, and let $\bar{D} = D \cup \gamma$. Prove that the boundary value problem

$$
\begin{array}{lll}
m u_{tt} = \tau (u_{xx} + u_{yy}) + q(x, y, t) & (x, y) \text{ in } D, & t > 0 \\
u(\gamma(s), t) = h(s) & & t > 0 \\
u(x, y, 0) = f(x, y) & (x, y) \text{ in } D \\
u_t(x, y, 0) = g(x, y) & (x, y) \text{ in } D
\end{array}
$$

corresponding to the forced vibrations of a membrane has a unique solution. Prove that the same is true if the boundary condition is replaced by $D_n u(\gamma(s), t) = 0$ for $t > 0$, where $D_n u$ is the directional derivative of u in the direction of the outward unit normal vector on γ. The first condition corresponds to a membrane that is attached to its boundary in a fixed manner, and the second, to one whose boundary is free to move up or down in time while its (x, y) coordinates remain on γ. Comment on the well posedness of these problems (**Hint:** show that, if u satisfies the wave equation, then the energy of the vibrating membrane, given by

$$E(t) = \int_D \left[\frac{m}{2} u_t^2 + \frac{\tau}{2} (u_x^2 + u_y^2) \right],$$

is constant by differentiating under the integral sign and then applying the two-dimensional version of the divergence theorem of Ostrogradski-Gauss to the vector field $u_t D_n u$).

---

[1] Originally found by Fourier without the divergence theorem and stated as Theorem IV in *The Analytical Theory of Heat*, page 112.

1.6 Use the divergence theorem of Ostrogradski-Gauss to prove Green's theorems: Let D be an open set in $\mathbb{R}^3$, let $u, v: D \to \mathbb{R}$ be functions of class C^2, define

$$\Delta u = u_{xx} + u_{yy} + u_{zz},$$

called the *Laplacian* of u (Green's own notation was δu) and, if p is any vector in $\mathbb{R}^3$, let $D_p u$ denote the directional derivative of u in the direction of p. Then, if S is a closed smooth surface in D bounding an open set V and if n denotes the outward normal unit vector field on S, we have

(*i*) $\displaystyle\int_V (v\Delta u + \langle \operatorname{grad} u, \operatorname{grad} v \rangle) = \int_S v D_n u$ and

(*ii*) $\displaystyle\int_V (v\Delta u - u\Delta v) = \int_S (v D_n u - u D_n v).$[1]

1.7 Prove *Gauss' Mean Value Theorem* in $\mathbb{R}^3$: if $D \subset \mathbb{R}^3$ is an open set, if $u: D \to \mathbb{R}$ is a harmonic function, and if $S \subset D$ is a sphere of radius R centered at p, then

$$u(p) = \frac{1}{4\pi R^2} \int_S u$$

(**Hint:** putting $v \equiv 1$ in Green's theorem (*i*) of Exercise 1.6 gives

$$\int_S D_n u = 0;$$

then, in $\mathbb{R}^3$, consider a sphere S_ϵ of radius $\epsilon < R$ also centered at p, and apply Green's theorem (*ii*) in the region between S_ϵ and S to u and the function $v: D \to \mathbb{R}$ defined by $v(q) = 1/r$, where r is the distance from p to q).

1.8 Derive again the equations of Exercises 1.2 and 1.3 but use Green's theorem (*i*) instead of the divergence theorem of Ostrogradski-Gauss.

2.1 Expand $f(x, y) = x|y|$ in a double Fourier series on R_π and also on the square $[0, 2\pi] \times [0, 2\pi]$.

2.2 Expand $f(x, y) = |x|y^2$ in a double Fourier series on $[-a, a] \times [-b, b]$.

2.3 Expand the function $f: R_\pi \to \mathbb{R}$ defined by

$$f(x, y) = \begin{cases} 1, & y < x \\ 0, & y \geq x \end{cases}$$

in a double Fourier series on R_π. Show that it converges at all points of R_π and find its sum at each point.

2.4 Define the *double Fourier sine series* of $f: (0, a) \times (0, b) \to \mathbb{R}$ to be the double Fourier series of the extension f_O of f to $(-a, a) \times (-b, b)$ such that $f_O(-x, y) = -f_O(x, y)$ and $f_O(x, -y) = -f_O(x, y)$, that is,

$$\sum_{m,n=1}^{\infty} d_{mn} \sin \frac{m\pi}{a} x \sin \frac{n\pi}{b} y,$$

---

[1] These equations are on pages 25 and 23, respectively, of Green's essay, quoted above.

where

$$d_{mn} = \frac{4}{ab} \int_0^a \int_0^b f(x, y) \sin \frac{m\pi}{a} x \sin \frac{n\pi}{b} y \, dx \, dy.$$

Find the double Fourier sine series of

(i) $f \equiv C$, where C is a constant, on $(0, a) \times (0, b)$ and

(ii) $f(x, y) = \sin(x + y)$ on $(0, \pi) \times (0, \pi)$.

2.5 Define the *complex double Fourier series* of $f: R_\pi \to \mathbb{R}$ to be

$$\sum_{m,n=-\infty}^{\infty} C_{mn} e^{i(mx+ny)},$$

where

$$C_{mn} = \frac{1}{4\pi^2} \int_{-\pi}^{\pi} \int_{-\pi}^{\pi} f(x, y) e^{-i(mx+ny)} \, dx \, dy,$$

and let S_{MN} be the partial sum

$$S_{MN} = \sum_{\substack{-M \le m \le M \\ -N \le n \le N}} C_{mn} e^{i(mx+ny)}.$$

Show that $S_{MN} = s_{MN}$, where s_{MN} is as stated before Definition 9.2. Thus, the complex double Fourier series of f on R_π is equivalent to the regular double Fourier series of f on R_π.

2.6 Find the complex double Fourier series of $f(x, y) = |x + y|$ on R_π.

2.7 Prove Lemma 9.1 (**Hint:** partition R into subrectangles and imitate the proof of Theorem 2.1).

2.8 The differentiability requirements in Theorem 9.1 can be relaxed if f is of the form $f(x, y) = g(x)h(y)$. In fact, prove the following convergence theorem for such an f: if g and h are piecewise continuous on $(-\pi, \pi)$, then the Fourier series of f on R_π converges to $f(x, y)$ at each point (x, y) in $\mathbb{R}^2$ at which f is continuous and g and h have right-hand and left-hand derivatives at x and y, respectively.

2.9 We shall define an orthogonal system on a rectangle R in the plane to be *complete* if and only if every continuous function that is orthogonal on R to all the elements of the system is identically zero. Show that, if $\{\varphi_m\}$ is a complete orthogonal system on $[a, b]$ and if $\{\phi_n\}$ is a complete orthogonal system on $[c, d]$, then $\{\varphi_m \phi_n\}$ is a complete orthogonal system on $[a, b] \times [c, d]$.

2.10 Use the result of Exercise 2.9 to show that the system of Example 9.2 is complete on R_π, and then that, if two continuous functions $f, g: R_\pi \to \mathbb{R}$ have the same Fourier coefficients on R_π, then $f \equiv g$ on R_π.

3.1 Prove that the double series

$$\sum_{m,n=1}^{\infty} \frac{1}{(m^2 + n^2)^2}$$

converges (**Hint:** use the integral test to show that the single series obtained by fixing m converges and that its Nth partial sum is no larger than

$$\frac{1}{m^4} + \frac{\pi - 2}{8m^3},$$

and then use the following theorem: if

$$\sum_{m,n=1}^{\infty} a_{mn}$$

is a double series of positive terms, then it converges if

$$\sum_{m=1}^{\infty}\left(\sum_{n=1}^{\infty} a_{mn}\right)$$

converges.[1]

3.2   Find a formal solution of the boundary value problem

$$
\begin{aligned}
u_t &= k(u_{xx} + u_{yy}) & 0 < x < a, && 0 < y < b, && t > 0 \\
u(0, y, t) &= 0 & && 0 \le y \le b, && t \ge 0 \\
u(a, y, t) &= 0 & && 0 \le y \le b, && t \ge 0 \\
u(x, 0, t) &= 0 & 0 \le x \le a, && && t \ge 0 \\
u(x, b, t) &= 0 & 0 \le x \le a, && && t \ge 0 \\
u(x, y, 0) &= f(x, y) & 0 \le x \le a, && 0 \le y \le b &&
\end{aligned}
$$

of the form

$$u(x, y, t) = \sum_{m,n=1}^{\infty} c_{mn} X_m(x) Y_n(y).$$

How is the solution to be modified if the conditions $u(0, y, t) = u(a, y, t) = 0$ are replaced with $u_x(0, y, t) = u_x(a, y, t) = 0$?

3.3   Propose a method of solution for the boundary value problem of Exercise 3.2 if the two-dimensional heat equation is replaced with $u_t = k(u_{xx} + u_{yy}) + q(x, y, t)$. Use the result of Exercise 2.4 (i) to formally solve in full the particular case in which $q \equiv Q$, a constant, and $f \equiv 0$ (**Hint:** assume that there is a solution of the form

$$u(x, y, t) = \sum_{m,n=1}^{\infty} c_{mn}(t) X_m(x) Y_n(y),$$

where X_m and Y_n are as in Exercise 3.2, and imitate the procedure suggested in §3.11 for the one-dimensional equation of heat propagation to find the function c_{mn}).

3.4   Propose a method of solution for the boundary value problem

$$
\begin{aligned}
u_t &= k(u_{xx} + u_{yy}) + q(x, y, t) & 0 < x < a, && 0 < y < b, && t > 0 \\
u(0, y, t) &= 0 & && 0 \le y \le b, && t \ge 0 \\
u(a, y, t) &= 0 & && 0 \le y \le b, && t \ge 0 \\
u(x, 0, t) &= 0 & 0 \le x \le a, && && t \ge 0 \\
u(x, b, t) &= g(x) & 0 \le x \le a, && && t \ge 0 \\
u(x, y, 0) &= f(x, y) & 0 \le x \le a, && 0 \le y \le b &&
\end{aligned}
$$

(**Hint:** think in terms of the steady-state solution of a related problem, and then the method of Exercise 3.3 will help at some point).

---

[1] See, for instance, E. W. Hobson, *The Theory of Functions of a Real Variable and the Theory of Fourier's Series*, Vol. 2, Dover, New York, 1957, pages 52–53.

3.5 Show that the boundary value problem

$$
\begin{aligned}
u_{tt} &= c^2(u_{xx} + u_{yy}) & 0 < x < a, \quad & 0 < y < b, & t > 0 \\
u(0, y, t) &= 0 & & 0 \le y \le b, & t \ge 0 \\
u(a, y, t) &= 0 & & 0 \le y \le b, & t \ge 0 \\
u(x, 0, t) &= 0 & 0 \le x \le a, & & t \ge 0 \\
u(x, b, t) &= 0 & 0 \le x \le a, & & t \ge 0 \\
u(x, y, 0) &= f(x, y) & 0 \le x \le a, & \quad 0 \le y \le b & \\
u_t(x, y, 0) &= g(x, y) & 0 \le x \le a, & \quad 0 \le y \le b, &
\end{aligned}
$$

corresponding to the vibrations of a membrane attached to a rectangular frame (see Exercise 1.3), has the formal solution

$$
u(x, y, t) = \sum_{m,n=1}^{\infty} (A_{mn} \cos \omega_{mn} t + B_{mn} \sin \omega_{mn} t) \sin \frac{m\pi}{a} x \sin \frac{n\pi}{b} y,
$$

where A_{mn} and B_{mn} are the coefficients of the double Fourier sine series of f and g on $(0, a) \times (0, b)$, respectively, and

$$
\omega_{mn} = \frac{\pi c}{ab} \sqrt{a^2 n^2 + b^2 m^2}.
$$

As in the case of the vibrating string, each of the component vibrations, which can be written in the form

$$
u_{mn} = c_{mn} \sin(\omega_{mn} t + \varphi_{mn}) \sin \frac{m\pi}{a} x \sin \frac{n\pi}{b} y,
$$

is called a *standing wave*, and the lines parallel to the coordinate axes on which it vanishes are called *nodal lines*. However, their individual frequencies $\omega_{mn}/2\pi$ are not integral multiples of a fundamental frequency, which renders the membrane useless for many musical purposes beyond keeping a beat. Moreover, different standing waves may have the same frequency such as, for instance, u_{13} and u_{31} if $a = b$. One can then consider compound standing waves, such as $u_{13} + u_{31}$, with more complex nodal structure.

3.6 Propose a method of solution for the boundary value problem of Exercise 3.5 if the two-dimensional wave equation is replaced with $u_{tt} = c^2(u_{xx} + u_{yy}) + q(x, y, t)$, corresponding to forced vibrations of the membrane. Use the result of Exercise 2.4 (i) to formally solve in full the particular case in which $q(x, y, t) = A \sin \omega t$, where A is a constant, and $f \equiv g \equiv 0$, and explore the possibility of resonance (**Hint:** assume that there is a solution of the form

$$
u(x, y, t) = \sum_{m,n=1}^{\infty} c_{mn}(t) \sin \frac{m\pi}{a} x \sin \frac{n\pi}{b} y
$$

and imitate the procedure suggested in §5.7 for the one-dimensional wave equation).

3.7 Use the double Fourier series of Exercise 2.1 to find a formal solution of the boundary value problem

$$
\begin{aligned}
u_{xx} + u_{yy} + u_{zz} &= 0 &\quad 0 < x < \pi, &\quad 0 < y < \pi, &\quad 0 < z < 1 \\
u(0, y, z) &= 0 & &\quad 0 \le y \le \pi, &\quad 0 \le z \le 1 \\
u(\pi, y, z) &= 0 & &\quad 0 \le y \le \pi, &\quad 0 \le z \le 1 \\
u_y(x, \pi, z) &= 0 &\quad 0 \le x \le \pi, & &\quad 0 \le z \le 1 \\
u_y(x, 0, z) &= 0 &\quad 0 \le x \le \pi, & &\quad 0 \le z \le 1 \\
u(x, y, 0) &= xy &\quad 0 \le x \le \pi, &\quad 0 \le y \le \pi \\
u(x, y, 1) &= 0 &\quad 0 \le x \le \pi, &\quad 0 \le y \le \pi.
\end{aligned}
$$

3.8 Find a formal solution of the boundary value problem of Exercise 3.7 if the Laplace equation is replaced with $u_{xx} + u_{yy} + u_{zz} = u$.

4.1 Find a formal solution of the boundary value problem

$$
\begin{aligned}
u_t &= k(u_{xx} + u_{yy} + u_{zz}) &\quad 0 < x < a, &\quad 0 < y < b, &\quad 0 < z < c, &\quad t > 0 \\
u(0, y, z, t) &= 0 & &\quad 0 \le y \le b, &\quad 0 \le z \le c, &\quad t \ge 0 \\
u(a, y, z, t) &= 0 & &\quad 0 \le y \le b, &\quad 0 \le z \le c, &\quad t \ge 0 \\
u(x, 0, z, t) &= 0 &\quad 0 \le x \le a, & &\quad 0 \le z \le c, &\quad t \ge 0 \\
u(x, b, z, t) &= 0 &\quad 0 \le x \le a, & &\quad 0 \le z \le c, &\quad t \ge 0 \\
u_z(x, y, 0, t) &= 0 &\quad 0 \le x \le a, &\quad 0 \le y \le b, & &\quad t \ge 0 \\
u_z(x, y, c, t) &= 0 &\quad 0 \le x \le a, &\quad 0 \le y \le b, & &\quad t \ge 0 \\
u(x, y, z, 0) &= f(x, y, z) &\quad 0 \le x \le a, &\quad 0 \le y \le b, &\quad 0 \le z \le c
\end{aligned}
$$

assuming that a suitable extension of f has a convergent triple Fourier series.

4.2 Find a formal solution of the boundary value problem

$$
\begin{aligned}
u_{tt} &= c_0^2(u_{xx} + u_{yy} + u_{zz}) &\quad 0 < x < a, &\quad 0 < y < b, &\quad 0 < z < c, &\quad t > 0 \\
u(0, y, z, t) &= 0 & &\quad 0 \le y \le b, &\quad 0 \le z \le c, &\quad t \ge 0 \\
u(a, y, z, t) &= 0 & &\quad 0 \le y \le b, &\quad 0 \le z \le c, &\quad t \ge 0 \\
u(x, 0, z, t) &= 0 &\quad 0 \le x \le a, & &\quad 0 \le z \le c, &\quad t \ge 0 \\
u(x, b, z, t) &= 0 &\quad 0 \le x \le a, & &\quad 0 \le z \le c, &\quad t \ge 0 \\
u(x, y, 0, t) &= 0 &\quad 0 \le x \le a, &\quad 0 \le y \le b, & &\quad t \ge 0 \\
u(x, y, c, t) &= 0 &\quad 0 \le x \le a, &\quad 0 \le y \le b, & &\quad t \ge 0 \\
u(x, y, z, 0) &= f(x, y, z) &\quad 0 \le x \le a, &\quad 0 \le y \le b, &\quad 0 \le z \le c \\
u_t(x, y, z, 0) &= f(x, y, z) &\quad 0 \le x \le a, &\quad 0 \le y \le b, &\quad 0 \le z \le c
\end{aligned}
$$

assuming that suitable extensions of f and g have convergent triple Fourier series.

5.1 Use the Fourier transform method rather than Duhamel's principle to solve the boundary value problem of Example 9.5.

5.2 Repeat Exercise 3.3 using Duhamel's principle for the heat equation.

5.3 If $r = (x, y, z)$, use the solution of the problems in Examples 9.4 and 9.5 to find a solution of the boundary value problem

$$
\begin{aligned}
u_t &= k(u_{xx} + u_{yy} + u_{zz}) + q(r, t) &\quad r \text{ in } \mathbb{R}^3, &\quad t > 0 \\
u(r, 0) &= f(r) &\quad r \text{ in } \mathbb{R}^3.
\end{aligned}
$$

5.4 Verify the solution of Example 9.6 assuming that f and g are functions of class C^2 (**Hint:** after differentiating with respect to t, under the integral sign for integrals, use Green's theorem (i) of Exercise 1.6 or the divergence theorem of Ostrogradski-Gauss to transform surface integrals into volume integrals).

5.5 Use the solution of Example 9.6 to solve the boundary value problem

$$\begin{aligned}
u_{tt} &= c^2(u_{xx} + u_{yy} + u_{zz}) & & r \text{ in } \mathbb{R}^3, & t > 0 \\
u(x, y, z, 0) &= x^2 + y^2 + z^2 & & r \text{ in } \mathbb{R}^3 \\
u_t(x, y, z, 0) &= x + y + z & & r \text{ in } \mathbb{R}^3.
\end{aligned}$$

5.6 Use the solution of Example 9.6 to solve the boundary value problem

$$\begin{aligned}
u_{tt} &= c^2(u_{xx} + u_{yy} + u_{zz}) & & 0 < x < a, \quad 0 < y < b, \quad 0 < z < c, \quad t > 0 \\
u(0, y, z, t) &= u(a, y, z, t) = 0 & & t \geq 0 \\
u(x, 0, z, t) &= u(x, b, z, t) = 0 & & t \geq 0 \\
u(x, y, 0, t) &= u(x, y, c, t) = 0 & & t \geq 0 \\
u(x, y, z, 0) &= 0 & & 0 \leq x \leq a, \quad 0 \leq y \leq b, \quad 0 \leq z \leq c \\
u_t(x, y, z, 0) &= f(x, y, z) & & 0 \leq x \leq a, \quad 0 \leq y \leq b, \quad 0 \leq z \leq c
\end{aligned}$$

after extending f to $\mathbb{R}^3$ in a suitable way. State conditions on f to guarantee the validity of the solution.

5.7 Use Poisson's formula of Example 9.6 to show that the boundary value problem

$$\begin{aligned}
u_{tt} &= c^2(u_{xx} + u_{yy} + u_{zz}) & & r \text{ in } \mathbb{R}^3, & t > 0 \\
u(x, y, z, 0) &= C_1 & & r \text{ in } \mathbb{R}^3 \\
u_t(x, y, z, 0) &= C_2 & & r \text{ in } \mathbb{R}^3,
\end{aligned}$$

where C_1 and C_2 are positive constants, has the solution

$$u(r, t) = \begin{cases}
C_1 + C_2 t, & \|r\| < R \text{ and } ct \leq R - \|r\| \\[2mm]
\dfrac{C_1}{2}\left(1 - \dfrac{ct}{\|r\|}\right) & \\[1mm]
\quad + \dfrac{C_2}{4c\|r\|}[R^2 - (\|r\| - ct)^2], & |R - \|r\|| < ct < R + \|r\| \\[2mm]
0, & \text{otherwise.}
\end{cases}$$

5.8 Solve the following boundary value problem for the *Klein-Gordon equation.*

$$\begin{aligned}
u_{tt} &= c^2(u_{xx} + u_{yy}) - m^2 u & & -\infty < x < \infty, & -\infty < y < \infty, & t > 0 \\
u(x, y, 0) &= 0 & & -\infty < x < \infty, & -\infty < y < \infty \\
u_t(x, y, 0) &= g(x, y) & & -\infty < x < \infty, & -\infty < y < \infty,
\end{aligned}$$

where m is a constant (**Hint:** solve a boundary value problem for the function defined by $v(x, y, z, t) = u(x, y, t)e^{imz/c}$, and make the substitution $\theta = \theta$ and $\rho = ct \sin \phi$ to deal with surface integrals in the Poisson form of the solution for v).

5.9 Use the three-dimensional solution of Example 9.6, but assume that the functions u, f, and g are independent of z, to show that, if $r = (x, y)$, then the boundary value problem

$$\begin{aligned}
u_{tt} &= c^2(u_{xx} + u_{yy}) & & -\infty < x < \infty, & -\infty < y < \infty, & t > 0 \\
u(x, y, 0) &= f(x, y) & & -\infty < x < \infty, & -\infty < y < \infty \\
u_t(x, y, 0) &= g(x, y) & & -\infty < x < \infty, & -\infty < y < \infty
\end{aligned}$$

has the solution

$$u(r,t) = \frac{d}{dt}\left[\frac{1}{2\pi c}\int_{D_{ct}}\frac{f(r+\sigma)}{\sqrt{(ct)^2-\|\sigma\|^2}}\,d\sigma\right] + \frac{1}{2\pi c}\int_{D_{ct}}\frac{g(r+\sigma)}{\sqrt{(ct)^2-\|\sigma\|^2}}\,d\sigma,$$

where D_{ct} is the disk of radius ct centered at the origin (**Hint:** make the substitution $\theta = \theta$ and $\rho = ct\sin\phi$ to deal with surface integrals).

5.10   Use the solution of Exercise 5.9 and the method of descent to show that the boundary value problem

$$\begin{aligned} u_{tt} &= c^2 u_{xx} & -\infty < x < \infty, \quad t > 0 \\ u(x,0) &= f(x) & -\infty < x < \infty \\ u_t(x,0) &= g(x) & -\infty < x < \infty \end{aligned}$$

has the solution

$$u(x,t) = \frac{1}{2}[f(x+ct) + f(x-ct)] + \frac{1}{2c}\int_{x-ct}^{x+ct} g.$$

The last integral shows that $u(x,t)$ depends on the values of g on the interval $(x-ct, x+ct)$, that is, that Huygens' principle is not valid for $n=1$ (**Hint:** write the integrals over D_{ct} in rectangular coordinates).

5.11   Prove the following version of *Duhamel's principle for the wave equation*: let D be a region in $\mathbb{R}^3$ with boundary ∂D (which may be empty), and for each function $q: D \times [0,\infty) \to \mathbb{R}$ and each fixed value of τ define a function $q_\tau : D \to \mathbb{R}$ by $q_\tau(x,y,z) = q(x,y,z,\tau)$. If u_{q_τ} is a solution of the boundary value problem

$$\begin{aligned} u_{tt} &= c^2(u_{xx} + u_{yy} + u_{zz}) & (x,y,z) \text{ in } D, \quad t > 0 \\ u(x,y,z,t) &= 0 & (x,y,z) \text{ in } \partial D, \quad t \ge 0 \\ u(x,y,z,0) &= 0 & (x,y,z) \text{ in } D \\ u_t(x,y,z,0) &= q_\tau(x,y,z) & (x,y,z) \text{ in } D, \end{aligned}$$

then the boundary value problem

$$\begin{aligned} u_{tt} &= c^2(u_{xx} + u_{yy} + u_{zz}) + q(x,y,z,t) & (x,y,z) \text{ in } D, \quad t > 0 \\ u(x,y,z,t) &= 0 & (x,y,z) \text{ in } \partial D, \quad t \ge 0 \\ u(x,y,z,0) &= 0 & (x,y,z) \text{ in } D \\ u_t(x,y,z,0) &= 0 & (x,y,z) \text{ in } D \end{aligned}$$

has the solution

$$u(x,y,z,t) = \int_0^t u_{q_\tau}(x,y,z,t-\tau)\,d\tau.$$

A similar statement holds for arbitrary dimension.

5.12   Use Duhamel's principle for the wave equation of Exercise 5.11 and assume that the solution of Exercise 3.5 can be integrated term by term to solve the boundary value problem

$$u_{tt} = c^2(u_{xx} + u_{yy}) + q(x, y, t) \qquad 0 < x < a, \qquad 0 < y < b, \qquad t > 0$$
$$u(0, y, t) = 0 \qquad\qquad\qquad\qquad\qquad 0 \le y \le b, \qquad t \ge 0$$
$$u(a, y, t) = 0 \qquad\qquad\qquad\qquad\qquad 0 \le y \le b, \qquad t \ge 0$$
$$u(x, 0, t) = 0 \qquad\qquad 0 \le x \le a, \qquad\qquad\qquad\qquad t \ge 0$$
$$u(x, b, t) = 0 \qquad\qquad 0 \le x \le a, \qquad\qquad\qquad\qquad t \ge 0$$
$$u(x, y, 0) = 0 \qquad\qquad 0 \le x \le a, \qquad 0 \le y \le b$$
$$u_t(x, y, 0) = 0 \qquad\qquad 0 \le x \le a, \qquad 0 \le y \le b.$$

5.13 Use Duhamel's principle for the wave equation of Exercise 5.11 to show that, if $r = (x, y, z)$, then the boundary value problem

$$u_{tt} = c^2(u_{xx} + u_{yy} + u_{zz}) + q(r, t) \qquad r \text{ in } \mathbb{R}^3, \qquad t > 0$$
$$u(r, 0) = 0 \qquad\qquad\qquad\qquad\qquad\qquad r \text{ in } \mathbb{R}^3$$
$$u_t(r, 0) = 0 \qquad\qquad\qquad\qquad\qquad\qquad r \text{ in } \mathbb{R}^3$$

has the solution

$$u(r, t) = \frac{1}{4\pi c^2} \int_{\|s - r\| \le ct} \frac{q(s, t - \|s - r\|/c)}{\|s - r\|} \, ds.$$

Because of the difference $t - \|s - r\|/c$, which shows a time delay in the amount $\|s - r\|/c$, and by comparison with the potential solution of Exercise 5.14, the solution above is called *retarded potential*.

5.14 Use the Fourier transform method to show that, if $r = (x, y, z)$, then the equation

$$u_{xx} + u_{yy} + u_{zz} = -q(x, y, z),$$

where q is a continuous, positive function on $\mathbb{R}^3$, has the formal solution

$$u(r) = \frac{1}{4\pi} \int_{\mathbb{R}^3} \frac{q(s)}{\|s - r\|} \, ds.$$

This represents the electrostatic potential created by a continuous distribution of charge in space or the Newtonian potential created by a continuous distribution of mass (**Hint:** after using the inversion formula, change the order of integration and switch to spherical coordinates to evaluate the inner integral).

5.15 Let u be the solution of Example 9.7. Show that
    (i) u is bounded and,
    (ii) if f vanishes outside some circle and if $r = (x, y, z)$, then $u(r) \to 0$ as $\|r\| \to \infty$.
Find an unbounded solution of the same problem by inspection, showing that the solution is not unique.

5.16 Find a solution of the boundary value problem

$$u_{xx} + u_{yy} + u_{zz} = 0 \qquad -\infty < x < \infty, \qquad -\infty < y < \infty, \qquad z > 0$$
$$u_z(x, y, 0) = f(x, y) \qquad -\infty < x < \infty, \qquad -\infty < y < \infty$$

such that $u(x, y, z) \to 0$ as $x^2 + y^2 + z^2 \to \infty$ (**Hint:** solve a boundary value problem for $v = u_z$).

6.1 Verify equations (35), (36), and 37).

6.2 Show that the double Laplace transform of

$$f(x, y) = \frac{x}{y\sqrt{\pi y}} e^{-x^2/y}$$

is

$$F(p, s) = \frac{1}{(p + 2\sqrt{s})}$$

(**Hint:** use the Laplace transform of Example 8.13).

6.3 Show that the double Laplace transform of

$$f(x, y) = \begin{cases} y^2, & y < \dfrac{x}{c} \\[2mm] y^2 - \left(y - \dfrac{x}{c}\right)^2, & y \geq \dfrac{x}{c}, \end{cases}$$

where c is a positive constant, is

$$F(p, s) = \frac{1}{ps^2(cp + s)}.$$

6.4 Use the double Laplace transform method to solve the boundary value problem

$$\begin{aligned} u_t &= ku_{xx} & x > 0, & \quad t > 0 \\ u(x, 0) &= 0 & x \geq 0 \\ u(0, t) &= g(t) & & \quad t \geq 0, \end{aligned}$$

where g has a one-dimensional Laplace transform (**Hint:** use the transform of Exercise 6.2).

6.5 Use the double Laplace transform method to solve the boundary value problem

$$\begin{aligned} u_{tt} &= c^2 u_{xx} - g & x > 0, & \quad t > 0 \\ u(0, t) &= 0 & & \quad t \geq 0 \\ u(x, 0) &= f(x) & x \geq 0 \\ u_t(x, 0) &= h(x) & x \geq 0 \end{aligned}$$

for a semi-infinite string whose initial position and velocity are known if g is the acceleration of gravity

(*i*) using the Laplace transform of Exercise 6.3 (this is the easy way out) and

(*ii*) without using the Laplace transform of Exercise 6.3.

6.6 Show that the boundary value problem

$$\begin{aligned} u_{xx} + u_{yy} &= 0 & x > 0, & \quad y > 0 \\ u(x, 0) &= f(x) & x \geq 0 \\ u(0, y) &= 0 & & \quad y \geq 0 \end{aligned}$$

cannot be solved by the double Laplace transform method.

10

BOUNDARY VALUE PROBLEMS
WITH CIRCULAR SYMMETRY

§10.1 Vibrations of a Circular Membrane

In 1759 Euler submitted a paper to the Academy of Saint Petersburg—which did not appear in print until 1766, dated 1764—containing a study of the vibrations of a drum and introducing, to that end, something new at the time: a partial differential equation involving more than one space variable. More precisely, Euler was able to show that the displacements of a membrane vibrating without friction satisfy the equation

$$u_{tt} = c^2(u_{xx} + u_{yy}),$$

called the *two-dimensional wave equation*, where $c^2 = \tau/m$, τ is the tensile force and m is the mass of the membrane per unit area (see Exercises 1.3 and 1.4 of Chapter 9). He started by studying a rectangular membrane (see Exercise 3.5 of Chapter 9) but then took up the case of a circular membrane, which he attacked by means of another original bold move, one that has become by now a standard routine in studying problems with circular symmetry: switching to polar coordinates. If v is a solution of the two-dimensional wave equation, if $x = r\cos\theta$ and $y = r\sin\theta$, and if we define $u(r, \theta, t) = v(x, y, t)$, then the chain rule gives $v_x = u_r r_x + u_\theta \theta_x$. Since

$$r = \sqrt{x^2 + y^2} \qquad \text{and} \qquad \theta = \arctan\frac{y}{x},$$

differentiating with respect to x, we have

$$r_x = \frac{x}{\sqrt{x^2 + y^2}} = \frac{x}{r}, \quad r_{xx} = \frac{r - xr_x}{r^2} = \frac{1}{r} - \frac{x^2}{r^3}, \quad \text{and} \quad \theta_x = -\frac{y}{x^2 + y^2} = -\frac{y}{r^2}.$$

Differentiating $v_x = u_r r_x + u_\theta \theta_x$ again with respect to x and using these derivatives, we obtain

$$v_{xx} = u_{rr} r_x^2 + u_{r\theta} \theta_x r_x + u_r r_{xx} + u_{\theta r} r_x \theta_x + u_{\theta\theta} \theta_x^2$$

$$= u_{rr} \frac{x^2}{r^2} - u_{r\theta} \frac{yx}{r^3} + u_r \left(\frac{1}{r} - \frac{x^2}{r^3} \right) - u_{\theta r} \frac{xy}{r^3} + u_{\theta\theta} \frac{y^2}{r^4}.$$

Similary, differentiating v twice with respect to y, we obtain

$$v_{yy} = u_{rr} \frac{y^2}{r^2} + u_{r\theta} \frac{xy}{r^3} + u_r \left(\frac{1}{r} - \frac{y^2}{r^3} \right) + u_{\theta r} \frac{yx}{r^3} + u_{\theta\theta} \frac{x^2}{r^4}.$$

Clearly, $v_{tt} = u_{tt}$, and then after substitution of these second-order partial derivatives into the two-dimensional wave equation it becomes

$$u_{tt} = c^2 \left(u_{rr} + \frac{1}{r} u_r + \frac{1}{r^2} u_{\theta\theta} \right).$$

Let us pose a specific boundary value problem, that of solving this equation for $0 < r < a$ and $-\pi \le \theta \le \pi$, where a is a positive constant, subject to the boundary and initial conditions

$$
\begin{array}{llll}
u(a, \theta, t) = 0 & & -\pi \le \theta \le \pi, & t \ge 0 \\
u(r, \theta, 0) = f(r, \theta) & r \le a, & -\pi \le \theta \le \pi & \\
u_t(r, \theta, 0) = 0 & r \le a, & -\pi \le \theta \le \pi, &
\end{array}
$$

where f is a known function that is periodic in the variable θ with period 2π and such that $f(a, \theta) = 0$. This problem corresponds to the vibrations of a membrane that is stretched over a fixed circular frame of radius a with an initial displacement given by $f(r, \theta)$ and no initial velocity.

We shall use the well known method of separation of variables and assume that there is a solution of the form $u(r, \theta, t) = R(r)\Theta(\theta)T(t)$ (Euler did the same, but assuming *a priori* that Θ and T are sine functions). Substitution into the two-dimensional wave equation and division by $c^2 R\Theta T$ leads to

$$\frac{T''}{c^2 T} = \frac{R''}{R} + \frac{R'}{rR} + \frac{\Theta''}{r^2\Theta} = -\lambda,$$

where $-\lambda$ is a constant. Putting $\Theta''/\Theta = -\mu$, another constant, and using the boundary conditions, we obtain

$$
\begin{array}{ll}
T'' + c^2\lambda T = 0 & T'(0) = 0 \\
\Theta'' + \mu\Theta = 0 & \Theta(-\pi) = \Theta(\pi) \\
r^2 R'' + r R' + (\lambda r^2 - \mu)R = 0 & R(a) = 0.
\end{array}
$$

We shall assume that $\lambda > 0$, or else the equation for T will not have nonconstant periodic solutions, contrary to an intuitive expectation of vibrations for a frictionless drumhead in motion. Similarly, an examination of the general solution for Θ and the requirement that Θ must be either constant or a periodic function of period 2π show that $\mu \geq 0$. Then, writing $\mu = p^2$, the equation for R, which was obtained by Euler, becomes

$$r^2 R'' + r R' + (\lambda r^2 - p^2) R = 0,$$

and it must be solved subject to the condition $R(a) = 0$. This is a singular Sturm-Liouville problem, as already remarked in Example 4.5, and the theory developed in §4.3 is not applicable. It will be easier to study this equation if we introduce a new variable $x = r\sqrt{\lambda}$ and a new function y of x defined by

$$(1) \qquad\qquad y(x) = R\left(\frac{x}{\sqrt{\lambda}}\right).$$

Differentiating we have

$$y'(x) = \frac{1}{\sqrt{\lambda}} R'(r) \qquad \text{and} \qquad y''(x) = \frac{1}{\lambda} R''(r),$$

so that the differential equation for R is transformed into

$$(2) \qquad\qquad x^2 y'' + x y' + (x^2 - p^2) y = 0,$$

which is independent of λ. Some particular cases of this equation appeared first in papers by the Bernoullis and Euler, then in the preceding work on the vibrating membrane, and also in some work of Fourier and Poisson. However, it is called the *Bessel differential equation of order p* for reasons that will be explained in §10.3.

The Bessel equation is the first ordinary differential equation that we encounter as a by-product of separation of variables that cannot be solved in closed form. It does not have constant coefficients and it is not of the Cauchy-Euler type (see Exercise 1.5 of Chapter 4), as the first two terms might suggest, because the coefficient of y is not a constant. Of course, there are other problems with circular symmetry that do not lead to anything new, but such problems are relegated to the exercises. To deal with this new type of situation, Euler revived a method already tried by Newton and Leibniz for some elementary equations, namely, to seek a solution in the form of an infinite series. That is, Euler assumed that there is a solution of the form

$$(3) \qquad\qquad y(x) = \sum_{k=0}^{\infty} a_k x^{\rho+k},$$

where ρ is a constant, and that the series is differentiable term by term. Then the coefficients a_k and the constant ρ, which is not necessarily an integer so that (3) may not be a power series, are to be determined by substituting (3) into (2) and then comparing terms with equal powers of x.[1] Substitution gives

$$\sum_{k=0}^{\infty}[(\rho + k)(\rho + k + 1) + (\rho + k) + (x^2 - p^2)]a_k x^{\rho+k} = 0,$$

that is,

$$\sum_{k=0}^{\infty}[(\rho + k)^2 - p^2]a_k x^{\rho+k} + \sum_{k=0}^{\infty} a_k x^{\rho+k+2}$$

$$= x^{\rho}\left[\sum_{k=0}^{\infty}[(\rho + k)^2 - p^2]a_k x^k + \sum_{k=2}^{\infty} a_{k-2} x^k\right]$$

$$= x^{\rho}\left[(\rho^2 - p^2)]a_0 + (\rho + 1 - p)(\rho + 1 + p)a_1 x\right.$$

$$\left. + \sum_{k=2}^{\infty}[(\rho + k - p)(\rho + k + p)a_k + a_{k-2}]x^k\right]$$

$$= 0.$$

Thus, for $x > 0$, the coefficients of each power of x in the series above must be zero. We shall assume that $a_0 \neq 0$; that is, the first nonzero coefficient in the series is labeled a_0, and then either $\rho = p$ or $\rho = -p$. We make the choice $\rho = p$, and then we choose $a_1 = 0$ so that $(2p + 1)a_1 = 0$ for any p. Finally, for $k \geq 2$, we must have

$$k(2p + k)a_k = -a_{k-2}.$$

Unless $2p$ is a negative odd integer, the fact that $a_1 = 0$ implies that $a_3 = 0$ and then that $a_{2j-1} = 0$ for every positive integer j. So we further choose $a_{2j-1} = 0$ for every positive integer j regardless of the value of p. Note that it is legitimate to make these choices so long as what we do leads to a solution. Now, if k is even and we write $k = 2j$, we have

$$4j(p + j)a_{2j} = -a_{2j-2},$$

and, multiplying the equations that result for values of $j = 1, \ldots, k$,

$$4^k k!(p + 1) \cdots (p + k)a_2 \cdots a_{2k} = (-1)^k a_0 \cdots a_{2k-2}.$$

That is, if we impose, at least for the time being, the restriction that p is not a negative integer,

$$a_{2k} = \frac{(-1)^k}{k!(p + 1) \cdots (p + k)2^{2k}} a_0.$$

---

[1] What Euler did in some examples was later expanded into a method for obtaining the general solution of certain types of nth-order differential equations by Lazarus Fuchs (1833–1902), a former student of Weierstrass and his successor at the University of Berlin, in 1866 and by Ferdinand Georg Frobenius (1849–1917), also of the University of Berlin, in 1873. This extension is now called *the method of Frobenius*.

This equation allows us to determine all the a_{2k} from a_0, which we are free to choose in any way we want. It is possible to simplify the expression on the right-hand side if p is a nonnegative integer—the most interesting case in applications—if we choose

$$a_0 = \frac{1}{p!2^p}.$$

Then

$$a_{2k} = \frac{(-1)^k}{k!(p+k)!}\frac{1}{2^{2k+p}},$$

and (3) becomes

(4) $$y(x) = \sum_{k=0}^{\infty} a_{2k}x^{2k+p} = \sum_{k=0}^{\infty} \frac{(-1)^k}{k!(p+k)!}\left(\frac{x}{2}\right)^{2k+p}.$$

This provides, subject to verification, a series solution of the Bessel differential equation of order p, but only if p is a nonnegative integer. This restriction on p is no obstacle in dealing with the vibrating membrane problem since we have already seen that $p^2 = \mu \geq 0$.

Equation (4) differs form the solution found by Euler only by a constant factor, a consequence of the fact that Euler did not assign a_0 the value that has by now become standard.[1] Two things remain in the study of the vibrating drumhead. The first is to find a second linearly independent solution of (2), which Euler could not do. The second is to deal with the boundary condition at $r = a$, a point on which Euler—although optimistic—was rather vague. We shall find a second solution in §10.5 and return to the boundary condition in §10.6 and §10.7.

§10.2 The Gamma Function

We have used the concept of factorial to simplify the product $(p+k)(p+k-1)\cdots 2 \cdot 1$, but this is possible for all k only if p is a nonnegative integer. In order to give an expression equivalent to (4) in the remaining cases we must first generalize the concept of factorial. This problem was considered by several mathematicians in the early eighteenth century—even before that of the vibrating membrane—and the first one to succeed was Euler, who presented his solution in a letter dated October 13, 1729, to Christian Goldbach—a competitor in this endeavor. We shall not be concerned with this early form of the solution involving infinite products, nor with several revisions and improvements, and simply mention that in 1781 Euler used integration by parts to obtain

$$\int_0^{\infty} t^k e^{-t}\, dt = k\int_0^{\infty} t^{k-1}e^{-t}\, dt,$$

---

[1]The problem of the vibrating membrane was solved independently by Poisson, who published the solution in 1829.

and then, if we repeat this procedure k times,

$$\int_0^\infty t^k e^{-t}\, dt = k(k-1)\cdots 3\cdot 2 \int_0^\infty e^{-t}\, dt = k!$$

for any positive integer k. But the integral on the left has meaning for any positive real value of k and we could define $x!$ simply by replacing k with x in this equation. However, this has not become a standard notation. Instead, the usual notation, introduced by Legendre in his *Exercises de calcul intégral* of 1811–1817, is

$$\Gamma(x) = \int_0^\infty t^{x-1} e^{-t}\, dt$$

for $x > 0$. This is called the *gamma function*.

It is easy to see that the last two equations imply that

(*i*) $\Gamma(1) = 1$,

(*ii*) $\Gamma\left(\frac{1}{2}\right) = \int_0^\infty t^{-1/2} e^{-t}\, dt = \int_0^\infty 2 e^{-\tau^2}\, d\tau = \int_{-\infty}^\infty e^{-\tau^2}\, d\tau = \sqrt{\pi}$,

and, if $x > 0$ and k is a positive integer,

(*iii*) $\Gamma(x+1) = x\Gamma(x)$,

(*iv*) $\Gamma(x+k) = (x+k-1)(x+k-2)\cdots x\,\Gamma(x)$, and

(*v*) $\Gamma(k+1) = k!$

Equation (*iii*) is obtained integrating by parts; (*iv*), by induction from (*iii*); and (*v*), from (*iv*) with $x = 1$ and (*i*).

We can extend the definition of Γ to any negative value of x other than an integer by choosing a positive integer k so large that $x + k > 0$ and then using (*iv*) to define

(5) $$\Gamma(x) = \frac{\Gamma(x+k)}{x(x+1)\cdots(x+k-1)}$$

for $x < 0$. It is easy to see using (*iv*) that the quotient on the right is independent of the choice of k.

By definition, $\Gamma(x) > 0$ on $(0, \infty)$. Also, Γ is continuous on $(0, \infty)$ since it is continuous on $(0, a)$ for every $a > 0$ by Theorem B.7. From (*ii*) and (*iii*) we obtain the quotients

$$\frac{\Gamma(x+1)}{\Gamma(x)} = x \quad \text{and} \quad \frac{\Gamma(x)}{\Gamma(x+1)} = \frac{1}{x},$$

which approach infinity as $x \to \infty$ and $x \to 0+$, respectively, and then

$$\Gamma(x) \to \infty \quad \text{as} \quad x \to \infty \quad \text{and} \quad \Gamma(x) \to \infty \quad \text{as} \quad x \to 0+.$$

These limits can be used to draw the graph of Γ for $x > 0$. Then, for each positive integer k, its graph for $-k < x < -k+1$ can be constructed from (5) since $x + k > 0$

and the values of Γ are already known for positive values of the argument. Note, in particular, that $\Gamma(x+k) \to \infty$ as $x \to -k$ from the right, but, if $x \to -k+1$ from the left, we have $\Gamma(x+k) \to \Gamma(1) = 1$ and $x+k-1 \to 0$. These facts and (5) imply that $|\Gamma(x)| \to \infty$ as $x < 0$ approaches any integer. The sign of $\Gamma(x)$ on each subinterval $-k < x < -k+1$ should be clear from (5).

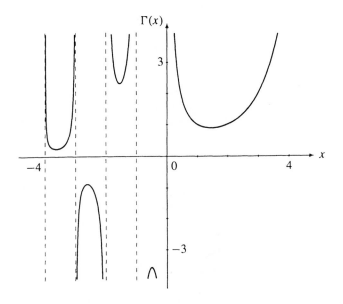

Figure 10.1

§10.3 Bessel Functions of the First Kind

On the basis of our study of the gamma function in §10.2 and if p is any real number other than a negative integer, we propose the following modification of (4) as a solution of the Bessel differential equation (2)

$$(6) \qquad y(x) = \sum_{k=0}^{\infty} \frac{(-1)^k}{k!\Gamma(p+k+1)} \left(\frac{x}{2}\right)^{2k+p}.$$

Of course, this coincides with (4) if p is a nonnegative integer because $\Gamma(p+k+1) = (p+k)!$ in that case. If $p \geq 0$, (6) is an infinite series that converges for all $x \geq 0$ (in fact, for all x in $\mathbb{R}$, but in applications to physical problems we are interested only in nonnegative values of x) by the ratio test because

$$\frac{k!\Gamma(p+k+1)2^{2k+p}|x|^{2k+2+p}}{(k+1)!\Gamma(p+k+2)2^{2k+2+p}|x|^{2k+p}} = \frac{|x|^2}{4(k+1)(p+k+1)}$$

approaches zero as $k \to \infty$. If $p < 0$, (6) contains two kinds of terms. Those with $2k + p < 0$ are finite in number and well defined for $x > 0$. The rest make up, as before, a convergent series. Therefore, this proposed solution of (2) is a well-defined function on $[0, \infty)$ for $p \geq 0$ and on $(0, \infty)$ for $p < 0$. The ratio test also shows, as above, that the first and second derived series of (6) converge for these values of x, and, by the Weierstrass M-test, the convergence is uniform on any bounded interval. Thus, this series can be differentiated term by term by Theorem A.6. This validates the steps in §10.1, so that the sum of (6) is indeed a solution of (2) if p is not a negative integer.

The first comprehensive study of the function defined by (6) for $p = 0$ was carried out by Fourier—while investigating the cooling of a cylinder—in Articles 116–139 of his 1807 memoir, including its zeros, some orthogonality properties (see §10.7), an integral form, its use in the terms of certain generalized Fourier series, and even its graph, hand drawn on page 193 of the manuscript. This function was also an essential component, but now for any integral value of p, of the extensive studies on the motion of the planets carried out by the German astronomer Friedrich Wilhelm Bessel (1784–1846), of the

FRIEDRICH WILHELM BESSEL
From H. S. Williams, *Modern Development of the Physical Sciences*, vol. 3, The Goodhue Company, 1912.

Könisberg observatory, and published in a series of papers from 1816 to 1824. Bessel, who wrote k instead of x and i instead of p, denoted the sum of (6) by $I_k{}^i$. One of his main contributions to the subject was establishing some recursion formulas relating the values of the function $I_k{}^i$ for different values of i (see §10.4). This function $I_k{}^i$ was named after Bessel by Carl Gustav Jacob Jacobi (1804–1851), of Berlin, in 1836, and also by Oskar Xavier Schlömilch (1823–1901), of Dresden, in 1857 and by Rudolph Lipschitz in 1859. In 1868, Heine proposed to name it *Fourier-Bessel function* to give credit to both researchers, but this term did not catch on. Neither did Bessel's notation, soon replaced by a variety of symbols that slowly evolved into the present notation: $J_p(x)$. It was first used by Isaac Todhunter (1820–1884) of Saint John's College, Cambridge, in 1875.

Definition 10.1 *For each real value of p that is not a negative integer the function J_p defined by*

$$J_p(x) = \sum_{k=0}^{\infty} \frac{(-1)^k}{k!\,\Gamma(p+k+1)}\left(\frac{x}{2}\right)^{2k+p}$$

on $[0, \infty)$ for $p \geq 0$ and on $(0, \infty)$ for $p < 0$ is called the **Bessel function of the first kind of order** p.

Example 10.1 If n is a nonnegative integer,

$$J_n(x) = \sum_{k=0}^{\infty} \frac{(-1)^k}{k!\,(n+k)!}\left(\frac{x}{2}\right)^{2k+n}$$

and, in particular,

$$J_0(x) = \sum_{k=0}^{\infty} \frac{(-1)^k}{(k!)^2}\left(\frac{x}{2}\right)^{2k}.$$

Example 10.2 Properties (*iv*), (*iii*), and (*ii*) of the gamma function, stated in the previous section, allow us to conclude that

$$\Gamma\left(\tfrac{3}{2}+k\right) = \left(\tfrac{3}{2}+k-1\right)\left(\tfrac{3}{2}+k-2\right)\cdots\tfrac{3}{2}\,\Gamma\left(\tfrac{3}{2}\right)$$

$$= \frac{1}{2^k}(2k+1)(2k-1)\cdots 3\cdot\tfrac{1}{2}\,\Gamma\left(\tfrac{1}{2}\right)$$

$$= \frac{1}{2^{k+1}}(2k+1)(2k-1)\cdots 3\cdot 1\cdot\sqrt{\pi}.$$

Then

$$J_{1/2}(x) = \sum_{k=0}^{\infty} \frac{(-1)^k}{k!\,\Gamma\left(\tfrac{3}{2}+k\right)}\left(\frac{x}{2}\right)^{2k+\frac{1}{2}}$$

$$= \sqrt{\frac{2}{\pi x}} \sum_{k=0}^{\infty} \frac{(-1)^k 2^{k+1}}{k!\,(2k+1)(2k-1)\cdots 3\cdot 1}\left(\frac{x}{2}\right)^{2k+1}$$

$$= \sqrt{\frac{2}{\pi x}} \sum_{k=0}^{\infty} \frac{(-1)^k}{2^k k(k-1)\cdots 2 \cdot 1 \cdot (2k+1)(2k-1)\cdots 3 \cdot 1} x^{2k+1}$$

$$= \sqrt{\frac{2}{\pi x}} \sum_{k=0}^{\infty} \frac{(-1)^k}{(2k)(2k-2)\cdots 4 \cdot 2 \cdot (2k+1)(2k-1)\cdots 3 \cdot 1} x^{2k+1}$$

$$= \sqrt{\frac{2}{\pi x}} \sum_{k=0}^{\infty} \frac{(-1)^k}{(2k+1)!} x^{2k+1}$$

$$= \sqrt{\frac{2}{\pi x}} \sin x.$$

Similarly,

$$J_{-1/2}(x) = \sqrt{\frac{2}{\pi x}} \cos x.$$

We shall see in §10.6 that for $x \gg 0$ any J_p behaves like a linear combination of $J_{1/2}$ and $J_{-1/2}$. In particular, this shows that J_p has infinitely many, arbitrarily large zeros.

The behavior of the Bessel functions at the origin will be found relevant throughout this chapter.

Lemma 10.1 *Let p be a real number that is not a negative integer. Then*
(i) $J_p(x) > 0$ *for sufficiently small $x > 0$,*
(ii) $J_0(0) = 1$,
(iii) *if $p > 0$, $J_p(0) = 0$, and*
(iv) *if $p < 0$, $|J_p(x)| \to \infty$ as $x \to 0$.*

Proof. Statements (ii) and (iii) are quite clear from the series in Definition 10.1, and the other two follow from the fact that, for very small x, this series shows that

$$J_p(x) \approx \frac{1}{\Gamma(p+1)} \left(\frac{x}{2}\right)^p,$$

Q.E.D.

Parts (ii) and (iii) of this lemma show that, if p is not an integer, then J_p and J_{-p} are linearly independent functions on $(0, \infty)$. This fact together with our findings in the opening paragraph of this section allows us to state the following result.

Theorem 10.1 *If p is not a negative integer, then J_p is a solution of (2) on $[0, \infty)$ for $p \geq 0$ and on $(0, \infty)$ for $p < 0$. If, in addition, p is not an integer the general solution of (2) on $(0, \infty)$ is*

$$y = AJ_p + BJ_{-p},$$

where A and B are arbitrary constants.

So far J_{-n} remains undefined if n is a positive integer. Note, however, that, if we let $p \to -n$ in the expression for J_p in Definition 10.1, then the terms of the series for $k = 0, \ldots, n-1$ approach zero because, as $p \to -n$, $p + k + 1$ approaches a negative integer or zero and then $|\Gamma(p + k + 1)| \to \infty$. This allows us to complete the definition of J_p as follows.

Definition 10.2 *If n is a positive integer, the* **Bessel function of the first kind of order** $-n$ *is the function* $J_{-n} \colon [0, \infty) \to \mathbb{R}$ *defined by*

$$J_{-n}(x) = \sum_{k=n}^{\infty} \frac{(-1)^k}{k!\,\Gamma(-n + k + 1)} \left(\frac{x}{2}\right)^{2k-n}.$$

As above, the series on the right, as well as its first two derived series, converges, and it is easy to see by direct substitution that J_{-n} is a solution of (2). However, it is not a very interesting solution because the series in this definition equals

$$\sum_{k=0}^{\infty} \frac{(-1)^{k+n}}{(k+n)!\,\Gamma(k+1)} \left(\frac{x}{2}\right)^{2k+n} = (-1)^n \sum_{k=0}^{\infty} \frac{(-1)^k}{\Gamma(k+n+1)k!} \left(\frac{x}{2}\right)^{2k+n} = (-1)^n J_n(x),$$

so that $J_{-n} = (-1)^n J_n$, and J_n and J_{-n} are not linearly independent.

§10.4 Recursion Formulas for Bessel Functions

The fact that a power series, such as that defining J_p, can be differentiated term by term and the equation $\Gamma(p+k+1) = (p+k)\Gamma(p+k)$ make it possible to derive a number of formulas relating some Bessel functions of different orders and their derivatives. It turns out that these formulas are extremely useful in establishing further properties of Bessel functions that we shall need later, and, although this may seem insufficient motivation at this point, we devote this short section to their study. All the formulas that we have in mind follow from the following two, which were proved by Bessel for integral p in 1824 and generalized to arbitrary p by Eugen Cornelius Joseph von Lommel (1837–1899), of Munich, in his *Studien über die Bessel'schen Functionen* of 1868.

Theorem 10.2 *If p is any real number and $x > 0$, then*

(i) $\dfrac{d}{dx}[x^p J_p(x)] = x^p J_{p-1}(x)$, *and*

(ii) $\dfrac{d}{dx}[x^{-p} J_p(x)] = -x^{-p} J_{p+1}(x).$

Proof. If p is not a negative integer, Definition 10.1 implies that

(7) $$x^p J_p(x) = \sum \frac{(-1)^k}{k!\,\Gamma(p+k+1)} \frac{x^{2k+2p}}{2^{2k+p}},$$

where the Σ on the right represents the addition of a finite sum over all k such that $2k + 2p < 0$ plus an infinite series over all k such that $2k + 2p \geq 0$. If, instead, p is a negative integer $-n$, then Definition 10.2 implies that

$$x^p J_p(x) = x^{-n} J_{-n}(x) = \sum_{k=n}^{\infty} \frac{(-1)^k}{k!\Gamma(-n+k+1)} \frac{x^{2k-2n}}{2^{2k-n}} = \sum \frac{(-1)^k}{k!\Gamma(p+k+1)} \frac{x^{2k+2p}}{2^{2k+p}},$$

which is identical in form to (7) but now Σ is an infinite series over all k from n to ∞. In either case, it is possible to differentiate term by term, and then

$$\frac{d}{dx}[x^p J_p(x)] = \sum \frac{(-1)^k (2k+2p)}{k!\Gamma(p+k+1)} \frac{x^{2k+2p-1}}{2^{2k+p}}$$

$$= \sum \frac{(-1)^k (k+p) x^p}{k!(p+k)\Gamma(p+k)} \left(\frac{x}{2}\right)^{2k+p-1}$$

$$= x^p \sum \frac{(-1)^k}{k!\Gamma(p+k)} \left(\frac{x}{2}\right)^{2k+p-1}$$

$$= x^p J_{p-1}.$$

This proves the first equation and the proof of the second is analogous, Q.E.D.

The equations in Theorem 10.2 can be rewritten as

$$x^p J_p'(x) + px^{p-1} J_p(x) = x^p J_{p-1}(x)$$

and

$$x^{-p} J_p'(x) - px^{-p-1} J_p(x) = -x^{-p} J_{p+1}(x).$$

Multiplying the first by x^{1-p} and the second by x^{1+p}, we obtain the following result.

Corollary 1 *For any real value of p and any $x > 0$,*

(i) $x J_p'(x) + p J_p(x) = x J_{p-1}(x)$ *and*

(ii) $x J_p'(x) - p J_p(x) = -x J_{p+1}(x).$

Subtracting and adding these equations we obtain two more relations.

Corollary 2 *For any real value of p and any $x > 0$,*

(i) $\dfrac{2p}{x} J_p(x) = J_{p-1}(x) + J_{p+1}(x)$ *and*

(ii) $J_p'(x) = \frac{1}{2}[J_{p-1}(x) - J_{p+1}(x)].$

Example 10.3 Putting $p = 0$ in Corollary 2 (ii), we obtain

$$J_0' = \frac{1}{2}(J_{-1} - J_1) = -J_1.$$

Example 10.4 Using Corollary 2(i) and the results of Example 10.2, we obtain

$$J_{3/2}(x) = \frac{1}{x} J_{1/2}(x) - J_{-1/2}(x) = \sqrt{\frac{2}{\pi x}} \left(\frac{\sin x}{x} - \cos x \right)$$

for $x > 0$.

Finally, the following integration formulas are an immediate consequence of Theorem 10.2 and the chain rule.

Corollary 3 *If p is any real number, $\alpha \neq 0$ is a constant, and $x > 0$, then*

(i) $\displaystyle \int x^p J_{p-1}(\alpha x) \, dx = \frac{x^p}{\alpha} J_p(\alpha x) + C$ *and*

(ii) $\displaystyle \int x^{-p} J_{p+1}(\alpha x) \, dx = -\frac{x^{-p}}{\alpha} J_p(\alpha x) + C,$

where C is an arbitrary constant.

§10.5 Bessel Functions of the Second Kind

In order to obtain the general solution of (2) if p is a nonnegative integer we still need a second linearly independent solution. We can try variation of constants and assume, following a classic method of d'Alembert, that it is of the form $y = c J_p$ where c is a function of x. Substitution into (2) yields[1]

$$c(x) = J_p(x) \int \frac{1}{x J_p^2(x)} \, dx.$$

Assuming now—as it turns out to be the case—that $1/J_p^2(x)$ can be expanded in a power series starting with a constant term, it follows that this second solution is a constant multiple of

$$J_p(x) \log x + w(x),$$

where $w(x)$ still has to be determined. This was done in 1867 by Carl Gottfried Neumann (1832–1925), a professor at Leipzig, for $p = 0$, and then, after deriving a recursion formula, he obtained the solution for any p by induction. However, a more convenient second solution was found independently by Hermann Hankel (1839–1873), of the University of Tübingen, in his 1869 memoir *Die Cylinderfunctionen erster und zweiter Art*. His definition was not valid if $2p$ is an odd integer, and for this reason it was modified in 1872 by Ernst Heinrich Weber (1842–1913), then at Heidelberg but soon to become a full professor at Könisberg, where he was one of Hilbert's teachers, and by Ludwig Schläfli (1814–1895) of the University of Bern (published in 1873 and 1875,

---

[1]For details see, for instance, D. G. Zill, *Differential Equations with Boundary Value Problems*, 2nd ed., PWS-KENT, Boston, 1989, pages 155–156.

respectively). Although Weber defined his solution by means of an integral, it turns out that if p is not an integer it can be written in the form

$$Y_p = \frac{J_p \cos p\pi - J_{-p}}{\sin p\pi}.$$

We shall omit a detailed account of the work that led to this function and simply observe that it is indeed a solution of (2) since it is a linear combination of solutions. The notation Y_p is due to Hankel and, except for a constant factor, this is the form proposed by Schläfli. Still, one may wonder about the reason for choosing this particular linear combination of J_p and J_{-p} since the stated quotient is undefined for every integral value of p. The reason will be stated at the end of §10.6, but note now that if n is an integer we can use l'Hôpital's rule to evaluate $\lim_{p \to n} Y_p$ from the expression above, and then define the resulting limit to be Y_n . We obtain

$$Y_n = \lim_{p \to n} \frac{-\pi J_p \sin p\pi + \cos p\pi \dfrac{\partial J_p}{\partial p} - \dfrac{\partial J_{-p}}{\partial p}}{\pi \cos p\pi} = \frac{1}{\pi} \left[\frac{\partial J_p}{\partial p} - (-1)^n \frac{\partial J_{-p}}{\partial p} \right]_{p=n}$$

since $\sin p\pi \to 0$ and $\cos p\pi \to (-1)^n$ as $p \to n$. The computations of the derivatives and limits in this expression were first carried out by Hankel, who also showed that the function so defined is a solution of (2) and obtained its expansion as an infinite series. However, these computations are rather long and tedious and we shall also omit them.[1] If we define

$$\varphi(k) = 1 + \frac{1}{2} + \frac{1}{3} + \cdots + \frac{1}{k}$$

for every positive integer k and then the so-called *Euler constant* by

$$\gamma = \lim_{k \to \infty} [\varphi(k) - \log k],$$

the results are

$$Y_0(x) = \frac{2}{\pi} \left(\log \frac{x}{2} + \gamma \right) J_0(x) - \frac{2}{\pi} \sum_{k=1}^{\infty} \frac{(-1)^k \varphi(k)}{(k!)^2} \left(\frac{x}{2} \right)^{2k}$$

and, if n is a positive integer,

$$Y_n(x) = \frac{2}{\pi} \left(\log \frac{x}{2} + \gamma \right) J_n(x) - \frac{1}{\pi} \sum_{k=0}^{n-1} \frac{(n-k-1)!}{k!} \left(\frac{x}{2} \right)^{2k-n}$$

$$- \frac{\varphi(n)}{\pi n!} \left(\frac{x}{2} \right)^n - \frac{1}{\pi} \sum_{k=1}^{\infty} \frac{(-1)^k [\varphi(k) + \varphi(k+n)]}{k!(k+n)!} \left(\frac{x}{2} \right)^{2k+n}$$

Then term-by-term differentiation and substitution into the Bessel differential equation of order n would show that Y_n is a solution for $n \geq 0$.

---

[1] See, for instance, G. N. Watson, *A Treatise on the Theory of Bessel Functions*, Cambridge University Press, Cambridge, 1958, pages 58–62.

Definition 10.3 *If n is a nonnegative integer, the function $Y_n : (0, \infty) \to \mathbb{R}$ defined by the two preceding equations is called the* **Bessel function of the second kind of order** *n.*

The behavior of Y_n near the origin is as follows.

Lemma 10.2 *If n is a nonnegative integer, then $Y_n(x) \to -\infty$ as $x \to 0$.*

Proof. For $n = 0$ this is clear from the expansion for $Y_0(x)$ since $J_0(x) \to 1$ as $x \to 0$. For $n > 0$ the well-known fact that $x^m \log x \to 0$ as $x \to 0$ for any positive integer m shows that the first term in the expansion for $Y_n(x)$ approaches zero as $x \to 0$. Then, for very small x,

$$Y_n(x) \approx -\frac{(n-1)!}{\pi}\left(\frac{x}{2}\right)^{-n},$$

and the result follows, Q.E.D.

Lemmas 10.1 and 10.2 show that J_n and Y_n are linearly inpendent for any nonnegative integer n. Since this is the only case we need to consider, we have found the general solution of (2) in the case excluded in Theorem 10.1.

Theorem 10.3 *If n is a nonnegative integer, then the general solution of (2) on $(0, \infty)$ is*

$$y = A J_n + B Y_n,$$

where A and B are arbitrary constants.

§10.6   The Zeros of Bessel Functions

At the end of §10.1 we asserted that Euler, in his solution of the vibrating membrane problem, was vague about the endpoint condition. Essentially, what he said is that there are infinitely many values of λ for which $J_p(a\sqrt{\lambda}) = 0$, so that—using (1), (6), and Definition 10.1—it is possible to have $R(a) = J_p(a\sqrt{\lambda}) = 0$, but he did not justify this statement. As we have already mentioned, Fourier showed in 1807 that J_0 has infinitely many zeros, a result that Bessel proved independently in 1824 and that was extended to J_p by von Lommel in 1868. We shall study the zeros of Bessel functions from the following formula, based on an idea proposed by Poisson in 1823 to approximate $J_0(x)$ for large values of x.

Theorem 10.4 *For any real value of p there are constants $A_p > 0$ and θ_p such that*

$$J_p(x) = \frac{A_p}{\sqrt{x}} \cos(x + \theta_p) + O(x^{-3/2})$$

for $x > 0$. An identical result holds for Y_p.[1]

---

[1] The Bachmann-Landau symbol O is used with the following meaning: $O(x^{-3/2})$ is a function of x such that $O(x^{-3/2}) \le Mx^{-3/2}$ for some constant $M > 0$ and all x large enough.

Proof. Let $x > 0$ and define a new function z by

$$z(x) = \sqrt{x}\, J_p(x).$$

Differentiating and using (2), we obtain

$$(8) \qquad z'' + z + \frac{1}{x^2}\left(\frac{1}{4} - p^2\right)z = \frac{\sqrt{x}}{x^2}\left[x^2 J_p'' + x J_p' + (x^2 - p^2)J_p\right] = 0.$$

For very large values of x this equation is like $z'' + z = 0$, whose solution is a linear combination of $\sin x$ and $\cos x$, that is, of the form $A\cos(x + \theta)$. This is what leads us to expect that $\sqrt{x}\, J_p(x) \approx A_p \cos(x + \theta_p)$, and to prove it we transform (8) into a first-order system

$$z' = y$$
$$y' = -z + \frac{1}{x^2}\left(p^2 - \frac{1}{4}\right)z.$$

Then, after the substitution $z = R\cos\theta$ and $y = -R\sin\theta$, where R and θ are functions of x, the first-order system becomes

$$-R\theta'\sin\theta + R'\cos\theta = -R\sin\theta$$
$$-R\theta'\cos\theta - R'\sin\theta = -R\cos\theta - \frac{R}{x^2}\left(p^2 - \frac{1}{4}\right)\cos\theta.$$

Solving for θ' and R' we obtain

$$\theta' = 1 + \frac{1}{x^2}\left(p^2 - \frac{1}{4}\right)\cos^2\theta$$

$$R' = \frac{R}{x^2}\left(p^2 - \frac{1}{4}\right)\sin\theta\cos\theta.$$

Integrating the first equation from 1 to x, we have

$$(9) \qquad \theta(x) = \theta(1) + x - 1 + \left(p^2 - \frac{1}{4}\right)\int_1^x \frac{1}{s^2}\cos^2\theta(s)\,ds,$$

and taking this expression for $\theta(x)$ to the second equation in the system gives a separable ordinary differential equation whose solution is

$$R(x) = R(1)e^{\left(p^2 - \frac{1}{4}\right)\int_1^x \frac{1}{s^2}\sin\theta(s)\cos\theta(s)\,ds}.$$

Now, the integral

$$\int_1^\infty \frac{1}{s^2}\cos^2\theta(s)\,ds$$

converges by Theorem B.3 because $0 \leq \cos^2 \theta(s) \leq 1$ and because the integral of s^{-2} over $(1, \infty)$ converges to 1. It follows that $\theta(x) - x$ has a limit as $x \to \infty$, which we denote by θ_p, and that (9) can be rewritten in terms of this limit as

$$\theta(x) = x + \theta_p - \left(p^2 - \frac{1}{4}\right) \int_x^\infty \frac{1}{s^2} \cos^2 \theta(s) \, ds.$$

Then, if we define

$$\vartheta_p(x) = -x\left(p^2 - \frac{1}{4}\right) \int_x^\infty \frac{1}{s^2} \cos^2 \theta(s) \, ds,$$

whose absolute value is bounded by

$$\left| x\left(p^2 - \frac{1}{4}\right) \int_x^\infty \frac{1}{s^2} \, ds \right| = \left| p^2 - \frac{1}{4} \right|,$$

we have

$$\theta(x) = x + \theta_p + \frac{\vartheta_p(x)}{x}.$$

Similarly, from the fact that $|\sin \theta(s) \cos \theta(s)| = \frac{1}{2}|\sin 2\theta(s)| \leq \frac{1}{2}$ for all s, it follows that the integral of $\sin \theta(s) \cos \theta(s)/s^2$ over $(1, \infty)$ converges, and then $R(x)$ has the limit

$$A_p = R(1) e^{\left(p^2 - \frac{1}{4}\right) \int_1^\infty \frac{1}{s^2} \sin \theta(s) \cos \theta(s) \, ds}$$

as $x \to \infty$. $A_p > 0$ because $R(1) > 0$, or else

$$0 = R^2(1) = z^2(1) + [z'(1)]^2$$

would imply that $z(1) = z'(1) = 0$, so that $z \equiv 0$ would be the unique solution of (8) corresponding to these initial conditions, and then $J_p \equiv 0$, which is a contradiction. Therefore, $R(x)$ is of the form

$$R(x) = A_p e^{-\left(p^2 - \frac{1}{4}\right) \int_x^\infty \frac{1}{s^2} \sin \theta(s) \cos \theta(s) \, ds}$$

with $A_p > 0$, and if we define

$$r_p(x) = -x\left(p^2 - \frac{1}{4}\right) \int_x^\infty \frac{1}{s^2} \sin \theta(s) \cos \theta(s) \, ds,$$

whose absolute value is bounded by $\frac{1}{2}\left|p^2 - \frac{1}{4}\right|$, we have

$$R(x) = A_p e^{r_p(x)/x}$$

and

$$(10) \qquad z(x) = R(x) \cos \theta(x) = A_p e^{r_p(x)/x} \cos\left(x + \theta_p + \frac{\vartheta_p(x)}{x}\right).$$

By the mean value theorem, there is a number c between zero and $r_p(x)/x$ such that

$$e^{r_p(x)/x} = 1 + \frac{r_p(x)}{x} e^{cr_p(x)/x}$$

and a number k between $x + \theta_p$ and $x + \theta_p + \vartheta_p(x)/x$ such that

$$\cos\left(x + \theta_p + \frac{\vartheta_p(x)}{x}\right) = \cos(x + \theta_p) - \frac{\vartheta_p(x)}{x} \sin\left(x + \theta_p + k\frac{\vartheta_p(x)}{x}\right).$$

Therefore, taking these expressions to (10),

$$J_p(x) = \frac{z(x)}{\sqrt{x}} = \frac{A_p}{\sqrt{x}} \cos(x + \theta_p) + \varphi(x),$$

where

$$\varphi(x) = \frac{A_p}{\sqrt{x}}\left[\frac{r_p(x)}{x} e^{cr_p(x)/x} \cos(x + \theta_p) \right.$$

$$\left. - \left(1 + \frac{r_p(x)}{x} e^{cr_p(x)/x}\right) \frac{\vartheta(x)}{x} \sin\left(x + \theta_p + k\frac{\vartheta_p(x)}{x}\right) \right],$$

which is $O(x^{-3/2})$.

The result is also valid for Y_p, since the only property of J_p that we have used in the preceding proof is that it is a solution of (2), Q.E.D.

Corollary $J_p(x) \to 0$ and $Y_p(x) \to 0$ as $x \to \infty$.

Theorem 10.5 *For any real value of p there is a real sequence $\{x_{pk}\}$ such that $0 < x_{p1} < x_{p2} \cdots < x_{pk} < \cdots$ and $x_{pk} \to \infty$ as $k \to \infty$ and such that J_p has a simple zero at each x_{pk} and no other positive zeros. Similarly this is true for Y_p.*

Proof. Let A_p and θ_p be as in Theorem 10.4 and let $0 < \epsilon < A_p$. By Theorem 10.4 we see that for large $x > 0$ the curve $y = \sqrt{x} \, J_p(x)$ is between the curve $y = A_p \cos(x + \theta_p) + \epsilon$ and the curve $y = A_p \cos(x + \theta_p) - \epsilon$. Since ϵ is arbitrarily small, it follows that J_p has infinitely many zeros, that for x large enough these are at points arbitrarily near $m\pi - \theta_p$, where m is a large positive integer, and, therefore, that there are arbitrarily large zeros.

The zeros are simple. Indeed, if J_p had a multiple zero at some $x_0 > 0$, then so would the function z in the proof of Theorem 10.4, and then $z(x_0) = z'(x_0) = 0$. But this would imply that $z \equiv 0$ is the unique solution of (8) that satisfies these initial conditions, and then $J_p \equiv 0$, which is a contradiction.

Finally, the zeros of J_p are isolated since this is true for any power series (usually proved in the complex case [1]). Thus, the zeros of J_p form a sequence. This is also the case for Y_p, Q.E.D.

---

[1] See, for instance, L. V. Ahlfors, *Complex Analysis*, 3rd ed., McGraw-Hill, 1979, page 127.

The information just gained about the zeros of either J_p or Y_p, together with the information about their behavior at the origin contained in Lemmas 10.1 and 10.2, allows us to construct their graphs. A few of these are shown in Figure 10.2. The graph of J_{-p} for $p > 0$ is considered in Exercise 6.4.

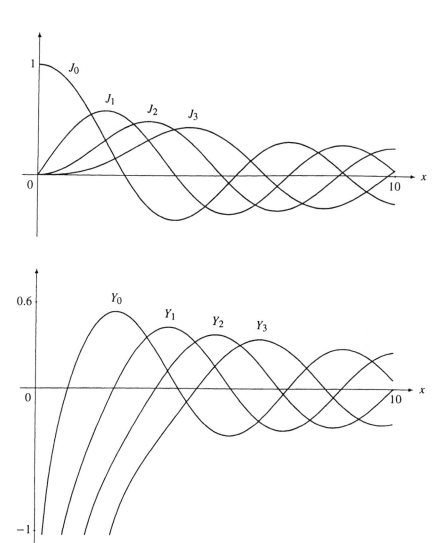

Figure 10.2

While the study of the zeros of J_p was motivated by the boundary condition of the problem in §10.1, other boundary conditions require knowledge of the zeros of $hJ_p(x) + xJ_p'(x)$, where h is a constant. The following result, a particular case of a theorem proved in 1903 by Alfred Cardew Dixon (1865–1936), will be useful in this respect.

Corollary *For any real value of p and any constant h there is a real sequence $\{x_{pk}\}$ such that $0 < x_{p1} < x_{p2} \cdots < x_{pk} < \cdots$ and $x_{pk} \to \infty$ as $k \to \infty$ and such that the function defined by*

$$F_p(x) = hJ_p(x) + xJ_p'(x)$$

has a zero at each x_{pk} and no other positive zeros. Similarly this is true for Y_p.

Proof. Let $r, s > 0$ be two consecutive zeros of J_p. This and the fact that zeros are simple implies that $J_p'(r)J_p'(s) < 0$. Then

$$F_p(r)F_p(s) = rsJ_p'(r)J_p'(s) < 0,$$

and the intermediate value theorem implies that $F_p(x) = 0$ for some x between r and s. The zeros of F_p are isolated for the same reason given in the proof of the theorem, Q.E.D.

The values of the constants A_p and θ_p in Theorem 10.4 are actually known. The precise formula is

$$(11) \qquad J_p(x) = \sqrt{\frac{2}{\pi x}} \cos\left(x - \frac{p\pi}{2} - \frac{\pi}{4}\right) + O(x^{-3/2}),$$

but there is no brief proof.[1] For $p = 0$ it was stated by Poisson in 1823 and proved by Lipschitz in 1859, for $p = 1$ it was stated by Peter Andreas Hansen (1795–1847), of Gotha, in 1843, and for any integral p it was stated by Jacobi in 1849. For arbitrary p, (11) and the companion formula

$$(12) \qquad Y_p(x) = \sqrt{\frac{2}{\pi x}} \sin\left(x - \frac{p\pi}{2} - \frac{\pi}{4}\right) + O(x^{-3/2})$$

were obtained and proved by Hankel in 1868. It is the likeness of equations (11) and (12) that motivates the selection of Y_p as the second linearly independent solution of the Bessel differential equation of integral order p.

§10.7 Orthogonal Systems of Bessel Functions

In §10.1 we posed the problem of finding a bounded solution of the equation $r^2R'' + rR' + (\lambda r^2 - p^2)R = 0$ on $(0, a)$, where $\lambda > 0$ and $p \geq 0$, subject to the condition

---

[1] For a proof see, for instance, J. L. Troutman, *Boundary Value Problems of Applied Mathematics*, PWS, Boston, 1994, pages 365–369.

$R(a) = 0$. Anticipating a wider range of applications (see Exercises 11.1 and 11.6), we now replace this last requirement with a more general boundary condition and also, to conform to the general notation of Chapter 4, we replace R and r with X and x. Thus, we pose the problem of finding bounded solutions of the singular Sturm-Liouville problem

$$x^2 X'' + x X' + (\lambda x^2 - p^2)X = 0 \qquad \text{on} \quad (0, a)$$
$$c_1 X(a) + c_2 X'(a) = 0,$$

where $\lambda > 0$, $p \geq 0$, and c_1 and c_2 are nonnegative constants such that $c_1 + c_2 > 0$. This restriction on the constants is due to reasons similar to those already explained in §3.7 and §5.6. If $c_2 = 0$, the boundary condition is $X(a) = 0$, so that this is the Sturm-Liouville problem posed in §10.1 for the vibrating membrane. If, on the other hand, $c_2 \neq 0$, we have a different problem for which the boundary condition can be writen as $h X(a) + a X'(a) = 0$, where $h = a c_1/c_2$. Equivalently, our Sturm-Liouville problem is

(13) $$(x X')' + \left(\lambda x - \frac{p^2}{x}\right) X = 0 \qquad \text{on} \quad (0, a)$$

subject to either

(14) $$X(a) = 0$$

or

(15) $$h X(a) + a X'(a) = 0, \qquad h \geq 0.$$

Now, the substitution $s = x\sqrt{\lambda}$, $y(s) = X(s/\sqrt{\lambda})$ transforms the differential equation into the Bessel equation of order p. The functions J_{-p} and Y_n in Theorems 10.1 and 10.3 are unbounded near the origin, and then any bounded solution of the Bessel equation is of the form

$$y(s) = C J_p(s),$$

where C is an arbitrary constant. Therefore,

$$X(x) = C J_p(x\sqrt{\lambda}).$$

The endpoint conditions at $x = a$ stated above imply that

$$J_p(a\sqrt{\lambda}) = 0$$

or

$$h J_p(a\sqrt{\lambda}) + a\sqrt{\lambda}\, J_p'(a\sqrt{\lambda}) = 0,$$

and then, according to Theorem 10.5 or its corollary, λ must be such that

$$a\sqrt{\lambda} = x_{pk},$$

where x_{pk} is a positive zero of $J_p(x)$ or of $hJ_p(x) + xJ_p'(x)$. Then

$$\sqrt{\lambda} = \alpha_{pk},$$

where $a\alpha_{pk} = x_{pk}$; that is, α_{pk} is a positive zero of either $J_p(ax)$ or $hJ_p(ax)+axJ_p'(ax)$. This means that our Sturm-Liouville problem has eigenvalues

$$\lambda_k = \alpha_{pk}^2$$

and corresponding eigenfunctions

$$J_p(x\sqrt{\lambda_k}) = J_p(\alpha_{pk}x)$$

for every positive integer k.

The condition $\lambda > 0$ was imposed in §10.1 on the basis of physical considerations about the motion of the membrane. However, λ may not be so restricted in other applications, and the existence of nonpositive eigenvalues of the stated Sturm-Liouville problem must be proved or disproved. It will be shown at the end of §10.10 that there are no negative eigenvalues. On the other hand, Ulisse Dini noticed in 1880 that $\lambda = 0$ is an eigenvalue in the particular case $p = h = 0$. Indeed, if $\lambda = 0$, the differential equation for X is of the Cauchy-Euler type and can be solved using the substitution $x = e^t$, as indicated in Exercise 1.5 of Chapter 4. Omitting the details, its general solution is

$$X(x) = \begin{cases} Ax^p + Bx^{-p} & \text{if} \quad p > 0 \\ A + B\log x & \text{if} \quad p = 0, \end{cases}$$

where A and B are arbitrary constants. For a bounded solution near the origin $B = 0$, and then the boundary condition $X(a) = 0$ implies that $A = 0$ and that $X \equiv 0$, so that in this case $\lambda = 0$ is not an eigenvalue. If $p > 0$, the boundary condition $hX(a) + aX'(a) = 0$ implies that

$$hAa^p + aApa^{p-1} = A(h + p)a^p = 0,$$

and then $A = 0$ and $X \equiv 0$ so that $\lambda = 0$ is not an eigenvalue in this case either. Finally, if $p = 0$, the boundary condition $hX(a) + aX'(a) = 0$ implies that $hA = 0$, so that $A \neq 0$ if and only if $h = 0$. This shows that $\lambda = 0$ is an eigenvalue, with corresponding eigenfunction $X \equiv 1$, if and only if $p = h = 0$. We have proved the following results.

Theorem 10.6 *Let $p \geq 0$, $a > 0$, and $h \geq 0$ be arbitrary constants and let $\{\alpha_{pk}\}$ be the sequence of positive zeros of $J_p(ax)$. Then the eigenvalues and eigenfunctions of the singular Sturm-Liouville problem* (13) *and* (14) *are*

$$\lambda_k = \alpha_{pk}^2 \qquad \text{and} \qquad X_k(x) = J_p(\alpha_{pk}x)$$

for all positive integral values of k.

Theorem 10.7 *Let $p \geq 0$, $a > 0$, and $h \geq 0$ are arbitrary constants and let $\{\alpha_{pk}\}$ be the sequence of positive zeros of $hJ_p(ax) + axJ_p'(ax)$. Then the eigenvalues and* *eigenfunctions of the singular Sturm-Liouville problem* (13) *and* (15) *are*

   (i) $\lambda_k = \alpha_{pk}^2$ *and* $X_k(x) = J_p(\alpha_{pk}x)$ *for all positive integral values of* k *if* $p + h > 0$

and

   (ii) $\lambda_0 = 0$ *and* $X_0 \equiv 1$ *plus* $\lambda_k = \alpha_{pk}^2$ *and* $X_k(x) = J_p(\alpha_{pk}x)$ *for all positive integral values of* k *if* $p = h = 0$.

According to Theorem 4.1 of Chapter 4, eigenfunctions corresponding to distinct eigenvalues are orthogonal. That is, using the weight function $r(x) = x$,

$$\int_0^a x J_p(\alpha_{pj}x) J_p(\alpha_{pk}x)\, dx = 0$$

for $j \neq k$, where $\{\alpha_{pk}\}$ is the sequence of positive zeros of $J_p(ax)$ or $h J_p(ax) + ax J_p'(ax)$. In the case of the additional eigenfunction of Theorem 10.7,

$$\int_0^a x J_0(\alpha_{0k}x)\, dx = 0,$$

where $\{\alpha_{0k}\}$ is the sequence of positive zeros of $J_1(ax)$, since $0 J_0(ax) + ax J_0'(ax) = ax J_0'(ax) = -ax J_1(ax)$.

While the problem of the vibrating membrane has led us to consider a singular Sturm-Liouville problem on $(0, a)$, other applications may lead to regular Sturm-Liouville problems on an interval of the form (a, b) with $0 < a < b$ (see Exercise 11.4). Then the general solution for X is of the form

$$X(x) = A J_p(x\sqrt{\lambda}) + B J_{-p}(x\sqrt{\lambda})$$

if p is not an integer or of the form

$$X(x) = A J_p(x\sqrt{\lambda}) + B Y_p(x\sqrt{\lambda})$$

if p is a nonnegative integer. The constants A and B are found from the endpoint conditions at $x = a$ and $x = b$, both of which must be present now.

§10.8 Fourier-Bessel Series and Dini-Bessel Series

We now consider the possibility of a generalized Fourier series expansion of a given function f with respect to the eigenfunctions of either Theorem 10.6 or Theorem 10.7,

$$f(x) = c_0 + \sum_{k=1}^{\infty} c_k J_p(\alpha_{pk}x),$$

where $c_0 = 0$ unless $p = h = 0$, in which case $\{\alpha_{0k}\}$ is the sequence of positive zeros of $J_0'(ax) = -J_1(ax)$. As we shall see in the next section when we return to the vibrating membrane, this question comes up when dealing with initial conditions in boundary value problems involving these eigenfunctions. It is easy to obtain the coefficients if we assume that equality holds and that integration term by term is permissible. Then multiplying both sides by $J_p(\alpha_{pj}x)$, integrating from zero to a, and using orthogonality,

$$\int_0^a xf(x)J_p(\alpha_{pj}x)\,dx = c_j \int_0^a xJ_p^2(\alpha_{pj}x)\,dx.$$

Since the integral on the right is the square of the norm of $J_p(\alpha_{pj}x)$, this leads us to belive that

$$c_k = \frac{1}{\|J_p(\alpha_{pk}x)\|^2} \int_0^a xf(x)J_p(\alpha_{pk}x)\,dx$$

for all k, a result that must be justified.

Our first step will be to compute the norms of $J_p(\alpha_{pk}x)$ with respect to the given inner product. The basic fact is the following lemma proved by von Lommel in 1879.

Lemma 10.3 *For every $p \geq 0$ and $\alpha > 0$,*

$$\int_0^a xJ_p^2(\alpha x)\,dx = \frac{a^2}{2}\left[(J_p')^2(a\alpha) + \left(1 - \frac{p^2}{a^2\alpha^2}\right)J_p^2(a\alpha)\right].$$

Proof. Since J_p satisfies the Bessel differential equation we have

$$(\alpha x)^2 J_p''(\alpha x) + \alpha x J_p'(\alpha x) + [(\alpha x)^2 - p^2]J_p(\alpha x) = 0,$$

and if we define $u(x) = J_p(\alpha x)$ to simplify the writing, this equation becomes

$$x^2 u'' + xu' + (\alpha^2 x^2 - p^2)u = 0.$$

Then, multiplying by $2u'$, this equation becomes

$$2x^2 u'' u' + 2x(u')^2 + 2(\alpha^2 x^2 - p^2)uu' = 0,$$

that is,

$$[(xu')^2]' + (\alpha^2 x^2 - p^2)(u^2)' = 0.$$

Since $J_p(\alpha x)$ is continuous and bounded on $[0, a]$ for $p \geq 0$ we can integrate from zero to a. Using integration by parts for the second term, we obtain

$$\left[(xu')^2 + (\alpha^2 x^2 - p^2)u^2\right]_0^a - 2\alpha^2 \int_0^a xu^2(x)\,dx,$$

that is,

$$2\alpha^2 \int_0^a xJ_p^2(\alpha x)\,dx = a^2[\alpha J_p'(\alpha a)]^2 + (\alpha^2 a^2 - p^2)J_p^2(\alpha a) + p^2 J_p^2(0).$$

But $pJ_p(0) = 0$ for $p \geq 0$, and the result follows, Q.E.D.

Then von Lommel applied the preceding result to compute the norms of the functions in the orthogonal system of Theorem 10.6 with respect to the given inner product as follows.

Theorem 10.8 *If $p \geq 0$, $a > 0$, and $h \geq 0$ are arbitrary constants and if $\{\alpha_{pk}\}$ is the sequence of positive zeros of*
(i) $J_p(ax)$, *then*

(16) $$\|J_p(\alpha_{pk}x)\|^2 = \tfrac{1}{2}a^2 J_{p+1}^2(a\alpha_{pk}) = \tfrac{1}{2}a^2 J_{p-1}^2(a\alpha_{pk}),$$

and of
(ii) $h J_p(ax) + ax J_p'(ax)$, *then*

(17) $$\|J_p(\alpha_{pk}x)\|^2 = \frac{h^2 - p^2 + a^2\alpha_{pk}^2}{2\alpha_{pk}^2} J_p^2(a\alpha_{pk}).$$

Proof. To prove (16) use Corollary 2 of Theorem 10.2 to obtain

$$J_p'(a\alpha_{pk}) = \frac{p}{a\alpha_{pk}} J_p(a\alpha_{pk}) - J_{p+1}(a\alpha_{pk}) = J_{p-1}(a\alpha_{pk}) - \frac{p}{a\alpha_{pk}} J_p(\alpha_{pk})$$

and note that in case (i) $J_p(a\alpha_{pk}) = 0$. Then (16) results from taking these values of $J_p'(a\alpha_{pk})$ to Lemma 10.3.

In case (ii) we have

$$J_p'(a\alpha_{pk}) = -\frac{h}{a\alpha_{pk}} J_p(a\alpha_{pk}),$$

and then Lemma 10.3 implies that

$$\|J_p(a\alpha_{pk})\|^2 = \frac{a^2}{2}\left(\frac{h^2}{a^2\alpha_{pk}^2} + 1 - \frac{p^2}{a^2\alpha_{pk}^2}\right) J_p^2(a\alpha_{pk}),$$

which proves (17), Q.E.D.

From now on we shall consider the two cases in this theorem separately.

Definition 10.4 *Let $p \geq 0$ and $a > 0$ be constants and let $f : (0, a) \to \mathbb{R}$ be an integrable function. If $\{\alpha_{pk}\}$ is the sequence of positive zeros of $J_p(ax)$, then the series*

$$\sum_{k=1}^{\infty} c_k J_p(\alpha_{pk}x),$$

where

$$c_k = \frac{2}{a^2 J_{p+1}^2(a\alpha_{pk})} \int_0^a x f(x) J_p(\alpha_{pk}x)\, dx,$$

is called the **Fourier-Bessel series** *of f of order p on $(0, a)$.*

It can be shown that the orthogonal system of Theorem 10.6 has the Parseval property and is, therefore, complete, and then the Fourier-Bessel series converges in the mean.[1] For pointwise convergence, the following result, first stated by Fourier in 1807 for $p = 0$, was proved by Hankel in 1869 and independently by Schläfli in 1876 using the theory of residues. The first proofs without the use of complex analysis were provided by Kneser in 1903 and 1907 and by Ernest William Hobson (1856–1933), of Christ's College, Cambridge, in 1909. All these proofs are beyond the scope of this book.[2]

Theorem 10.9 *Let $p \geq 0$ and $a > 0$ be constants and let $f : (0, a) \to \mathbb{R}$ be a piecewise continuous function with a piecewise continuous derivative. Then the Fourier-Bessel series of f order p on $(0, a)$ converges to the value*

$$\tfrac{1}{2}[f(x+) + f(x-)]$$

at each point x in $(0, a)$.

In the next section we shall use this result to complete the solution of the boundary value problem of §10.1.

Example 10.5 If p is a nonnegative real number the coefficients of the Fourier-Bessel series of $f(x) = x^p$ of order p on $(0, a)$ are

$$c_k = \frac{2}{a^2 J_{p+1}^2(a\alpha_{pk})} \int_0^a x^{p+1} J_p(\alpha_{pk} x) \, dx.$$

Using now Corollary 3 (ii) of Theorem 10.2, we obtain

$$\int_0^a x^{p+1} J_p(\alpha_{pk} x) \, dx = \frac{x^{p+1}}{\alpha_{pk}} J_{p+1}(\alpha_{pk} x) \Big|_0^a = \frac{a^{p+1}}{\alpha_{pk}} J_{p+1}(a\alpha_{pk}).$$

Therefore, according to Definition 10.4, the Fourier-Bessel series of x^p of order p on $(0, a)$ is

$$x^p = 2a^{p-1} \sum_{k=1}^{\infty} \frac{1}{\alpha_{pk} J_{p+1}(a\alpha_{pk})} J_p(\alpha_{pk} x).$$

Equality holds because f is of class C^1 on $(0, a)$.

In particular, for $p = 0$ we obtain the representation

$$1 = \frac{2}{a} \sum_{k=1}^{\infty} \frac{1}{\alpha_{0k} J_1(a\alpha_{0k})} J_0(\alpha_{0k} x).$$

---

[1]See, for instance, K. Yosida, *Lectures on Differential and Integral Equations*, Interscience, New York, 1960, page 204.

[2]See, for instance, G. N. Watson, *op. cit.*, Cambridge University Press, Cambridge, 1958, pages 591–592.

The orthogonal system of Theorem 10.7 includes the function $X_0 \equiv 1$ if $p = h = 0$. The generalized Fourier coefficient of f corresponding to X_0 is

$$c_0 = \frac{\displaystyle\int_0^a x f(x)\, dx}{\displaystyle\int_0^a x\, dx} = \frac{2}{a^2} \int_0^a x f(x)\, dx.$$

Definition 10.5 *Let* $p \geq 0, a > 0,$ *and* $h \geq 0$ *be constants and let* $f : (0, a) \to \mathbb{R}$ *be an integrable function. If* $\{\alpha_{pk}\}$ *is the sequence of positive zeros of* $h J_p(ax) + ax J_p'(ax),$ *then the series*

$$c_0 + \sum_{k=1}^{\infty} c_k J_p(\alpha_{pk} x),$$

where

$$c_0(x) = \begin{cases} \dfrac{2}{a^2} \displaystyle\int_0^a x f(x)\, dx & \text{if } p = h = 0 \\[4mm] 0 & \text{if } p + h > 0 \end{cases}$$

and

$$c_k = \frac{2\alpha_{pk}^2}{(h^2 - p^2 + a^2 \alpha_{pk}^2) J_p^2(a\alpha_{pk})} \int_0^a x f(x) J_p(\alpha_{pk} x)\, dx$$

for $k \geq 1,$ *is called the* **Dini-Bessel series** *of* f *of order* p *on* $(0, a)$.

The proof of the following theorem can be reduced to that of Theorem 10.9.[1]

Theorem 10.10 *Let* $p \geq 0,\ a > 0,$ *and* $h \geq 0$ *be constants and let* $f : (0, a) \to \mathbb{R}$ *be a piecewise continuous function with a piecewise continuous derivative. Then the Dini-Bessel series of* f *of order* p *on* $(0, a)$ *converges to the value*

$$\tfrac{1}{2}[f(x+) + f(x-)]$$

at each point x *in* $(0, a)$.

Example 10.6 The coefficients of the Dini-Bessel series of $f(x) = x^2$ of order zero on $(0, a)$ with $h = 0$ are

$$c_0 = \frac{2}{a^2} \int_0^a x^3\, dx = \frac{a^2}{2}$$

and, for $k \geq 1,$

$$c_k = \frac{2}{a^2 J_0^2(a\alpha_{0k})} \int_0^a x^3 J_0(\alpha_{0k} x)\, dx.$$

---

[1] See, for instance, Watson *op. cit.*, pages 601–602.

Upon integration by parts and using Corollary 3 (i) twice (or see Exercise 4.6) and then Corollary 2 (i) of Theorem 10.2, the integral on the right becomes

$$
\int_0^a x^2 [x J_0(\alpha_{0k} x)] \, dx = \left[x^2 \frac{x}{\alpha_{0k}} J_1(\alpha_{0k} x) \right]_0^a - \frac{2}{\alpha_{0k}} \int_0^a x^2 J_1(\alpha_{0k} x) \, dx
$$

$$
= \frac{a^3}{\alpha_{0k}} J_1(a\alpha_{0k}) - \frac{2}{\alpha_{0k}} \left[\frac{x^2}{\alpha_{0k}} J_2(\alpha_{0k} x) \right]_0^a
$$

$$
= \frac{a^3}{\alpha_{0k}} J_1(a\alpha_{0k}) - \frac{2a^2}{\alpha_{0k}^2} J_2(a\alpha_{0k})
$$

$$
= \frac{a^3}{\alpha_{0k}} J_1(a\alpha_{0k}) - \frac{2a^2}{\alpha_{0k}^2} \left[\frac{2}{a\alpha_{0k}} J_1(a\alpha_{0k}) - J_0(a\alpha_{0k}) \right].
$$

But $\{\alpha_{0k}\}$ is the sequence of positive zeros of $J_1(ax)$ because $h J_0(ax) + ax J_0'(ax) = ax J_0'(ax) = -ax J_1(ax)$, and then

$$
c_k = \frac{2}{a^2 J_0^2(a\alpha_{0k})} \frac{2a^2}{\alpha_{0k}^2} J_0(a\alpha_{0k}) = \frac{4}{\alpha_{0k}^2 J_0(a\alpha_{0k})}.
$$

Therefore,

$$
x^2 = \frac{a^2}{2} + 4 \sum_{k=1}^\infty \frac{1}{\alpha_{0k}^2 J_0(a\alpha_{0k})} J_0(\alpha_{0k} x)
$$

because f is of class C^1 on $(0, a)$.

§10.9   Return to the Vibrating Membrane

In §10.1 we posed the boundary value problem

$$
\begin{array}{llll}
u_{tt} = c^2 \left(u_{rr} + \dfrac{1}{r} u_r + \dfrac{1}{r^2} u_{\theta\theta} \right) & r < a, & -\pi < \theta < \pi, & t > 0 \\[2mm]
u(a, \theta, t) = 0 & & -\pi \le \theta \le \pi, & t \ge 0 \\[1mm]
u(r, \theta, 0) = f(r, \theta) & r \le a, & -\pi \le \theta \le \pi \\[1mm]
u_t(r, \theta, 0) = 0 & r \le a, & -\pi \le \theta \le \pi
\end{array}
$$

corresponding to the displacements of a circular membrane that is stretched over a fixed circular frame of radius a and vibrating without friction after an initial displacement given by $f(r, \theta)$ but with no initial velocity. We found that, if we assume a solution of the form $u(r, \theta, t) = R(r)\Theta(\theta)T(t)$, then

$$
\begin{array}{ll}
T'' + c^2 \lambda T = 0 & T'(0) = 0 \\[1mm]
\Theta'' + p^2 \Theta = 0 & \Theta(-\pi) = \Theta(\pi) \\[1mm]
r^2 R'' + r R' + (\lambda r^2 - p^2) R = 0 & R(a) = 0,
\end{array}
$$

where $\lambda > 0$.

The periodic Sturm-Liouville problem for Θ has the eigenvalue $p = 0$ with eigen-function $\frac{1}{2}$ (in fact, any constant will do) and the additional eigenvalues $p = n$ for each positive integer n with corresponding eigenfunctions $\cos n\theta$ and $\sin n\theta$. Then, in §10.7, we showed that the equation for R, now with $p = n \geq 0$, has eigenvalues $\lambda_k = \alpha_{nk}^2$ and corresponding eigenfunctions $J_n(\alpha_{nk}r)$, $k \geq 1$, where $\{\alpha_{nk}\}$ is the sequence of positive values of r such that $J_n(ar) = 0$. Incorporating the solutions

$$T(t) = A \cos \sqrt{\lambda_k}\, ct = A \cos \alpha_{nk} ct$$

of the initial value problem for T into the product $R(r)\Theta(\theta)T(t)$ and trying an infinite series of such solutions we propose

$$(18) \quad u(r, \theta, t) = \tfrac{1}{2} \sum_{k=1}^{\infty} a_{0k} J_0(\alpha_{0k}r) \cos \alpha_{0k} ct$$
$$+ \sum_{n=1}^{\infty} \left[\sum_{k=1}^{\infty} J_n(\alpha_{nk}r)(a_{nk} \cos n\theta + b_{nk} \sin n\theta) \cos \alpha_{nk} ct \right],$$

where a_{nk} and b_{nk} are arbitrary constants. Then, in order to satisfy the initial condition $u(r, \theta, 0) = f(r, \theta)$, we must have

$$f(r, \theta) = \tfrac{1}{2} \sum_{k=1}^{\infty} a_{0k} J_0(\alpha_{0k}r)$$
$$+ \sum_{n=1}^{\infty} \left[\left(\sum_{k=1}^{\infty} a_{nk} J_n(\alpha_{nk}r) \right) \cos n\theta + \left(\sum_{k=1}^{\infty} b_{nk} J_n(\alpha_{nk}r) \right) \sin n\theta \right].$$

Thus, for each fixed r, the right-hand side must be the Fourier series of $f(r, \cdot)$ on $(-\pi, \pi)$, that is,

$$\sum_{k=1}^{\infty} a_{nk} J_n(\alpha_{nk}r) = \frac{1}{\pi} \int_{-\pi}^{\pi} f(r, \theta) \cos n\theta\, d\theta, \qquad n = 0, 1, 2, \ldots$$

$$\sum_{k=1}^{\infty} b_{nk} J_n(\alpha_{nk}r) = \frac{1}{\pi} \int_{-\pi}^{\pi} f(r, \theta) \sin n\theta\, d\theta, \qquad n = 1, 2, 3, \ldots .$$

Next, according to Theorem 10.9, the constants a_{nk} and b_{nk} must be chosen so that the series on the left are the Fourier-Bessel series of order n of the functions of r on the right on $(0, a)$. That is, using Definition 10.4,

$$(19) \qquad a_{nk} = \frac{2}{\pi a^2 J_{n+1}^2(a\alpha_{nk})} \int_0^a r \left(\int_{-\pi}^{\pi} f(r, \theta) \cos n\theta\, d\theta \right) J_n(\alpha_{nk}r)\, dr$$

$$(20) \qquad b_{nk} = \frac{2}{\pi a^2 J_{n+1}^2(a\alpha_{nk})} \int_0^a r \left(\int_{-\pi}^{\pi} f(r, \theta) \sin n\theta\, d\theta \right) J_n(\alpha_{nk}r)\, dr.$$

This completes the construction of a formal solution, which we shall be unable to verify.[1]

Two things can easily be deduced from this formal solution. The first is that the individual frequencies of vibration of the different components, $\alpha_{nk}c/2\pi$, are not integral multiples of a fundamental frequency, not even close. This is clear from an examination of some of the zeros of Bessel functions compiled in the following table for $0 \leq n \leq 4$ and $1 \leq k \leq 5$.[2] What this means is that the sound of a drum is not as good in tone

k	1	2	3	4	5
$a\alpha_{0k}$	2.40483	5.52008	8.65373	11.79153	14.93092
$a\alpha_{1k}$	3.83171	7.01559	10.17347	13.32369	16.47063
$a\alpha_{2k}$	5.13562	8.41724	11.61984	14.79595	17.95982
$a\alpha_{3k}$	6.38016	9.76102	13.01520	16.22347	19.40942
$a\alpha_{4k}$	7.58834	11.06471	14.37254	17.61597	20.82693

quality as the sound of a string instrument. On the other hand, there is something in common between the individual vibrations of a drum and those of a string. In the same manner as there were nodes in each of the component vibrations of a string, there are nodal lines in the vibrations of a drum, that is, lines that do not move at all with the passing of time. The amplitudes of the individual vibrations for the drum of our problem are constant multiples of the functions defined by

$$A_{nkc}(r, \theta) = J_n(\alpha_{nk}r) \cos n\theta \qquad \text{or} \qquad A_{nks}(r, \theta) = J_n(\alpha_{nk}r) \sin n\theta,$$

and they vanish wherever one of the factors on the corresponding right-hand side is zero. Note that $J_n(\alpha_{nk}r) = 0$ for

$$r = \frac{a\alpha_{nj}}{\alpha_{nk}},$$

which happens for $r < a$ if $1 \leq j < k$. This is represented (not to scale) for a few values of n and k in Figure 10.3. The plus and minus signs indicate the regions in which $A_{nkc}(r, \theta)$ or $A_{nkc}(r, \theta)$ is positive or negative.

If the given initial conditions are replaced with

$$u(r, \theta, 0) = 0 \qquad r \leq a, \qquad -\pi \leq \theta \leq \pi$$
$$u_t(r, \theta, 0) = g(r, \theta) \qquad r \leq a, \qquad -\pi \leq \theta \leq \pi,$$

the new problem is solved as above but replacing $\cos \alpha_{nk}ct$ with $\sin \alpha_{nk}ct$ in (18). If this series can be differentiated term by term, then, in order to satisfy the initial condition,

---

[1] For a verification if f is of class C^4 see, for instance, H. F. Weinberger, *A First Course in Partial Differential Equations*, Ginn, Waltham, Massachusetts, 1965, pages 182–184.

[2] From M. Abramowitz and I. A. Stegun, Eds., *Handbook of Mathematical Functions*, Dover, New York, 1965, page 409.

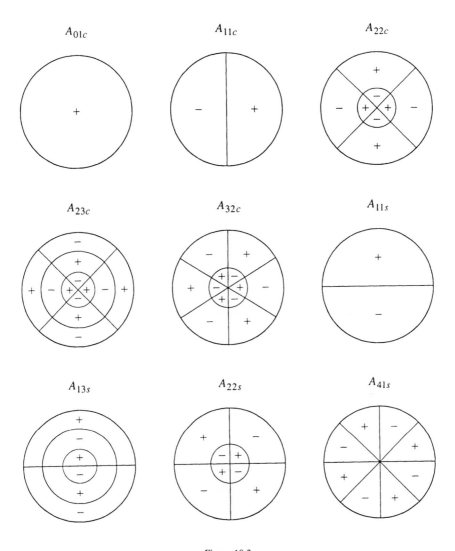

Figure 10.3

the coefficients a_{nk} and b_{nk} are obtained from (19) and (20) after replacing f by g and dividing the right-hand sides by $-\alpha_{nk}c$. The details are left to Exercise 9.2. A more general problem in which the initial position and velocity are nonvanishing is solved by superposition.

Up to this moment we have considered the free vibrations of a membrane only if its mass per unit area is constant and in the absence of any external forces. If such forces are present, the wave equation must be replaced with

$$
(21) \qquad mu_{tt} = \tau\left(u_{rr} + \frac{1}{r}u_r + \frac{1}{r^2}u_{\theta\theta}\right) + q(r,\theta,t).
$$

The simplest case to deal with is that in which q is independent of θ and t. First we find a bounded solution U that depends only on r and satisfies the boundary condition $U(a) = 0$. That is,

$$
(U')' + \frac{1}{r}U' = q(r),
$$

and U is found by integrating the general solution of this first-order equation for U' and then determining the constants of integration from the fact that U must be bounded and $U(a) = 0$. Then, if u is a solution of the modified problem with external forces, the function $v = u - U$ satisfies the wave equation plus the boundary and initial conditions

$$
\begin{aligned}
v(a,\theta,t) &= 0 \\
v(r,\theta,0) &= f(r,\theta) - U(r) \\
v_t(r,\theta,0) &= g(r,\theta)
\end{aligned}
$$

and is found as above. Finally, $u = U + v$. As an example, Exercise 9.5 considers the particular case in which the only external force is due to gravity.

The general case in which q depends on all three variables is considerably more complex. Using the same idea that was successful in §3.11 and §5.7, we shall try a solution of the form (18) but with arbitrary time-varying coefficients, that is,

$$
(22) \ u(r,\theta,t)
$$

$$
= \frac{1}{2}\sum_{k=1}^{\infty} a_{0k}(t)J_0(\alpha_{0k}r) + \sum_{n=1}^{\infty}\left[\sum_{k=1}^{\infty} J_n(\alpha_{nk}r)[a_{nk}(t)\cos n\theta + b_{nk}(t)\sin n\theta]\right].
$$

This works well if q can be expanded in a series of the same form

$$
q(r,\theta,t) = \frac{1}{2}\sum_{k=1}^{\infty} A_{0k}(t)J_0(\alpha_{0k}r) + \sum_{n=1}^{\infty}\left[\sum_{k=1}^{\infty} J_n(\alpha_{nk}r)[A_{nk}(t)\cos n\theta + B_{nk}(t)\sin n\theta]\right]
$$

and if we assume that the series for u can be differentiated term by term as desired. Then, taking these derivatives of u and the series for q to (21), multiplying the result by $\cos j\theta$, where j is any nonnegative integer, integrating from $-\pi$ to π, and dividing by π we obtain

$$
m\sum_{k=1}^{\infty} a_{jk}'' J_j(\alpha_{jk}r)
$$

$$
= \tau\sum_{k=1}^{\infty}\left[\frac{d^2}{dr^2}J_j(\alpha_{jk}r) + \frac{1}{r}\frac{d}{dr}J_j(\alpha_{jk}r) - \frac{j^2}{r^2}J_j(\alpha_{jk}r)\right]a_{jk} + \sum_{k=1}^{\infty} A_{jk}J_j(\alpha_{jk}r).
$$

Given the fact that α_{jk}^2 is an eigenvalue of the differential equation for R if $p = j$ with corresponding eigenfunction $J_j(\alpha_{jk}r)$, as shown in §10.7, the sum in the brackets equals $-\alpha_{jk}^2 J_j(\alpha_{jk}r)$, and then

$$\sum_{k=1}^{\infty}(ma_{jk}'' + \tau\alpha_{jk}^2 a_{jk} - A_{jk})J_j(\alpha_{jk}r) = 0.$$

Multiplying this series by $rJ_j(\alpha_{jl}r)$, where l is any positive integer, integrating term by term from 0 to a, and using the orthogonality of the system $\{J_j(\alpha_{jk}r)\}$ gives

$$(ma_{jl}'' + \tau\alpha_{jl}^2 a_{jk} - A_{jl})\,\|J_j(\alpha_{jl}r)\|^2 = 0,$$

and then, dividing by $\|J_j(\alpha_{jl}r)\|^2$ and replacing j and l with n and k, we obtain

$$ma_{nk}'' + \tau\alpha_{nk}^2 a_{nk} - A_{nk} = 0, \qquad n = 0, 1, 2, \ldots .$$

Similarly,

$$mb_{nk}'' + \tau\alpha_{nk}^2 b_{nk} - A_{nk} = 0, \qquad n = 1, 2, 3 \ldots .$$

The initial conditions for these equations are obtained by putting $t = 0$ both in (23) and in its derived series with respect to t. From the condition $u(r, \theta, 0) = f(r, \theta)$ it follows, as above, that $a_{nk}(0)$ and $b_{nk}(0)$ are given by the right-hand sides of (19) and (20), while the condition $u_t(r, \theta, 0) = g(r, \theta)$ implies that $a_{nk}'(0)$ and $b_{nk}'(0)$ are also obtained from (19) and (20) after replacing f by g.

In any given situation in which the stated expansion for $q(r, \theta, t)$ is known, the formal solution is assembled as follows. First $a_{nk}(0)$, $b_{nk}(0)$, $a_{nk}'(0)$ and $b_{nk}'(0)$ are obtained from (19) and (20), then the differential equations for a_{nk} and b_{nk} are solved using these initial conditions, and the results are taken to (23). In view of the unjustified assumptions made in the outlined procedure and if at all possible, this solution must be verified *a posteriori*. If it is verified, then it is the only solution as shown in Exercise 1.5 of Chapter 9.

§10.10 Modified Bessel Functions

Consider now the solution of the Dirichlet problem for a finite circular cylinder of radius a, with its axis along the z-axis and located between $z = 0$ and $z = c$. Such a problem arises, for example, when trying to determine the electrostatic potential inside the cylinder if it is known on the lateral surface and on the top and bottom or if we seek a steady-state solution of the heat equation

$$u_t = k(u_{xx} + u_{yy} + u_{zz})$$

since a solution independent of time automatically satisfies the Laplace equation

$$u_{xx} + u_{yy} + u_{zz} = 0.$$

We shall impose the boundary conditions

$$
\begin{aligned}
u(r, 0, \theta) &= f(r, \theta) & r \le a, & & -\pi \le \theta \le \pi \\
u(r, c, \theta) &= g(r, \theta) & r \le a, & & -\pi \le \theta \le \pi \\
u(a, z, \theta) &= h(z, \theta) & 0 \le z \le c, & & -\pi \le \theta \le \pi,
\end{aligned}
$$

where f, g, and h are known functions that are periodic in the variable θ of period 2π, and we assume that u is continuous on the boundary.

The best way to deal with this problem is to find its solution as the sum of those of three separate problems, in each of which only one of the boundary conditions is as stated and the other two are zero. Two of these simpler problems involve nothing new and are assigned as Exercise 10.1. Here we shall find a formal solution of the third, in which $f \equiv g \equiv 0$.

As in §10.1 we rewrite the Laplace equation in polar coordinates

$$
u_{rr} + \frac{1}{r} u_r + \frac{1}{r^2} u_{\theta\theta} + u_{zz} = 0,
$$

and if we assume a solution of the form $u(r, z, \theta) = R(r)Z(z)\Theta(\theta)$, we must have

$$
\frac{R''}{R} + \frac{R'}{rR} + \frac{\Theta''}{r^2\Theta} = -\frac{Z''}{Z} = \lambda,
$$

where λ is a constant. Again as in §10.1, but using now the boundary conditions $u(r, 0, \theta) = u(r, c, \theta) = 0$, this leads to the Sturm-Liouville problems

$$
\begin{aligned}
Z'' + \lambda Z &= 0 & Z(0) = Z(c) = 0 \\
\Theta'' + \mu\Theta &= 0 & \Theta(-\pi) = \Theta(\pi),
\end{aligned}
$$

the first of which is regular and the second, periodic. As we already know, the eigenvalues are

$$
\lambda_m = \frac{m^2\pi^2}{c^2},
$$

where m is any positive integer, for the first problem and $\mu_n = n^2$, where n is any nonnegative integer, for the second. The corresponding eigenfunctions are

$$
Z_m(z) = \sin\frac{m\pi}{c} z,
$$

$\Theta_0 \equiv \frac{1}{2}$, and the linear combination

$$
\Theta_n(\theta) = A_n \cos n\theta + B_n \sin n\theta,
$$

corresponding to $\mu_n = n^2 > 0$.

Finally, since $\Theta''/\Theta = -n^2$ we obtain the following differential equation for R

$$
r^2 R'' + r R' - (\lambda r^2 + n^2)R = 0.
$$

Once more, as in §10.1, if we introduce a new variable $x = r\sqrt{\lambda}$ and a new function y of x defined by

$$y(x) = R\left(\frac{x}{\sqrt{\lambda}}\right)$$

and if we replace n with an arbitrary real number p, the differential equation for R is transformed into

(23) $$x^2 y'' + xy' - (x^2 + p^2)y = 0.$$

This is called the *modified Bessel equation of order p*. The only difference between (23) and (2) is the sign of the term $x^2 y$.

This equation had already been used by Euler, as far back as 1736, in his study of the motion of a hanging chain. Then it reappeared, all this in a slightly different form, in papers by J. Nicolas (1882) and A. Basset (1886), who proposed certain forms of its solution. But it was in *A Treatise on Hydrodynamics* of 1888 (Vol. II, page 17), in which Alfred Barnard Basset (1854–1930), at one time of Trinity College, Cambridge, introduced the following solution in its current form and notation, $I_p : (0, \infty) \to \mathbb{R}$, defined by

$$I_p(x) = i^{-p} J_p(ix),$$

where i is the imaginary unit. Note that $J_p(ix)$ is well defined by the series in either Definition 10.1 or Definition 10.2, and it is easily seen by differentiation that I_p is a solution of the modified Bessel equation of order p. It is called the *modified Bessel function of the first kind of order p*. We should say in passing that Basset was not concerned with the Dirichlet problem in a cylinder, but rather with more sophisticated applications such as, for example, the study of the rotational motion of a liquid in a closed vessel bounded by two coaxial circular cylinders, a topic that we cannot explore here.[1]

Note that replacing x with ix in the series in either Definition 10.1 or Definition 10.2, we obtain

$$I_p(x) = i^{-p} \sum \frac{(-1)^k}{k!\,\Gamma(p+k+1)} \left(\frac{ix}{2}\right)^{2k+p} = \sum \frac{1}{k!\,\Gamma(p+k+1)} \left(\frac{x}{2}\right)^{2k+p},$$

where Σ is an infinite series whose first term is for $k = 0$ if p is not a negative integer or for $k = n$ if $p = -n$ and n is a positive integer. Therefore, I_p has only real values and the following additional properties:

1. if $p > 0$, $I_0(0) = 1$ and $I_p(0) = 0$,

2. if $p \geq 0$, $I_p(x) > 0$ and $I_p'(x) > 0$ for $x > 0$,

3. if $p < 0$ is not an integer, $|I_p(x)| \to \infty$ as $x \to 0$ because, if $x > 0$ is very small,

$$I_p(x) \approx \frac{1}{\Gamma(p+1)} \left(\frac{x}{2}\right)^p,$$

---

[1] See Basset's treatise, Vol. II, page 94.

and

4. $I_p(x) \to \infty$ as $x \to \infty$.

It is left to the reader to approximately draw the graph of I_p for $p \geq 0$. Note also that, if p is not an integer, the first and third properties imply that I_p and I_{-p} are linearly independent, and then the general solution of (23) is

$$y = A I_p + B I_{-p},$$

where A and B are arbitrary constants.

If p is an integer, we need a second linearly independent solution of (23) because it is easy to verify using Definition 10.2 that $I_n = I_{-n}$ for every integer n. Such a solution was found by Basset, who denoted it by K_n, in 1888, and was extended to arbitrary p by Hector Munro Macdonald (1865–1935) in 1899. This is the function $K_p \colon (0, \infty) \to \mathbb{R}$ defined by

$$K_p = \begin{cases} \dfrac{\pi}{2} \dfrac{I_{-p} - I_p}{\sin p\pi} & \text{if } p \text{ is not an integer} \\[4mm] \displaystyle\lim_{\alpha \to p} \dfrac{\pi}{2} \dfrac{I_{-\alpha} - I_\alpha}{\sin \alpha\pi} & \text{if } p \text{ is an integer} \end{cases}$$

and is called the *modified Bessel function of the second kind of order* p. It can be shown that it is a solution of (23), that

$$K_0(x) = -\left(\log \frac{x}{2} + \gamma \right) I_0(x) + \sum_{k=1}^{\infty} \frac{\varphi(k)}{(k!)^2} \left(\frac{x}{2} \right)^{2k},$$

and that, if n is a positive integer,

$$K_n(x) = (-1)^{n+1} \left(\log \frac{x}{2} + \gamma \right) I_n(x) + \frac{1}{2} \sum_{k=0}^{n-1} \frac{(-1)^k (n-k-1)!}{k!} \left(\frac{x}{2} \right)^{2k-n}$$

$$+ (-1)^n \frac{\varphi(n)}{2n!} \left(\frac{x}{2} \right)^n + \frac{(-1)^n}{2} \sum_{k=1}^{\infty} \frac{\varphi(k) + \varphi(k+n)}{k!(k+n)!} \left(\frac{x}{2} \right)^{2k+n}.$$

The logarithmic term shows that $K_0(x) \to \infty$ as $x \to 0$. Also, for $n > 0$ and very small values of $x > 0$,

$$K_n(x) \approx \frac{(n-1)!}{2} \left(\frac{x}{2} \right)^{-n},$$

which is unbounded near $x = 0$. This shows that, for every nonnegative integer n, I_n and K_n are linearly independent and then that the general solution of (23) is

$$y = A I_n + B K_n,$$

where A and B are arbitrary constants.

We now return to the Dirichlet problem for the cylinder. For each value of $m > 0$ and $p = n \geq 0$ the general solution of (23) gives the following solution for R

$$R_{mn}(r) = y(r\sqrt{\lambda}) = y\left(\frac{m\pi}{c}r\right) = C_{mn}I_n\left(\frac{m\pi}{c}r\right) + D_{mn}K_n\left(\frac{m\pi}{c}r\right),$$

and $D_{mn} = 0$ since our solution must be bounded but K_n is unbounded on $(0, a)$. Therefore, we have found product solutions $R_{mn}(r)Z_m(z)\Theta_n(\theta)$ of the Laplace equation, each of which is periodic in θ of period 2π and vanishes for $z = 0$ and $z = c$. Then we try a series solution

$$u(r, z, \theta) = \frac{1}{2}\sum_{m=1}^{\infty} c_{m0}I_0\left(\frac{m\pi}{c}r\right)\sin\frac{m\pi}{c}z$$

$$+ \sum_{m,n=1}^{\infty} I_n\left(\frac{m\pi}{c}r\right)\left(\sin\frac{m\pi}{c}z\right)(c_{mn}\cos n\theta + d_{mn}\sin n\theta),$$

where $c_{mn} = C_{mn}A_n$ and $d_{mn} = C_{mn}B_n$ are arbitrary constants, in order to satisfy the boundary condition $u(a, z, \theta) = h(z, \theta)$. We must have

$$h(z, \theta) = \frac{1}{2}\sum_{m=1}^{\infty} c_{m0}I_0\left(\frac{m\pi}{c}a\right)\sin\frac{m\pi}{c}z$$

$$+ \sum_{m,n=1}^{\infty}\left[c_{mn}I_n\left(\frac{m\pi}{c}a\right)\sin\frac{m\pi}{c}z\cos n\theta + d_{mn}I_n\left(\frac{m\pi}{c}a\right)\sin\frac{m\pi}{c}z\sin n\theta\right],$$

and then, referring to Definition 9.1, the right-hand side must be the double Fourier series of the extension of h to the rectangle $[-c, c] \times [-\pi, \pi]$ that is odd in z; that is,

$$c_{mn} = \frac{2}{\pi c I_n(m\pi a/c)}\int_0^c\int_{-\pi}^{\pi} h(z, \theta)\sin\frac{m\pi}{c}z\cos n\theta\, dz\, d\theta, \qquad m \geq 1, \ n \geq 0,$$

and

$$d_{mn} = \frac{2}{\pi c I_n(m\pi a/c)}\int_0^c\int_{-\pi}^{\pi} h(z, \theta)\sin\frac{m\pi}{c}z\sin n\theta\, dz\, d\theta, \qquad m, n \geq 1.$$

The work done in this section can also be used to explain why the Sturm-Liouville problem of §10.7 has no negative eigenvalues. If $\lambda < 0$, we define $\gamma = -\lambda > 0$, and then equation (13) becomes $x^2X'' + xX' - (\gamma x^2 + p^2)X = 0$. After a substitution that transforms it into the modified Bessel equation of order p, we would see that the only bounded solutions are of the form $I_p(x\sqrt{\gamma})$. But I_p is positive and increasing for $x > 0$ if $p \geq 0$ and cannot satisfy either the boundary condition $X(a) = 0$ or $hX(a) + aX'(a) = 0$ if $h \geq 0$.

§10.11   The Skin Effect

In a paper read on March 17, 1890, Sir William Thomson related the strange circumstances of an incident at Cragside, where Lord Armstrong had recently electrified his manor house. Lord Armstrong was holding a steel bar that was accidentally allowed to come in contact with the poles of a working dynamo, and later, in a letter to Thomson of March 7, he described his experience in the following manner:

> The sudden pain caused me to dash the bar instantaneously to the ground, and an attendant *immediately* picked it up and found it quite cold. Three of my fingers and my thumb were blistered . . . but nothing was scorched except the skin at the points of grasp.

Thomson went on to quote the work of Lord Rayleigh, who, in a paper of 1886, had shown that an alternating current through a cylindrical conductor has a tendency to condense on the outer part and that the penetration of the current into the conductor may be only skin deep. Near the end of his own paper Thomson summed up the incident with the folowing conclusion:

> We have thus no difficulty in understanding that there should have been amply sufficient current through an exceedingly thin shell of the bar to produce very suddenly the high temperature of the surface which Lord Armstrong perceived, and yet that the total amount of heat generated was insufficient to heat the bar to any sensible degree after the second or two required for the thermal diffusion, to spread it nearly uniformly through the body of the bar.[1]

The discussion of this incident poses two mathematical problems. The first is the diffusion of heat in a solid cylinder, which had already been considered by Fourier and is taken up in the exercises.[2] The second is the distribution of electric current inside the conductor, and it is to its study that we shall devote the rest of this section. This matter was investigated by Thomson himself, who presented the solution—introducing some further modifications of Bessel functions—as part of his Presidential Address to the Institution of Electrical Engineers of January 10, 1890.

But we must go back in time to find the fundamental principles of electromagnetism that will allow us to set up the partial differential equation governing the density of electric current in a cylindrical conductor. Of course, the reader that is not interested in this derivation can proceed directly to equation (26). In 1820 Hans Christian Ørsted (1777–1851), of Denmark, showed that when an electric current flows through a wire it induces a magnetic field around it.[3] This is evidenced by the fact that a magnetic needle

---

[1] These quotations are taken from *Mathematical and Physical Papers by Sir William Thomson*, Vol. III, Cambridge University Press Warehouse, 1890, pages 474 and 482–483.

[2] *The Analytical Theory of Heat*, Chapter VI.

[3] This field was later defined as follows: if a positive unit charge passes through a point P near the wire with velocity v and if a sidewise force F acts on the charge (in the absence of electric or mechanical forces), then there is a magnetic induction vector B at P such that F is the cross product of v and B.

placed near such a current, in a plane perpendicular to the wire, deviates from its position until it becomes tangent to a circle in the stated plane with center on the wire. Thus, the magnetic field is tangent to such a circle at any point. It is found experimentally that, if the current in the wire has intensity I and if the circle has radius r, then the magnitude of the induced field at any point of this circle is proportional to the quotient I/r.

This experiment was repeated in France by André Marie Ampère (1775–1836), a professor at the *École Polytechnique*, who developed a mathematical theory of electro-dynamic action through the early 1820s. To state its fundamental principle, assume that a magnetic field B exists in a region of space, created in whatever manner, and that its field lines—not necessarily circles now—are contained in planes perpendicular to a certain line l. It is an experimental fact that, if γ is a simple, closed curve in a plane perpendicular to l, then the line integral along γ of the inner product of B and the tangent vector to γ is proportional to the total current passing through the interior of γ in the direction of l. That is, if γ is given in parametric form by $\gamma(\theta) = (x(\theta), y(\theta))$ for $0 \leq \theta \leq 2\pi$ and if $B(\gamma(\theta))$ is the value of B at $\gamma(\theta)$, then

$$\int_0^{2\pi} \langle B(\gamma(\theta)), \gamma'(\theta) \rangle \, d\theta = \mu I,$$

where μ is a constant called the *permeability* of the medium. This is known as *Ampère's law*, although Ampère himself did not formulate his work in terms of vector fields.

Now consider a cylindrical conductor of radius a, let $u(r, t)$ be the electric current flowing parallel to its axis per unit cross area at a distance $r \leq a$ from the axis and at time t, let γ be a circle of radius r centered on the axis of the conductor and contained in a plane perpendicular to it, and let $B(r, t)$ denote now the magnitude of the magnetic field induced at any point of γ at time t. Since $\gamma(\theta) = (r \cos \theta, r \sin \theta)$ and since the magnetic field at this point is $(-B(r, t) \sin \theta, B(r, t) \cos \theta)$, its inner product with $\gamma'(\theta)$ is easily computed to be $r B(r, t)$. Using polar coordinates, the total current through the portion of the conductor inside γ is

$$I = \int_0^r \int_0^{2\pi} u(s, t) s \, ds \, d\theta = 2\pi \int_0^r s u(s, t) \, ds,$$

and then Ampère's law gives $2\pi r B(r, t) = \mu I$, and we conclude that

(24)
$$B(r, t) = \frac{\mu}{r} \int_0^r s u(s, t) \, ds.$$

While Ampère studied the magnetism produced by a current, Michael Faraday (1791–1867) conducted a sequence of experiments in England to show that the reverse of Ørsted's discovery is also true, that is, that magnetism could produce an electric current. It was not easy, but he finally succeeded on August 29, 1831, when he discovered something that was far from obvious at the time: that magnetism must be moving to produce electricity.[1] More specifically, consider a small plane area A bounded by a conducting

---

[1] If a stationary magnet could produce an electric current that, in turn, could run a machine, we would have a violation of the principle of conservation of energy.

thin wire and placed in a magnetic field of magnitude B, perpendicular to the field lines. The following experimental fact is known as *Faraday's law*: the current created on the bounding wire by variations in B with time is produced by an electromotive force of magnitude

$$E = \frac{\partial}{\partial t} \int_A B.$$

Returning to our cylindrical conductor of radius a, let r and h be positive numbers such that both r and $r + h$ are no larger than a. Now consider a plane containing the cylinder's axis, and in that plane a rectangular area A bounded by two segments of length l parallel to the conductor's axis and located at distances r and $r + h$ from it and by the two segments of length h perpendicular to the axis (see Figure 10.4). Applying

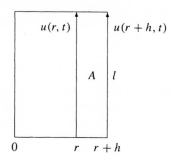

Figure 10.4

Faraday's law to this area A gives

$$E = \frac{\partial}{\partial t} \int_r^{r+h} \int_0^l B(s, t) \, ds \, dz = l \int_r^{r+h} B_t(s, t) \, ds.$$

Note that the current flows parallel to the conductor's axis and then, if ρ denotes the resistance of the material per unit cross area and unit length, Kirchhoff's voltage law applied to the boundary of A gives

$$E = \rho l[u(r + h, t) - u(r, t)].$$

Eliminating E from the preceding equations, dividing by lh, letting letting $h \to 0$, and assuming that all necessary derivatives exist, we obtain

(25) $$\rho u_r(r, t) = \lim_{h \to 0} \frac{1}{h} \int_r^{r+h} B_t(s, t) \, ds = \frac{\partial}{\partial r} \int_0^r B_t(s, t) \, ds = B_t(r, t).$$

But (24) implies that

$$B_r(r, t) = \frac{\mu}{r} r u(r, t) - \frac{\mu}{r^2} \int_0^r s u(s, t) \, ds = \mu u(r, t) - \frac{1}{r} B(r, t),$$

and then, differentiating with respect to t and using (25),

$$B_{rt}(r, t) = \mu u_t(r, t) - \frac{1}{r} B_t(r, t), = \mu u_t(r, t) - \frac{\rho}{r} u_r(r, t).$$

Also, differentiating (25) with respect to r shows that $B_{tr} = \rho u_{rr}$, and, assuming that $B_{tr} = B_{rt}$, we arrive at

(26) $$\rho u_{rr} = \mu u_t - \frac{\rho}{r} u_r.$$

Here u is the intensity of electric current per unit cross area flowing parallel to the conductor's axis, μ is a proportionality constant called the permeability of the medium, and ρ is the electric resistance of the material per unit cross area and unit length.

Assuming now that an alternating current flows on the surface of the conductor as a consequence of accidentally coming in contact with the poles a working dynamo, we pose the problem consisting of solving this equation subject to the boundary condition

$$u(a, t) = C \sin \omega t,$$

where C and ω are given constants. As usual, we try separation of variables. If (26) has a solution of the form $u(r, t) = R(r)T(t)$, then substitution into (26) and division by ρRT yield

$$\frac{\mu T'}{\rho T} = \frac{R''}{R} + \frac{R'}{rR} = i\lambda,$$

where λ is a real constant and i is the imaginary unit. Then

$$T' - i\frac{\rho}{\mu}\lambda T = 0$$

and

(27) $$r^2 R'' + r R' - i\lambda r^2 R = 0.$$

Note that $i\lambda$ has to be a purely imaginary constant because, if it had a real part α, then the solution for T would contain a factor of the form $e^{\alpha \rho t/\mu}$ and it would be impossible to satisfy the boundary condition. But with λ real, the equation for T does not have real coefficients and its solution

$$T(t) = T(0)e^{i\lambda \rho t/\mu}$$

is complex valued. Since the boundary value is the imaginary part of $Ce^{i\omega t}$, this suggests that we should find first a complex valued solution of (26) of the form $R(r)T(t)$ subject to the boundary condition

$$R(a)T(t) = Ce^{i\omega t},$$

and then take

$$u(r, t) = \text{Im}\,[R(r)T(t)]$$

as the solution of (26). Then, in view of the solution obtained for T, we must choose

$$\lambda = \frac{\omega \mu}{\rho} > 0$$

and use it to solve (27).

As in §10.1 we introduce a new variable $x = r\sqrt{\lambda}$ and a new function $y(x) = R(x/\sqrt{\lambda})$. Then (27) is transformed into

(28) $$x^2 y'' + xy' - ix^2 y = 0,$$

and it is easy to see by direct substitution that $I_0(\sqrt{i}\,x)$ and $K_0(\sqrt{i}\,x)$, defined by the series expansions of the preceding section but replacing x with $\sqrt{i}\,x$, are solutions. They are linearly independent because $I_0(0) = 1$ but, as a consequence of the logarithmic term in the series for $K_0(x)$ on page 386, $|K_0(\sqrt{i}\,x)| \to \infty$ as $x \to 0$. Since we are looking for a bounded solution, it must be of the form $y(x) = A I_0(\sqrt{i}\,x)$ for some constant A, and then

$$R(r) = A I_0(\sqrt{i\lambda}\,r).$$

Using the solution already found for T and the boundary condition $R(a)T(t) = Ce^{i\omega t}$, we obtain

$$A T(0) I_0(\sqrt{i\lambda}\,a) = C,$$

and thus we arrive at the product solution

$$R(r)T(t) = C \frac{I_0(\sqrt{i\lambda}\,r)}{I_0(\sqrt{i\lambda}\,a)} e^{i\omega t}.$$

To complete the solution of this problem Sir William Thomson introduced in 1889 two new real valued functions, which he denoted by *ber* and *bei* for *Bessel real* and *Bessel imaginary*, defined by the equation

$$\text{ber}(x) + i\,\text{bei}(x) = I_0(\sqrt{i}\,x).[1]$$

These are called the *Kelvin functions*, in honor of Sir William Thomson, who would become Baron Kelvin of Largs in 1892. We can use them to write our product solution in the form

$$R(r)T(t) = C \frac{\text{ber}(\sqrt{\lambda}\,r) + i\,\text{bei}(\sqrt{\lambda}\,r)}{\text{ber}(\sqrt{\lambda}\,a) + i\,\text{bei}(\sqrt{\lambda}\,a)} (\cos\omega t + i\sin\omega t).$$

Finally,

$$u(r,t) = \text{Im}[R(r)T(t)]$$

$$= \frac{C}{\text{ber}^2(\sqrt{\lambda}\,a) + \text{bei}^2(\sqrt{\lambda}\,a)} \left\{ \left[\text{bei}(\sqrt{\lambda}\,r)\text{ber}(\sqrt{\lambda}\,a) - \text{ber}(\sqrt{\lambda}\,r)\text{bei}(\sqrt{\lambda}\,a) \right] \cos\omega t \right.$$

$$\left. + \left[\text{ber}(\sqrt{\lambda}\,r)\text{ber}(\sqrt{\lambda}\,a) + \text{bei}(\sqrt{\lambda}\,r)\text{bei}(\sqrt{\lambda}\,a) \right] \sin\omega t \right\}$$

$$= C\sqrt{\frac{\text{ber}^2(\sqrt{\lambda}\,r) + \text{bei}^2(\sqrt{\lambda}\,r)}{\text{ber}^2(\sqrt{\lambda}\,a) + \text{bei}^2(\sqrt{\lambda}\,a)}} \cos(\omega t - \varphi_r),$$

---

[1] In a similar manner, but using K_0 instead of I_0, A. Russell introduced in 1909 the functions *ker* and *kei*, which we shall have no occasion to use.

where

$$\tan \varphi_r = \frac{\operatorname{ber}(\sqrt{\lambda}\,r)\operatorname{ber}(\sqrt{\lambda}\,a) + \operatorname{bei}(\sqrt{\lambda}\,r)\operatorname{bei}(\sqrt{\lambda}\,a)}{\operatorname{bei}(\sqrt{\lambda}\,r)\operatorname{ber}(\sqrt{\lambda}\,a) - \operatorname{ber}(\sqrt{\lambda}\,r)\operatorname{bei}(\sqrt{\lambda}\,a)}.$$

Note that the current in the interior of the wire is not in phase with the current at the surface. But we are more interested in the quotient

$$Q(r,a) = \frac{|u(r,t)|}{|u(a,t)|} = \sqrt{\frac{\operatorname{ber}^2(\sqrt{\lambda}\,r) + \operatorname{bei}^2(\sqrt{\lambda}\,r)}{\operatorname{ber}^2(\sqrt{\lambda}\,a) + \operatorname{bei}^2(\sqrt{\lambda}\,a)}}.$$

Lord Armstrong's bar was 1.4 centimeters in diameter, that is, $a = 0.007$ meters. If it had been made of copper, for which $\mu = 3.9996\pi \times 10^{-7}$ henrys per meter and $1/\rho = 5.88 \times 10^7$ mhos per meter, then, with $\omega = 2\pi f$, we would have $\sqrt{\lambda}\,a = 0.1508\sqrt{f}$. In a typical alternating-current installation of electric light in late nineteenth century England the frequency was about 80 periods per second, and then $\sqrt{\lambda}\,a = 1.349$. Tabulated values for the Kelvin functions, originally computed by Thomson's assistant, Mr. Magnus Maclean, show that, if $0 \le r \le a$,

$$\operatorname{ber}(\sqrt{\lambda}\,r) \approx \operatorname{ber}(\sqrt{\lambda}\,a) \approx 1$$

and

$$\operatorname{bei}(\sqrt{\lambda}\,r) \approx \operatorname{bei}(\sqrt{\lambda}\,a) \approx 0,$$

that is, $Q(r,a) \approx 1$. Therefore, for this low value of the frequency the current would be uniformly distributed throughout a copper conductor. However, Lord Armstrong's bar was made of iron, for which—as for other ferromagnetic materials—the value of the permeability is considerably higher than that for copper. We cannot assign an exact value to it because it depends nonlinearly on the magnetic induction B and on the previous magnetic history of the material. However, a typical value of $20000\pi \times 10^{-7}$ henrys per meter is reasonable. With a value of $1/\rho = 10^7$ mhos per meter at room temperature, $\sqrt{\lambda}\,a \approx 39.339$, which is a much higher value than that for copper. Then, using estimates for the values of

$$\operatorname{ber}^2(\sqrt{\lambda}\,x) + \operatorname{bei}^2(\sqrt{\lambda}\,x)$$

for $x > 10^1$ we can construct the following table.

r/a	0.90	0.92	0.94	0.96	0.98	1
$Q(r,a)$	0.07	0.11	0.19	0.34	0.58	1

Figure 10.5 represents $Q(r,a)$ versus r/a. It shows that most of the current flows through a thin shell near the surface of the bar, from $r = 0.9a$ to $r = a$. This phenomenon is known as the *skin effect*. It is present in iron at low frequencies, but for nonferromagnetic conductors such as copper it is strictly a high-frequency affair.

---

[1] See, for instance, M. Abramowitz and I. A. Stegun, Eds., *op. cit.*, pages 382–383.

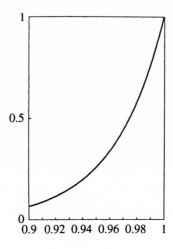

Figure 10.5

EXERCISES

Exercises 1.1 to 1.8 refer to the Dirichlet problem on an open disc D in $\mathbb{R}^2$ of radius a, centered at the origin and bound by a circumference C. The boundary values are given by a piecewise continuous function $f : [-\pi, \pi] \to \mathbb{R}$ such that $f(-\pi) = f(\pi)$, and we shall use the notation $\bar{D} = D \cup C$. By performing a translation it is easily seen that these results are valid or easily adapted to the case of a disc with arbitrary center.

1.1 Use separation of variables to show that the Dirichlet problem on D has the formal series solution

$$u(r, \theta) = \tfrac{1}{2} a_0 + \sum_{n=1}^{\infty} \left(\frac{r}{a} \right)^n (a_n \cos n\theta + b_n \sin n\theta),$$

where a_0, a_n, and b_n are the Fourier coefficients of f on $(-\pi, \pi)$ (**Hint:** use the polar expression for $u_{xx} + u_{yy}$ derived in the text).

1.2 Verify that the series in Exercise 1.1 converges to a function $u : \bar{D} \to \mathbb{R}$ that is harmonic in D and that, if f is continuous and has a piecewise continuous derivative, then u is continuous on $\bar{D}$ (this result will be improved in Exercise 1.5).

1.3 By replacing the coefficients in the series solution of Exercise 1.1 with their integral representations, interchanging the order of summation and integration (justify this step), and then writing the cosine in terms of exponentials to sum the resulting series, derive

Poisson's integral formula for the disc:

$$u(r, \theta) = \frac{1}{2\pi} \int_{-\pi}^{\pi} \frac{a^2 - r^2}{a^2 - 2ar \cos(\theta - \varphi) + r^2} f(\varphi) \, d\varphi$$

for $r < a$.

1.4 Consider the function $P^r : [-\pi, \pi] \to \mathbb{R}$ defined by

$$P^r(\alpha) = \frac{a^2 - r^2}{a^2 - 2ar \cos \alpha + r^2}$$

for each $r < a$ (evaluated at $\alpha = \theta - \varphi$ in the formula of Exercise 1.3). The quotient $P^r(\alpha)/2\pi$ is called *Poisson's kernel for the disc*. Show that $P^r \geq 0$ and that

$$\int_{-\pi}^{0} P^r = \int_{0}^{\pi} P^r = \pi.$$

1.5 Show that the function u defined by the Poisson integral formula of Exercise 1.3 is harmonic in D and then use the result of Exercise 1.4 to show that

$$u(r, \theta) \to \tfrac{1}{2}[f(\theta+) + f(\theta-)]$$

as $r \to a$. Show also that, if f is continuous, then the extension of u to $\bar{D}$ such that $u(a, \theta) = f(\theta)$ is continuous (**Hint:** this proof is almost identical to that of Theorems 7.7 and 7.8).

1.6 If u is the function defined by the Poisson integral formula of Exercise 1.3, then

$$u(0, \cdot) = \frac{1}{2\pi} \int_{-\pi}^{\pi} f;$$

that is, the value of u at the center of D is the average value of f on C. This is known as *Gauss' Mean Value Theorem* for the disc. Use this result to prove that, if $u : \bar{D} \to \mathbb{R}$ is a continuous function that is harmonic in D, then u attains its maximum and minimum values on C and, unless u is constant, only on C.

1.7 Use the result of Exercise 1.6 to prove that, if f is continuous, then the Dirichlet problem for the disc is well posed in the sense of Hadamard. Is this still true if the Laplace equation is replaced by the Poisson equation $u_{xx} + u_{yy} = -q(x, y)$?

1.8 Use the results of Exercises 1.3, 1.6, and 1.7 to prove the following theorem of Axel Harnack (1851–1888): if u is a nonnegative function that is harmonic on an open set containing D, then for any point (x, y) at a distance $r < a$ from the origin we have

$$\frac{a - r}{a + r} u(0, 0) \leq u(x, y) \leq \frac{a + r}{a - r} u(0, 0).$$

These are called *Harnack's inequalities*.

1.9 Use Harnack's inequalities of Exercise 1.8 to prove the following theorem of Charles Émile Picard (1856–1941) of Paris: if u is harmonic in $\mathbb{R}^2$ and either bounded above or below by a constant, then u is a constant (**Hint:** consider first the case $u \geq 0$ and use a disc of arbitrarily large radius).

1.10 Use Poisson's integral formula and Gauss' mean value theorem (for a disc of arbitrary center) of Exercises 1.3 and 1.6 to prove the strong form of the maximum principle for the Laplace equation (compare this form with the weak form of Theorem 9.6): if D is a connected bounded open set in $\mathbb{R}^2$, ∂D denotes the boundary of D, $\bar{D} = D \cup \partial D$, and if $u : \bar{D} \to \mathbb{R}$ is a continuous function that is harmonic in D, then u attains its maximum and minimum values on ∂D and, unless u is constant, only on ∂D (**Hint:** you need to know the following facts from point set topology: (*i*) connected means that D is not the union of disjoint nonempty open subsets or, in short, that it is in one piece, and (*ii*) the inverse image of an open set under a continuous function is open).[1]

1.11 Solve the exterior Dirichlet problem for the disc, that is, find a harmonic function for $r > a$ such that $u(a, \theta) = f(\theta)$, where we assume that f is the sum of its Fourier series on $(-\pi, \pi)$, and show that

$$u(r, \theta) = \frac{1}{2\pi} \int_{-\pi}^{\pi} \frac{r^2 - a^2}{a^2 - 2ar\cos(\theta - \varphi) + r^2} f(\varphi)\, d\varphi$$

for $r > a$.

1.12 Solve the Dirichlet problem on an annulus:

$$u_{rr} + \frac{1}{r}u_r + \frac{1}{r^2}u_{\theta\theta} = 0 \qquad a < r < b, \qquad -\pi < \theta < \pi$$
$$u(a, \theta) = f(\theta) \qquad\qquad\qquad -\pi \le \theta \le \pi$$
$$u(b, \theta) = g(\theta) \qquad\qquad\qquad -\pi \le \theta \le \pi,$$

where f and g are continuous functions with piecewise continuous derivatives and such that $f(-\pi) = f(\pi)$ and $g(-\pi) = g(\pi)$.

1.13 Solve the boundary value problem

$$u_{rr} + \frac{1}{r}u_r + \frac{1}{r^2}u_{\theta\theta} = 0 \qquad 0 < r < a, \qquad -\pi < \theta < \pi$$
$$u_r(a, \theta) = f(\theta) \qquad\qquad\qquad -\pi \le \theta \le \pi,$$

where f is a continuous function that has a piecewise continuous derivative and such that $f(-\pi) = f(\pi)$. What additional condition must f satisfy? Is the solution of this problem unique?

1.14 Find a formal solution of the following boundary value problem on an annular wedge,

$$u_{rr} + \frac{1}{r}u_r + \frac{1}{r^2}u_{\theta\theta} = 0 \qquad 0 < r < a, \qquad 0 < \theta < \theta_0$$
$$u(r, 0) = 0 \qquad\qquad\qquad 0 \le r \le a,$$
$$u(r, \theta_0) = T \qquad\qquad\qquad 0 \le r \le a,$$
$$u_r(1, \theta) = 0 \qquad\qquad\qquad\qquad 0 \le \theta \le \theta_0$$
$$u(b, \theta) = f(\theta) \qquad\qquad\qquad\qquad 0 \le \theta \le \theta_0,$$

---

[1]There is an extension of this result to the case in which u is only piecewise continuous on ∂D. It was discovered in 1904 by Edvard Phragmén (1863–1937), of Stockholm, and generalized by him and Ernst Lindelöf (1870–1946) in 1908. For a statement and proof of the *Phragmén-Lindelöf principle* see, for instance, E. Hille, *Analytic Function Theory*, Vol. II, Chelsea, New York, 1977, page 394.

where θ_0 and b are constants such that $0 < \theta_0 < \pi$ and $b > 1$ and f is a continuous function that has a piecewise continuous derivative such that $f(-\pi) = f(\pi)$. The solution represents the steady-state temperature distribution on the wedge if the straight sides are maintained at constant temperatures 0 and T, the smaller curved side is insulated, and the temperature on the larger curved side is given by f (**Hint:** you need the eigenvalues and eigenfunctions of Example 4.1).

2.1   Show that

$$\int_0^\infty e^{-t^2} t^x \, dt = \frac{1}{2} \Gamma\left(\frac{x+1}{2}\right)$$

for any $x > 0$.

2.2   Use the result of Exercise 2.1 to show that for any pair of nonnegative integers m and n

$$\int_0^{\pi/2} \cos^m \theta \sin^n \theta \, d\theta = \frac{\Gamma\left(\dfrac{m+1}{2}\right)\Gamma\left(\dfrac{n+1}{2}\right)}{2\Gamma\left(\dfrac{m+n+2}{2}\right)}$$

(**Hint:** evaluate the double integral

$$\int_0^\infty \int_0^\infty e^{-(x^2+y^2)} x^m y^n \, dx \, dy$$

in two ways: (i) as a product of single integrals and (ii) switching to polar coordinates first).

3.1   Prove the following inequality discovered by Cauchy in 1841: for any nonnegative integer n

$$|J_n(x)| \le \frac{x^n}{2^n n!} e^{x^2/4}.$$

Deduce that

$$\lim_{n \to \infty} x^m J_n(x) = 0$$

for every pair of nonnegative integers m and n. What is the corresponding inequality for J_p if p is any real number greater than -1?

3.2   Prove the following theorem discovered by Schlömilch in 1857: for any $x > 0$ and any complex number $z \ne 0$

$$e^{x(z^2-1)/2z} = \sum_{m=-\infty}^\infty J_m(x) z^m,$$

where m is an integer. The left-hand side is known as *the generating function* of the Bessel functions of integral order (**Hint:** use the Maclaurin series for the exponential and then the binomial thorem to expand each term, and rearrange the terms of the resulting series).

3.3   Put $z = \pm i e^{-i\theta}$ in the identity of Exercise 3.2 and use the result to show that for any real x

$$\cos(x \cos \theta) = J_0(x) + 2 \sum_{n=1}^\infty (-1)^n J_{2n}(x) \cos 2n\theta$$

and

$$\sin(x \cos \theta) = 2 \sum_{n=1}^{\infty} (-1)^{n+1} J_{2n-1}(x) \cos(2n-1)\theta.$$

These expansions were obtained by Jacobi in 1836 by a different method. State their particular forms for $\theta = 0$.

3.4　Putting $z = e^{i\theta}$ in the equation of Exercise 3.2 and then comparing with the complex Fourier series of $e^{ix \sin \theta}$ on $(-\pi, \pi)$ (refer to Example 6.4) suggest that

$$J_n(x) = \frac{1}{2\pi} \int_{-\pi}^{\pi} e^{i(x \sin \theta - n\theta)} \, d\theta = \frac{1}{\pi} \int_0^{\pi} \cos(x \sin \theta - n\theta) \, d\theta$$

for any nonnegative integer n. Use the Maclaurin series for $e^{ix \sin \theta}$ and integration term by term to prove this identity, discovered by Hansen in 1843. The right-hand side was used by Bessel as the definition of $J_n(x)$ in 1824 and is called *Bessel's integral form* of $J_n(x)$ (**Hint:** show that the integral

$$\int_{-\pi}^{\pi} (\sin^m \theta) \, e^{-in\theta} \, d\theta$$

has value zero unless $m = n + 2j$, where j is a nonnegative integer, and compute its value in this case).

3.5　Use the result of Exercise 3.4 and differentiation under the integral sign (which you must justify) to show that

$$\left| J_n^{(k)}(x) \right| \leq 1$$

for any pair of nonnegative integers k and n.

3.6　Prove that for any real number x

$$[J_0(x)]^2 + 2 \sum_{n=1}^{\infty} [J_n(x)]^2 = 1.$$

This identity was given by Hansen in 1843 and it shows, in particular, that

$$|J_n(x)| \leq \tfrac{1}{2}$$

for any positive integer n (**Hint:** apply Parseval's identity of Exercise 4.12 of Chapter 6 to the function $e^{ix \sin \theta}$ of Exercise 3.4).

3.7　Use the result of Exercise 3.4 to prove that

$$J_n(x) = \frac{1}{\pi} \int_0^{\pi} \cos(x \sin \theta) \, \cos 2n\theta \, d\theta$$

if n is an even nonnegative integer and

$$J_n(x) = \frac{1}{\pi} \int_0^{\pi} \sin(x \sin \theta) \, \sin n\theta \, d\theta$$

if n is an odd positive integer.

3.8   Use the result of Exercise 3.7 to prove the following formulas

$$J_{2n}(x) = \frac{(-1)^n}{\pi} \int_0^\pi \cos(x \cos \theta) \cos 2n\theta \, d\theta, \qquad n = 0, 1, 2, \ldots$$

and

$$J_{2n-1}(x) = \frac{(-1)^{n-1}}{\pi} \int_0^\pi \sin(x \cos \theta) \cos(2n-1)\theta \, d\theta, \qquad n = 1, 2, 3, \ldots$$

which were obtained by Jacobi in 1836.

3.9   Use the formulas of Exercise 3.8 to show that

$$\lim_{x \to \infty} J_n(x) = 0$$

for every nonnegative integer n (**Hint:** for $n = 0$ use the substitution $u = \cos \theta$ and the Riemann-Lebesgue theorem, and for $n > 0$ use de Moivre's formula and the binomial theorem to show first that

$$\cos m\theta = \tfrac{1}{2} \sum_{k=0}^m \binom{2n}{k} i^{m-k} [1 + (-1)^{m-k}] \cos^k \theta \, \sin^{m-k} \theta$$

for every positive integer m).

3.10   Use the Maclaurin series for the cosine and integration term by term to show that

$$J_n(x) = \frac{1}{\Gamma\left(n + \frac{1}{2}\right)\Gamma\left(\frac{1}{2}\right)} \left(\frac{x}{2}\right)^n \int_0^\pi \cos(x \cos \theta) \sin^{2n} \theta \, d\theta$$

for every nonnegative integer n. This is known as *Poisson's integral* for $J_n(x)$, since it was discovered in 1823 by Poisson while he was studying heat propagation in a cylinder, and it was used by von Lommel as the definition of $J_n(x)$ for all nonintegral values of $n > -\frac{1}{2}$ (**Hint:** evaluate the integral

$$\int_0^\pi \cos^{2m} \theta \, \sin^{2n} \theta \, d\theta$$

using a reduction formula).

4.1   Show that

$$J_{-p}(x)J_p'(x) - J_p(x)J_{-p}'(x) = J_p(x)J_{-(p-1)}(x) + J_{-p}(x)J_{p-1}(x)$$

and that for each fixed value of p and $x > 0$ the expression

$$x[J_p(x)J_{-(p-1)}(x) + J_{-p}(x)J_{p-1}(x)]$$

is independent of x.

4.2   Prove that, if n is a positive integer, then

$$\left(\frac{1}{x}\frac{d}{dx}\right)^n [x^{-p}J_p(x)] = (-1)^n x^{-p-n} J_{p+n}(x),$$

a formula due to Bessel for integral p (the operation on the left consists of repeating n times the sequence of differentiation followed by division by x).

4.3   Use the result of Exercise 4.2 to prove the following formula for Bessel functions of positive half-integral order

$$J_{n+1/2}(x) = (-1)^n \sqrt{\frac{2}{\pi x}} x^{n+1} \left(\frac{1}{x} \frac{d}{dx} \right)^n \left(\frac{\sin x}{x} \right),$$

as stated by Arnold J. W. Sommerfeld in 1909. Then use this formula to evaluate $J_{5/2}$.

4.4   Show that J_{p+1}, J_{p-1} and all their derivatives are of the form

$$P \left(\frac{1}{x} \right) J_p(x) + Q \left(\frac{1}{x} \right) J'_p(x),$$

where $P(x)$ and $Q(x)$ are polynomials in x.

4.5   Use the result of Exercise 4.4 to show that J_p and all its derivatives are of the form

$$P \left(\frac{1}{x} \right) J_{p-n}(x) + Q \left(\frac{1}{x} \right) J'_{p-n}(x),$$

where $P(x)$ and $Q(x)$ are polynomials in x and n is any integer. In particular, if p is an integer then J_p and all its derivatives are of the form

$$P \left(\frac{1}{x} \right) J_0(x) + Q \left(\frac{1}{x} \right) J_1(x).$$

4.6   Use integration by parts to derive the reduction formula

$$\int_0^x s^n J_0(\alpha s) \, ds = \frac{x^n}{\alpha} J_1(\alpha x) + \frac{(n-1)x^{n-1}}{\alpha^2} J_0(\alpha x) - \frac{(n-1)^2}{\alpha^2} \int_0^x s^{n-2} J_0(\alpha s) \, ds$$

for $n \geq 2$, where $\alpha \neq 0$ is a constant. Thus, by repeated application of the reduction formula, the integral on the left can be evaluated for n odd or reduced to the integral of J_0, which is tabulated, for n even.

The equations in Exercises 4.7 and 4.8 are particular cases of a more general formula proved by Nikolai J. Sonine (1849–1915) in 1880.

4.7   Show that, if $a > 0$, if $\alpha \neq 0$ is a constant, and if n is a nonnegative integer, then

$$\int_0^a x^{p+1} (a^2 - x^2)^n J_p(\alpha x) \, dx = \frac{2^n n! a^{p+n+1}}{\alpha^{n+1}} J_{p+n+1}(a\alpha)$$

(**Hint:** use induction on n).

4.8   Show that, if $a > 0$, if $\alpha \neq 0$ is a constant, and if n is a nonnegative integer, then

$$\int_0^a x^{p+1} (a^2 - x^2)^{n+1/2} J_p(\alpha x) \, dx = \frac{2^{n+1/2} \Gamma \left(n + \frac{3}{2} \right) a^{p+n+3/2}}{\alpha^{n+3/2}} J_{p+n+3/2}(a\alpha)$$

(**Hint:** for $n = 0$ integrate the series in the integrand term by term, make the substitution $x = a \cos \theta$, and use the result of Exercise 2.2; then use induction on n).

5.1 Show that

$$Y_n = \frac{1}{\pi} \lim_{p \to n} \frac{J_p - (-1)^n J_{-p}}{p - n}$$

for every integer n, and then use this identity to prove that

$$Y_{-n} = (-1)^n Y_n,$$

a result esssentially due to von Lommel.

5.2 Prove that the recursion formulas in Theorem 10.2, and as a consequence those in its corollaries, are also valid for Y_p, a result due to von Lommel (**Hint:** it is enough to do it for nonintegral p, and then the continuity of Y_p with respect to p implies the full result).

5.3 Use the result of Exercise 5.2 to show that

$$Y_p(x) J_p'(x) - J_p(x) Y_p'(x) = J_p(x) Y_{p+1}(x) - J_{p+1}(x) Y_p(x)$$

and that for each fixed value of p and $x > 0$ the expression

$$x[J_p(x) Y_{p+1}(x) - J_{p+1}(x) Y_p(x)]$$

is independent of x.

5.4 In 1902 Niels Nielsen (1865–1931) defined the *Bessel functions of the third kind*, for which he used the letter H in honor of Hankel, by $H_p^\pm = J_p \pm i Y_p$. Prove that

(i) $H_p^+ = \dfrac{J_{-p} - e^{-ip\pi} J_p}{i \sin p\pi}$

(ii) $H_p^- = \dfrac{e^{ip\pi} J_p - J_{-p}}{i \sin p\pi}$

(iii) $H_{-p}^\pm = e^{\pm ip\pi} H_p^\pm$

(iv) $H_{1/2}^\pm = \sqrt{\dfrac{2}{\pi x}} \, e^{\pm i(x - \pi/2)}$.

6.1 Use the Bessel differential equation of order p to show that the smallest positive zero of
(i) J_p,
(ii) J_p', and
(iii) $h J_p(x) + x J_p'(x)$,
where h is a constant, is larger than p (**Hint:** show first that, for $p > 0$, $J_p(x)$ and $J_p'(x)$ are positive for small $x > 0$).

6.2 Let $\{x_{pn}\}$ be the sequence of positive zeros of J_p. Use the functions z and θ introduced in the proof of Theorem 10.4 to show that

$$\lim_{n \to \infty} \left[x_{p(n+1)} - x_{pn} \right] = \pi.$$

6.3 Let $\{x_{pn}\}$ be the sequence of positive zeros of $h J_p(x) + x J_p'(x)$, where h is a constant. Use the graph of the tangent to show that, if θ is the function introduced in the proof of Theorem 10.4, then

$$\left(n - \tfrac{1}{2} \right) \pi < \theta(x_{pn}) < \left(n + \tfrac{1}{2} \right) \pi.$$

Then show that

$$\lim_{n \to \infty} \left[x_{p(n+1)} - x_{pn} \right] = \pi.$$

6.4 Examine the behavior of J_{-p} for $p > 0$ and small $x > 0$, and show how it depends on the greatest integer no larger than $-p$ being even or odd. Use this information and Theorem 10.4 to draw an approximate sketch of the graph of J_{-p}.

6.5   Use Theorem 10.2 to prove the following theorem (interlacing of the zeros): between every two consecutive positive zeros of J_p there is a unique zero of J_{p+1}, and between every two consecutive positive zeros of J_p there is a unique zero of J_{p+1}.

6.6   Use formula (11) to prove that

$$Y_p(x) = \sqrt{\frac{2}{\pi x}} \sin\left(x - \frac{p\pi}{2} - \frac{\pi}{4}\right) + O(x^{-3/2})$$

and

$$J_p'(x) = -\sqrt{\frac{2}{\pi x}} \sin\left(x - \frac{p\pi}{2} - \frac{\pi}{4}\right) + O(x^{-3/2})$$

(**Hint:** use the definition of Y_p to prove the first formula for nonintegral p, and then the more general result follows from the fact that Y_p is a continuous function of p by definition).

6.7   Use the second formula in Exercise 6.6 to show that, if x_{pk} is a zero of $J_p(x)$, then

$$J_{p+1}(x_{pk}) \approx \pm\sqrt{\frac{2}{\pi x_{pk}}}$$

for k large enough.

6.8   Use the results of Exercises 4.1 and 5.3, together with the asymptotic formulas for J_p and Y_p (Exercise 6.6), to show that

$$J_p(x)J_{-(p-1)}(x) + J_{-p}(x)J_{p-1}(x) = \frac{2\sin p\pi}{\pi x}$$

and

$$J_p(x)Y_{p+1}(x) - J_{p+1}(x)Y_p(x) = -\frac{2}{\pi x}$$

for all p and $x > 0$. These formulas were originally proved by von Lommel in 1871.

8.1   Show that for any fixed p and if α sufficiently large, there are constants A and B such that

$$\frac{A}{\alpha} \le \int_0^a x J_p^2(\alpha x)\, dx \le \frac{B}{\alpha}$$

(**Hint:** use the asymptotic formula of Theorem 10.4.)

8.2   Show that, if $f : (0, a) \to \mathbb{R}$ is an integrable function and either (a) $\{\alpha_{pn}\}$ is the sequence of positive zeros of $J_p(ax)$ and c_n are the coefficients of the Fourier-Bessel series of f of order p on $(0, a)$ or (b) $\{\alpha_{pn}\}$ is the sequence of positive zeros of $hJ_p(ax) + axJ_p'(ax)$ and c_n are the coefficients of the Dini-Bessel series of f of order p on $(0, a)$, then

   (i) $\displaystyle \lim_{n \to \infty} \frac{c_n}{\sqrt{\alpha_{pn}}} = 0$ and

   (ii) $\displaystyle \lim_{n \to \infty} c_n J_p(\alpha_{pn}x) = 0$

(**Hint:** for the first limit use Bessel's inequality and the result of Exercise 8.1).

8.3   Use the results of Exercises 8.1 and 8.2 to prove the following analogue of the Riemann-Lebesgue theorem for Fourier-Bessel or Dini-Bessel series: if $f : (0, a) \to \mathbb{R}$ is an integrable function and if $\{\alpha_{pn}\}$ is the sequence of positive zeros of $J_p(ax)$ or of $hJ_p(ax) + axJ_p'(ax)$, then

$$\lim_{n \to \infty} \int_0^a x f(x) J_p(\alpha_{pn}x)\, dx = 0.$$

8.4 Prove that, if $p \geq 0$, if $\{\alpha_{pn}\}$ is the sequence of positive zeros of $J_p(ax)$ or of $h J_p(ax) + ax J_p'(ax)$, and if there are constants $C > 0$ and $q > 1$ such that

$$|c_n| \leq \frac{C}{\alpha_{pn}^q},$$

then the series

$$\sum_{n=1}^{\infty} c_n J_p(\alpha_{pn} x)$$

converges absolutely and uniformly on $[0, a]$ (**Hint:** use the results of Exercises 6.2 and 6.3 to show that $x_{pn} \geq n/2$ for n large enough).

8.5 Show that the product of $\sqrt{x}$ by the Fourier-Bessel series of $f(x)$ of order $1/2$ on $(0, a)$ equals the Fourier sine series of $\sqrt{x} f(x)$ on $(0, a)$.

8.6 Show that the product of $\sqrt{x}$ by the Fourier-Bessel series of $f(x)$ of order $-1/2$ on $(0, a)$ equals a Fourier cosine series. Of what function and on what interval is this series valid?

8.7 Use the result of Exercise 4.6 to show that the Fourier-Bessel expansion of $f(x) = x$ of order zero on $(0, a)$ is

$$x = \sum_{k=1}^{\infty} \frac{2}{\alpha_{0k} J_1(a\alpha_{0k})} \left[1 - \frac{1}{(a\alpha_{0k})^2 J_1(a\alpha_{0k})} \int_0^{a\alpha_{0k}} J_0(x)\, dx \right] J_0(\alpha_{0k} x).$$

8.8 Find the Fourier-Bessel expansions of

(i) $f(x) = \begin{cases} 1, & 0 \leq x < a/2 \\ 0, & a/2 \leq x < a \end{cases}$ (ii) $f(x) = \begin{cases} x, & 0 \leq x < a/2 \\ 0, & a/2 \leq x < a \end{cases}$

of order zero and of order 1, respectively, on $(0, a)$.

8.9 If $a > 0$ and n is a nonnegative integer, use the result of Exercise 4.7 to show that the Fourier-Bessel expansion of order $p \geq 0$ of $f(x) = x^p (a^2 - x^2)^n$ on $(0, a)$ is

$$x^p (a^2 - x^2)^n = 2^{n+1} n! a^{p+n-1} \sum_{k=1}^{\infty} \frac{J_{p+n+1}(a\alpha_{pk})}{\alpha_{pk}^{n+1} J_{p+1}^2(a\alpha_{pk})} J_p(\alpha_{pk} x).$$

8.10 Use the result of Exercise 8.9 to find the Fourier-Bessel expansions of
(i) $a^2 - x^2$ and
(ii) x^2
of order zero on $(0, a)$.

8.11 Use the result of Exercise 8.9 to show that

$$(a^2 - x^2)^2 = \frac{16}{a} \sum_{k=1}^{\infty} \frac{8 - (a\alpha_{0k})^2}{\alpha_{0k}^5 J_1(a\alpha_{0k})} J_0(\alpha_{0k} x)$$

and that

$$(a^2 - x^2)^3 = \frac{768}{a} \sum_{k=1}^{\infty} \frac{6 - (a\alpha_{0k})^2}{\alpha_{0k}^7 J_1(a\alpha_{0k})} J_0(\alpha_{0k} x).$$

8.12   If $a > 0$, use the result of Exercise 4.8 to find the Fourier-Bessel series of

(i) $\sqrt{a^2 - x^2}$ and

(ii) $x\sqrt{a^2 - x^2}$

of order zero and of order 1, respectively, on $(0, a)$ (**Hint:** for (ii) use the result of Exercise 4.3).

8.13   Use the result of Exercise 4.6 to show that the Dini-Bessel expansion of $f(x) = x$ of order zero on $(0, a)$ is

$$x = 2a \sum_{k=1}^{\infty} \frac{1}{(h^2 + a^2\alpha_{0k}^2)J_0(a\alpha_{0k})}\left[1 - \frac{1}{a\alpha_{0k}J_0(a\alpha_{0k})}\int_0^{a\alpha_{0k}} J_0(x)\,dx\right]J_0(\alpha_{0k}x)$$

for $h > 0$ and find the corresponding expansion for $h = 0$.

8.14   Find the Dini-Bessel expansions of

(i) $f(x) = \begin{cases} 1, & 0 \le x < a/2 \\ 0, & a/2 \le x < a \end{cases}$ (ii) $f(x) = \begin{cases} x, & 0 \le x < a/2 \\ 0, & a/2 \le x < a \end{cases}$

of order zero and of order 1, respectively, on $(0, a)$. Write your results using Bessel functions of order no higher than 1.

8.15   If $a > 0$ and n is a nonnegative integer, use the result of Exercise 4.7 to show that the Dini-Bessel expansion of order $p \ge 0$ of $f(x) = x^p(a^2 - x^2)^n$ on $(0, a)$ is

$$x^p(a^2 - x^2)^n = 2^{n+1}n!a^{p+n+1}\sum_{k=1}^{\infty} \frac{J_{p+n+1}(a\alpha_{pk})}{\alpha_{pk}^{n-1}(h^2 - p^2 + a^2\alpha_{pk}^2)J_p^2(a\alpha_{pk})} J_p(\alpha_{pk}x)$$

if $p + h > 0$ and

$$(a^2 - x^2)^n = \frac{a^{2n}}{n+1} + 2^{n+1}n!a^{n-1}\sum_{k=1}^{\infty} \frac{J_{n+1}(a\alpha_{0k})}{\alpha_{0k}^{n+1}J_0^2(a\alpha_{0k})} J_0(\alpha_{0k}x)$$

if $p = h = 0$.

8.16   Use the result of Exercise 8.15 to find the Dini-Bessel expansion of $f(x) = x^p$ of order $p \ge 0$ on $(0, a)$ if $p + h > 0$. Obtain, in particular, the representation

$$1 = 2h \sum_{k=1}^{\infty} \frac{1}{(h^2 + a^2\alpha_{0k}^2)J_0(a\alpha_{0k})} J_0(\alpha_{0k}x),$$

where $\{\alpha_{0k}\}$ is the sequence of positive zeros of $hJ_0(ax) - axJ_1(ax) = 0$.

8.17   Use the result of Exercise 8.15 to find the Dini-Bessel expansions of

(i) $a^2 - x^2$ and

(ii) x^2

of order zero on $(0, a)$.

8.18   If $a > 0$, use the result of Exercise 4.8 to find the Dini-Bessel series of

(i) $\sqrt{a^2 - x^2}$ and

(ii) $x\sqrt{a^2 - x^2}$

of order zero and of order 1, respectively, on $(0, a)$ (**Hint:** for (ii) use the result of Exercise 4.3).

9.1   Find the particular form of the motion of the membrane if it has no initial velocity and its initial displacement $f(r, \theta)$ is independent of θ. Use the result of Exercise 8.11 to find the solution if $f(r, \theta) = (a^2 - r^2)^2$.

9.2   Find a formal solution of the boundary value problem

$$u_{tt} = c^2 \left(u_{rr} + \frac{1}{r} u_r + \frac{1}{r^2} u_{\theta\theta} \right) \qquad r < a, \qquad -\pi < \theta < \pi, \qquad t > 0$$

$$u(a, \theta, t) = 0 \qquad\qquad\qquad\qquad -\pi \leq \theta \leq \pi, \qquad t \geq 0$$

$$u(r, \theta, 0) = 0 \qquad\qquad r \leq a, \qquad -\pi \leq \theta \leq \pi$$

$$u_t(r, \theta, 0) = g(r, \theta) \qquad r \leq a, \qquad -\pi \leq \theta \leq \pi$$

corresponding to the vibrations of a circular membrane with initial velocity given by a known function g that is periodic of period 2π and such that $g(a, \theta) = 0$. Then solve the same problem if $u(r, \theta, 0) = 0$ is replaced with $u(r, \theta, 0) = f(r, \theta)$, where f is periodic of period 2π and such that $f(a, \theta) = 0$.

9.3   Consider a circular membrane attached to a rigid frame that is falling vertically inside a snuggly fitting cylinder. Assume that at all times the frame remains on a horizontal plane and that at $t = 0$ it hits a circular lower lip in the cylinder with constant velocity $v_0 < 0$ and instantly comes to a stop. If only the frame is constrained by the cylinder's lip, so that the membrane is free to move vertically and if it is initially on a horizontal plane, use the solution of Exercise 9.2 to find its motion for $t > 0$.

9.4   A circular membrane, attached to a rigid frame at $r = a$ and initially at rest, is struck by a hammer of mass M with velocity v_0. The hammer hits only the points of the membrane for $r < \epsilon < a$, in such a way that the hammer's momentum is evenly distributed over all these points. Use the solution of Exercise 9.2 to find the motion of the membrane and compute its limit as $\epsilon \to 0$.

9.5   Find the motion of a circular membrane attached to a rigid frame if it moves under the force of gravity, that is, if the wave equation is replaced by

$$m u_{tt} = \tau \left(u_{rr} + \frac{1}{r} u_r + \frac{1}{r^2} u_{\theta\theta} \right) - mg,$$

where g is the acceleration of gravity, if
(i) m is a constant and
(ii) m is a function of r of the form $m = M - r/a$ where $M > 1$ is a constant.
Assume, in each case, that the membrane has no initial velocity and satisfies an initial condition of the form $u(r, \theta, 0) = f(r, \theta)$.

9.6   Consider a circular membrane that is initially at rest, attached to a rigid frame and subject to an external force $F(t) = A \sin \omega t$ per unit area for $t > 0$. Since this situation is independent of θ, the wave equation is replaced by

$$u_{tt} = c^2 \left(u_{rr} + \frac{1}{r} u_r \right) + A \sin \omega t,$$

where A is a constant. Show that, if $\{\alpha_{0k}\}$ is the sequence of positive zeros of $J_0(ax)$ and if $\omega \neq c\alpha_{0k}$ for any k, then the motion of the membrane is formally given by

$$u(r, t) = \frac{2A}{ac} \sum_{k=1}^{\infty} \frac{c\alpha_{0k} \sin \omega t - \omega \sin c\alpha_{0k} t}{\alpha_{0k}^2 (c^2 \alpha_{0k}^2 - \omega^2)} \frac{J_0(\alpha_{0k} r)}{J_1(a\alpha_{0k})}.$$

Discuss the possibility of resonance if $\omega = c\alpha_{0k}$ for some k (**Hint:** use the result of Example 10.5).

9.7   Find a formal solution of the boundary value problem

$$u_{tt} = c^2\left(u_{rr} + \frac{1}{r}u_r\right) - 2hu_t \qquad r < a, \qquad t > 0$$

$$u(a, t) = 0 \qquad\qquad\qquad\qquad\qquad t \geq 0$$

$$u(r, 0) = f(r) \qquad\qquad\qquad\qquad r \leq a$$

$$u_t(r, 0) = g(r) \qquad\qquad\qquad\qquad r \leq a,$$

where h is a positive constant, corresponding to the vibrations of a circular membrane moving in a medium in which the friction force is proportional to the velocity of the string. Assume that $h < c\alpha_{01}$, where α_{01} is the first positive zero of $J_0(ax)$. Since the initial position and velocity of the membrane are independent of θ so is its motion. Show that in the particular case of the falling membrane of Exercise 9.3

$$u(r, t) = \frac{2v_0}{a}e^{-ht}\sum_{k=1}^{\infty}\frac{1}{\alpha_{0k}\sqrt{c^2\alpha_{0k}^2 - h^2}\, J_1(a\alpha_{0k})} J_0(\alpha_{0k}r)\sin\sqrt{c^2\alpha_{0k}^2 - h^2}\, t$$

(**Hint:** make the substitution $v = e^{ht}u$ and then use separation of variables).

9.8   Repeat Exercise 9.6 if the membrane moves in a medium with friction, that is, if the equation of motion is replaced with

$$u_{tt} = c^2\left(u_{rr} + \frac{1}{r}u_r\right) - 2hu_t + A\sin\omega t$$

(**Hint:** make the substitution $v = e^{ht}u$, then assume a solution of the form

$$v(r, t) = \sum_{k=1}^{\infty} a_{0k}(t) J_0(\alpha_{0k}r),$$

and modify the result of Exercise 9.6 as needed).

10.1   Solve the Dirichlet problem

$$u_{rr} + \frac{1}{r}u_r + \frac{1}{r^2}u_{\theta\theta} + u_{zz} = 0 \qquad r < a, \qquad -\pi \leq \theta \leq \pi, \qquad 0 < z < c$$

$$u(r, \theta, 0) = f(r, \theta) \qquad\qquad r \leq a, \qquad -\pi \leq \theta \leq \pi$$

$$u(r, \theta, c) = g(r, \theta) \qquad\qquad r \leq a, \qquad -\pi \leq \theta \leq \pi$$

$$u(a, \theta, z) = 0 \qquad\qquad\qquad\qquad -\pi \leq \theta \leq \pi, \qquad 0 \leq z \leq c$$

(**Hint:** obtain the solution by the superposition of those of the particular cases (i) $g \equiv 0$ and (ii) $f \equiv 0$).

10.2   How is the solution of Exercise 10.1 to be modified to become a bounded solution of the Laplace equation in the semi-infinite cylinder defined by $r < a$ and $z > 0$ if the boundary condition at $z = c$ is dropped? Show that, if $f \equiv C$, a constant, this solution is

$$u(r, z) = \frac{2C}{a}\sum_{k=1}^{\infty}\frac{1}{\alpha_{0k}J_1(a\alpha_{0k})} J_0(\alpha_{0k}r)e^{-\alpha_{0k}z}.$$

10.3   How is the solution of Exercise 10.1 to be modified if the Laplace equation is replaced
with the Poisson equation

$$u_{rr} + \frac{1}{r}u_r + \frac{1}{r^2}u_{\theta\theta} + u_{zz} = -q,$$

where q is a constant? Complete the details of the following outlined procedure. Add
to the solution of Exercise 10.1 a solution of the equation

$$v_{rr} + \frac{1}{r}v_r + v_{zz} = -q$$

that is independent of θ and satisfies zero boundary conditions. To find v use Example 10.5 to expand q in a Fourier Bessel series on $(0, a)$ and then, using variation of
constants, try a solution of the form

$$v(r, \theta, z) = \sum_{k=1}^{\infty} a_k(z) J_0(\alpha_{0k} r).$$

Imitate, if necessary, some of the steps at the end of §10.9.

10.4   Solve the Dirichlet problem

$$u_{rr} + \frac{1}{r}u_r + \frac{1}{r^2}u_{\theta\theta} + u_{zz} = 0 \qquad r < a, \qquad 0 \le \theta \le \pi, \qquad 0 < z < c$$

$$u_z(r, \theta, 0) = 0 \qquad\qquad r \le a, \qquad 0 \le \theta \le \pi$$

$$u_z(r, \theta, c) = 0 \qquad\qquad r \le a, \qquad 0 \le \theta \le \pi$$

$$u(r, 0, z) = 0 \qquad\qquad r \le a, \qquad\qquad\qquad 0 \le z \le c$$

$$u(r, \pi, z) = 0 \qquad\qquad r \le a, \qquad\qquad\qquad 0 \le z \le c$$

$$u(a, \theta, z) = f(\theta, z) \qquad\qquad 0 \le \theta \le \pi, \qquad 0 \le z \le c$$

for a half-cylinder that is insulated at the top and bottom if the temperature is prescribed
to be zero on its flat side and equal to $f(\theta, z)$ on the semicircular surface, where we
assume that $f_z(\theta, 0) = f_z(\theta, c) = 0$.

10.5   Find a formal bounded solution of the Laplace equation in the semi-infinite cylinder
defined by $r < a$ and $z > 0$ if $u_z(r, 0) = 0$ for $r \le a$ and $u(a, z) = f(z)$ for $z \ge 0$,
using
   (*i*) separation of variables and
   (*ii*) the Fourier cosine transform method as outlined before Exercise 6.5 of Chapter 7.
Show that, if

$$f(z) = \begin{cases} c - z, & 0 \le z \le c \\ 0, & z > c, \end{cases}$$

then the solution is

$$u(r, z) = \frac{2}{\pi} \int_0^\infty \frac{I_0(\omega r)}{\omega^2 I_0(\omega a)} (1 - \cos \omega c) \cos \omega z \, d\omega$$

(**Hint:** refer to Exercises 2.3 and 3.9(*iii*) of Chapter 7).

10.6   Use the Fourier transform method to show that the formal solution of the Laplace equation in the infinite cylinder $r < a$ such that $u(a, z) = f(z)$ for all z is

$$u(r, z) = \frac{1}{\pi} \int_0^\infty \frac{I_0(\omega r)}{I_0(\omega a)} \left(\int_{-\infty}^{\infty} f(s) \cos \omega(z - s) \, ds \right) d\omega.$$

Exercises 11.1 to 11.6 ask for solutions of the three-dimensional heat equation

$$u_t = k\left(u_{rr} + \frac{1}{r}u_r + \frac{1}{r^2}u_{\theta\theta} + u_{zz}\right)$$

in cylindrical coordinates or of some of its simplified versions when the temperature can be assumed to be independent of one or more of the variables.

11.1   Consider a circular cylindrical bar of radius a centered on the z-axis and located between $z = 0$ and $z = c$. If the temperature is independent of θ, find a formal solution of the heat equation that satisfies the boundary and initial conditions

$$
\begin{aligned}
u(r, 0, t) &= T & r &\leq a, & t &\geq 0 \\
u(r, c, t) &= T & r &\leq a, & t &\geq 0 \\
u_r(a, z, t) &= C[T - u(a, z, t)] & & 0 \leq z \leq c, & t &\geq 0 \\
u(r, z, 0) &= f(r, z) & r &\leq a, & 0 &\leq z \leq c,
\end{aligned}
$$

where T and C are nonnegative constants and f is a given function such that $f(r, 0) = f(r, c) = T$. Thus, we assume that for $t > 0$ the ends of the bar are at the temperature T of the outside medium, but that there is heat loss by convection on the lateral cylindrical surface.

11.2   Consider a very long circular cylinder, or a very thin circular plate, in which the temperature is independent of θ and z. Find a formal solution of the heat equation that satisfies the boundary and initial conditions $u(a, t) = Ce^{-\gamma t}$ for $t \geq 0$ and $u(r, 0) = f(r)$ for $r \leq a$, where C and γ are positive constants and f is a given function (**Hint:** find first a solution $U(r, t)$ that satisfies the boundary condition only, and then consider the function $v = u - U$).

11.3   How should the solution of Exercise 11.2 be modified if the heat equation is replaced with

$$u_t = k\left(u_{rr} + \frac{1}{r}u_r\right) + q(r, t),$$

where $q(r, t)$ represents internal heat generation? Consider the particular cases
   (*i*) q is independent of r and
   (*ii*) $q(r, t) = Q(a^2 - r^2)$, where Q is a positive constant
(**Hint:** for general q imitate some of the steps at the end of §10.9, for part (*i*) use the result of Example 10.5, and for part (*ii*) use the result of Exercise 8.10 (*i*)).

11.4   Consider a long hollow cylindrical tube from $r = a > 0$ to $r = b > a$. Find a formal solution of the heat equation inside this tube that satisfies the boundary and initial conditions

$$
\begin{aligned}
u(a, t) &= T_1 & t &\geq 0 \\
u(b, t) &= T_2 & t &\geq 0 \\
u_t(r, 0) &= f(r) & r &\leq a,
\end{aligned}
$$

where T_1 and T_2 are nonnegative constants and f is a given function that has the values $f(a) = T_1$ and $f(b) = T_2$.

11.5   Consider a wedge of a circular cylindrical bar of radius a located along the z-axis between $\theta = 0$ and $\theta = \theta_0$, where $0 < \theta_0 < 2\pi$. If its temperature is independent of z

and is maintained at a constant value T for $t > 0$ on the lateral surface, find a formal solution of the heat equation that satisfies the initial condition

$$u(r, \theta, 0) = f(r, \theta),$$

where f is a given function. Find the solution in the particular cases of a quarter-round ($\theta_0 = \pi/2$) and a half-round ($\theta_0 = \pi$).

11.6 Consider a quarter-round wedge of a circular cylindrical bar of radius a located along the z-axis between $z = 0$ and $z = c$. If its lateral surface is insulated, find a formal solution of the heat equation that satisfies the initial condition

$$u(r, \theta, z, 0) = f(r, \theta, z),$$

where f is a given function.

11

BOUNDARY VALUE PROBLEMS
WITH SPHERICAL SYMMETRY

§11.1 The Potbellied Stove

We consider the problem of finding the temperature in a large room outside a potbellied stove, which we shall take to be perfectly spherical as a simplifying assumption and whose surface temperature is known. As usual, the temperature is a solution of the heat equation

$$u_t = k(u_{xx} + u_{yy} + u_{zz}),$$

which was derived in Exercises 1.2 and 1.4 of Chapter 9. In the current situation it is convenient to express it in spherical coordinates, r, θ, and ϕ, that are related to rectangular coordinates by the equations

$$
\begin{aligned}
x &= r \cos \theta \sin \phi \\
y &= r \sin \theta \sin \phi \\
z &= r \cos \phi
\end{aligned}
$$

as shown in Figure 11.1. This figure also shows the distance from the origin to the point $(x, y, 0)$, labeled ρ. In order to compute $u_{xx} + u_{yy} + u_{zz}$, we first proceed as in §10.1 in cylindrical coordinates, but with ρ replacing r, to obtain

(1) $$u_{xx} + u_{yy} = u_{\rho\rho} + \frac{1}{\rho} u_\rho + \frac{1}{\rho^2} u_{\theta\theta}.$$

Since the two pairs of equations

$$
\begin{aligned}
z &= r \cos \phi, & \rho &= r \sin \phi \\
x &= \rho \cos \theta, & y &= \rho \sin \theta
\end{aligned}
$$

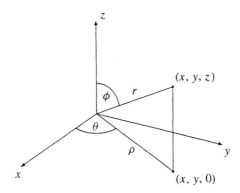

Figure 11.1

show that the coordinates (z, ρ) are related to (r, ϕ) in exactly the same manner as (x, y) are related to (ρ, θ), (1) implies that

$$(2) \qquad u_{zz} + u_{\rho\rho} = u_{rr} + \frac{1}{r}u_r + \frac{1}{r^2}u_{\phi\phi}.$$

Adding (1) and (2) and canceling $u_{\rho\rho}$, we obtain

$$(3) \qquad u_{xx} + u_{yy} + u_{zz} = u_{rr} + \frac{1}{r}u_r + \frac{1}{r^2}u_{\phi\phi} + \frac{1}{\rho}u_\rho + \frac{1}{\rho^2}u_{\theta\theta}.$$

Now, ρ depends on r and ϕ but is independent of θ, and then the chain rule gives

$$(4) \qquad u_\rho = u_r r_\rho + u_\phi \phi_\rho.$$

The remaining derivatives are computed using the equations

$$r = \sqrt{\rho^2 + z^2} \qquad \text{and} \qquad \phi = \arctan\frac{\rho}{z},$$

from which we obtain

$$r_\rho = \frac{\rho}{\sqrt{\rho^2 + z^2}} = \frac{\rho}{r} \qquad \text{and} \qquad \phi_\rho = \frac{1}{1 + (\rho/z)^2}\frac{1}{z} = \frac{z}{z^2 + \rho^2} = \frac{\cos\phi}{r}.$$

Taking these derivatives to (4) and then the resulting expression to (3) gives

$$u_{xx} + u_{yy} + u_{zz} = u_{rr} + \frac{1}{r}u_r + \frac{1}{r^2}u_{\phi\phi} + \frac{1}{\rho}\left(u_r\frac{\rho}{r} + u_\phi\frac{\cos\phi}{r}\right) + \frac{1}{\rho^2}u_{\theta\theta}.$$

Finally, simplifying and using the fact that $\rho = r\sin\phi$ lead to the following expression for the heat equation in spherical coordinates

$$(5) \qquad u_t = k\left(u_{rr} + \frac{2}{r}u_r + \frac{1}{r^2}u_{\phi\phi} + \frac{\cot\phi}{r^2}u_\phi + \frac{1}{r^2\sin^2\phi}u_{\theta\theta}\right).$$

Let us assume that we are interested only in the steady state solution, that is, in the temperature after the stove has had ample time to heat the room. Then we are looking for a solution independent of time, and the heat equation becomes the Laplace equation. Assume also that the surface temperature of the stove is the same all the way around, that is, that it is independent of θ. Then, if this temperature is a given function f of ϕ and if the stove has radius a we impose the boundary condition

$$u(a, \phi) = f(\phi).$$

Using the method of separation of variables, we look for a bounded solution of the form $u(r, \phi, \theta) = R(r)\Phi(\phi)\Theta(\theta)$ valid for $r \geq a$. Since the initial condition implies that $R(a)\Phi(\phi)\Theta(\theta) = f(\phi)$, Θ must be a constant function. This means that the proposed separated solution must be of the form $u(r, \phi) = R(r)\Phi(\phi)$. Substituting into the Laplace equation, multiplying by r^2, and dividing by $R\Phi$, we have

$$r^2 \frac{R''}{R} + 2r \frac{R'}{R} + \frac{\Phi''}{\Phi} + \cot \phi \frac{\Phi'}{\Phi} = 0,$$

which leads to the ordinary differential equations

$$r^2 R'' + 2r R' - \lambda R = 0$$

and

$$\Phi'' + (\cot \phi)\Phi' + \lambda \Phi = 0.$$

The equation for R is of the Cauchy-Euler type and easily solvable using the substitution $r = e^t$, as indicated in Exercise 1.5 of Chapter 4. Omitting the details, its general solution is

$$R(r) = Ar^{-\frac{1}{2}+\sqrt{\frac{1}{4}+\lambda}} + Br^{-\frac{1}{2}-\sqrt{\frac{1}{4}+\lambda}},$$

where A and B are arbitrary constants. The equation for Φ can be rewritten by defining a new variable $x = \cos \phi$, which varies from 1 to -1 as ϕ varies from 0 to π, and a new function X of x defined by

$$X(x) = \Phi(\arccos x).$$

Differentiating we have

$$X'(x) = -\frac{1}{\sqrt{1 - x^2}}\Phi'(\phi)$$

and

$$X''(x) = \frac{1}{1 - x^2}\Phi''(\phi) - \frac{x}{(1 - x^2)\sqrt{1 - x^2}}\Phi'(\phi),$$

and then the equation for Φ becomes

$$(1 - x^2)X'' - 2xX' + \lambda X = 0.$$

This is known as the *Legendre equation* for reasons to be explained later. It is not of a type previously encountered, and we shall devote the next section to its solution.

§11.2 Solutions of the Legendre Equation

Our task is to solve the Legendre equation on $[-1, 1]$. Note that it is of the form

$$(pX')' + (q + \lambda r)X = 0$$

with $p(x) = 1 - x^2, q \equiv 0$ and $r \equiv 1$. Since p vanishes at $x = \pm 1$, this equation is by itself a singular Sturm-Liouville problem on $[-1, 1]$. We shall see at the end of §11.4 that the only eigenvalues are $\lambda = n(n + 1)$, where n is a nonnegative integer, so we shall take λ to be of this form. There is no loss of generality in doing this even if we wanted to solve the Legendre equation on another interval, because λ is always of this form for an appropriate—but not necessarily integral—choice of n. If we try a series solution of the form

$$X(x) = \sum_{k=0}^{\infty} a_k x^k,$$

which we assume to be twice differentiable term by term, substitution into the Legendre equation gives

$$\sum_{k=0}^{\infty} a_k[k(k - 1)(1 - x^2)x^{k-2} - 2kx^k + \lambda x^k] = 0,$$

and then, noting that $k(k - 1) = 0$ for $k = 1, 2$ and regrouping terms,

$$\sum_{k=2}^{\infty} k(k - 1)a_k x^{k-2} + \sum_{k=0}^{\infty} [-k(k - 1) - 2k + \lambda]a_k x^k$$

$$= \sum_{k=0}^{\infty} \{(k + 2)(k + 1)a_{k+2} - [k(k + 1) - \lambda]a_k\} x^k$$

$$= 0.$$

Equating the coefficient of each power of x to zero and then putting $\lambda = n(n + 1)$ yield the recurrence relation

$$(6) \qquad a_{k+2} = \frac{k(k + 1) - n(n + 1)}{(k + 2)(k + 1)} a_k = -\frac{(n - k)(n + k + 1)}{(k + 2)(k + 1)} a_k$$

for every nonnegative integer k. This means that a_0 and a_1 are arbitrary, and then the remaining coefficients are chosen using (6). If we make the choice $a_0 = 0$ and $a_1 \neq 0$, then $a_{k+2} = 0$ for all even k and we obtain a solution of the form

$$(7) \qquad X(x) = \sum_{k=1}^{\infty} a_{2k-1} x^{2k-1}.$$

If, instead, we choose $a_0 \neq 0$ and $a_1 = 0$, then $a_{k+2} = 0$ for all odd k and we obtain a solution of the form

$$(8) \qquad X(x) = \sum_{k=0}^{\infty} a_{2k} x^{2k}.$$

It is easy to use the ratio test to show that these series converge for $|x| < 1$ and diverge for $|x| > 1$. Then, by the Weierstrass M-test, they converge uniformly on any closed subinterval of $(-1, 1)$. The same is true of their first and second derived series, and then they can be differentiated term by term. This validates the steps above and shows that (7) and (8) are solutions of the Legendre equation on $(-1, 1)$. Since one of these solutions contains only even powers of x and the other only odd powers, they are linearly independent.

If n is a nonnegative integer in (6), we see that $a_{n+2} = a_{n+4} = a_{n+6} = \cdots = 0$. Therefore, if n is odd, (7) becomes a polynomial and, if n is even, (8) becomes a polynomial. For every nonnegative integer n, this polynomial solution is then an eigenfunction of the stated Sturm-Liouville problem corresponding to the eigenvalue $\lambda = n(n+1)$. If we define

$$N = \begin{cases} \dfrac{n}{2} & \text{if } n \text{ is even} \\[2mm] \dfrac{n-1}{2} & \text{if } n \text{ is odd,} \end{cases}$$

then the polynomial solution given by either (7) or (8) is of the form

$$X(x) = \sum_{m=0}^{N} a_{n-2m} x^{n-2m},$$

whose first term is $a_n x^n$ and whose last term is $a_1 x$ if n os odd and a_0 if n is even. It is customary to choose a_n arbitrarily, rather than a_0 or a_1, and then to determine the remaining coefficients using (6). Indeed, let m be a positive integer no larger than N (so that $2m \le n$), multiply the m equations that result from substituting $k = n - 2$, $n - 4, \ldots, n - 2m$ in (6), and cancel common factors to obtain

$$a_n = (-1)^m \frac{2 \cdot 4 \cdots 2m}{n(n-2) \cdots (n-2m+2)} \frac{(2n-1)(2n-3) \cdots (2n-2m+1)}{(n-1)(n-3) \cdots (n-2m+1)} a_{n-2m}.$$

After combination of some of the terms on the right-hand side into the appropriate factorials, this equation becomes

$$a_n = (-1)^m \frac{m!(n-2m)!(2n)!(n-m)!}{(2n-2m)!(n!)^2} a_{n-2m}$$

(see Exercise 2.1). Note that it is also valid for $m = 0$, and then, if we choose

$$a_n = \frac{(2n)!}{2^n (n!)^2},$$

we arrive at the value

$$a_{n-2m} = (-1)^m \frac{(2n-2m)!}{2^n m!(n-2m)!(n-m)!}.$$

With this choice of coefficients the eigenfunction corresponding to the eigenvalue $\lambda = n(n + 1)$, n a nonnegative integer, is denoted by P_n and is given by

$$(9) \qquad P_n(x) = \frac{1}{2^n} \sum_{m=0}^{N} \frac{(-1)^m}{m!} \frac{(2n - 2m)!}{(n - 2m)!(n - m)!} x^{n-2m}.$$

Clearly, P_n is an even function if n is even and an odd function if n is odd, that is,

$$P_n(-x) = (-1)^n P_n(x).$$

The P_n are called the *Legendre polynomials of degree n* since they were introduced by Adrien Marie Legendre (1752–1833), of the *Académie des Sciences*. Not too much

ADRIEN MARIE LEGENDRE
David Eugene Smith Collection.
Rare Book and Manuscript Library.
Columbia University.
Reprinted with permission.

is known of his private life, nor did he wish to be known for any reason but his work. Uninterested in notoriety and even in making the right career moves, he devoted his entire life to mathematical research on a very wide range of subjects. The polynomials in question appear in his papers *Recherches sur la figure des planètes* and *Recherches sur l'attraction des sphéroïdes*, published in 1783 and 1785, respectively, at which time he limited himself to polynomials of even order and proved (9) for P_{2n}. Later, in another paper on the same subject written in 1787, he proved some properties of the polynomials P_{2n} including their orthogonality with respect to the inner product

$$(f, g) = \int_{-1}^{1} fg.$$

Then, in a fourth paper dated 1789 but published in 1793, he introduced P_n for n odd, proving (9) and the orthogonality property for any n. It is in this paper that he stated the Legendre equation for the first time and showed that the P_n are solutions. Actually, this equation is a particular case of another that appears in a paper by Laplace, also on the attraction of spheroids, dated 1782 but published in 1785.

There is an alternate expression for $P_n(x)$ that is often convenient. Since $2N \leq n$,

$$\frac{d^n}{dx^n} x^{2n-2m} = \begin{cases} \dfrac{(2n - 2m)!}{(n - 2m)!} x^{n-2m} & \text{if } 0 \leq m \leq N \\ 0 & \text{if } m > N \end{cases}$$

and then the right-hand side of (9) can be written in the form

$$\frac{1}{2^n n!} \frac{d^n}{dx^n} \sum_{m=0}^{n} (-1)^m \frac{n!}{m!(n - m)!} x^{2n-2m}.$$

But the sum on the right is the binomial expansion of $(x^2 - 1)^n$, and then

$$P_n(x) = \frac{1}{2^n n!} \frac{d^n}{dx^n} (x^2 - 1)^n.$$

This equation was originally found in 1816 by the French mathematician, banker, and social and religious reformer Olinde Rodrigues (1794–1851), also working on the attraction of spheroids, and it is called the *Rodrigues formula*.

Using this formula and the Leibniz rule for the nth derivative of a product,

$$(fg)^{(n)} = \sum_{k=0}^{n} \binom{n}{k} f^{(k)} g^{(n-k)},$$

it is easy to show that

$$P_n(x) = \frac{1}{2^n n!} \frac{d^n}{dx^n} [(x + 1)^n (x - 1)^n] = \frac{1}{2^n n!} [(n!)(x + 1)^n + \cdots],$$

where the dots indicate a sum of n terms each of which containing the factor $x - 1$. It follows at once that

$$P_n(1) = 1$$

for all n, and then that

$$P_n(-1) = (-1)^n P_n(1) = (-1)^n.$$

If n is a nonnegative integer, P_n is just one solution of the Legendre equation with $\lambda = n(n + 1)$. The other linearly independent solution is of one of the forms (7) or (8) and does not terminate; that is, it remains an infinite series whose precise form is given in Exercise 2.8 for n even or in Exercise 2.9 for n odd. It easy to see that these series diverge at $x = \pm 1$, and then they cannot be eigenfunctions of our singular Sturm-Liouville problem because they are not defined on $[-1, 1]$. In fact, we have already mentioned above, and we shall show it at the end of §11.4, that the only eigenvalues are $\lambda = n(n + 1)$, where n is a nonnegative integer (this also follows from Exercise 2.12), and then the only eigenfunctions are P_n. As we have already shown in a more general situation in Theorem 4.1, the P_n form an orthogonal system.

§11.3 The Norms of the Legendre Polynomials

We shall now consider expanding a given function in a generalized Fourier series with respect to the orthogonal system of the Legendre polynomials. The first step is to compute the norm of P_n. This was originally done by Legendre in his 1793 paper. As in the case of Bessel functions, we start with a recursion formula.

Theorem 11.1 *For any positive integer n,*

$$(n + 1)P_{n+1}(x) - (2n + 1)x P_n(x) + n P_{n-1}(x) = 0.$$

Proof. The polynomial

$$(n + 1)P_{n+1}(x) - (2n + 1)x P_n(x)$$

is odd because $P_{n+1}(x)$ is odd and $P_n(x)$ is even. Then it is of degree $n - 1$ because the coefficient of x^{n+1} is

$$(n+1)\frac{1}{2^{n+1}}\frac{(2n + 2)!}{[(n + 1)!]^2} - (2n+1)\frac{1}{2^n}\frac{(2n)!}{(n!)^2} = \left[\frac{(n + 1)(2n + 2)}{2(n + 1)^2} - 1\right]\frac{2n + 1}{2^n}\frac{(2n)!}{(n!)^2} = 0.$$

Then it is of the form

$$(n + 1)P_{n+1}(x) - (2n + 1)x P_n(x) = \sum_{i=0}^{n-1}c_i P_i(x),$$

where the c_i are constants. Let m be an integer such that $0 \le m \le n - 2$, multiply the equation above by P_m, and use orthogonality to conclude that

$$-(2n + 1)\int_{-1}^1 x P_m(x)P_n(x)\,dx = c_m \int_{-1}^1 P_m^2(x)\,dx.$$

Since $x P_m(x)$ is a polynomial of degree no larger than $n - 1$, it is a linear combination of $P_0, \ldots, P_{n-1}$ and, again by orthogonality, the left-hand side is zero. This means that $c_m = 0$ for $0 \le m \le n - 2$, and then

$$(n + 1)P_{n+1}(x) - (2n + 1)x P_n(x) = c_{n-1}P_{n-1}(x).$$

Putting $x = 1$ and using the fact that $P_n(1) = 1$ for all n, we obtain

$$n + 1 - (2n + 1) = c_{n-1},$$

that is, $c_{n-1} = -n$, and the result follows, Q.E.D.

Theorem 11.2 *For any nonnegative integer* n,

$$\| P_n \|^2 = \frac{2}{2n + 1}.$$

Proof. The result follows by direct computation for $P_0 \equiv 1$ and $P_1(x) = x$. If $n \geq 2$, Theorem 11.1, with n replaced by $n - 1$ in the first of the next two equations, gives

$$n P_n(x) - (2n - 1)x P_{n-1}(x) + (n - 1)P_{n-2}(x) = 0$$

$$(n + 1)P_{n+1}(x) - (2n + 1)x P_n(x) + n P_{n-1}(x) = 0.$$

Multiplying these equations by $P_n(x)$ and $P_{n-1}(x)$, respectively, and integrating from -1 to 1, we obtain

$$n \int_{-1}^{1} P_n^2(x)\, dx = (2n - 1) \int_{-1}^{1} x P_{n-1}(x) P_n(x)\, dx$$

$$n \int_{-1}^{1} P_{n-1}^2(x)\, dx = (2n + 1) \int_{-1}^{1} x P_n(x) P_{n-1}(x)\, dx,$$

that is,

$$(2n + 1) \int_{-1}^{1} P_n^2(x)\, dx = (2n - 1) \int_{-1}^{1} P_{n-1}^2(x)\, dx = \cdots = 3 \int_{-1}^{1} P_1^2(x)\, dx = 2,$$

Q.E.D.

§11.4   Fourier-Legendre Series

If $f : (-1, 1) \to \mathbb{R}$ is integrable, it has a generalized Fourier series

$$f \sim \sum_{n=0}^{\infty} c_n P_n$$

with respect to the Legendre polynomials P_n, where

$$c_n = \frac{2n + 1}{2} \int_{-1}^{1} f P_n.$$

The series above and these coefficients are called the **Fourier-Legendre series** and the **Fourier-Legendre coefficients** of f on $(-1, 1)$, respectively.

In order to prove the pointwise convergence of the Fourier-Legendre series of a sufficiently nice function we need two preliminary results. The following lemma, giving an integral representation of P_n, is due to Laplace and appears in the second chapter of the fifth volume of his *Mécanique Céleste, Livre XI*, of 1825.

Lemma 11.1 *For $-1 \leq x \leq 1$ and any nonnegative integer n,*

$$P_n(x) = \frac{2}{\pi} \int_0^{\pi/2} (x + i\sqrt{1 - x^2} \cos s)^n \, ds.$$

Proof. If n is a nonnegative integer

$$\int_{-\pi}^{\pi} e^{ins} = \begin{cases} 0 & \text{if } n > 0 \\ 2\pi & \text{if } n = 0 \end{cases}$$

and then, for any real number x and any real or complex constant c,

$$\frac{1}{2\pi} \int_{-\pi}^{\pi} (x + ce^{is})^n \, ds = \frac{1}{2\pi} \int_{-\pi}^{\pi} \sum_{k=0}^{n} \binom{n}{k} x^k c^{n-k} e^{i(n-k)s} \, ds = x^n.$$

Thus, any polynomial P in x satisfies

$$P(x) = \frac{1}{2\pi} \int_{-\pi}^{\pi} P(x + ce^{is}) \, ds.$$

But, according to the Rodrigues formula, $P_n(x) = g^{(n)}(x)$, where

$$g(x) = \frac{1}{2^n n!} (x^2 - 1)^n$$

for $-1 \leq x \leq 1$, so that

(10) $$P_n(x) = \frac{1}{2\pi} \int_{-\pi}^{\pi} g^{(n)}(x + ce^{is}) \, ds.$$

Now, if k is an integer such that $0 \leq k \leq n - 1$, we have

$$\int_{-\pi}^{\pi} e^{-iks} g^{(n-k)}(x + ce^{is}) \, ds$$

$$= \int_{-\pi}^{\pi} e^{-i(k+1)s} \frac{d}{ds} \left[\frac{1}{ic} g^{(n-k-1)}(x + ce^{is}) \right] ds$$

$$= e^{-i(k+1)s} \frac{1}{ic} g^{(n-k-1)}(x + ce^{is}) \Big|_{-\pi}^{\pi}$$

$$\qquad\qquad - \int_{-\pi}^{\pi} \frac{-i(k+1)}{ic} e^{-i(k+1)s} g^{(n-k-1)}(x + ce^{is}) \, ds$$

$$= \frac{k+1}{c} \int_{-\pi}^{\pi} e^{-i(k+1)s} g^{(n-k-1)}(x + ce^{is}) \, ds.$$

Starting with (10) and applying the preceding formula for $k = 0, \ldots n - 1$ lead to

$$(11) \qquad P_n(x) = \frac{n!}{2\pi c^n} \int_{-\pi}^{\pi} e^{-ins} g(x + ce^{is}) \, ds.$$

Using the definition of g and putting $c = \sqrt{x^2 - 1}$ we obtain

$$
\begin{aligned}
2^n n! g(x + ce^{is}) &= [(x + ce^{is})^2 - 1]^n \\
&= (x^2 + c^2 e^{2is} + 2xce^{is} - 1)^n \\
&= (2xce^{is} + c^2 e^{2is} + c^2)^n \\
&= c^n e^{ins}(2x + ce^{is} + ce^{-is})^n \\
&= 2^n c^n e^{ins}(x + c \cos s)^n \\
&= 2^n c^n e^{ins}(x + i\sqrt{1 - x^2} \cos s)^n,
\end{aligned}
$$

and then the result follows from (11) and the fact that the cosine is symmetric with respect to the line $x = \pi/2$, Q.E.D.

Laplace's integral formula can be used to prove two useful inequalities for the Legendre polynomials. Both are based on the fact that

$$
\begin{aligned}
|x + i\sqrt{1 - x^2} \cos s| &= \sqrt{x^2 + (1 - x^2) \cos^2 s} \\
&= \sqrt{x^2 + (1 - x^2)(1 - \sin^2 s)} \\
&= \sqrt{1 - (1 - x^2) \sin^2 s}.
\end{aligned}
$$

Since the right-hand side is no larger than 1 it follows form Lemma 11.1 that

$$|P_n(x)| \le \frac{2}{\pi} \int_0^{\pi/2} |x + i\sqrt{1 - x^2} \cos s|^n \, ds \le 1$$

for $-1 \le x \le 1$. The second inequality, which shall be of help in proving the convergence of a Fourier-Legendre series, is of a type first obtained by Thomas Jan Stieltjes (1856–1894), of the University of Toulouse, in 1890, but the particular one that we present below was proved by Sergeĭ Natanovich Bernstein (1880–1968) in 1931.

Lemma 11.2 *For any fixed x in $(-1, 1)$ and any positive integer n*

$$|P_n(x)| < \frac{\sqrt{\pi}}{\sqrt{2n(1 - x^2)}}.$$

Proof. First note that

$$\sin s \geq \frac{2s}{\pi}$$

for $0 \leq s \leq \pi/2$ and also that $1 - t < e^{-t}$ for $t > 0$ because both sides have value 1 for $t = 0$, but $1 - t$ has a smaller derivative than e^{-t} for $t > 0$. Therefore,

$$|x + i\sqrt{1 - x^2}\cos s|^2 = 1 - (1 - x^2)\sin^2 s \leq 1 - \frac{4s^2}{\pi^2}(1 - x^2) < e^{-4s^2(1-x^2)/\pi^2}$$

for $0 \leq s \leq \pi/2$. Then, using Laplace's formula in Lemma 11.1 and the preceding inequalities, we obtain

$$|P_n(x)| \leq \frac{2}{\pi}\int_0^{\pi/2} |x + i\sqrt{1 - x^2}\cos s|^n \, ds < \frac{2}{\pi}\int_0^{\pi/2} e^{-2ns^2(1-x^2)/\pi^2} \, ds.$$

Choosing now a new variable of integration,

$$\sigma = \frac{\sqrt{2n(1 - x^2)}}{\pi} s,$$

we obtain

$$|P_n(x)| < \frac{2}{\sqrt{2n(1 - x^2)}}\int_0^{\sqrt{n(1-x^2)/2}} e^{-\sigma^2} \, d\sigma < \frac{2}{\sqrt{2n(1 - x^2)}}\int_0^{\infty} e^{-\sigma^2} \, d\sigma,$$

which proves the result bacause the integral on the right has value $\sqrt{\pi}/2$, Q.E.D.

The first to attempt a proof of the convergence of a Fourier-Legendre series was Poisson in 1829 and 1831, but Dirichlet later pointed out some errors in this proof. In *Elementary Principles of the Theories of Electricity, Heat and Molecular Actions*, published in 1833 by Robert Murphy (1806–1843), a contemporary of George Green and a Fellow at Gonville and Caius College, Cambridge University, he provided a sketch of proof based on term-by-term integration and orthogonality. After some work by other investigators, Hobson succeeded in proving some general sufficient conditions for convergence in 1908. Hobson's theorem is beyond the scope of this book,[1] but we can prove a simpler theorem without much effort. Let $s_N(x)$ denote the $(N + 1)$-st partial sum

$$\sum_{n=0}^{N} c_n P_n$$

of the Fourier-Legendre series of f on $(-1, 1)$. By the definition of the c_n and the identity $P_0 \equiv 1$,

$$s_N(x) = \frac{1}{2}\int_{-1}^{1} f(s)\left[1 + \sum_{n=1}^{N}(2n + 1)P_n(s)P_n(x)\right] ds.$$

---

[1]See, for instance, E. W. Hobson, *The Theory of Spherical and Ellipsoidal Harmonics*, Chelsea, New York, 1965, pages 318–329, or G. Sansone, *Orthogonal Functions*, Dover, New York, 1991, pages 220–234.

What is needed now is a summation formula for the term in the brackets, such as the one resulting in the Dirichlet kernel in Chapter 2. It was found by Elwin Bruno Christoffel (1829–1900), a professor at Zurich and later at Strasbourg, in 1858.

Lemma 11.3 *For every pair of real numbers s and x,*

$$1 + \sum_{n=1}^{N}(2n+1)P_n(s)P_n(x) = (N+1)\frac{P_{N+1}(s)P_N(x) - P_{N+1}(x)P_N(s)}{s-x}.$$

Proof. If we multiply the recursion formulas

$$(2n+1)s\,P_n(s) = (n+1)P_{n+1}(s) + nP_{n-1}(s)$$
$$(2n+1)x\,P_n(x) = (n+1)P_{n+1}(x) + nP_{n-1}(x),$$

established by Theorem 11.1, by $P_n(x)$ and $P_n(s)$, respectively, and subtract we obtain

$$(2n+1)(s-x)P_n(s)P_n(x) = (n+1)[P_{n+1}(s)P_n(x) - P_{n+1}(x)P_n(s)]$$
$$- n[P_n(s)P_{n-1}(x) - P_n(x)P_{n-1}(s)].$$

Adding the equations resulting from substituting $n = 1, \ldots, N$ in this identity yields

$$\sum_{n=1}^{N}(2n+1)(s-x)P_n(s)P_n(x) = (N+1)[P_{N+1}(s)P_N(x) - P_{N+1}(x)P_N(s)]$$
$$- [P_1(s)P_0(x) - P_1(x)P_0(s)].$$

Dividing both sides by $s-x$ and recalling that $P_0 \equiv 1$ and $P_1(x) = x$ give the Christoffel formula. Note that the numerator on the right-hand side is zero for $s = x$ and therefore contains $s - x$ as a factor. Hence, the quotient is well defined for all real s and x, Q.E.D.

We are now ready to give sufficient conditions for the convergence of a Fourier-Legendre series.

Theorem 11.3 *Let $f : (-1, 1) \to \mathbb{R}$ be piecewise continuous. Then the Fourier-Legendre series of f on $(-1, 1)$ converges to*

$$\tfrac{1}{2}[f(x+) + f(x-)]$$

at each point x in $(-1, 1)$ at which f has one-sided derivatives.

Proof. Using the Christoffel formula, we obtain

$$s_N(x) = \frac{N+1}{2}\int_{-1}^{1}f(s)\frac{P_{N+1}(s)P_N(x) - P_{N+1}(x)P_N(s)}{s-x}\,ds.$$

In particular, if $f \equiv 1$, then $f \equiv P_0$ and $s_N \equiv 1$ for all N, and then

$$1 = \frac{N+1}{2}\int_{-1}^{1}\frac{P_{N+1}(s)P_N(x) - P_{N+1}(x)P_N(s)}{s-x}\,ds.$$

Hence,

$$s_N(x) - \tfrac{1}{2}[f(x+) + f(x-)] = \frac{N+1}{4} \int_{-1}^{1} F(s)[P_{N+1}(s)P_N(x) - P_{N+1}(x)P_N(s)]\,ds,$$

where

$$F(s) = \frac{f(s) - f(x+)}{s - x} + \frac{f(s) - f(x-)}{s - x}.$$

If f has one-sided derivatives at x, then F is piecewise continuous and therefore integrable. Hence its Fourier-Legendre coefficients with respect to the Legendre polynomials,

$$C_n = \frac{1}{\|P_n\|^2} \int_{-1}^{1} F P_n,$$

are well defined, and then

$$s_N(x) - \tfrac{1}{2}[f(x+) + f(x-)] = \frac{N+1}{4}[P_N(x)\|P_{N+1}\|^2 C_{N+1} - P_{N+1}(x)\|P_N\|^2 C_N].$$

Using Lemma 11.2 and writing $M(x) = \sqrt{\pi/2(1-x^2)}$, we obtain

$$|P_N(x)| < \frac{M(x)}{\sqrt{N}} \qquad \text{and} \qquad |P_{N+1}(x)| < \frac{M(x)}{\sqrt{N+1}} < \frac{M(x)}{\sqrt{N}},$$

and, by Theorem 11.2,

$$\|P_N\| < \sqrt{\frac{2}{2N+1}} < \frac{1}{\sqrt{N}} \qquad \text{and} \qquad \|P_{N+1}\| < \sqrt{\frac{2}{2N+3}} < \frac{1}{\sqrt{N}}.$$

Therefore

$$|s_N(x) - \tfrac{1}{2}[f(x+) + f(x-)]| < M(x)\frac{N+1}{4N}(\|P_{N+1}\|C_{N+1} + \|P_N\|C_N).$$

Since the series on the left-hand side of Bessel's inequality for F,

$$\sum_{n=0}^{\infty} C_N^2 \|P_N\|^2 < \|F\|^2,$$

must be convergent (refer to Corollary 2 of Theorem 6.2), $\|P_N\|C_N \to 0$ as $N \to \infty$ and

$$s_N(x) \to \tfrac{1}{2}[f(x+) + f(x-)]$$

as $N \to \infty$, Q.E.D.

In a follow-up theorem of 1909, Hobson also gave conditions for convergence at $x = \pm 1$.[1] As a particular case of his result and under the stated conditions on f, its

---

[1] See, for instance, E. W. Hobson, *op. cit.*, pages 331–335, or G. Sansone, *op. cit.*, pages 236–239.

Fourier-Legendre series converges at $x = -1$ to the value $f(-1+)$ and at $x = 1$ to the value $f(1-)$. If, in addition, f is continuous on a closed subinterval I of $(-1, 1)$, it is a consequence of Hobson's first theorem that the Fourier-Legendre series of f converges uniformly on any interval interior to I. It is also possible to establish the convergence in the mean of the Fourier-Legendre series of an integrable function with a proof very similar to that of Theorem 6.5 (see Exercise 4.9).

Corollary   *The only eigenvalues of the singular Sturm-Liouville problem consisting of the Legendre equation*

$$(1 - x^2)X'' - 2xX' + \lambda X = 0$$

on $[-1, 1]$ *are* $\lambda = n(n + 1)$ *for any nonnegative integer* n, *with* P_n *as corresponding eigenfuctions.*

Proof. Assume that λ is an eigenvalue that is not of the stated form and that X is a corresponding eigenfunction. Then X is orthogonal to all the P_n and its Fourier-Legendre coefficients are zero. Since X is of class C^1, the theorem shows that it is the sum of its Fourier-Legendre series, that is, that $X \equiv 0$. But this is impossible by the definition of eigenfunction. Therefore, λ cannot be an eigenvalue, Q.E.D.

Sometimes it is necessary to solve the Legendre equation on the interval $(0, 1)$ with a boundary condition at $x = 0$. We shall state the results that are relevant in this situation, but leaving most of the details as exercises. First observe that each of the sets $\{P_{2n+1}\}$ and $\{P_{2n}\}$, consisting of all the Legendre polynomials of odd and even order, respectively, is an orthogonal system on $(0, 1)$ (Exercise 2.3). Then we define the *odd* and *even Fourier-Legendre series* of an integrable function $f : (0, 1) \to \mathbb{R}$ to be the Fourier series of f, in the sense of Definition 6.2, with respect to the orthogonal systems $\{P_{2n+1}\}$ and $\{P_{2n}\}$, respectively. The following result can be deduced directly from Theorem 11.3 (Exercise 4.10).

Theorem 11.4 *Let* $f : (0, 1) \to \mathbb{R}$ *be piecewise continuous. Then the odd and even Fourier-Legendre series of* f *on* $(0, 1)$ *converge to*

$$\tfrac{1}{2}[f(x+) + f(x-)]$$

at each point x *in* $(0, 1)$, *at which* f *has one-sided derivatives.*

Corollary   *The only eigenvalues of the singular Sturm-Liouville problem consisting of the Legendre equation on* $[0, 1]$ *and the boundary condition*
   (i) $X(0) = 0$ *are* $\lambda = (2n + 1)(2n + 2)$ *for any nonnegative integer* n, *with* P_{2n+1} *as corresponding eigenfuctions; or the boundary condition*
   (ii) $X'(0) = 0$ *are* $\lambda = 2n(2n + 1)$ *for any nonnegative integer* n, *with* P_{2n} *as corresponding eigenfuctions.*

The proof is entirely analogous to that of the previous corollary. Note that the solutions P_{2n} cannot be used in case (i) because $P_{2n}(0) \neq 0$ (Exercise 2.4) and that the solutions P_{2n+1} cannot be used in case (ii) because $P'_{2n+1}(0) \neq 0$ (Exercise 2.5).

§11.5   Return to the Potbellied Stove

In §11.1 we were in the process of finding the temperature u outside a spherical stove of radius a subject to the boundary condition

$$u(a, \phi) = f(\phi),$$

where f is the surface temperature. We found that, if we assume a solution of the form $u(r, \phi) = R(r)\Phi(\phi)$ valid for $r \geq a$, and if $X(x) = \Phi(\arccos x)$, then X must be an eigenfunction of the singular Sturm-Liouville problem

$$(1 - x^2)X'' - 2xX' + \lambda X = 0$$

on $[-1, 1]$. According to the corollary of Theorem 11.3, $\lambda = n(n + 1)$, where n is a nonnegative integer, and then the solution for R found in §11.1 becomes

$$R(r) = Ar^n + B\frac{1}{r^{n+1}}.$$

Since the temperature cannot increase with the distance from the stove, we must choose $A = 0$, and then

$$R(r) = B\frac{1}{r^{n+1}}.$$

We also know that the eigenfunctions for X are the Legendre polynomials, so that the solutions for Φ are $P_n(\cos \phi)$. In order to satisfy the initial condition we shall try an infinite series of the form

$$u(r, \phi) = \sum_{n=0}^{\infty} \frac{c_n}{r^{n+1}} P_n(\cos \phi),$$

where c_n is a constant, and then the boundary condition $u(a, \phi) = f(\phi)$ requires that

$$f(\arccos x) = \sum_{n=0}^{\infty} \frac{c_n}{a^{n+1}} P_n(x).$$

If f is as in Theorem 11.3, this equation holds at every point of continuity if we choose c_n/a^{n+1} to be the nth Fourier-Legendre coefficient of $f(\arccos x)$ on $(-1, 1)$, that is,

$$\frac{c_n}{a^{n+1}} = \frac{2n + 1}{2} \int_{-1}^{1} f(\arccos x) P_n(x) \, dx = \frac{2n + 1}{2} \int_{0}^{\pi} f(\sigma) P_n(\cos \sigma) \sin \sigma \, d\sigma.$$

This leads to the formal solution

$$(12) \quad u(r, \phi) = \sum_{n=0}^{\infty} \left[\frac{2n + 1}{2} \int_{0}^{\pi} f(\sigma) P_n(\cos \sigma) \sin \sigma \, d\sigma \right] \left(\frac{a}{r} \right)^{n+1} P_n(\cos \phi).$$

It is possible to verify this solution by a sequence of steps very similar to that used to prove Propositions 3.1 and 3.2 and Theorem 3.1, and we shall sketch this verification soon. The proofs given in Chapter 3 use the fact that the sine and cosine functions, as well as their first and second derivatives, are bounded by 1 in absolute value. We have shown above, after Lemma 11.1, that this is also true for the Legendre polynomials. For their first and second derivatives we have

$$|P_n'(x)| \leq \frac{n(n+1)}{2} \quad \text{and} \quad |P_n''(x)| \leq \frac{n^4}{2}$$

for $-1 \leq x \leq 1$ (see Exercise 3.3). Then the proposed solution is verified as follows. If $|f| \leq M$, each term of the series in (12) is bounded by

$$(2n+1)M\left(\frac{a}{r}\right)^{n+1}$$

because $|P_n| \leq 1$. Either the ratio or the root test shows that the series converges absolutely for any ϕ and any $r \geq r_0 > a$, and the Weierstrass M-test shows that it is uniformly convergent on $[r_0, \infty) \times [0, \pi]$. Its sum u is continuous there by Theorem A.4 and then on $(a, \infty) \times [0, \pi]$ since r_0 is arbitrary. The same is true of the first and second derived series with respect to r or with respect to ϕ, but using for the last two the inequalities for P' and P'' stated above. Comparison of all these series shows that u is a solution of the Laplace equation on $(a, \infty) \times [0, \pi]$. The choice of coefficients guarantees that u satisfies the boundary condition at every point at which f is continuous. If f is only piecewise continuous, then at each point of discontinuity we have

$$u(r, \phi) \to \tfrac{1}{2}[f(\phi+) + f(\phi-)]$$

as $r \to a$ because, for each ϕ, the series in (12) converges uniformly with respect to r on $[a, \infty)$ by the corollary of Theorem A.3 with

$$a_n = \left[\frac{2n+1}{2} \int_0^\pi f(\sigma) P_n(\cos\sigma) \sin\sigma \, d\sigma\right] P_n(\cos\phi)$$

and

$$g_n(r) = \left(\frac{a}{r}\right)^{n+1} \leq 1.$$

If, in addition, f is continuous, then the Fourier-Legendre series involved in this problem converges uniformly on any closed subinterval I of $(-\pi, \pi)$, as explained at the end of the previous section. Then the series in (12) converges uniformly on $[a, \infty) \times I$ by Theorem A.3 if we define $f_n(r, \phi)$ and $g_n(r, \phi)$ as a_n and $g_n(r)$ were defined above. Such a solution is continuous on $[a, \infty) \times I$ by Theorem A.4 and then on $[a, \infty) \times (-\pi, \pi)$ since I is arbitrary.

Since our boundary value problem is posed for an unbounded set—the exterior of the sphere—and since the maximum principle for the Laplace equation (Theorem 9.5) is valid only on a bounded set, the questions of uniqueness and continuous dependence of solutions on boundary values cannot be answered through this principle. However, it can be shown that, if u satisfies the additional condition that ru, ru_{xx}, ru_{yy}, and ru_{zz}

are bounded in absolute value for all sufficiently large r, then the maximum principle holds on the region bounded by and exterior to the sphere.[1] Then this problem is shown to be well posed in the sense of Hadamard as in the proof of Theorem 9.6.

§11.6   The Dirichlet Problem for the Sphere

In §11.1 we posed an exterior problem for the sphere. We now pose the problem of finding the temperature at any interior point if the temperature on the surface is known. Since the heat equation reduces to the Laplace equation for the steady-state temperature, this is just the Dirichlet problem for the sphere. We could just as well ask for the electrostatic potential inside the sphere if the potential on the surface is known. But now we consider a more general boundary condition than in §11.1, namely,

$$u(a, \phi, \theta) = f(\phi, \theta).$$

This will take us, once again, into new mathematical territory.

We shall start, as usual, looking for a product solution of the form $u(r, \phi, \theta) = R(r)\Phi(\phi)\Theta(\theta)$. Substituting into the Laplace equation, that is, the right-hand side of (5) equated to zero, and dividing by $R\Phi\Theta$, we obtain

$$\frac{rR'' + 2R'}{rR} + \frac{\Phi'' + (\cot\phi)\Phi'}{r^2\Phi} + \frac{1}{r^2\sin^2\phi}\frac{\Theta''}{\Theta} = 0.$$

If we multiply this equation by r^2, it follows that there is a constant λ such that

$$\frac{r^2R'' + 2rR'}{R} = \lambda = -\frac{\Phi'' + (\cot\phi)\Phi'}{\Phi} - \frac{1}{\sin^2\phi}\frac{\Theta''}{\Theta}.$$

In particular, this means that $r^2R'' + 2rR' - \lambda R = 0$, which is the same equation as in §11.1. Also, there is a constant μ such that

$$\left(\lambda + \frac{\Phi'' + (\cot\phi)\Phi'}{\Phi}\right)\sin^2\phi = \mu = -\frac{\Theta''}{\Theta},$$

which gives the differential equations $\Theta'' + \mu\Theta = 0$ and

$$(13) \qquad\qquad \Phi'' + (\cot\phi)\Phi' + \left(\lambda - \frac{\mu}{\sin^2\phi}\right)\Phi = 0.$$

The equation for Θ subject to the boundary conditions $\Theta(-\pi) = \Theta(\pi)$ and $\Theta'(-\pi) = \Theta'(\pi)$, which are imposed by circular symmetry, has solutions of the form

$$\Theta(\theta) = A\cos m\theta + B\sin m\theta,$$

---

[1] See, for instance, O. D. Kellogg, *Foundations of Potential Theory*, Dover, New York, 1954, pages 216–218.

where $m = \sqrt{\mu}$ is a nonnegative integer. Finally, as in §11.1, the choice of a new variable $x = \cos \phi$ and a new function

$$X(x) = \Phi(\arccos x),$$

plus the fact that $\mu = m^2$, transforms the equation for Φ into

$$(1 - x^2)X'' - 2xX' + \left(\lambda - \frac{m^2}{1 - x^2}\right)X = 0,$$

which is known as the *associated Legendre equation* of order m.

In order for the solution of this problem to coincide with that of §11.5 when f is independent of θ, we shall take λ to be of the form $n(n + 1)$, where n is a nonnegative integer, in the equations for R and Φ above. Then, in the next section we shall investigate the solutions of

$$(14) \qquad (1 - x^2)X'' - 2xX' + \left[n(n + 1) - \frac{m^2}{1 - x^2}\right]X = 0.$$

§11.7   The Associated Legendre Functions

The associated Legendre equation of order $m = 0$ becomes the usual Legendre equation, and its solutions are already known. On the other hand, solving (14) for $m > 0$ is not a straightforward matter. In fact, it can be accomplished only with a little bit of luck. Suppose that, for lack of anything better to do, we put $\mu = 0$ and $\lambda = n(n + 1)$ in (13) and differentiate. We obtain

$$\Phi''' + (\cot \phi)\Phi'' + \left[n(n + 1) - \frac{1}{\sin^2 \phi}\right]\Phi' = 0,$$

which shows that $\pm\Phi'$ are solutions of (13) for $\mu = 1$, or $m = 1$, if Φ is a solution of the Legendre equation on $[-1, 1]$, that is, if $\Phi(\phi) = P_n(\cos \phi)$. Therefore,

$$-\Phi'(\phi) = P_n'(\cos \phi) \sin \phi.$$

Restoring the variable $x = \cos \phi$, we have found a solution of the associated Legendre equation (14) of order $m = 1$ that, if we denote it by P_{n1}, is defined by

$$P_{n1}(x) = \sqrt{1 - x^2}\, P_n'(x).$$

Unfortunately, differentiating (13) again does not immediately lead to a solution for $m = 2$. Instead, we shall use the defining equation for P_{n1} to conjecture that for any positive integer m the function P_{nm} defined by

$$(15) \qquad P_{nm}(x) = (1 - x^2)^{m/2} P_n^{(m)}(x),$$

where the superscript (m) indicates mth derivative, is a solution of (14). To verify this conjecture we substitute P_{nm} in the left-hand side of (14). After simplification we obtain

$$(1 - x^2)P_{nm}'' - 2x P_{nm}' + \left[n(n+1) - \frac{m^2}{1-x^2}\right]P_{nm} =$$

$$(1 - x^2)^{m/2}\left[(1-x^2)P_n^{(m+2)} - 2(m+1)x P_n^{(m+1)} + [n(n+1) - m(m+1)]P_n^{(m)}\right].$$

Then P_{nm} would be a solution of (14) if we show that the expression in the larger brackets on the right-hand side is zero. To do this we shall replace X with P_n and λ with $n(n+1)$ in the Legendre equation and then differentiate m times. Using the Leibniz rule (see page 416) with $f(x) = 1 - x^2$ and $g = P_n''$ and noting that $f^{(k)} \equiv 0$ for $k > 2$, we have

$$\frac{d^m}{dx^m}\left[(1-x^2)P_n''\right] = \sum_{k=0}^{2} \frac{m!}{k!(m-k)!}\left[\frac{d^k}{dx^k}(1-x^2)\right]P_n^{(m+2-k)}$$

$$= (1-x^2)P_n^{(m+2)} - 2mx P_n^{(m+1)} - m(m-1)P_n^{(m)}.$$

Similarly,

$$\frac{d^m}{dx^m}(-2x P_n') = \sum_{k=0}^{1} \frac{m!}{k!(m-k)!}\left[\frac{d^k}{dx^k}(-2x)\right]P_n^{(m+1-k)} = -2x P_n^{(m+1)} - 2m P_n^{(m)},$$

and then

$$0 = \frac{d^m}{dx^m}\left[(1-x^2)P_n'' - 2x P_n' + n(n+1)P_n\right]$$

$$= (1-x^2)P_n^{(m+2)} - 2(m+1)x P_n^{(m+1)} + [n(n+1) - m(m+1)]P_n^{(m)}.$$

This completes the proof that P_{nm} is a solution of (14) for $m \geq 1$. Note that this solution is nontrivial only if $m \leq n$ because P_n is a polynomial of degree n and all its derivatives of an order higher than n vanish identically. We complete the definition of the P_{nm} (for which the standard but less convenient notation is P_n^m, where m is used as a superscript rather than as an exponent) with

$$P_{n0} = P_n.$$

These solutions of (14) are called the *associated Legendre functions*. They were introduced in 1877 by Norman M. Ferrers (1829–1903), of Cambridge University, and for this reason they are sometimes called the *Ferrers functions*.[1]

Since the P_{nm} are solutions of the singular Sturm-Liouville problem (14) on $[-1, 1]$, they are orthogonal by Theorem 4.1, that is, for each fixed nonnegative integer m and all $k, n \geq 0$ with $k \neq n$,

$$\int_{-1}^{1} P_{km}(x)P_{nm}(x)\, dx = 0.$$

Next we evaluate the norm of P_{nm} since it will be necessary in the next section.

---

[1] See Ferrers' book *Spherical Harmonics*, London, 1877, page 76.

Theorem 11.5 *For any nonnegative integer n and any integer m with $0 \le m \le n$,*

$$\| P_{nm} \|^2 = \frac{(n+m)!}{(n-m)!} \frac{2}{2n+1}.$$

Proof. For $m = 0$ this is true by Theorem 11.2 since $P_{n0} = P_n$. To find the norm of P_{nm} for $0 < m \le n$ we start by replacing m with $m - 1$ in (15) and then differentiate to obtain

$$(16) \quad P'_{n(m-1)} = \frac{d}{dx} \left[(1 - x^2)^{(m-1)/2} P_n^{(m-1)} \right]$$

$$= -(m - 1)x(1 - x^2)^{(m-3)/2} P_n^{(m-1)} + (1 - x^2)^{(m-1)/2} P_n^{(m)}$$

$$= -\frac{(m-1)x}{1 - x^2} P_{n(m-1)} + (1 - x^2)^{-1/2} P_{nm}.$$

This leads to the equation

$$P_{nm} = \sqrt{1 - x^2}\, P'_{n(m-1)} + \frac{(m-1)x}{\sqrt{1 - x^2}} P_{n(m-1)},$$

and then, if we square both sides and integrate from -1 to 1,

$$\| P_{nm} \|^2 =$$

$$\int_{-1}^{1} \left[(1 - x^2)[P'_{n(m-1)}]^2 + 2(m - 1)x P_{n(m-1)} P'_{n(m-1)} + \frac{(m-1)^2 x^2}{1 - x^2} P^2_{n(m-1)} \right] dx.$$

The first two terms are integrated by parts since they are of the form fg' with $g = P_{n(m-1)}$. In the first term $f(x) = (1 - x^2) P'_{n(m-1)}(x)$ or, using (16) and (15),

$$f(x) = (1 - m)x P_{n(m-1)}(x) + \sqrt{1 - x^2}\, P_{nm}(x)$$

$$= (1 - m)x(1 - x^2)^{(m-1)/2} P_n^{(m-1)}(x) + \sqrt{1 - x^2}\, P_{nm}(x).$$

Performing the integration by parts, noting that $f(1) = f(-1) = 0$ for $m \ge 1$, and recalling that $P_{n(m-1)}$ is a solution of the associated Legendre equation of order $m - 1$, we obtain

$$\int_{-1}^{1} (1 - x^2)[P'_{n(m-1)}]^2 = -\int_{-1}^{1} \left[(1 - x^2) P'_{n(m-1)} \right]' P_{n(m-1)}\, dx$$

$$= -\int_{-1}^{1} \left[(1 - x^2) P''_{n(m-1)} - 2x P'_{n(m-1)} \right] P_{n(m-1)}\, dx$$

$$= \int_{-1}^{1} \left[n(n + 1) - \frac{(m-1)^2}{1 - x^2} \right] P^2_{n(m-1)}\, dx.$$

To evaluate the second integral we define

$$f(x) = (m-1)x\, P_{n(m-1)}(x) = (m-1)x(1-x^2)^{(m-1)/2} P_n^{(m-1)}(x),$$

and note again that $f(1) = f(-1) = 0$ for $m \geq 1$. Then

$$\int_{-1}^{1} (m-1)x\, P_{n(m-1)} P_{n(m-1)}' \, dx = -\int_{-1}^{1} [(m-1)x\, P_{n(m-1)}]'\, P_{n(m-1)} \, dx$$

$$= -\int_{-1}^{1} (m-1)[x\, P_{n(m-1)} P_{n(m-1)}' + P_{n(m-1)}^2]\, dx,$$

that is,

$$2\int_{-1}^{1} (m-1)x\, P_{n(m-1)} P_{n(m-1)}' \, dx = -\int_{-1}^{1} (m-1) P_{n(m-1)}^2 \, dx.$$

Therefore,

$$\|P_{nm}\|^2 = \int_{-1}^{1} \left[n(n+1) - \frac{(m-1)^2}{1-x^2} - (m-1) + \frac{(m-1)^2 x^2}{1-x^2} \right] P_{n(m-1)}^2 \, dx$$

$$= [n(n+1) - m(m-1)] \int_{-1}^{1} P_{n(m-1)}^2 \, dx$$

$$= (n+m)(n-m+1)\|P_{n(m-1)}\|^2.$$

Similarly,

$$\|P_{nk}\|^2 = (n+k)(n-k+1)\|P_{n(k-1)}\|^2$$

for $k = m-1, m-2, \ldots, 1$, and then repeated use of this equation and the fact that $P_{n0} = P_n$ lead to

$$\|P_{nm}\|^2 = (n+m) \cdots (n+2)(n+1)(n-m+1) \cdots (n-1)n\|P_{n0}\|^2$$

$$= (n+m) \cdots (n+2)(n+1)n(n-1) \cdots (n-m+1)\|P_{n0}\|^2$$

$$= \frac{(n+m)!}{(n-m)!}\|P_n\|^2$$

$$= \frac{(n+m)!}{(n-m)!} \frac{2}{2n+1},$$

Q.E.D.

§11.8   Solution of the Dirichlet Problem for the Sphere

We can now assemble the solution of the hot sphere problem. In §11.6 we proposed the product solution $u(r, \phi, \theta) = R(r)\Phi(\phi)\Theta(\theta)$ and found that $\Theta(\theta) = A \cos m\theta + B \sin m\theta$. Since the equation for R is the same as that in §11.1 and since $\lambda = n(n+1)$, it has a general solution of the form

$$R(r) = Cr^{-\frac{1}{2}+\sqrt{\frac{1}{4}+\lambda}} + Br^{-\frac{1}{2}-\sqrt{\frac{1}{4}+\lambda}} = Cr^n + \frac{B}{r^{n+1}},$$

and we must take $B = 0$ for a bounded solution inside the sphere. Finally, the solutions P_{nm} of (14) found in the preceding section provide the solutions $\Phi(\phi) = P_{nm}(\cos\phi)$ of (13) for $\lambda = n(n+1)$. That is, we have found the product solutions

$$u_{mn}(r, \phi, \theta) = r^n(a_{mn} \cos m\theta + b_{mn} \sin m\theta)P_{nm}(\cos\phi)$$

for $n \geq 0$ and $0 \leq m \leq n$ (recall that $P_{nm} \equiv 0$ if $m > n$). Then, in order to satisfy the boundary condition $u(a, \phi, \theta) = f(\phi, \theta)$, we shall try a series solution of the form

(17) $$u(r, \phi, \theta) = \sum_{n=0}^{\infty} \sum_{m=0}^{n} r^n(a_{mn} \cos m\theta + b_{mn} \sin m\theta)P_{nm}(\cos\phi)$$

and require that for $r = a$

$$f(\phi, \theta) = \sum_{n=0}^{\infty} \sum_{m=0}^{n} a^n(a_{mn} \cos m\theta + b_{mn} \sin m\theta)P_{nm}(\cos\phi).$$

In order to find the coefficients a_{mn} and b_{mn} assume that the series on the right-hand side converges and that equality holds, then multiply both sides by $\cos j\theta$, $0 \leq j \leq n$, and assume that integration term by term from $-\pi$ to π is permissible. We obtain

(18) $$\int_{-\pi}^{\pi} f(\phi, \theta) \cos j\theta \, d\theta = c\pi \sum_{n=0}^{\infty} a^n a_{jn} P_{nj}(\cos\phi), \qquad 0 \leq j \leq n,$$

where $c = 2$ if $j = 0$ and $c = 1$ if $j > 0$, and then, since the orthogonality of the P_{nj} on $[-1, 1]$ implies that

$$\int_{0}^{\pi} P_{kj}(\cos\phi)P_{nj}(\cos\phi) \sin\phi \, d\phi = \int_{-1}^{1} P_{kj}(x)P_{nj}(x) \, dx = 0$$

for $k \neq n$, we can multiply (18) by $P_{kj}(\cos\phi) \sin\phi$ and integrate term by term from 0 to π. We find that

$$\frac{1}{c\pi} \int_{0}^{\pi} \left(\int_{-\pi}^{\pi} f(\phi, \theta) \cos j\theta \, d\theta \right) P_{kj}(\cos\phi) \sin\phi \, d\phi$$

$$= a^k a_{jk} \int_{0}^{\pi} P_{kj}^2(\cos\phi) \sin\phi \, d\phi = a^k a_{jk} \frac{2}{2k+1} \frac{(k+j)!}{(k-j)!}.$$

for $k \geq 0$ and $0 \leq j \leq k$. Solving for a_{jk}, replacing the subscripts j and k with m and n, and replacing the varibales ϕ and θ with σ and τ, we obtain

$$a_{mn} = \frac{2n+1}{2c\pi a^n} \frac{(n-m)!}{(n+m)!} \int_0^\pi \left(\int_{-\pi}^\pi f(\sigma,\tau) \cos m\tau \, d\tau \right) P_{nm}(\cos\sigma) \sin\sigma \, d\sigma$$

for $n \geq 0$ and $0 \leq m \leq n$, where $c = 2$ if $m = 0$ and $c = 1$ if $m > 0$. Similarly,

$$b_{mn} = \frac{2n+1}{2\pi a^n} \frac{(n-m)!}{(n+m)!} \int_0^\pi \left(\int_{-\pi}^\pi f(\sigma,\tau) \sin m\tau \, d\tau \right) P_{nm}(\cos\sigma) \sin\sigma \, d\sigma$$

for $n \geq 0$ and $1 \leq m \leq n$ (b_{0n} is actually irrelevant since $\sin mx = 0$ for $m = 0$). Taking these coefficients to (17) and writing $\cos m(\theta - \tau)$ for $\cos m\theta \cos m\tau + \sin m\theta \sin m\tau$ lead to the formal solution

$$u(r,\phi,\theta) = \sum_{n=0}^\infty \sum_{m=0}^n \frac{2n+1}{2c\pi} \frac{(n-m)!}{(n+m)!} \left(\frac{r}{a}\right)^n P_{nm}(\cos\phi)$$

$$\times \int_0^\pi \left(\int_{-\pi}^\pi f(\sigma,\tau) \cos m(\theta - \tau) \, d\tau \right) P_{nm}(\cos\sigma) \sin\sigma \, d\sigma,$$

where $c = 2$ for $m = 0$ and $c = 1$ for $m > 0$. If we recall that $P_{n0} = P_n$, this can be rewritten as

$$(19) \quad u(r,\phi,\theta) = \frac{1}{4\pi} \sum_{n=0}^\infty \int_0^\pi \left(\int_{-\pi}^\pi f(\sigma,\tau)(2n+1)\left(\frac{r}{a}\right)^n \left[P_n(\cos\phi) P_n(\cos\sigma) \right.\right.$$

$$\left.\left. + 2\sum_{m=1}^n \frac{(n-m)!}{(n+m)!} P_{nm}(\cos\phi) P_{nm}(\cos\sigma) \cos m(\theta - \tau) \right] d\tau \right) \sin\sigma \, d\sigma.$$

The verification of this solution is not a straightforward matter. In the next section we shall find an alternative form that can be verified and shown to be identical to this series expansion.

The separated solutions

$$r^n (a_{mn} \cos m\theta + b_{mn} \sin m\theta) P_{nm}(\cos\phi)$$

of the Laplace equation from which (19) was built were called *spherical harmonics* by William Thompson, or *solid spherical harmonics* to distinguish them from the *surface spherical harmonics*

$$(a_{mn} \cos m\theta + b_{mn} \sin m\theta) P_{nm}(\cos\phi).$$

An interesting, and perhaps unexpected, property of the solid spherical harmonics is that they become homogeneous polynomials if expressed in the Cartesian coordinates x, y, and z (see Exercise 8.1). Recall that a polynomial $P(x, y, z)$ is said to be homogeneous of degree n if for any constant α

$$P(\alpha x, \alpha y, \alpha z) = \alpha^n P(x, y, z).$$

In fact, sometimes this is taken as the starting point of a more general theory in which a (*solid*) *spherical harmonic of degree n* is defined to be any solution of the Laplace equation that is homogeneous in x, y, z of degree n.[1] From this point of view the spherical harmonics used in this section are just a special kind. The special surface harmonics vanish on equally spaced meridians, on which the trigonometric factor does, and parallel circles $\phi = \arccos x_0$, where x_0 is any of the zeros of P_{nm} in $[-1, 1]$ (see Exercises). Since this divides the surface of the sphere into curvilinear rectangles, these spherical harmonics are sometimes called *tesseral harmonics*, from the Greek *tessares*, meaning four.

§11.9 Poisson's Integral Formula for the Sphere

Consider the simplest problem in potential theory: to determine the attraction force between two point masses m_1 and m_2, the first of which is fixed in space and the second is rotating in a circle of radius $r < 1$ about a fixed point P located at a distance 1 from m_1. If ϕ denotes the angle between the lines joining P to m_1 and to m_2, as shown in Figure 11.2, and if the distance between the two masses at any given instant is denoted

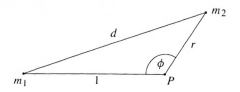

Figure 11.2

by d, then the force of attraction of m_1 on m_2 is given by Newton's law of gravitation

$$F = -m_2 \frac{\partial}{\partial d}\left(\frac{Gm_1}{d}\right),$$

where G is the gravitation constant and the quotient Gm_1/d is called the *potential* for m_1. If the law of cosines is used to compute d, the potential for m_1 is proportional to

(20)
$$\frac{1}{[1 - (2r\cos\phi - r^2)]^{1/2}}.$$

In his first paper of 1872, Legendre studied a more sophisticated gravitational problem and was able to find that the force of attraction exerted by a solid of revolution on

---

[1]For an exposition of such a theory, see, for instance, E. W. Hobson, *op. cit.*, or T. M. MacRobert, *Spherical Harmonics; An Elementary Treatise on Harmonic Functions with Applications*, 3rd ed., Pergamon Press, Oxford, 1967.

an external point mass was given by an integral contaning a factor of the form

$$\frac{1}{[1 - (2r \cos \phi - r^2)]^{3/2}},$$

where $r < 1$. In order to perform the integration he expanded this quotient in a power series in r and found the Legendre polynomials in the coefficients of the expansion. We have already found these polynomials by an alternate method, but in solving the Dirichlet problem for the sphere it will be useful to establish their relationship with Legendre's own expansion or, rather, with the series expansion of (20), which is an easier task and from which we could then obtain the expansion of the last quotient by differentiation.

It is easy to see using the ratio test that the Taylor series of $f(t) = (1 - t)^{-1/2}$ about $t = 0$ converges absolutely for $|t| < 1$. This guarantees the validity of a power series expansion of the form

$$\frac{1}{[1 - (2xr - r^2)]^{1/2}} = \sum_{n=0}^{\infty} C_n (2xr - r^2)^n$$

for $|2xr - r^2| < 1$ and an appropriate choice of the coefficients C_n. Note that this expansion is valid for any fixed value of x with $|x| \le 1$ if $|r| < \sqrt{2} - 1$ because these inequalities imply that

$$|2xr - r^2| = |r||2x - r| \le |r|(2 + |r|) < (\sqrt{2} - 1)(\sqrt{2} + 1) = 1.$$

The factor $(2xr - r^2)^n$ in the preceding series can be expanded using the binomial theorem, and then the terms of this absolutely convergent series can be rearranged in order of ascending powers of r. This leads to a power series expansion of the form

$$(21) \qquad \frac{1}{(1 - 2xr + r^2)^{1/2}} = \sum_{n=1}^{\infty} c_n(x) r^n,$$

where the coefficients, which depend on the choice of x, are still to be found. It is easily shown using Theorem A.6 that differentiation term by term with respect to r is permissible for a power series inside the interval of convergence, and then

$$\frac{x - r}{(1 - 2xr + r^2)^{3/2}} = \sum_{n=0}^{\infty} n c_n(x) r^{n-1}.$$

Then, if we multiply both sides of this equation by $1 - 2xr + r^2$ and use (21), we obtain

$$(x - r) \sum_{n=0}^{\infty} c_n(x) r^n = (1 - 2xr + r^2) \sum_{n=1}^{\infty} n c_n(x) r^{n-1}.$$

Equating the coefficients of equal powers of r on both sides gives $x c_0(x) = c_1(x)$ and

$$x c_n(x) - c_{n-1}(x) = (n + 1) c_{n+1}(x) - 2xn c_n(x) + (n - 1) c_{n-1}(x).$$

The last equation can be more conveniently rearranged as

$$(n + 1)c_{n+1}(x) - (2n + 1)xc_n(x) + nc_{n-1}(x) = 0,$$

a recursion formula valid for $n \geq 1$. Since putting $r = 0$ in (21) shows that $c_0(x) = 1 = P_0(x)$, we obtain $c_1(x) = x = P_1(x)$, and then the remaining c_n can be obtained from the recursion formula. But this is exactly the equation in Theorem 11.1 for the Legendre polynomials, and it follows that $c_n(x) = P_n(x)$ for all n. This completes the determination of the coefficients on the right-hand side of (21) and establishes this series expansion for sufficiently small r. However, we can prove that it is valid for any r with $|r| < 1$.

Theorem 11.6 *For any numbers x and r such that $-1 \leq x \leq 1$ and $-1 < r < 1$,*

$$(22) \qquad \frac{1}{\sqrt{1 - 2xr + r^2}} = \sum_{n=0}^{\infty} P_n(x)r^n.$$

Proof. The inequality $|P_n(x)| \leq 1$, which we deduced from Lemma 11.1, and comparison with the geometric series show that $\sum_{n=0}^{\infty} P_n(x)r^n$ converges for $|x| \leq 1$ and $|r| < 1$. Now, it is easy to see by differentiation with respect to r that, for any fixed x, the left-hand side of (22) is a solution of the first-order differential equation

$$(23) \qquad (1 - 2xr + r^2)\frac{dy}{dr} + (r - x)y = 0.$$

But $\sum_{n=0}^{\infty} P_n(x)r^n$ is also a solution of this equation for the same value of x. Indeed, differentiating term by term, we obtain

$$(1 - 2xr + r^2)\sum_{n=0}^{\infty} nP_n(x)r^{n-1} + (r - x)\sum_{n=0}^{\infty} P_n(x)r^n = \sum_{n=0}^{\infty} nP_n(x)r^{n-1}$$

$$+ \sum_{n=0}^{\infty} nP_n(x)r^{n+1} - 2x\sum_{n=0}^{\infty} nP_n(x)r^n + \sum_{n=0}^{\infty} P_n(x)r^{n+1} - x\sum_{n=0}^{\infty} P_n(x)r^n.$$

The first term on the right-hand side can be rewritten as

$$\sum_{n=1}^{\infty} nP_n(x)r^{n-1} = \sum_{n=0}^{\infty} (n + 1)P_{n+1}(x)r^n,$$

while the second and fourth can be combined into

$$\sum_{n=0}^{\infty} (n + 1)P_n(x)r^{n+1} = \sum_{n=1}^{\infty} nP_{n-1}(x)r^n = \sum_{n=0}^{\infty} nP_{n-1}(x)r^n.$$

Therefore,

$$(1 - 2xr + r^2)\sum_{n=0}^{\infty} nP_n(x)r^{n-1} + (r - x)\sum_{n=0}^{\infty} P_n(x)r^n$$

$$= \sum_{n=0}^{\infty} \left[(n + 1)P_{n+1}(x) - (2n + 1)xP_n(x) + nP_{n-1}(x)\right]r^n,$$

which equals zero by Theorem 11.1. Then both sides of (22) are solutions of (23) that satisfy the initial condition $y(0) = 1$. Since x is arbitrary, the uniqueness of solutions for the initial value problem consisting of (23) and this initial condition proves (22) for $|x| \leq 1$ and $|r| < 1$, Q.E.D.

In view of this result, the quotient on the left-hand side of (22) is usually known as the *generating function* for the Legendre polynomials. It can be used to obtain some Fourier-Legendre expansions (see Exercises 9.1 to 9.3), but here we shall use it to transform the formal series solution of §11.8 into an integral solution, which then can be proved to be the unique solution of the Dirichlet problem for the sphere.

We start by evaluating (19) for $\phi = 0$, that is, for a point on the z-axis. Then the right-hand side becomes a much simpler expression because $P_n(1) = 1$ and, as the definition of P_{nm} shows, $P_{nm}(1) = 0$ for $m \geq 1$. If we assume that the order of summation and integration can be reversed, (19) becomes

$$(24) \quad u(r, 0, \cdot) = \frac{1}{4\pi} \int_0^\pi \left(\int_{-\pi}^\pi f(\sigma, \tau) \sum_{n=0}^\infty (2n+1) \left(\frac{r}{a} \right)^n P_n(\cos \sigma) \, d\tau \right) \sin \sigma \, d\sigma,$$

where the dot replacing θ expresses the fact that the θ coordinate is undefined on the z-axis. Now replace r in (22) with r/a, $r \leq a$, differentiate term by term with respect to r, multiply by r/a, and simplify to obtain

$$\sum_{n=0}^\infty n P_n(x) \left(\frac{r}{a} \right)^n \frac{1}{a} = \frac{r(ax - r)}{(a^2 - 2axr + r^2)^{3/2}}.$$

Then, using this equation and (22), we obtain

$$(25) \qquad \sum_{n=0}^\infty (2n+1) P_n(x) \left(\frac{r}{a} \right)^n = \frac{a(a^2 - r^2)}{(a^2 - 2axr + r^2)^{3/2}},$$

and this can be used to rewrite (24) as

$$u(r, 0, \cdot) = \frac{a}{4\pi} \int_0^\pi \left(\int_{-\pi}^\pi f(\sigma, \tau) \frac{a^2 - r^2}{(a^2 - 2ar \cos \sigma + r^2)^{3/2}} \, d\tau \right) \sin \sigma \, d\sigma.$$

Note that σ is just the angle between the vector $p = (r, 0, \cdot)$ and the vector $q = (a, \sigma, \tau)$ and also that this equation must hold for any point p on a sphere of radius r because we can always choose to position the z-axis through p. Therefore,

$$(26) \quad u(r, \phi, \theta) = \frac{a}{4\pi} \int_0^\pi \left(\int_{-\pi}^\pi f(\sigma, \tau) \frac{a^2 - r^2}{(a^2 - 2ar \cos \alpha + r^2)^{3/2}} \, d\tau \right) \sin \sigma \, d\sigma,$$

where α is the angle between $p = (r, \phi, \theta)$ and $q = (a, \sigma, \tau)$. Rewriting the integral on the right as a surface integral, the preceding discussion leads to the conjecture that the following result holds.

Theorem 11.7 *Let S be a sphere of radius a centered at the origin and let D denote the interior of S. If $f : S \to \mathbb{R}$ is a continuous function, define a function $u : D \cup S \to \mathbb{R}$ by $u \equiv f$ on S and by*

(27)
$$u(p) = \frac{1}{4\pi a} \int_S f(q) \frac{a^2 - \|p\|^2}{(a^2 - 2a\|p\|\cos\alpha + \|p\|^2)^{3/2}} \, dS_q$$

in D, where α is the angle between the vector $p \in D$ and the vector $q \in S$ and the notation dS_q indicates that this is a surface integral and that the variable is q. Then u is continuous on $D \cup S$ and harmonic in D; that is, it satisfies the Laplace equation in D.

It is possible to rewrite (27) in a more compact form if we recall that

$$\langle p, q \rangle = \|p\| \|q\| \cos\alpha = a\|p\|\cos\alpha,$$

so that

$$\|p - q\|^2 = \langle p - q, p - q \rangle = \|p\|^2 - 2\langle p, q \rangle + \|q\|^2 = \|p\|^2 - 2a\|p\|\cos\alpha + a^2.$$

Then, if we define a function $P^q : D \to \mathbb{R}$ by

$$P^q(p) = \frac{\|q\|^2 - \|p\|^2}{\|p - q\|^3},$$

(27) can be written in the form

(28)
$$u(p) = \frac{1}{4\pi a} \int_S f(q) P^q(p) \, dS_q.$$

The transformation of (24) into (26) shown above was first carried out by Poisson in a paper of 1823, and for this reason any of the Equations (26), (27), or (28) is called *Poisson's integral formula for the sphere* and the quotient $P^q(p)/4\pi$ is called the *Poisson kernel for the sphere*. This 1823 paper is actually an addition to an 1820 paper in which Poisson used similar integral formulas in an attempt to prove the convergence of Fourier series.

The proof of Theorem 11.7 is quite similar to those of Theorems 7.7 and 7.8, but it is also based on Gauss' mean value theorem for the sphere (Exercise 1.7 of Chapter 9).

Proof. First we can differentiate $P^q(p)$, in either its Cartesian or its spherical coordinate form, to verify that it is a harmonic function of p in D, and the same is true for u by differentiation under the integral sign. In particular, u is continuous in D and then, since f is continuous on S, it remains to be shown that

$$\lim_{p \to q_0} u(p) = f(q_0)$$

for every q_0 in S.

We shall show below that the surface integral of $P^q(p)$ itself over S is independent of p and that, in fact,[1]

(29)
$$\int_S P^q(p)\, dS_q = 4\pi a.$$

Then, using (28) and (29),

$$4\pi a |u(p) - f(q_0)| = \left| \int_S f(q) P^q(p)\, dS_q - \int_S f(q_0) P^q(p)\, dS_q \right|$$

$$\leq \int_S |f(q) - f(q_0)| P^q(p)\, dS_q$$

$$= \int_{\|q - q_0\| < \delta} |f(q) - f(q_0)| P^q(p)\, dS_q$$

$$\qquad\qquad + \int_{\|q - q_0\| \geq \delta} |f(q) - f(q_0)| P^q(p)\, dS_q.$$

Making δ small enough, the first integral on the right-hand side can be made arbitrarily small by the continuity of f and (29). Having chosen δ, if $\|q - q_0\| \geq \delta$ and if $\|q_0 - p\| < \delta/2$, then

$$\|p - q\| = \|q - p\| \geq \|q - q_0\| - \|q_0 - p\| \geq \delta - \frac{\delta}{2} = \frac{\delta}{2}$$

and

$$P^q(p) \leq \frac{8(\|q\|^2 - \|p\|^2)}{\delta^3}.$$

Since f is continuous and its domain S is closed and bounded, then f is bounded (see the footnote to Theorem 3.4). It follows that the last integral can be made arbitrarily small by choosing p close enough to q_0, since this implies that $\|p\|$ is close enough to $\|q_0\| = \|q\|$. This proves the desired limit.

It now remains to prove (29). To this end, we shall transform the integral in question into one that allows a direct application of Gauss' mean value theorem by means of the substitution $s = rq/a$, where $\|s\| = r$. Note that

$$P^q(p) = P^{ap/r}(s)$$

because the identity $\|v\|^2 = \langle v, v \rangle$, valid for any vector v in $\mathbb{R}^3$, can be easily used to verify that

$$\left\| \frac{a}{r} p - s \right\| = \|p - q\|,$$

---

[1] This result is analogous to Equations (11) of §7.5 and to the integrals in Exercise 1.4 of Chapter 10.

and then, since $\|p\| = r$ and $\|q\| = a$,

$$P^q(p) = \frac{a^2 - r^2}{\|p - q\|^3} = \frac{\left\|\frac{a}{r}p\right\|^2 - \|s\|^2}{\left\|\frac{a}{r}p - s\right\|^3} = P^{ap/r}(s).$$

Therefore, making the substitution $s = rq/a$ and noting that s is on the sphere S_r of radius $r < a$ and centered at the origin, the integral in (29) becomes

$$\int_S P^q(p)\,dS_q = \int_S P^{ap/r}(s)\,dS_q = \int_{S_r} P^{ap/r}(s)\frac{a^2}{r^2}\,dS_s = \frac{a^2}{r^2}\int_{S_r} P^{ap/r}(s)\,dS_s.$$

Since $\|ap/r\| = a > \|s\|$, the last integrand is a harmonic function of s on and inside S_r and then, according to Gauss' mean value theorem, its value at the center of S_r is the average of its values on the surface. That is,

$$\int_S P^q(p)\,dS_q = \frac{a^2}{r^2}\,4\pi r^2 P^{ap/r}(0) = 4\pi a^2 \frac{1}{\|ap/r\|} = 4\pi a,$$

as stated above, Q.E.D.

Using the maximum principle as in the proof of Theorem 9.6, the Dirichlet problem for the sphere is then seen to be well posed in the sense of Hadamard.

It is possible, *a posteriori*, to justify the series solution (19) of §11.8. The series in (25) is uniformly convergent with respect to x by the Weierstrass M-test and the inequality $|P_n(x)| \le 1$. Using this series to replace the quotient in the integrand in (27), with x replaced by $\cos\alpha$, and changing the order of summation and integration we obtain

$$u(r, \phi, \theta) = \frac{a}{4\pi}\sum_{n=0}^{\infty}\int_0^\pi \left(\int_{-\pi}^\pi f(\sigma, \tau)(2n + 1)P_n(\cos\alpha)\left(\frac{r}{a}\right)^n d\tau\right)\sin\sigma\,d\sigma.$$

The series solution (19) then follows from the identity

$$P_n(\cos\alpha) = P_n(\cos\phi)P_n(\cos\sigma) + 2\sum_{m=1}^{n}\frac{(n - m)!}{(n + m)!}P_{nm}(\cos\phi)P_{nm}(\cos\sigma)\cos m(\theta - \tau),$$

where α is the angle between the vectors (r, ϕ, θ) and (r, σ, τ). It was originally proved by Legendre in his fourth paper of 1789 in a somewhat different form, but we shall omit the proof.[1]

---

[1] See, for instance, J. S. Walker, *Fourier Analysis*, Oxford University Press, New York, 1988, pages 361–363.

§11.10 The Cooling of a Sphere

In §11.1 and §11.6 we posed some heat problems, but in asking for a steady-state solution we actually posed boundary value problems for the Laplace equation. We shall now pose a problem for the heat equation: find the temperature of a sphere of radius a if its initial temperature is known and if the surface temperature is maintained at zero degrees for $t > 0$. A simpler variant of this problem, in which the temperature is assumed to depend only on r, was originally considered by Fourier. In our case, we have initial and boundary conditions of the form

$$u(r, \phi, \theta, 0) = f(r, \phi, \theta) \qquad \text{and} \qquad u(a, \phi, \theta, t) = 0$$

for $t \geq 0$.

Once more, we shall look for a product solution $u(r, \phi, \theta, t) = R(r)\Phi(\phi)\Theta(\theta)T(t)$. Substituting into the heat equation (5) and dividing by $kR\Phi\Theta T$, we obtain

$$\frac{T'}{kT} = \frac{rR'' + 2R'}{rR} + \frac{\Phi'' + (\cot \phi)\Phi'}{r^2\Phi} + \frac{1}{r^2 \sin^2 \phi} \frac{\Theta''}{\Theta}.$$

Then each side of this equation must be equal to a constant $-\gamma$, which gives the differential equation $T' + \gamma kT = 0$ and shows the existence of a constant λ such that

$$\frac{r^2 R'' + 2r R'}{R} + \gamma r^2 = \lambda = -\frac{\Phi'' + (\cot \phi)\Phi'}{\Phi} - \frac{1}{\sin^2 \phi} \frac{\Theta''}{\Theta}.$$

Then we obtain the same differential equations for Φ and Θ as those in §11.6. Well-known solutions for T, Θ, and Φ are

$$T(t) = e^{-\gamma kt}, \quad \Theta(\theta) = A \cos m\theta + B \sin m\theta, \quad \text{and} \quad \Phi(\phi) = P_{nm}(\cos \phi),$$

where m and n are nonnegative integers and λ must be of the form $n(n+1)$. The equation for R is

$$r^2 R'' + 2r R' + [\gamma r^2 - n(n+1)]R = 0,$$

and it is easy to see that the substitution $X(r) = \sqrt{r}\, R(r)$ transforms it into the Bessel differential equation

$$r^2 X'' + r X' + \left[\gamma r^2 - \left(n + \tfrac{1}{2}\right)^2\right]X = 0.$$

Together with the boundary condition, $X(a) = 0$, it makes a singular Sturm-Liouville problem on $(0, a)$. Its eigenvalues and eigenfunctions were found in §10.6, and if we define

$$p_n = n + \tfrac{1}{2},$$

they are

$$\gamma_l = \alpha_{nl}^2 \qquad \text{and} \qquad X_l(r) = J_{p_n}(\alpha_{nl}r),$$

respectively, where $\{\alpha_{nl}\}$, $l = 1, 2, \ldots$, is the sequence of positive zeros of $J_{p_n}(ar)$. Corresponding to the stated eigenfunctions we have the following separated solutions for R,

$$R_{nl}(r) = \frac{J_{p_n}(\alpha_{nl}r)}{\sqrt{r}},$$

and then the product solutions

$$R_{nl}(r)(a_{lmn} \cos m\theta + b_{lmn} \sin m\theta) P_{nm}(\cos\phi)e^{-\alpha_{nl}^2 kt}$$

for $l \geq 1$, $n \geq 0$, and $0 \leq m \leq n$ (recall that $P_{nm} \equiv 0$ if $m > n$). In order to satisfy the initial condition $u(r, \phi, \theta, 0) = f(r, \phi, \theta)$ we shall try a series solution of the form

$$(30) \quad u(r, \phi, \theta, t) = \sum_{l=1}^{\infty} \sum_{n=0}^{\infty} \sum_{m=0}^{n} R_{nl}(r)(a_{lmn} \cos m\theta + b_{lmn} \sin m\theta) P_{nm}(\cos\phi)e^{-\alpha_{nl}^2 kt}$$

and require that for $t = 0$

$$f(r, \phi, \theta) = \sum_{l=1}^{\infty} \sum_{n=0}^{\infty} \sum_{m=0}^{n} R_{nl}(r)(a_{lmn} \cos m\theta + b_{lmn} \sin m\theta) P_{nm}(\cos\phi).$$

In order to determine the coefficients a_{lmn} and b_{lmn} assume that the series on the right-hand side converges and that equality holds, then multiply both sides by $\cos j\theta$, $0 \leq j \leq n$, and assume that integration term by term from $-\pi$ to π is permissible. We obtain

$$(31) \quad \int_{-\pi}^{\pi} f(r, \phi, \theta) \cos j\theta \, d\theta = c\pi \sum_{l=1}^{\infty} \sum_{n=0}^{\infty} R_{nl}(r) a_{ljn} P_{nj}(\cos\phi), \qquad 0 \leq j \leq n,$$

where $c = 2$ if $j = 0$ and $c = 1$ if $j > 0$, and then, since the orthogonality of the P_{nj} on $[-1, 1]$ implies that

$$\int_{0}^{\pi} P_{kj}(\cos\phi) P_{nj}(\cos\phi) \sin\phi \, d\phi = \int_{-1}^{1} P_{kj}(x) P_{nj}(x) \, dx = 0$$

for $k \neq n$, we can multiply (31) by $P_{kj}(\cos\phi) \sin\phi$ and integrate term by term from zero to π. We find that

$$(32) \quad \frac{1}{c\pi} \int_{0}^{\pi} \left(\int_{-\pi}^{\pi} f(r, \phi, \theta) \cos j\theta \, d\theta \right) P_{kj}(\cos\phi) \sin\phi \, d\phi$$

$$= \sum_{l=1}^{\infty} R_{kl}(r) a_{ljk} \int_{0}^{\pi} P_{kj}^2(\cos\phi) \sin\phi \, d\phi = \| P_{kj} \|^2 \sum_{l=1}^{\infty} R_{kl}(r) a_{ljk}$$

for $k \geq 0$ and $0 \leq j \leq k$. Finally, we can use the orthogonality of Bessel functions (Theorem 10.5) to establish the fact that

$$\int_{0}^{a} r^2 R_{ki}(r) R_{kl}(r) \, dr = \int_{0}^{a} r J_{k+(1/2)}(\alpha_{ki}r) J_{k+(1/2)}(\alpha_{kl}r) \, dr = 0$$

for $i \neq l$. Then if we multiply both sides of (32) by $r^2 R_{ki}(r)$, integrate with respect to r from 0 to a, use Theorem 10.6(i) to evaluate

$$\int_0^a r^2 R_{ki}^2(r)\, dr = \int_0^a r J_{k+(1/2)}^2(\alpha_{ki} r)\, dr = \tfrac{1}{2} a^2 J_{k+(3/2)}^2(a\alpha_{ki}),$$

solve for a_{ijk}, and replace the subscripts i, j, and k with l, m, and n, we obtain

$$a_{lmn} = \frac{2}{c\pi a^2 J_{n+(3/2)}^2(a\alpha_{nl}) \| P_{nm} \|^2}$$

$$\times \int_0^a \left(\int_0^\pi \left(\int_{-\pi}^\pi r^2 R_{nl}(r) f(r, \phi, \theta) \cos m\theta\, d\theta \right) P_{nm}(\cos\phi) \sin\phi\, d\phi \right) dr$$

for $l \geq 1$, $n \geq 0$, and $0 \leq m \leq n$, where $c = 2$ if $m = 0$ and $c = 1$ if $m > 0$, and $\| P_{nm} \|$ is given by Theorem 11.5. Similarly,

$$b_{lmn} = \frac{2}{\pi a^2 J_{n+(3/2)}^2(a\alpha_{nl}) \| P_{nm} \|^2}$$

$$\times \int_0^a \left(\int_0^\pi \left(\int_{-\pi}^\pi r^2 R_{nl}(r) f(r, \phi, \theta) \sin m\theta\, d\theta \right) P_{nm}(\cos\phi) \sin\phi\, d\phi \right) dr$$

for $l \geq 1$, $n \geq 0$, and $1 \leq m \leq n$. Using these coefficients in (30) and the definition of $R_{nl}(r)$ leads to the formal solution

$$u(r, \phi, \theta, t) = \sum_{l=1}^\infty \sum_{n=0}^\infty \sum_{m=0}^n \frac{J_{n+(1/2)}(\alpha_{nl} r)}{\sqrt{r}}$$

$$\times (a_{lmn} \cos m\theta + b_{lmn} \sin m\theta) P_{nm}(\cos\phi) e^{-\alpha_{nl}^2 kt}.$$

Because of the decreasing exponentials and the boundedness of the remaining factors in each term of this series, the validity of the solution can be established in a manner analogous to that for the heated bar, the first problem ever considered in this book, but the precise details would require some knowledge of double series.[1]

EXERCISES

1.1   Assume that the temperature anywhere in the universe is a function of the distance r from a certain point and of the time t and that it satisfies the simplified heat equation

---

[1] For the basic definitions and theorems relating to double series see, for instance, T. M. Apostol, *Mathematical Analysis*, 2nd ed., Addison-Wesley, Reading, Massachusetts, 1974, pages 200–203.

$ru_t = k(ru_{rr} + 2u_r)$. Find a solution of the boundary value problem

$$\begin{array}{lll} ru_t = k(ru_{rr} + 2u_r) & r > 0, & t > 0 \\ u(r,t) \to 0 \quad \text{as} \quad t \to 0 & r > 0 \\ u(r,t) \to 0 \quad \text{as} \quad t \to \infty & r > 0 \\ u(r,t) \to 0 \quad \text{as} \quad r \to \infty & t > 0 \end{array}$$

(**Hint:** make the substitution $v = ru$ and use the solution of the boundary value problem of Exercise 3.2 of Chapter 1).

1.2   Find a solution of the boundary value problem of Exercise 1.1 that satisfies the additional condition that the total amount of heat in the entire universe is a constant Q.

1.3   Given that X is a nontrivial solution of the Legendre equation on $[-1, 1]$, find a solution that is even and a solution that is odd (**Hint:** consider the function Y defined by $Y(x) = X(-x)$ for all x in $[-1, 1]$).

1.4   Show that the initial value problem consisting of the Legendre equation and the initial conditions $X(1) = c_1$ and $X'(1) = c_2$, where c_1 and c_2 are arbitrary constants, need not have a solution. What condition must c_1 and c_2 satisfy for the solution to exist?

2.1   Use the identities

$$n! = n(n-1)\cdots(n-2m+1)(n-2m)!$$
$$(2n)! = 2n(2n-1)\cdots(2n-2m+1)(2n-2m)!$$

to verify the equation

$$a_n = (-1)^m \frac{m!(n-2m)!(2n)!(n-m)!}{(2n-2m)!(n!)^2} a_{n-2m}$$

for the coefficients of the Legendre polynomials.

2.2   Show that the first five Legendre polynomials are $P_0(x) = 1$, $P_1(x) = x$,

$$P_2(x) = \tfrac{1}{2}(3x^2 - 1), \quad P_3(x) = \tfrac{1}{2}(5x^3 - 3x) \quad \text{and} \quad P_4(x) = \tfrac{1}{8}(35x^4 - 30x^2 + 3).$$

Draw their graphs, indicating the location of maxima, minima, and points of inflection.

2.3   Show that each of the sets $\{P_{2n}\}$ and $\{P_{2n+1}\}$, $0 \le n < \infty$, is an orthogonal system on $(0, 1)$ and that their norms are

$$\|P_{2n}\| = \frac{1}{\sqrt{4n + 1}} \qquad \text{and} \qquad \|P_{2n+1}\| = \frac{1}{\sqrt{4n + 3}}.$$

2.4   Show that[1]

$$P_{2n}(0) = \frac{(-1)^n (2n-1)!!}{2^n n!} \quad (n > 0) \qquad \text{and} \qquad P_{2n+1}(0) = 0.$$

2.5   Show that

$$P'_{2n}(0) = 0 \qquad \text{and} \qquad P'_{2n+1}(0) = \frac{(-1)^n (2n+1)!!}{2^n n!}.$$

---

[1]Note that, if p is an odd integer, $p!!$ means $p(p-2)(p-4)\cdots 3 \cdot 1$. This notation is also used in Exercises 2.5, 2.8, and 2.9.

2.6 Use the Rodrigues formula and the mean value theorem, or Rolle's theorem, to show that P_n has exactly n distinct zeros in the interval $(-1, 1)$ (**Hint:** recall that, according to the fundamental theorem of algebra, a polynomial of degree n has exactly n zeros).

2.7 If X_1 is a solution of a second-order linear differential equation

$$a_2(x)X''(x) + a_1(x)X'(x) + a_0(x)X(x) = 0,$$

it can be shown that a second linearly independent solution is given by[1]

$$X_2(x) = X_1(x) \int \frac{1}{[X_1(x)]^2} e^{-\int [a_1(x)/a_2(x)]dx}\, dx.$$

Apply this method to find, for each $P_n, n = 0, 1, 2$ and 3, a second linearly independent solution Q_n of the Legendre equation. Such a solution is called the *Legendre function of the second kind*.

2.8 Let m be a nonnegative integer. Use Equations (6) with $\lambda = n(n + 1)$ and (7) with

$$a_1 = \frac{(-1)^m 2^{2m}(m!)^2}{(2m)!}$$

to show that, if $n = 2m$, then a second solution of the Legendre equation is

$$Q_n(x) = \frac{(-1)^m 2^{2m}(m!)^2}{(2m)!} \left[x + \sum_{1 \le k \le m} (-1)^k \right.$$

$$\left. \times \frac{2^k(2m - 1)(2m - 3) \cdots (2m - 2k + 1)(k + m)!}{m!(2k + 1)!} x^{2k+1} \right]$$

$$+ \sum_{k=m+1}^{\infty} \frac{2^{k+m}(2k - 2m - 1)!!(k + m)!}{(2k + 1)!} x^{2k+1}.$$

2.9 Let m be a nonnegative integer. Use Equations (6) with $\lambda = n(n + 1)$ and (8) with

$$a_0 = \frac{(-1)^{m+1} 2^{2m}(m!)^2}{(2m + 1)!}$$

to show that, if $n = 2m + 1$, then a second solution of the Legendre equation is

$$Q_n(x) \stackrel{\cdot}{=} \frac{(-1)^{m+1} 2^{2m}(m!)^2}{(2m + 1)!} \left[1 + \sum_{1 \le k \le m+1} (-1)^k \right.$$

$$\left. \times \frac{2^k(2m + 1)(2m - 1) \cdots (2m - 2k + 3)(k + m)!}{m!(2k)!} x^{2k} \right]$$

$$+ \sum_{k=m+2}^{\infty} \frac{2^{k+m}(2k - 2m - 3)!!(k + m)!}{(2k)!} x^{2k}.$$

---

[1] See, for instance, D. G. Zill, *Differential Equations with Boundary Value Problems*, 2nd ed., PWS-KENT, Boston, 1989, pages 155–156.

2.10   Show that the solutions Q_0, Q_1, Q_2, and Q_3 obtained in Exercises 2.8 and 2.9 are the same as those obtained in Exercise 2.7 (**Hint:** use the fact that

$$\log \frac{1+x}{1-x} = 2 \sum_{k=0}^{\infty} \frac{x^{2k+1}}{2k+1}$$

for $-1 < x < 1$).

2.11   Show that the series in Exercises 2.8 and 2.9 diverge at $x = \pm 1$.

2.12   Prove that every solution of the Legendre equation corresponding to a value of λ not of the form $n(n+1)$, where n is a nonnegative integer, is unbounded on $(-1, 1)$ (**Hint:** use Theorem 4.1, the fact that every polynomial is a linear combination of Legendre polynomials, and the Weierstrass approximation theorem of Exercise 7.3 (*iii*) of Chapter 2).

3.1   Use the Rodrigues formula and Theorem 11.1 to prove the recursion formulas
    (*i*) $P_{n+1}'(x) - x P_n'(x) = (n+1) P_n(x)$ and
    (*ii*) $P_{n+1}'(x) - P_{n-1}'(x) = (2n+1) P_n(x)$
for $n \geq 1$.

3.2   Use the results of Exercise 3.1 to show that

$$(1 - x^2) P_n'(x) = n P_{n-1} - n x P_n$$

for $n \geq 1$. Then use this equation to show that, if a is a point in $(-1, 1)$ such that $P_{n-1}(a) = P_{n+1}(a) = 0$, then P_n has a maximum or a minimum at a.

3.3   Use the result of Exercise 3.1 (*ii*) to show that

$$|P_n'(x)| \leq \frac{n(n+1)}{2} \quad \text{and} \quad |P_n''(x)| \leq \frac{n^4}{2}$$

for any x in $[-1, 1]$ and any nonnegative integer n.

3.4   Use the result of Exercise 3.1 (*ii*) to show that for any number a in $[-1, 1]$ and $n \geq 1$

$$\int_a^1 P_n = \frac{1}{2n+1} \left[P_{n-1}(a) - P_{n+1}(a) \right] = - \int_{-1}^a P_n.$$

4.1   Use the Rodrigues formula to show that, if $f : [-1, 1] \to \mathbb{R}$ is a function of class C^n, then its Fourier-Legendre coefficients are

$$c_n = \frac{(-1)^n (2n+1)}{2^{n+1} n!} \int_{-1}^1 f^{(n)}(x)(x^2 - 1)^n \, dx.$$

4.2   Show that,
    (*i*) for every pair of nonnegative integers m and n with $m < n$, x^m is orthogonal to $P_n(x)$ and
    (*ii*) every polynomial of degree n that is orthogonal to 1, x, x^2, ... x^{n-1} is of the form $C P_n(x)$, where C is a constant
(**Hint:** to prove (*i*) use the result of Exercise 4.1).

4.3   Let p be a positive even integer. Use the result of Exercise 4.1 and the Fourier-Legendre series of $f(x) = x^p$ on $(-1, 1)$ to show that

$$x^p = \sum_{m=0}^{p/2} \frac{4m+1}{2^{2m}} \binom{p}{2m} \left[\sum_{k=0}^{2m} \binom{2m}{k} \frac{(-1)^k}{p+1+2m-2k} \right] P_{2m}(x).$$

4.4   Find and prove an identity similar to that of Exercise 4.3 for p odd.

4.5   Expand
$$f(x) = \begin{cases} 0, & -1 \leq x \leq 0 \\ 1, & 0 < x \leq 1 \end{cases}$$
in a Fourier-Legendre series (**Hint:** use the results of Exercises 3.4 and 2.4).

4.6   Expand
$$f(x) = \begin{cases} -1, & -1 \leq x \leq 0 \\ 1, & 0 < x \leq 1 \end{cases}$$
in a Fourier-Legendre series (**Hint:** use the results of Exercises 3.4 and 2.4).

4.7   Consider the function
$$f(x) = \begin{cases} 0, & -1 \leq x \leq 0 \\ x, & 0 \leq x \leq< 1. \end{cases}$$

Show that
   (*i*) $c_{2m+1} = 0$ for $m \geq 1$ and
   (*ii*) for every x in $[-1, 1]$

$$f(x) = \tfrac{1}{4} + \tfrac{1}{2}x + \sum_{m=1}^{\infty} \left[\frac{P_{2m+2}(0)}{2(4m + 3)} - \frac{(4m + 1)P_{2m}(0)}{(4m + 3)(4m - 1)} + \frac{P_{2m-2}(0)}{2(4m - 1)} \right] P_{2m}(x)$$

(**Hint:** for (*i*) use Exercise 4.2 (*i*) use and for (*ii*) use Exercises 3.1 (*ii*) and 3.4).

4.8   Prove or disprove that every function $f(\phi)$ of class C^1 on $[-\pi, \pi]$ can be expanded in a series of the form
$$\sum_{n=0}^{\infty} c_n P_n(\cos \phi)$$
on $(-\pi, \pi)$.

4.9   Prove that, if $f : [-1, 1] \to \mathbb{R}$ is integrable, then its Fourier-Legendre series converges to f in the mean, and deduce from this fact the completeness of the orthogonal system $\{P_n\}$, $0 \leq n < \infty$ (**Hint:** follow the steps in the proof of Theorem 6.5 but use the Weierstrass approximation theorem of Exercise 7.3 (*iii*) of Chapter 2 in step 3).

4.10   Find the Fourier series, in the sense of Definition 6.2, of an integrable function $f : (0, 1) \to \mathbb{R}$ with respect to the orthogonal systems $\{P_{2n+1}\}$ and $\{P_{2n}\}$ of Exercise 2.3. Show that these series are the same as the Fourier-Legendre series of the odd and even extensions of f to $(-1, 1)$, and prove Theorem 11.4.

5.1   Find the particular solution of the potbellied stove problem for
$$f(\phi) = \begin{cases} 1, & 0 \leq \phi \leq \pi/2 \\ -1, & \pi/2 \leq x \leq \pi \end{cases}$$

(**Hint:** use the solution of Exercise 4.6).

5.2   Find the steady-state temperature distribution inside a sphere of radius a subject to the boundary condition $u(a, \phi) = f(\phi)$, where f is a piecewise continuous function. Find the particular solution for
$$f(\phi) = \begin{cases} \cos \phi, & 0 \leq \phi \leq \pi/2 \\ 0, & \pi/2 \leq x \leq \pi \end{cases}$$

(**Hint:** use the Fourier-Legendre series of Exercise 4.7).

5.3   Find the steady-state temperature distribution inside a sphere of radius a subject to the boundary condition $u_r(a, \phi) = f(\phi)$ if $u(0, \cdot) = T_0$. Assume that, for the temperature to be in a steady state, the rate of heat flow through the entire spherical surface is zero, that is, that

$$\int_0^\pi f(\phi) \sin \phi \, d\phi = 0.$$

5.4   Find the steady-state temperature distribution inside a spherical shell $a \le r \le b, a > 0$, subject to the boundary conditions $u(a, \phi) = 0$ and $u(b, \phi) = f(\phi)$.

5.5   Find the steady-state temperature distribution inside a hemisphere $r \le a, 0 \le \phi \le \pi/2$, subject to the boundary conditions $u(a, \phi) = f(\phi)$ for $0 \le \phi \le \pi/2$ and $u(r, \pi/2) = 0$ for $r < a$. Find the particular solution for $f \equiv 1$ (**Hint:** use the corollary of Theorem 11.4 and the results of Exercises 3.4 and 2.4).

5.6   Find the steady-state temperature distribution inside the hemisphere of Exercise 5.5 if the base is insulated instead of kept at zero temperature (**Hint:** write the boundary condition at the base in cylindrical coordinates and use the equation $u_\phi = u_z z_\phi + u_\rho \rho_\phi$ to transform it into spherical coordinates).

5.7   Find the steady-state temperature distribution outside the hemisphere of Exercise 5.6.

5.8   Find the steady-state temperature distribution inside the hemisphere of the preceding exercises if the base is insulated, the boundary condition on the curved surface is $u_r(a, \phi) = f(\phi)$ for $0 \le \phi \le \pi/2$, and $u(0, \cdot) = T_0$. Assume that, for the temperature to be in a steady state, the rate of heat flow through the entire hemispherical surface is zero, that is, that

$$\int_0^{\pi/2} f(\phi) \sin \phi \, d\phi = 0.$$

7.1   Evaluate P_{nm} for $0 \le m \le n \le 2$.

7.2   Prove that $P_{nn}(\cos \phi) = C \sin^n \phi$, where C is a constant, and find the value of C for $n = 1, 2$, and 3.

7.3   Show that, if $1 \le m \le n$, then P_{nm} has exactly $n - m$ zeros in the interval $(-1, 1)$ (**Hint:** use the result of Exercise 2.6).

7.4   Prove or disprove that

$$\int_{-1}^1 P_{kl}(x) P_{nm}(x) \, dx = 0$$

for
   (i) $k = n$ and $l \ne m$ and for
   (ii) $k \ne n$ and $l \ne m$.

7.5   Prove the recursion formula

$$P_{n(m+2)}(x) - \frac{2(m + 1)x}{\sqrt{1 - x^2}} P_{n(m+1)}(x) + [n(n + 1) - m(m + 1)]P_{nm}(x) = 0.$$

7.6   Prove the recursion formula

$$(n - m + 1)P_{(n+1)m}(x) - (2n + 1)x P_{nm}(x) + (m + n)P_{(n-1)m}(x) = 0$$

for $n \ge 1$ (**Hint:** differentiate the equation in Theorem 11.1 m times with respect to x and then use Exercise 3.1 (ii)).

8.1   Find a particular solution of the Dirichlet problem for the sphere subject to the boundary condition $u(a, \phi, \theta) = \sin \phi \cos \theta (1 + \sin \phi \cos \theta)$ (**Hint:** use the results of Exercises 7.1 and 7.2).

8.2   Find a formal bounded solution for the Dirichlet problem outside a sphere of radius a subject to the boundary condition $u(a, \phi, \theta) = f(\phi, \theta)$.

8.3   Find a formal solution of the Dirichlet problem inside a spherical shell $a \leq r \leq b, a > 0$, subject to the boundary conditions $u(a, \phi, \theta) = f(\phi, \theta)$ and $u(b, \phi, \theta) = g(\phi, \theta)$ (**Hint:** solve separately the cases $f \equiv 0$ and $g \equiv 0$).

8.4   Show that the solid spherical harmonics are homogeneous polynomials if expressed in the Cartesian coordinates x, y, and z (**Hint:** note that $a \cos m\theta + b \sin m\theta$ is a linear combination of $e^{im\theta}$ and $e^{-im\theta}$, and use the definition of spherical coordinates).

9.1   Expand $f(x) = (5 - 3x)^{-1/2}$ in a Fourier-Legendre series on $(-1, 1)$.

9.2   Expand

$$f(x) = \log \frac{1 + x}{1 - x}$$

in a Fourier-Legendre series on $(-1, 1)$ (**Hint:** integrate (22) with respect to r from 0 to x).

9.3   Expand $f(x) = \sqrt{1 - x}$ in a Fourier-Legendre series on $(-1, 1)$ (**Hint:** multiply (22) by $\sqrt{1 - x}$, integrate with respect to x from -1 to 1, and then use the identity in the hint for Exercise 2.10).

9.4   Find an integral solution for the Dirichlet problem outside a sphere (**Hint:** start with the series solution of Exercise 8.2).

9.5   Prove the following theorem of Axel Harnack (1851–1888): if u is a nonnegative function that is harmonic on an open set of $\mathbb{R}^3$ containing a sphere S of radius a and its interior D, then for any point p at a distance $r < a$ from the origin 0 we have

$$a \frac{a - r}{(a + r)^2} u(0) \leq u(p) \leq a \frac{a + r}{(a - r)^2} u(0).$$

These are called *Harnack's inequalities* (**Hint:** use the fact that $-1 \leq \cos \alpha \leq 1$ to find upper and lower bounds for the function $P^q(p)$ defined after the statement of Theorem 11.7, multiply the resulting inequalities by $f(q)/4\pi a$, integrate over S, and then use Equation (29) and Gauss' mean value theorem of Exercise 1.7 of Chapter 9).

9.6   Use Harnack's inequalities of Exercise 9.5 to prove the following theorem of Charles Émile Picard (1856–1941) of Paris: if u is harmonic in $\mathbb{R}^3$ and either bounded above or below by a constant, then u is a constant (**Hint:** consider first the case $u \geq 0$ and use a sphere of arbitrarily large radius).

9.7   Prove *Harnack's first convergence theorem*: if $\{u_n\}$ is a sequence of harmonic functions on an open set $O \subset \mathbb{R}^3$ that converges uniformly to a function u, then u is harmonic in O (**Hint:** if S is an arbitrary sphere that together with its interior D is in O, show that each u_n is represented in D by the Poisson integral formula of Theorem 11.7, and then let $n \to \infty$).

9.8   Prove the following simplified form of *Harnack's second convergence theorem*: if $u_1 \leq u_2 \leq u_3 \leq \cdots \leq u_n \leq \cdots$ is an increasing sequence of harmonic functions on an open set $O \subset \mathbb{R}^3$ that converges to a function u, then u is harmonic in O (**Hint:** if S is an arbitrary sphere that together with its interior D is in O, use Harnack's inequalities of Exercise 9.5 to prove that the convergence is uniform in D and then apply Harnack's first convergence theorem of Exercise 9.7).

10.1 Find the particular solution of the cooling sphere problem if the initial temperature is a constant T_0 (**Hint:** refer to Examples 10.2 and 10.4).

10.2 How is the solution of the cooling sphere problem to be modified if the boundary condition is replaced by $u_r(a, \phi, \theta, t) = -Ku(a, \phi, \theta, t)$, where K is a nonnegative constant? This is the boundary condition used by Fourier (**Hint:** define $h = aK - \frac{1}{2}$).

10.3 Find the general solution of the three-dimensional wave equation

$$u_{tt} = c^2\left(u_{rr} + \frac{2}{r}u_r + \frac{1}{r^2}u_{\phi\phi} + \frac{\cot\phi}{r^2}u_\phi + \frac{1}{r^2\sin^2\phi}u_{\theta\theta}\right)$$

for $r < a$ subject to the boundary condition $u(a, r\phi, \theta) = 0$.

10.4 Find the particular solution of the wave equation problem of Exercise 10.3 corresponding to initial conditions of the form

(i) $u(r, \phi, \theta, 0) = f(r, \phi, \theta)$ and $u_t(r, \phi, \theta, 0) = 0$ and of the form
(ii) $u(r, \phi, \theta, 0) = 0$ and $u_t(r, \phi, \theta, 0) = f(r, \phi, \theta)$.

12

DISTRIBUTIONS AND
GREEN'S FUNCTIONS

§12.1 Historical Prologue

In Chapter 5 we witnessed a controversy over the kind of function that can be allowed to describe the initial position of the vibrating string. It must be of class C^2 for d'Alembert's solution of Theorem 5.2 to actually be a twice-differentiable solution at all points. If instead, as Euler proposed, a function with corners is allowed to describe the string's initial position, then d'Alembert's solution cannot be differentiable, and therefore is not a solution, on certain characteristics. This state of affairs represents a limitation in some applied problems. A different kind of limitation was found in the application of the Fourier transform method because some elementary functions, such as a constant, the exponential, the sine, and the cosine, do not have Fourier transforms.

It must be accepted that not all functions have derivatives or Fourier transforms. However, one could think—at least as a flight of fancy—of a way to deal with the aforementioned limitations: widen the concept of function, and along with it those of derivative and Fourier transform, and generally do whatever it takes so that d'Alembert's solution becomes a twice-differentiable object that satisfies the wave equation at all points and so that the elementary functions stated above have transforms. But, if this could be done, it would necessarily turn mathematical analysis topsy-turvy, and it can be argued that we lack sufficient motivation to embark on such a major undertaking. As it turns out, the first step in this course of action had already been taken by Fourier himself and is implicitly contained in his proof of the equation

$$(1) \qquad f(x) = \int_{-\infty}^{\infty} \hat{f}(\omega) e^{i\omega x} \, d\omega.$$

Fourier's proof contains a number of unrigorous steps, but is interesting because it contains the germ of further discoveries, and we shall examine it next. If we replace x

with s in

$$\hat{f}(\omega) = \frac{1}{2\pi} \int_{-\infty}^{\infty} f(x)e^{-i\omega x}\, dx,$$

substitute the result into the right-hand side of (1), reverse the order of integration, and simplify, we obtain

$$(2) \quad \int_{-\infty}^{\infty} f(s)\left(\frac{1}{\pi}\int_{0}^{\infty} \cos\omega(s-x)\,d\omega\right) ds = \int_{-\infty}^{\infty} f(s) \lim_{p\to\infty} \frac{\sin p(s-x)}{\pi(s-x)}\, ds.$$

At this point Fourier stated that the right-hand side is equal to

$$(3) \quad \int_{-\infty}^{\infty} f(s)\frac{\sin p(s-x)}{\pi(s-x)}\, ds$$

where, he said, $p = \infty$.[1] Now, this is mathematically impossible, but typical of Fourier's determination to go ahead and obtain results. Let us just say that $p > 0$ is fixed and very large and that (3) is an approximation to the right-hand side of (1). The quotient in (3) was well known to Fourier. It is an even function of $s - x$ with zeros at $s = x + k\pi/p$. For p very large the length of any interval over which $\sin p(s - x)$ undergoes a complete oscillation is very small, and then the quotient $f(s)/(s - x)$ remains approximately constant over any of these intervals except for the one enclosing x. Thus the contribution to the previous integral is negigible over these intervals. On the one containing x, $f(s) \approx f(x)$ and then (3) can be approximated by

$$f(x) \int_{x-\pi/p}^{x+\pi/p} \frac{\sin p(s-x)}{\pi(s-x)}\, ds.$$

But, as above, the integral of this quotient over the rest of the real line is negigible, and this expression can be further approximated by

$$(4) \quad f(x) \int_{-\infty}^{\infty} \frac{\sin p(s-x)}{\pi(s-x)}\, ds = \frac{f(x)}{\pi} \int_{-\infty}^{\infty} \frac{\sin u}{u}\, du = f(x)$$

for $p \gg 0$. If we allow $p = \infty$, this is Fourier's proof of (1).

It is not hard to imagine how this proof can be tightened to make it valid under current standards. But the point we want to make is along different lines. If we examine Fourier's argument, it seems that he would have us believe that there is a function δ defined by

$$(5) \quad \delta(x) = \lim_{p\to\infty} \frac{\sin px}{\pi x}$$

and such that

$$(6) \quad \int_{-\infty}^{\infty} f(s)\delta(s-x)\, ds = f(x),$$

---

[1] See *The Analytical Theory of Heat*, pages 426–429.

as suggested by (2) and (4), while (4) itself shows that

(7)
$$\int_{-\infty}^{\infty} \delta = 1.$$

Furthermore, from the argument following (3) we deduce that δ has a vanishing integral over any interval $[a, b]$ that does not contain the origin. Since a and b are arbitrary, δ vanishes outside the origin, and since δ is, by definition, unbounded in any neighborhood of the origin, we could think that

(8)
$$\delta(x) = \begin{cases} \infty & \text{if} \quad x = 0 \\ 0 & \text{if} \quad x \neq 0. \end{cases}$$

There is, of course, no such function. But we wish there were because it would be very useful in applications. For instance, in his 1828 essay, already mentioned in Chapter 9, George Green considered the problem of solving the equation

(9)
$$u_{xx} + u_{yy} + u_{zz} = f$$

in a bounded region of space that, we shall assume without loss of generality, contains the origin. Here u is the electrostatic potential created by a charge distribution given by f. Green showed that he could solve this problem if he could first solve the problem for the restricted case in which there is just one point charge—infinite charge density—at the origin and no charge elsewhere.[1] Now, let us say that there is a δ function on $\mathbb{R}^3$ with the properties (6)–(8) except that the integrals are three dimensional. Since, in particular, this δ function is infinite at the origin and zero elsewhere, we can rephrase Green's claim as follows: a solution of (9) can be obtained from a solution of

(10)
$$u_{xx} + u_{yy} + u_{zz} = \delta.$$

Indeed, this is very simple to see with the machinery at our disposal. Let u^δ be a solution of (10), denote the function of x defined by the left-hand side of (6)—or rather its three-dimensional equivalent—by $f * \delta$, which is in keeping with the definition of convolution, and define $f * u^\delta$ in the same way but replacing δ with u^δ in the integrand. Then $u = f * u^\delta$ is a solution of (9) because, if differentiation under the integral sign were permitted, we would have

$$u_{xx} + u_{yy} + u_{zz} = f * (u^\delta_{xx} + u^\delta_{yy} + u^\delta_{zz}) = f * \delta = f,$$

where the last equality is just (6).

Green, of course, proceeded along different lines since he did not have the δ function. Nor do we, for that matter, since we have relied on some unrigorous implicit definition of δ arising from Fourier's work. The first explicit definition was given by Kirchhoff

---

[1] *An Essay on the Application of Mathematical Analysis to the Theories of Electricity and Magnetism*, Nottingham, 1828, pages 32–33.

in 1882 while working on Huygens' principle for the wave equation. Starting with Green's theorem (*i*) of Exercise 1.6 of Chapter 9, his work was similar to Green's for the Laplace equation, but at a certain point he needed and defined a function that vanishes for each nonzero value of its argument, is never negative, and whose integral from a negative value to a positive value is 1. This is the same δ implicity defined in Fourier's proof but, once again, there is no such function.

The next definition of δ was given by Heaviside. In *EMT*II, pages 54 to 57, he considered a line without self-induction (while issuing a warning that *it would not at all be desirable to bring a practical telegraph circuit to such a state*), and, under certain conditions that are not relevant to our story, he found the current entering the cable to be

$$p[Q1],$$

where $p = d/dt$, Q is a constant, and **1** is the function defined, as we know, to vanish for $t < 0$ and to have value 1 for $t \geq 0$. So what is the current? Heaviside, who wrote pQ rather than $p[Q1]$, found it as follows (page 55):

> We have to note that if Q is any function of the time, then pQ is its rate of increase. If, then, as in the present case, Q is zero before and constant after $t = 0$, pQ is zero except when $t = 0$. It is then infinite. But its total amount is Q. That is to say, $p\mathbf{1}$ means a function of t which is wholly concentrated at the moment $t = 0$, of total amount 1. It is an impulsive function, so to speak.

Even though unnamed and unrigorously defined, this is, once more, the δ function— this time as the derivative of the Heaviside function—and the current entering the cable is $Q\delta$. From this moment on there was no turning back. The δ function, whatever it may be, since engineers did not seem to care as long as they obtained correct results, became indispensable in electrical engineering. Then it became prominent in the development of quantum mechanics by the British physicist Paul Adrien Maurice Dirac (1902–1987), who defined it by (7) and the second part of (8), that is, in the same manner as Heaviside. Dirac is the man responsible for the notation δ because it reminded him of the Kronecker delta.[1]

For quite some time mathematicians were amazed that manipulations with such "functions" would lead to correct results about ordinary functions. Then, in 1936, Sergeĭ L'vovich Sobolev (1908–1989), working on the solution of certain partial differential equations, pointed out that these illegitimate functions never appear in the final results, but only in equations such as (6), and that the intermediate manipulations lead to the definition of perfectly legitimate *linear functionals*, that is, real valued functions whose domain is a certain space of ordinary functions. We cannot describe this space, whose definition is highly technical and relevant to a theorem that Sobolev was considering on the existence of solutions, but note that, if f is in that space, then the integral in (6) is an example of such a functional: to each function f it assigns the real number $f(x)$. If $x = 0$, (6) becomes

$$\int_{-\infty}^{\infty} \delta(s) f(s)\, ds = f(0),$$

---

[1]This is a number denoted by δ_{ij} and defined to be 1 if $i = j$ and zero if $i \neq j$.

which is reinterpreted as follows: delta is the functional defined by

$$\delta(f) = f(0).$$

This is the current definition of the "delta function."

But before the preceding facts could be fully accepted by the mathematical community it was necessary to develop a rigorous theory of Sobolev's functionals. This is precisely what Laurent Schwartz (1915–), working in isolation at Grenoble as Fourier had done so many years before, set out to accomplish. During the period 1945–1948 he

LAURENT SCHWARTZ IN 1978
Photograph by Paul Halmos.
Reprinted from *I Have a Photographic Memory* by Paul R. Halmos,
by permission of the American Mathematical Society.

developed a complete theory of these functionals, which he called *distributions*, starting with a paper entitled *Généralisation de la notion de fonction, de dérivation, de transformation de Fourier, et applications mathématiques et physiques*—thereby fulfilling the entire wish list contained in our earlier flight of fancy—and culminating in the publication of his two-volume work *Théorie des distributions*, which appeared in Paris in 1950 and 1951.[1]   Once the theory of distributions had been developed, Leon Ehrenpreis

---

[1] A detailed account of all the events leading to this theory can be found in J. Lützen, *The Prehistory of the Theory of Distributions*, Springer-Verlag, New York, 1972.

in 1954 and independently Bernard Malgrange in 1955 were able to prove that the solution of an equation of the form

$$P[u] = \delta,$$

where P denotes any linear partial differential operator with constant coefficients— that is, the left-hand side of an equation such as (10) or any of those in the preceding chapters—always exists as a distribution. It is called a *fundamental solution* because, as shown above in a particular case and later in general, the solution of $P[u] = f$ can be obtained from it.[1] A modification of a fundamental solution designed to satisfy some given boundary conditions is usually called a *Green's function*, after Riemann's name for the function that Green actually used for u^δ above.

Unfortunately, the search for fundamental solutions or Green's functions in specific situations is not a simple matter and requires a preliminary development of some of the theory of distributions, especially those aspects related to δ. Before we plunge into such a task, it is interesting to note, once again, that the work of Fourier and Heaviside— these *enfants terribles* of science who insisted on using concepts and theories as yet undeveloped or underdeveloped—would provide some of the initial impulse that resulted in new advances in mathematics and its applications.

We shall present the essential part of the Schwartz theory of distributions in the next five sections, but the reader must be alerted to the fact that it is not entirely possible to develop it with the limited mathematical background assumed so far. Some concepts and facts from mathematical analysis will be used at certain points, giving precise statements and appropriate references. Then, in the next four sections, we consider some applications to problems in partial differential equations.

§12.2   Distributions

In the previous section we defined the delta function as a linear functional, that is, a real-valued linear function whose domain is a certain space of ordinary functions. But we did not describe this space. It turns out that it is both convenient and sufficient to define these functionals on a space of very smooth ordinary functions called *test functions*, precisely defined as follows.

Definition 12.1 *Let $\mathcal{D}^n$ be the set of all functions $\varphi : \mathbb{R}^n \to \mathbb{R}$ of class C^∞ with bounded support (defined as the closure of the set of all points x in $\mathbb{R}^n$ such that $\varphi(x) \neq 0$). The elements of $\mathcal{D}^n$ are called* **test functions**.

Before we make another move we must make sure that test functions exist.

---

[1] It was first called an *elementary solution* by Hadamard, who originated this definition in a paper of 1908 and also included it in the same 1920 set of lectures at Yale in which he introduced the concept of a well-posed problem. Incidentally, Hadamard was *Oncle Jacques* to Schwartz since Hadamard's wife was the sister of the mother of Schwartz' mother.

Example 12.1 For each $a > 0$ the function $\varphi_a : \mathbb{R}^n \to \mathbb{R}$ defined by

$$\varphi_a(x) = \begin{cases} e^{-a^2/(a^2 - \|x\|^2)} & \text{if } \|x\| \le a \\ 0 & \text{if } \|x\| > a \end{cases}$$

is a test function.

Example 12.2 If φ is in $\mathcal{D}^n$ and if $f : \mathbb{R}^n \to \mathbb{R}$ is of class C^∞, then the product $f\varphi$ is in $\mathcal{D}^n$.

The reason why functionals defined on test functions are sufficient for our purposes is best illustrated by considering the case of a continuous function f. We claim that, if the values of the functional defined by

$$\int_{\mathbb{R}^n} f\varphi$$

are known for all test functions φ, then the values of f are known for all x. If so, we can define f by giving either its value at each x or the value of the stated functional at each test function φ. To prove the claim consider a text function φ such that

$$\int_{\mathbb{R}^n} \varphi = 1$$

(which is always possible if we multiply any φ by an appropriate constant). Then, if we define

$$\varphi_t(x) = \frac{1}{t^n} \varphi\left(\frac{1}{t}x\right)$$

for each $t > 0$, it is possible to show rather simply that

$$\lim_{t \to 0} \int_{\mathbb{R}^n} f(s)\varphi_t(x - s)\, ds = f(x)$$

(see Exercise 2.3); that is, the value of f is determined for each x by its functional values on test functions. It is possible to prove a similar statement if f is only locally integrable, but then its values can be determined only for almost all x.

In order to develop a simple but accurate notation for higher order partial derivatives we shall establish the following terminology. A *multi-index* is an n-tuple of nonnegative integers $(m_1, m_2, \ldots, m_n)$. If $m = (m_1, m_2, \ldots, m_n)$ is a multi-index and if $(x_1, x_2, \ldots, x_n)$ are the components of x in $\mathbb{R}^n$, we define

$$D^m \varphi = \frac{\partial^{m_1 + m_2 + \cdots + m_n} \varphi}{\partial x_1^{m_1} \partial x_2^{m_2} \cdots \partial x_n^{m_n}}.$$

When any one of the m_i is zero then the corresponding differentiation is not present.

Definition 12.2 *We say that a sequence* $\{\varphi_k\}$ *of test functions* **converges in** $\mathcal{D}^n$ *to a test function* φ, *and then we write* $\varphi_k \to \varphi$ *in* $\mathcal{D}^n$ *as* $k \to \infty$, *if and only if there is a bounded set* E *in* $\mathbb{R}^n$ *such that*

(i) φ *and all the* φ_k *vanish outside* E, *and*

(ii) *for any multi-index* m *the sequence* $\{D^m \varphi_k\}$ *converges uniformly on* E *to* $D^m \varphi$ *as* $k \to \infty$.

We are now ready to define the linear functionals that are the object of our study.

Definition 12.3 *A* **distribution** *is a function* $T : \mathcal{D}^n \to \mathbb{R}$ *such that*

(i) T *is linear, that is, if* φ *and* ϕ *are in* $\mathcal{D}^n$ *and if* c *is a real constant, then* $T(\varphi + \phi) = T(\varphi) + T(\phi)$ *and* $T(c\varphi) = cT(\varphi)$, *and*

(ii) $T(\varphi_k) \to T(\varphi)$ *in* $\mathbb{R}$ *if* $\varphi_k \to \varphi$ *in* $\mathcal{D}^n$ *as* $k \to \infty$.

If a function $T : \mathcal{D}^n \to \mathbb{R}$ satisfies the second requirement in Definition 12.3, then we say that T is *continuous*. Thus, a distribution is a continuous linear functional on $\mathcal{D}^n$. It must be remarked that although this continuity requirement looks impressive it is actually rather weak. Almost any linear functional that one can explicitely define on $\mathcal{D}^n$ would automatically be continuous.

Example 12.3 If $f : \mathbb{R}^n \to \mathbb{R}$ is locally integrable, define $T_f : \mathcal{D}^n \to \mathbb{R}$ by

$$T_f(\varphi) = \int_{\mathbb{R}^n} f\varphi.$$

Clearly, T_f is linear since

$$\int_{\mathbb{R}^n} f \cdot (\varphi + \phi) = \int_{\mathbb{R}^n} f\varphi + \int_{\mathbb{R}^n} f\phi$$

and, if c is a constant,

$$\int_{\mathbb{R}^n} f \cdot (c\varphi) = c \int_{\mathbb{R}^n} f\varphi.$$

Also, T_f is continuous because, if $\varphi_k \to \varphi$ in $\mathcal{D}^n$, then $\varphi_k \to \varphi$ as $k \to \infty$ uniformly on some bounded set E in $\mathbb{R}^n$. Then, if M_k denotes the maximum value of $|\varphi_k - \varphi|$ on E,

$$|T_f(\varphi_k) - T_f(\varphi)| \le \int_{\mathbb{R}^n} |f||\varphi_k - \varphi| \le M_k \int_E |f|.$$

Since $M_k \to 0$ as $k \to \infty$, so does the left-hand side, showing that T_f is continuous.

The integral defining T_f can be viewed as a "distribution" of the values of f over those of each φ. Hence the name adopted in Definition 12.3 for the functionals that are the object of our study.

Example 12.4 For each x in $\mathbb{R}^n$ define $\delta^x : \mathcal{D}^n \to \mathbb{R}$ by

$$\delta^x(\varphi) = \varphi(x)$$

(compare this definition with Equation (6)). Then δ^x is linear because

$$\delta^x(\varphi + \phi) = (\varphi + \phi)(x) = \varphi(x) + \phi(x) = \delta^x(\varphi) + \delta^x(\phi)$$

for any test functions φ and ϕ and, if c is a constant,

$$\delta^x(c\varphi) = (c\varphi)(x) = c\varphi(x) = c\delta^x(\varphi).$$

Now, if $\varphi_k \to \varphi$ in $\mathcal{D}^n$, then $\varphi_k(x) \to \varphi(x)$ in $\mathbb{R}$ as $k \to \infty$, and it follows that

$$\delta^x(\varphi_k) = \varphi_k(x) \to \varphi(x) = \delta^x(\varphi),$$

which shows that δ^x is also continuous and, therefore, a distribution. In particular, if $x = 0$, we define $\delta = \delta^0$, a distribution called the *delta function*.

Distributions can be classified into two broad categories, and each of the preceding examples illustrates one of them.

Definition 12.4 *A distribution* $T : \mathcal{D}^n \to \mathbb{R}$ *is called* **regular** *if and only if there is a locally integrable function* f *such that*

$$T(\varphi) = \int_{\mathbb{R}^n} f\varphi$$

for all φ *in* $\mathcal{D}^n$, *that is,* $T = T_f$ *in the notation of the previous example. A distribution is called* **singular** *if and only if it is not regular.*

Example 12.5 The distribution δ^x is singular for any x in $\mathbb{R}^n$. Indeed, assume, contrary to the assertion, that there is a locally integrable function f such that

$$\delta^x(\varphi) = \int_{\mathbb{R}^n} f\varphi$$

for all φ in $\mathcal{D}^n$, and choose the test function defined by $\varphi(s) = \varphi_a(s - x)$ for all s in $\mathbb{R}^n$, where φ_a is as in Example 12.1. Then

$$\delta^x(\varphi) = \varphi(x) = \varphi_a(x - x) = \varphi_a(0) = e^{-1} > 0$$

for any $a > 0$, while

$$\int_{\mathbb{R}^n} f\varphi = \int_{\mathbb{R}^n} f(s)\varphi_a(s - x)\,ds \to 0$$

as $a \to 0$, a contradiction.

Additional singular distributions can be found in the exercises for this section.

§12.3   Basic Properties of Distributions

We start by studying the local behavior of distributions, which will allow us to verify that δ vanishes outside the origin.

Definition 12.5 *A distribution T is said to* **vanish on an open set** *O in $\mathbb{R}^n$ if and only if $T(\varphi) = 0$ for every test function φ with support in O. T is said to* **vanish at a point** *x in $\mathbb{R}^n$ if and only if it vanishes on some open set that contains x. If U is the union of all open sets on which T vanishes, then the complement of U in $\mathbb{R}^n$ is called the* **support** *of T.*

Example 12.6 δ^x vanishes at every $s \neq x$, and, in particular, δ vanishes outside the origin. In fact, let O be an open set in $\mathbb{R}^n$ that does not contain x and let φ be a test function with support in O. Then

$$\delta^x(\varphi) = \varphi(x) = 0.$$

Assume now that O is an open set containing x, let $a > 0$ be so small that the closed disc D with center x and radius a is in O, and define $\varphi(s) = \varphi_a(s - x)$ for all s in $\mathbb{R}^n$, where φ_a is the test function of Example 12.1. Then φ vanishes outside D but

$$\delta^x(\varphi) = \varphi(x) = \varphi_a(x - x) = \varphi_a(0) = e^{-1} > 0,$$

which shows that the set consisting of the single point x is the support of δ^x. Thus, the support of δ is the origin.

It may be tempting to conclude, as an immediate consequence of Definition 12.5, that a distribution vanishes on the complement of its support. However, this is a fact that requires a very delicate proof, one based on the following result from mathematical analysis, whose proof is beyond our reach.[1]

Lemma 12.1 *Let $\mathcal{O}$ be a collection of open sets in $\mathbb{R}^n$ and let U be their union. Then there is a sequence $\{\phi_k\}$ of real-valued functions of class C^∞ on $\mathbb{R}^n$ such that*
   (i) $0 \leq \phi_k \leq 1$ *for all k,*
   (ii) *for each ϕ_k there is an open set O in $\mathcal{O}$ such that ϕ_k has support in O,*
   (iii) *we have*

$$\sum_{k=1}^{\infty} \phi_k(x) = 1$$

for each x in U, and,
   (iv) *if E is a closed and bounded subset of U, then there is an open set O in U that contains E and a positive integer K such that*

$$\sum_{k=1}^{K} \phi_k(x) = 1$$

for all x in O.

---

[1] For a proof see, for instance, W. Rudin, *Functional Analysis*, McGraw-Hill, New York, 1973, pages 147–148.

The sequence $\{\phi_k\}$ defined above is called a *partition of unity for U subordinate to $\mathcal{O}$*. If we accept this lemma, then we can prove the following result that we shall need later.

Theorem 12.1 *If the support of a test function φ does not intersect the support of a distribution T, then $T(\varphi) = 0$.*

Proof. Let $\mathcal{O}$ be the collection of open sets in $\mathbb{R}^n$ on which T vanishes, let U be their union, and let $\{\phi_k\}$ be a partition of unity for U subordinate to $\mathcal{O}$. If φ is a test function with support in U, then, since this support is closed and bounded by definition, Lemma 12.1 (iv) guarantees the existence of a partition of unity $\{\phi_k\}$ for U subordinate to $\mathcal{O}$ and a positive integer K such that

$$1 = \sum_{k=1}^{K} \phi_k(x)$$

on the support of φ. If we multiply both sides of this identity by φ, apply T, and use linearity, we obtain

$$T(\varphi) = \sum_{k=1}^{K} T(\varphi \phi_k) = 0$$

since each ϕ_k has its support in some O in $\mathcal{O}$, Q.E.D.

Definition 12.6 *Two distributions S and T are said to be* **equal** *on an open set O in $\mathbb{R}^n$ if and only if $S(\varphi) = T(\varphi)$ for every test function φ with support in O. Then we write $S = T$ on O.*

Example 12.7 For any x in $\mathbb{R}^n$, $\delta^x = \delta$ on the open set O obtained from $\mathbb{R}^n$ by removing the origin and x. Indeed, if φ is a test function with support in O, then

$$\delta^x(\varphi) = \varphi(x) = 0 = \varphi(0) = \delta(\varphi).$$

It is customary in many books to identify a locally integrable function f with the regular distribution T_f, perhaps to create the illusion that the set of distributions contains the ordinary functions as a proper subset. We shall not do so here because—even if we choose to ignore the fact that the domains of these functions are radically different—this identification cannot be free of inconsistencies. For instance, let $n = 1$ and consider the locally integrable function defined by $f(x) = |x|$. Clearly, f vanishes at the origin, but T_f does not. To see this let O be a neighborhood of the origin and let $a > 0$ be so small that the interval $[-a, a]$ is in O. Then φ_a vanishes outside $[-a, a]$ but

$$T_f(\varphi_a) = \int_{-\infty}^{\infty} f \varphi_a > 0.$$

Next we shall be concerned with adding distributions and multiplying them by constants and functions. There is, of course nothing new about adding distributions since this is just adding functions. In the same way, multiplication of a constant by a function is a well-known operation, but in the case of distributions linearity shows that

$$(cT)(\varphi) = cT(\varphi) = T(c\varphi),$$

and this paves the way to define, by imitation, the product of a function by a distribution. Note, in this respect, that if φ is a test function, then so is $f\varphi$ by Example 12.2. Altogether, we have the following definition.

Definition 12.7 *If S and T are distributions, c is a real constant, and $f : \mathbb{R}^n \to \mathbb{R}$ is a function of class C^∞, then we define $S + T, cT$, and fT by*
   (i) $(S + T)(\varphi) = S(\varphi) + T(\varphi)$,
   (ii) $(cT)(\varphi) = T(c\varphi)$, *and*
   (iii) $(fT)(\varphi) = T(f\varphi)$
for all test functions φ.

From now on we shall write $-T$ instead of $(-1)T$ and $T - S$ instead of $T + (-S)$.

Theorem 12.2 *If S and T are distributions, c is a real constant, and $f : \mathbb{R}^n \to \mathbb{R}$ is a function of class C^∞, then $S + T, cT$, and fT are distributions.*

Proof. $T + S, cT$, and fT are clearly linear. To show that they are also continuous let $\varphi_k \to \varphi$ in $\mathcal{D}^n$ as $k \to \infty$. Then

$$(S + T)(\varphi_k) = S(\varphi_k) + T(\varphi_k) \to S(\varphi) + T(\varphi) = (S + T)(\varphi)$$

and

$$(cT)(\varphi_k) = T(c\varphi_k) \to T(c\varphi) = (cT)(\varphi),$$

which prove that $S + T$ and cT are continuous. To prove that fT is also continuous recall first that $f\varphi_k$ and $f\varphi$ are in $\mathcal{D}^n$ by Example 12.2. Then let E be a bounded set in $\mathbb{R}^n$ such that $f\varphi$ and all the $f\varphi_k$ vanish outside E. Such a set exists according to Definition 12.2. Now, if $l = (l_1, l_2, \ldots, l_n)$ and $m = (m_1, m_2, \ldots, m_n)$ are multi-indices, we shall write $l \leq m$ if and only if $l_i \leq m_i$ for $i = 1, 2, \ldots, n$ and define $m - l = (m_1 - l_1, m_2 - l_2, \ldots, m_n - l_n)$. It can be shown using induction that Leibniz rule

$$D^m(f\varphi_k) = \sum_{l \leq m} \binom{m_1}{l_1}\binom{m_2}{l_2}\cdots\binom{m_n}{l_n} D^{m-l} f \, D^l \varphi_k$$

holds for any derivative D^m. Then the fact that $D^l\varphi_k \to D^l\varphi$ uniformly on E implies that $D^l(f\varphi_k) \to D^l(f\varphi)$ uniformly on E, so that $f\varphi_k \to f\varphi$ in $\mathcal{D}^n$ as $k \to \infty$, and the continuity of T shows that

$$(fT)(\varphi_k) = T(f\varphi_k) \to T(f\varphi) = (fT)(\varphi),$$

Q.E.D.

The proof of the following result for regular distributions is left as an exercise.

Theorem 12.3 *Let* $f, g : \mathbb{R}^n \to \mathbb{R}$ *be locally integrable functions, let* T_f *and* T_g *be the regular distributions corresponding to* f *and* g, *respectively, and let* c *be a real constant. Then*

(i) $T_f + T_g = T_{f+g}$,

(ii) $c T_f = T_{cf}$, *and*

(iii) *if* f *is of class* C^∞, $f T_g = T_{fg}$.

We should remark at this point on the fact that there is no natural way to define the product of distributions in the general case. In fact, the usual product of functions $(TS)(\varphi) = T(\varphi)S(\varphi)$ is not linear and cannot be a distribution.

§12.4 Differentiation of Distributions

In this section and the next we shall begin to appreciate the advantages of distributions over functions. Every distribution has derivatives of all orders and every convergent sequence or series of distributions can be differentiated term by term any number of times, always and unconditionally.

Of course, we have to define the derivatives of a distribution before the predecing assertions can be verified, and for motivation we look at the case of regular distributions for $n = 1$. We would like the formula

$$(11) \qquad\qquad\qquad D^m T_f = T_{f^{(m)}}$$

to be valid if f is of class C^m on $\mathbb{R}$. Now, if φ is a test function and if $a > 0$ is so large that the interval $[-a, a]$ contains the support of φ, we have

$$T_{f^{(m)}}(\varphi) = \int_{-\infty}^{\infty} f^{(m)} \varphi = \int_{-a}^{a} f^{(m)} \varphi = f^{(m-1)}\varphi \Big|_{-a}^{a} - \int_{-a}^{a} f^{(m-1)}\varphi' = -\int_{-\infty}^{\infty} f^{(m-1)}\varphi'.$$

Repeated application of integration by parts yields

$$T_{f^{(m)}}(\varphi) = (-1)^m \int_{-\infty}^{\infty} f\varphi^{(m)} = (-1)^m T_f(\varphi^{(m)}).$$

Then (11) is valid if we define $D^m T_f$ by

$$(D^m T_f)(\varphi) = (-1)^m T_f(\varphi^{(m)}).$$

This motivates the following definition in the general case.

Definition 12.8 *Let* $m = (m_1, m_2, \ldots, m_n)$ *be a multi-index and define*

$$|m| = m_1 + m_2 + \cdots + m_n.$$

If T *is a distribution on* $\mathcal{D}^n$, *we define its* **mth derivative** $D^m T$ *by*

$$(12) \qquad\qquad\qquad D^m T(\varphi) = (-1)^{|m|} T(D^m \varphi)$$

for all φ *in* $\mathcal{D}^n$.

In particular, this justifies (11) if $n = 1$. In the general case of a regular distribution corresponding to a function f of class C^m on $\mathbb{R}^n$, integration by parts shows that

$$(-1)^{|m|} \int_{\mathbb{R}^n} f D^m \varphi = \int_{\mathbb{R}^n} (D^m f) \varphi,$$

that is, that

(13) $$D^m T_f = T_{D^m f}.$$

It is easy to show that $D^m T$ is linear and continuous, that is, a distribution (Exercise 4.1). Therefore, we can sum up as follows.

Theorem 12.4 *Every distribution T has derivatives of all orders given by (12) and, if $T = T_f$, where f is a function of class C^m on $\mathbb{R}^n$, by (13).*

Example 12.8 If m is a multi-index, (12) implies that

$$D^m \delta^x (\varphi) = (-1)^{|m|} \delta^x (D^m \varphi) = (-1)^{|m|} D^m \varphi(x).$$

In particular, when $n = 1$ and if we denote first derivative with a prime,

$$(\delta^x)'(\varphi) = -\varphi'(x)$$

and

$$\delta'(\varphi) = -\varphi'(0).$$

Example 12.9 If x is a real number, if $H^x : \mathbb{R} \to \mathbb{R}$ is defined by

$$H^x(s) = \begin{cases} 0 & \text{if} \quad s < x \\ 1 & \text{if} \quad s \geq x, \end{cases}$$

and if φ is a test function, we have

$$(T_{H^x})'(\varphi) = -T_{H^x}(\varphi') = -\int_{-\infty}^{\infty} H^x \varphi' = -\int_x^{\infty} \varphi'.$$

But φ vanishes on $[a, \infty)$ for some $a > x$, and then

$$(T_{H^x})'(\varphi) = -\int_x^a \varphi' = \varphi(x) - \varphi(a) = \varphi(x)$$

for any test function φ. That is, by comparison with Example 12.4,

$$(T_{H^x})' = \delta^x$$

and, in particular, if H is the Heaviside funtion of Example 8.6 (that is, Heaviside's **1**),

$$(T_H)' = \delta.$$

Definition 12.9 *If* $f : \mathbb{R}^n \rightarrow \mathbb{R}$ *is a locally integrable function and m is a multi-index, we say that* $D^m T_f$ *is the* **mth derivative of** f **in the distribution sense.**

Example 12.9 shows that δ is the derivative of H in the distribution sense, which justifies Heaviside's quotation in §12.1. Every function, differentiable or not, has a derivative in the distribution sense, but Equation (13) applies only when f is of class C^m.

Having defined differentiation we are in a position to exhibit a fundamental solution of a partial differential equation. We shall consider the wave equation.

Example 12.10 Define $u : \mathbb{R}^2 \rightarrow \mathbb{R}$ by

$$
u(x, t) = \begin{cases} \dfrac{1}{2c}[H(x + ct) - H(x - ct)] & \text{if } t > 0 \\ 0 & \text{if } t < 0, \end{cases}
$$

where $c > 0$ is a constant and H is the Heaviside function. Thus, u has value $1/2c$ in the region bound by the characteristics $x = ct$ and $x = -ct$, $t \geq 0$, containing the positive t-axis, and value zero elsewhere. It is discontinuous on these characteristics, but we shall show that it is a fundamental solution of the wave equation, that is, that it satisfies the partial differential equation

$$
u_{tt} - c^2 u_{xx} = \delta
$$

in the distribution sense. Indeed, according to Definition 12.8 and if we denote the partial derivatives of test functions with subscripts in the usual way,

$$
D^{(2,0)} T_u(\varphi) = (-1)^2 T_u(\varphi_{xx})
$$

$$
= \int_{\mathbb{R}^2} u(x, t) \varphi_{xx}(x, t)\, dx\, dt
$$

$$
= \frac{1}{2c} \int_0^\infty \left(\int_{-ct}^{ct} \varphi_{xx}(x, t)\, dx \right) dt
$$

$$
= \frac{1}{2c} \int_0^\infty [\varphi_x(ct, t) - \varphi_x(-ct, t)]\, dt.
$$

Also,

$$
D^{(0,2)} T_u(\varphi) = (-1)^2 T_u(\varphi_{tt})
$$

$$
= \int_{\mathbb{R}^2} u(x, t) \varphi_{tt}(x, t)\, dt\, dx
$$

$$
= \frac{1}{2c} \int_{-\infty}^0 \left(\int_{-x/c}^\infty \varphi_{tt}(x, t)\, dt \right) dx + \frac{1}{2c} \int_0^\infty \left(\int_{x/c}^\infty \varphi_{tt}(x, t)\, dt \right) dx
$$

$$= \frac{1}{2c} \int_{-\infty}^{0} [-\varphi_t(x, -x/c)] \, dx + \frac{1}{2c} \int_{0}^{\infty} [-\varphi_t(x, x/c)] \, dx$$

$$= \frac{1}{2} \int_{\infty}^{0} \varphi_t(-ct, t) \, dt - \frac{1}{2} \int_{0}^{\infty} \varphi_t(ct, t) \, dt.$$

But use of the chain rule gives

$$\frac{d}{dt} \varphi(-ct, t) = -c \varphi_x(-ct, t) + \varphi_t(-ct, t)$$

and

$$\frac{d}{dt} \varphi(ct, t) = c \varphi_x(ct, t) + \varphi_t(ct, t),$$

and then

$D^{(0,2)} T_u(\varphi)$

$$= \frac{1}{2} \int_{\infty}^{0} \left[\frac{d}{dt} \varphi(-ct, t) + c \varphi_x(-ct, t) \right] dt - \frac{1}{2} \int_{0}^{\infty} \left[\frac{d}{dt} \varphi(ct, t) - c \varphi_x(ct, t) \right] dt$$

$$= \frac{1}{2} \varphi(0, 0) + \frac{c}{2} \int_{\infty}^{0} \varphi_x(-ct, t) \, dt - \frac{1}{2} [-\varphi(0, 0)] + \frac{c}{2} \int_{0}^{\infty} \varphi_x(ct, t) \, dt$$

$$= \varphi(0, 0) + \frac{c}{2} \left[-\int_{0}^{\infty} \varphi_x(-ct, t) \, dt + \int_{0}^{\infty} \varphi_x(ct, t) \, dt \right]$$

$$= \delta(\varphi) + c^2 D^{(2,0)} T_u(\varphi).$$

Therefore,

$$D^{(0,2)} T_u - c^2 D^{(2,0)} T_u = \delta,$$

that is, $u_{tt} - c^2 u_{xx} = \delta$ in the distribution sense.

The basic rules of differentiation are very easy to establish.

Theorem 12.5 *Let S and T be distributions, let $f : \mathbb{R}^n \to \mathbb{R}$ be a function of class C^∞, let c be a real constant, and let $l = (l_1, l_2, \ldots, l_n)$ and $m = (m_1, m_2, \ldots, m_n)$ be multi-indices. Then*

(i) $D^m(S + T) = D^m S + D^m T,$

(ii) $D^m(cT) = c D^m T,$

(iii) $D^m(fT) = \sum_{l \le m} \binom{m_1}{l_1} \binom{m_2}{l_2} \cdots \binom{m_n}{l_n} D^{m-l} f \, D^l T,$ *and*

(iv) $D^l(D^m T) = D^{l+m} T = D^m(D^l T).$

Proof. The first two properties are easy consequences of Definition 12.8. The third is a tedious but simple computation using the Leibniz rule, and we shall prove only the one-dimensional form

$$(fT)' = f'T + fT'$$

as a sample.[1] If φ is an arbitrary test function, we have

$$(fT)'(\varphi) = -(fT)(\varphi') = (fT)(-\varphi') = T(-f\varphi')$$
$$= T[-(f\varphi)' + f'\varphi] = T[-(f\varphi)'] + T(f'\varphi)$$
$$= T'(f\varphi) + T(f'\varphi) = fT'(\varphi) + f'T(\varphi)$$
$$= (fT' + f'T)(\varphi).$$

Also,

$$[D^l(D^m T)](\varphi) = (-1)^{|l|} (D^m T)(D^l \varphi) = (-1)^{|l|} (-1)^{|m|} T(D^m(D^l \varphi))$$
$$= (-1)^{|l|+|m|} T(D^{m+l}\varphi) = (-1)^{|l+m|} T(D^{m+l}\varphi) = D^{l+m} T(\varphi)$$

for any test function φ, which proves (iv), Q.E.D.

It is now possible to differentiate a wide class of discontinuous functions in the distribution sense.

Theorem 12.6 *Let $f : \mathbb{R} \to \mathbb{R}$ be a bounded piecewise continuous function with a finite number of discontinuities at points $x_1, x_2, \ldots, x_k$, and assume that f is differentiable on any interval that includes none of the x_i. Then*

$$(T_f)' = T_{f'} + \sum_{i=1}^{k} [f(x_i+) - f(x_i-)] \delta^{x_i}.$$

Proof. The function $g : \mathbb{R} \to \mathbb{R}$ defined by

$$g = f - \sum_{i=1}^{k} [f(x_i+) - f(x_i-)] H^{x_i}$$

is continuous and such that $g' = f'$ except at the x_i. By Theorem 12.3,

$$T_f = T_g + \sum_{i=1}^{k} [f(x_i+) - f(x_i-)] T_{H^{x_i}},$$

and then, using Theorem 12.5 and the result of Example 12.9,

$$(T_f)' = (T_g)' + \sum_{i=1}^{k} [f(x_i+) - f(x_i-)] \delta^{x_i}.$$

Since g is continuous and has a piecewise continuous derivative, it is easy to see that $(T_g)' = T_{g'}$ (Exercise 4.2), Q.E.D.

---

[1] For the general case see, for instance, W. Rudin, *Functional Analysis*, McGraw-Hill, New York, 1973, page 145.

§12.5 Sequences and Series of Distributions

We start by defining convergence.

Definition 12.10 *A sequence of distributions* $\{T_k\}$ *is said to converge to a distribution* T *if and only if* $T_k(\varphi) \to T(\varphi)$ *in* $\mathbb{R}$ *as* $k \to \infty$ *for all test functions* φ. *Then we write* $T_k \to T$.

The same definition is valid for a series

$$\sum_{k=1}^{\infty} T_k,$$

which is defined in the usual manner as the sequence of its partial sums, and we write

$$\sum_{k=1}^{\infty} T_k = T.$$

Example 12.11 The sum

$$\sum_{i=-k}^{k} \delta^i$$

converges as $k \to \infty$. In fact, for any test function φ,

$$\left(\sum_{i=-k}^{k} \delta^i \right)(\varphi) = \sum_{i=-k}^{k} \delta^i(\varphi) = \sum_{i=-k}^{k} \varphi(i).$$

But there is a positive integer K such that φ vanishes outside $[-K, K]$, and then, for $k > K$,

$$\left(\sum_{i=-k}^{k} \delta^i \right)(\varphi) = \sum_{i=-K}^{K} \varphi(i) = \sum_{i=-\infty}^{\infty} \varphi(i).$$

Since the right-hand side is a constant the left-hand side converges to this constant. Since φ is arbitrary the given sum converges, and we denote its limit by

$$\sum_{k=-\infty}^{\infty} \delta^k.$$

It is easily shown that it is a distribution.

The proof of the following theorem is left as an exercise.

Theorem 12.7 *Let* $\{S_k\}$ *and* $\{T_k\}$ *be sequences of distributions that converge to distributons* S *and* T, *respectively, and let* c *be a real constant. Then*
  (i) $S_k + T_k \to S + T$ *and*
  (ii) $c T_k \to c T$
as $k \to \infty$.

We now pose the following question. If a sequence $\{f_k\}$ of locally integrable functions converges to a locally integrable function f, does the sequence $\{T_{f_k}\}$ converge to T_f? The answer is yes under a wide range of conditions but the proof uses of the n-dimensional version of the Lebesgue dominated convergence theorem of Exercise 9.15 of Chapter 2.[1]

Theorem 12.8 *If* $\{f_k\}$ *is a sequence of locally integrable functions on* $\mathbb{R}^n$ *that converges to a locally integrable function* f *(except, possibly, at a finite number of points) and if there is a constant* M *such that* $|f_k| \leq M$ *on* $\mathbb{R}^n$ *for all* k, *then* $T_{f_k} \to T_f$ *as* $k \to \infty$.

Proof. If φ is a test function and if R is a closed rectangle (an interval if $n = 1$) that contains the support of φ, then, as $k \to \infty$,

$$T_{f_k}(\varphi) = \int_R f_k \varphi \to \int_R f\varphi = T_f(\varphi)$$

by the n-dimensional version of the theorem in Exercise 9.15 of Chapter 2 since $f_k\varphi$ is bounded by $M \max \varphi$, Q.E.D.

Example 12.12 Fix a positive integer k and then, for each integer i such that $-k \leq i \leq k$, define $f_i : \mathbb{R} \to \mathbb{R}$ by

$$f_i(x) = \begin{cases} -\dfrac{\pi + (x - 2i\pi)}{2} & \text{if} \quad (2i-1)\pi < x < 2i\pi \\[2ex] \dfrac{\pi - (x - 2i\pi)}{2} & \text{if} \quad 2i\pi < x < (2i+1)\pi \\[2ex] 0 & \text{otherwise} \end{cases}$$

as represented in Figure 12.1. Then define

$$F_k = \sum_{i=-k}^{k} f_i.$$

Clearly, $|F_k| \leq \pi/2$ and, if f is the periodic extension of f_0 to $\mathbb{R}$ of period 2π, $F_k \to f$ on $\mathbb{R}$ as $k \to \infty$. Therefore, $T_{F_k} \to T_f$ as $k \to \infty$ by Theorem 12.8 with $M = \pi/2$.

---

[1] If $n = 1$, the only case in which we shall use this result, the following elementary theorem proved in 1885 by Cesare Arzelà (1847–1912), of the University of Bologna, can be used instead: if $\{f_k\}$ is a sequence of Riemann integrable functions on a closed interval I that converges to a Riemann integrable function f on I, if there is a constant M such that $|f_k| \leq M$ for all k, and if there is a finite number of points in I such that the convergence of $\{f_k\}$ is uniform on any subset of I not containing any of those points (this last condition, included to facilitate the proof, is actually superfluous and not part of Arzelà's theorem), then

$$\lim_{k\to\infty} \int_I f_k = \int_I f.$$

The reader should attempt a proof of this fact by enclosing the set of points at which the convergence of $\{f_k\}$ fails to be uniform with a set of arbitrarily small total length, but a proof can be found in T. M. Apostol, *Mathematical Analysis*, 2nd ed., Addison-Wesley, Reading, Massachusetts, 1974, pages 227–228.

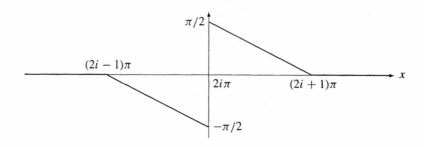

Figure 12.1

The fact that any distribution has derivatives of all orders already shows that distributions are pretty well behaved—much better than ordinary functions—and this is confirmed by the following result.

Theorem 12.9 *Any convergent sequence or series of distributions can be differentiated term by term any number of times; that is, if* $\{T_k\}$ *is a sequence of distributions that converges to a distribution* T, *then* $D^m T_k \to D^m T$ *as* $k \to \infty$ *for any multi-index m.*

Proof. For any test function φ,

$$D^m T_k(\varphi) = (-1)^{|m|} T_k(D^m \varphi) \to (-1)^{|m|} T(D^m \varphi) = D^m T(\varphi),$$

Q.E.D.

Example 12.13 If F_k and f are as in Example 12.12, then

$$(T_{F_k})' \to (T_f)'$$

as $k \to \infty$ and, according to the result of Exercise 4.7,

$$(T_{F_k})' = -\tfrac{1}{2} T_{H-(2k+1)\pi} + \tfrac{1}{2} T_{H(2k+1)\pi} + \pi \sum_{i=-k}^{k} \delta^{2i\pi}.$$

But

$$T_{H-(2k+1)\pi}(\varphi) = \int_{-(2k+1)\pi}^{\infty} \varphi \to \int_{-\infty}^{\infty} \varphi = T_1(\varphi),$$

while

$$T_{H(2k+1)\pi}(\varphi) = \int_{(2k+1)\pi}^{\infty} \varphi \to 0$$

as $k \to \infty$ for any test function φ. These limits together with the result of Example 12.11 show that

$$(T_f)' = \lim_{k \to \infty} (T_{F_k})' = -\tfrac{1}{2} T_1 + \pi \sum_{k=-\infty}^{\infty} \delta^{2k\pi}.$$

Example 12.14 The sequence $\{f_k\}$, where

$$f_k(x) = \frac{\sin kx}{\pi x},$$

diverges in the ordinary sense, but $f_k \to \delta$ in the distribution sense. To see this, fix $a < 0$ and define

$$F_k(x) = \int_a^x \frac{\sin kt}{\pi t}\, dt = \frac{1}{\pi} \int_{ka}^{kx} \frac{\sin u}{u}\, du.$$

It is known that

$$\int_0^x \frac{\sin u}{u}\, du$$

is bounded by its value at $x = \pi$ (see §2.6 of Chapter 2) and then, since the integrand is an even function,

$$|F_k(x)| \le \frac{2}{\pi} \int_0^\pi \frac{\sin u}{u}\, du$$

for all k. Now, if $x < 0$,

$$F_k(x) = \frac{1}{\pi} \int_a^x \frac{1}{t} \sin kt\, dt \to 0$$

as $k \to \infty$ by the Riemann-Lebesgue Theorem and, if $x > 0$,

$$F_k(x) \to \frac{1}{\pi} \int_{-\infty}^\infty \frac{\sin u}{u}\, du = 1$$

as $k \to \infty$. Therefore, F_k converges to the Heaviside function H as $k \to \infty$ (except at the origin), and $T_{F_k} \to T_H$ as $k \to \infty$ by Theorem 12.8. Then

$$T_{f_k} = T_{F_k'} = (T_{F_k})' \to (T_H)' = \delta$$

as $k \to \infty$ by Theorems 12.6 and 12.9 and the result of Example 12.9. This justifies Equation (5) for the delta function.

The preceding example shows that we can prove the convergence of a sequence of distributions by first integrating each term, then showing convergence, and finally differentiating each term.

§12.6 Convolution

Equation (6) and the solution already outlined for equation (9) suggest that convolution is an important operation when distributions are involved. As a guide to its definition we consider the case of regular distributions, and our goal is a definition that makes the equation $T_f * T_g = T_{f*g}$ true.[1]

---

[1] Our presentation of convolution is simpler than that of Schwartz, avoiding the introduction of the direct product of distributions.

If φ is a test function, we have

$$
\begin{aligned}
T_{f*g}(\varphi) &= \int_{\mathbb{R}^n} \left(\int_{\mathbb{R}^n} f(s)g(x-s)\,ds \right) \varphi(x)\,dx \\
&= \int_{\mathbb{R}^n} \left(\int_{\mathbb{R}^n} f(s)g(x-s)\varphi(x)\,ds \right) dx \\
&= \int_{\mathbb{R}^n} f(s) \left(\int_{\mathbb{R}^n} g(x-s)\varphi(x)\,dx \right) ds \\
&= \int_{\mathbb{R}^n} f(s) \left(\int_{\mathbb{R}^n} g(\sigma)\varphi(\sigma+s)\,d\sigma \right) ds,
\end{aligned}
$$

where the change of order of integration is justified by an extension of Theorem B.9 proved in 1907 by Guido G. Fubini (1879–1943).[1] If we define a test function φ^s by

$$
\varphi^s(x) = \varphi(x+s)
$$

and then $\varphi_{T_g} : \mathbb{R}^n \to \mathbb{R}$ by

$$
\varphi_{T_g}(s) = T_g(\varphi^s),
$$

we conclude that

$$
T_{f*g}(\varphi) = \int_{\mathbb{R}^n} f(s)T_g(\varphi^s)\,ds = \int_{\mathbb{R}^n} f(s)\varphi_{T_g}(s)\,ds = T_f(\varphi_{T_g}),
$$

provided, of course, that φ_{T_g} is a test function. We shall prove this below, and then, keeping in mind our goal to make the equation $T_f * T_g = T_{f*g}$ true, we would like to define the convolution of distributions in such a manner that

$$
T_f * T_g(\varphi) = T_f(\varphi_{T_g})
$$

for regular distributions. Then, in the general case of two distributions S and T, we would like to define

$$
S*T(\varphi) = S(\varphi_T),
$$

where $\varphi_T(x) = T(\varphi^x)$, provided, once again, that φ_T is a test function. We shall prove this fact, under sufficiently general conditions, in Lemma 12.3 below, but we need some preparatory work.

Lemma 12.2 *If φ is a test function, if s is fixed in $\mathbb{R}^n$, and if we define $\varphi^s(x) = \varphi(x+s)$ for all x in $\mathbb{R}^n$, then φ^s is of class C^∞ and, if m is a multi-index,*

$$
D^m \varphi^s = (D^m \varphi)^s.
$$

---

[1] Any proof requires knowledge of measure theory. See, for instance, H. L. Royden, *Real Analysis*, 2nd ed., Macmillan, New York, page 269.

Proof. Note that φ^s is of class C^∞ since it is just a translation of φ. The last equation is obvious for $|m| = 0$. Then, since any higher order derivative is an iteration of first-order derivatives, it is enough to prove the stated equation for one of these, say the ith first-order partial derivative, which we shall denote by D_i. Using the chain rule,

$$D_i \varphi^s(x) = D_i[\varphi(x+s)] = \sum_{j=1}^{n} (D_j \varphi)(x+s) D_i(x_j + s_j) = (D_i \varphi)(x+s) = (D_i \varphi)^s(x),$$

Q.E.D.

Definition 12.11 *For any test function φ and any distribution T, define a function $\varphi_T : \mathbb{R}^n \to \mathbb{R}$ by*

$$\varphi_T(x) = T(\varphi^x).$$

Example 12.15 For any test function φ and any s in $\mathbb{R}^n$ we have $\varphi_{\delta^s} = \varphi^s$ since

$$\varphi_{\delta^s}(x) = \delta^s(\varphi^x) = \varphi^x(s) = \varphi(s+x) = \varphi(x+s) = \varphi^s(x).$$

Lemma 12.3 *For any test function φ and any distribution T, φ_T is of class C^∞ on $\mathbb{R}^n$ and for any multi-index m with $|m| \neq 0$*

$$D^m \varphi_T = (D^m \varphi)_T = (-1)^{|m|} \varphi_{D^m T}.$$

If, in addition, T has bounded support, then φ_T is a test function.

Proof. Let x be fixed but arbitrary in $\mathbb{R}^n$. We shall show first that φ_T is continuous at x. If h is any vector in the unit ball in $\mathbb{R}^n$ centered at the origin, then the supports of the test functions φ^{x+h} are translations of the support of φ by no more than a certain finite amount and are, therefore, contained in a certain closed disc D. According to a theorem proved by Heine in 1872, a function that is continuous on a closed and bounded set is uniformly continuous there,[1] and then each φ^{x+h} is uniformly continuous on the larger disc consisting of all points of the form $s + h$ with s in D and $h \leq 1$. Hence, given any $\epsilon > 0$, we have

$$|\varphi^{x+h}(s) - \varphi^x(s)| = |\varphi^x(s+h) - \varphi^x(s)| < \epsilon$$

for all s in D if $\|h\| < 1$. This means that φ^{x+h} converges to φ^x uniformly on D as $h \to 0$. The same type of result is true, by an entirely analogous argument, if these functions are replaced with any of their partial derivatives, which means that the convergence just shown holds in $\mathcal{D}^n$. Then, by the continuity of T,

$$\varphi_T(x+h) - \varphi_T(x) = T(\varphi^{x+h}) - T(\varphi^x) = T(\varphi^{x+h} - \varphi^x) \to T(0) = 0$$

---

[1] For a proof see, for instance, W. Rudin, *Principles of Mathematical Analysis*, 3rd ed., McGraw-Hill, New York, 1976, page 91.

as $h \to 0$. This proves that φ_T is continuous at x.

As in the proof of Lemma 12.2, it is enough to prove the remaining statements about partial derivatives for any of the first order, say D_i. If e_i denotes the ith vector in the standard basis of $\mathbb{R}^n$ and if h denotes now any positive real number smaller than 1, then $D_i\varphi_T(x)$ is defined as the limit as $h \to 0$ of the quotient

$$\frac{\varphi_T(x + he_i) - \varphi_T(x)}{h} = \frac{T(\varphi^{x+he_i}) - T(\varphi^x)}{h} = T\left(\frac{\varphi^{x+he_i} - \varphi^x}{h}\right).$$

But as $h \to 0$

$$(14) \quad \frac{\varphi^{x+he_i}(s) - \varphi^x(s)}{h} = \frac{\varphi(s + x + he_i) - \varphi(s + x)}{h} \to D_i\varphi(s+x) = (D_i\varphi)^x(s)$$

for any s in $\mathbb{R}^n$, where the last equality follows from Lemma 12.2. Now, if D is as above, the convergence in (14) is uniform in D. Indeed, if D_{ii} denotes a second partial derivative in the ith direction, then Taylor's theorem with Lagrange's form of the remainder shows that

$$\varphi(s + x + he_i) = \varphi(s + x) + D_i\varphi(s + x)h + \tfrac{1}{2}D_{ii}\varphi(c)h^2$$

for some point c in the segment between $s + x$ and $s + x + he_i$. Since $D_{ii}\varphi$ is continuous and has bounded support, there is a positive constant M such that

$$\left|\frac{\varphi(s + x + he_i) - \varphi(s + x)}{h} - D_i\varphi(s + x)\right| = \tfrac{1}{2}|D_{ii}\varphi(c)h| \leq M|h|.$$

This shows that the left-hand side can be made smaller than any given $\epsilon > 0$ for all s in D by choosing h small enough; that is, the convergence in (14) is uniform on D. Since the same result can be shown after replacing φ with any of its partial derivatives, the convergence just shown holds in $\mathcal{D}^n$. Then, by the continuity of T,

$$D_i\varphi_T(x) = T[(D_i\varphi)^x] = (D_i\varphi)_T(x)$$

for any x in $\mathbb{R}^n$. This proves the first equality in the statement of the lemma for $m = 1$. To prove the second we use Lemma 12.2 and Definitions 12.8 and 12.11 to show that

$$T[(D_i\varphi)^x] = T(D_i\varphi^x) = -D_iT(\varphi^x) = -\varphi_{D_iT}(x).$$

It remains to show that, if T has bounded support, then so does φ_T. The support of φ^x is a translation of the support of φ, and it will not intersect the support of T if $\|x\|$ is large enough. Then, for such an x,

$$\varphi_T(x) = T(\varphi^x) = 0$$

by Theorem 12.1, showing that φ_T vanishes outside a bounded set, Q.E.D.

Lemma 12.3 implies that, if T is a distribution with bounded support, then $S(\varphi_T)$ is well defined for any distribution S and any test function φ. If T does not have bounded support, but S does, it is still possible to attach a meaning to the expression $S(\varphi_T)$ as follows. Since φ_T is of class C^∞ it is enough to show that any distribution S with bounded support can be extended to the set of all real-valued functions of class C^∞ on $\mathbb{R}^n$, which we denote by $C^\infty(\mathbb{R}^n)$, as a continuous linear functional. To do this let supp(S) denote the support of S, let R be the radius of the smallest closed disc containing supp(S) and centered at the origin (we take $R = 0$ if supp(S) is the origin), and then define a test function ϕ by

$$\phi(x) = \begin{cases} 1 & \text{if } \|x\| \le R+1 \\ e^{1-(2R+3)/[(R+2)^2-\|x\|^2]} & \text{if } R+1 \le \|x\| < R+2 \\ 0 & \text{if } \|x\| \ge R+2 \end{cases}$$

so that $\phi \equiv 1$ on supp(S). Note that ϕ depends on S although we have chosen not to encumber the notation with this fact. Also, if φ is a test function, then $\varphi - \phi\varphi \equiv 0$ on supp(S) and, according to Theorem 12.1, $S(\varphi) - S(\phi\varphi) = S(\varphi - \phi\varphi) = 0$. That is, $S(\varphi) = S(\phi\varphi)$ for any test function φ, and this allows us to extend the definition of S to $C^\infty(\mathbb{R}^n)$ by

$$S(f) = S(\phi f)$$

for every f in $C^\infty(\mathbb{R}^n)$. We shall show next that this extension is linear and continuous.

Lemma 12.4 *The extension of a distribution S with bounded support to $C^\infty(\mathbb{R}^n)$ defined by*

$$S(f) = S(\phi f),$$

where ϕ is as above and f is in $C^\infty(\mathbb{R}^n)$, is linear. Also, if $\{f_k\}$ is a sequence in $C^\infty(\mathbb{R}^n)$ such that $D^m f_k \to D^m f$ as $k \to \infty$ for every multi-index m, we have

$$S(f_k) \to S(f)$$

as $k \to \infty$.

Proof. This extension is clearly linear because if f and g are in $C^\infty(\mathbb{R}^n)$, then

$$S(f+g) = S(\phi(f+g)) = S(\phi f + \phi g) = S(\phi f) + S(\phi g) = S(f) + S(g)$$

and

$$S(cf) = S(\phi(cf)) = S(c(\phi f)) = cS(\phi f) = cS(f).$$

Now, each $D^m f_k \to D^m f$ uniformly on supp(S). It is enough to prove this for $|m| = 0$ and the same proof applies to the mth derivatives. If, contrary to the assertion, the convergence of $f_k \to f$ on supp(S) is not uniform, then, for every $\epsilon > 0$, there is a subsequence $\{f_{k_i}\}$ of $\{f_k\}$ and corresponding points x_i in supp(S) such that

$$|f_{k_i}(x_i) - f(x_i)| \ge \epsilon.$$

But supp(S) is bounded and then the x_i have a limit point x in supp(S) (refer to Cantor's definition in §2.9) according to the *Bolzano-Weierstrass theorem*, which states that every bounded infinite set has a limit point. The one-dimensional version of this result was established by Bolzano in 1817 in the process of proving the intermediate value theorem (see the comments preceding Lemma A.1), but it was reproved in its general form, with due credit to Bolzano, by Weierstrass in the 1860s.[1] We assume, by taking a subsequence if necessary, that $x_i \to x$ as $i \to \infty$. Then

$$|f_{k_i}(x_i) - f(x_i)| \le |f_{k_i}(x_i) - f_{k_i}(x)| + |f_{k_i}(x) - f(x)| + |f(x) - f(x_i)|.$$

Now, $f_{k_i} \to f$ and $x_i \to x$ as $i \to \infty$, and the last two terms on the right approach zero. So does the first term because the mean value theorem implies that there is a point c_i in the segment from x_i to x such that

$$f_{k_i}(x_i) - f_{k_i}(x) = \sum_{j=1}^{k} D_j f_{k_i}(c_i)(x_i - x)_j,$$

where $(x_i - x)_j$ is the jth component of $x_i - x$. The right-hand side in the preceding equation approaches zero as $x_i \to x$ unless some partial derivative is unbounded near x, and this is not possible because $D_j f_{k_i} \to D_j f$ as $i \to \infty$ for each j and $D_j f$ is bounded near x since it is continuous on a closed disc centered at x. This proves that $f_k \to f$ uniformly on supp(S) and, more generally, that $D^m f_k \to D^m f$ uniformly on supp(S) for any multi-index m. Then the Leibniz rule shows that $D^m(\phi f_k) \to D^m(\phi f)$ uniformly on supp(S). That is, $\phi f_k \to \phi f$ in $\mathcal{D}^n$, and then the continuity of S implies that

$$S(f_k) = S(\phi f_k) \to S(\phi f) = S(f),$$

Q.E.D.

Example 12.16   The support of δ^s consists of the single point s, and then $\phi(s) = 1$ implies that the extension of δ^s to $C^\infty(\mathbb{R}^n)$ is given by

$$\delta^s(f) = \delta^s(\phi f) = (\phi f)(s) = \phi(s) f(s) = f(s).$$

Definition 12.12   *If S and T are distributions such that at least one of them has bounded support, then the **convolution** of S and T is the function $S * T : \mathcal{D}^n \to \mathbb{R}$ defined by*

$$S * T(\varphi) = S(\varphi_T),$$

where it is implicitly assumed that S is replaced with its extension to $C^\infty(\mathbb{R}^n)$ if the support of T is not bounded.

---

[1] For a proof see, for instance, W. Rudin, *Principles of Mathematical Analysis*, 3rd ed., McGraw-Hill, New York, 1976, page 40.

This definition and the discussion at the beginning of this section show that

$$T_f * T_g(\varphi) = T_f(\varphi_{T_g}) = T_{f*g}(\varphi)$$

for every test function φ, that is, that $T_f * T_g = T_{f*g}$.

We need only prove some additional elementary properties of the function φ_T to show that $S * T$ is a distribution.

Lemma 12.5 *Let φ and ϕ be test functions, let c be a real constant, let S and T be distributions, and let $\{\varphi_k\}$ be a sequence of test functions such that $\varphi_k \to \varphi$ in $\mathcal{D}^n$ as $k \to \infty$. Then*

(i) $(\varphi + \phi)_T = \varphi_T + \phi_T$,

(ii) $(c\varphi)_T = c\varphi_T = \varphi_{cT}$,

(iii) $\varphi_{S+T} = \varphi_S + \varphi_T$, *and*

(iv) *for any multi-index m, $D^m(\varphi_k)_T \to D^m \varphi_T$ as $k \to \infty$.*

Proof. Parts (i) to (iii) are straightforward from the definition of φ_T and the linearity of T. To prove (iv) note that the convergence of $\varphi_k \to \varphi$ in $\mathcal{D}^n$ implies the convergence of their translations $\varphi_k^x \to \varphi^x$ in $\mathcal{D}^n$, and then Lemmas 12.3 and 12.2 and the continuity of T imply that

$$D^m(\varphi_k)_T(x) = (D^m \varphi_k)_T(x) = T[(D^m \varphi_k)^x] = T(D^m \varphi_k^x) \to T(D^m \varphi^x)$$

as $k \to \infty$ for any multi-index m and any x in $\mathbb{R}^n$. Similar steps applied to φ show that

$$T(D^m \varphi^x) = D^m \varphi_T(x),$$

which shows that $D^m(\varphi_k)_T \to D^m \varphi_T$ as $k \to \infty$ for any multi-index m, Q.E.D.

Theorem 12.10 *If S and T are distributions such that at least one of them has bounded support, then their convolution $S * T$ is a distribution.*

Proof. Lemma 12.5(i) and the linearity of S give

$$S * T(\varphi + \phi) = S[(\varphi + \phi)_T] = S(\varphi_T + \phi_T) = S(\varphi_T) + S(\phi_T) = S * T(\varphi) + S * T(\phi)$$

for all test functions φ and ϕ. Similarly,

$$S * T(c\varphi) = c(S * T)(\varphi)$$

for any real constant c and any test function φ. This shows that $S * T$ is linear. Also, if $\varphi_k \to \varphi$ in $\mathcal{D}^n$, then $D^m(\varphi_k)_T \to D^m \varphi_T$ as $k \to \infty$ by Lemma 12.5(iv) and then Lemma 12.4 implies that

$$S * T(\varphi_k) = S[(\varphi_k)_T] \to S(\varphi_T) = S * T(\varphi)$$

for any test function φ, which shows that $S * T$ is continuous, Q.E.D.

It is possible to prove that the convolution of any two distributions is commutative if at least one of them has bounded support. This is far from trivial for arbitrary distributions, but we shall not need such a general result in our presentation.[1] For regular distributions it follows from the fact that the convolution of functions is commutative, and then

$$T_f * T_g = T_{f*g} = T_{g*f} = T_g * T_f.$$

In addition, it is easily shown as part of the next example that the convolution of δ^x with any distribution is commutative.

Example 12.17 For any x in $\mathbb{R}^n$, any distribution T, and any test function φ, we have

$$T * \delta^x(\varphi) = T(\varphi_{\delta^x}) = T(\varphi^x)$$

using the result of Example 12.15. If we define T^x by $T^x(\varphi) = T(\varphi^x)$, we have

$$T * \delta^x = T^x.$$

Also,

$$\delta^x * T(\varphi) = \delta^x(\varphi_T) = \varphi_T(x) = T(\varphi^x)$$

using Example 12.16. But the right-hand side equals $T^x(\varphi)$, and then

$$\delta^x * T = T^x.$$

This shows that δ^x commutes with any distribution. In particular, putting $x = 0$ we obtain

$$\delta * T = T * \delta = T$$

for every distribution T. Thus, we can think of δ as the convolution identity.

We shall need, and can easily prove, some additional properties of the convolution of distributions.

Theorem 12.11 *Let S, T, and U be distributions such that at least two of them have bounded supports. Then*
(i) $(S + T) * U = S * U + T * U$,
(ii) $S * (T + U) = S * T + S * U$, *and*
(iii) *for any real constant c, $c(S * T) = (cS) * T = S * (cT)$.*

Proof. The hypothesis on the supports of the distributions implies that all the convolutions stated above are well defined. Now, for any test function φ,

$$(S + T) * U(\varphi) = (S + T)(\varphi_U) = S(\varphi_U) + T(\varphi_U) = S * U(\varphi) + T * U(\varphi),$$

which proves (i), and the proof of (ii) is similar. Also,

$$c(S * T)(\varphi) = c[S(\varphi_T)] = (cS)(\varphi_T) = (cS) * T(\varphi)$$

---

[1] For a proof, way beyond the level of this book, see W. Rudin, *Functional Analysis*, McGraw-Hill, New York, 1973, pages 160–161.

and, using Lemma 12.5 (ii),

$$c(S * T)(\varphi) = (S * T)(c\varphi) = S[(c\varphi)_T] = S(\varphi_{cT}) = S * (cT)(\varphi),$$

which proves (iii), Q.E.D.

To prove our next result we need to extend the definition of $D^m S$ to $C^\infty(\mathbb{R}^n)$ when S has compact support. Since $D^m S$ is a distribution it can be extended in the usual way: if f is in $C^\infty(\mathbb{R}^n)$ and ϕ is as defined above before Lemma 12.4,

$$D^m S(f) = D^m S(\phi f) = (-1)^{|m|} S[D^m(\phi f)].$$

According to the Leibniz rule there are constants c_{ml} such that

$$D^m(\phi f) = \sum_{l \leq m} c_{ml} D^{m-l}\phi D^l f.$$

But $\phi \equiv 1$ on the support of S and then $D^{m-l}\phi \equiv 0$ there for all $l < m$. Thus,

$$D^m S(f) = (-1)^{|m|} S(\phi D^m f) = (-1)^{|m|} S(D^m f),$$

which coincides with the definition of $D^m S$ on $\mathcal{D}^n$.

Theorem 12.12 *If S and T are distributions, at least one of which has bounded support, and if m is a multi-index,*

$$D^m(S * T) = D^m S * T = S * D^m T.$$

Proof. For any test function φ,

$$D^m(S * T)(\varphi) = (-1)^{|m|}(S * T)(D^m \varphi) = (-1)^{|m|} S[(D^m \varphi)_T].$$

Using the first equality in Lemma 12.3, we see that the right-hand side equals

$$(-1)^{|m|} S(D^m \varphi_T) = D^m S(\varphi_T) = D^m S * T(\varphi),$$

proving that $D^m(S * T) = D^m S * T$. Using now the second equality in Lemma 12.3,

$$(-1)^{|m|} S[(D^m \varphi)_T] = (-1)^{|m|} S[(-1)^{|m|} \varphi_{D^m T}] = S(\varphi_{D^m T}) = S * D^m T(\varphi),$$

which proves that $D^m(S * T) = S * D^m T$, Q.E.D.

Example 12.18 For any distribution T and any multi-index m, the result of Example 12.17 and the preceding theorem give

$$D^m \delta * T = \delta * D^m T = D^m T * \delta = T * D^m \delta.$$

§12.7   The Poisson Equation on the Sphere

We are now ready to apply our knowledge of distributions to some problems in partial differential equations. Let $L(u)$ denote a linear partial differential operator with constant coefficients. For example,

$$L(u) = u_{xx} + u_{yy} + u_{zz}$$

gives the Laplace operator in $\mathbb{R}^3$. In general, if $u : \mathbb{R}^n \to \mathbb{R}$, $L(u)$ is of the form

$$L(u) = \sum_{|m| \le M} c_m D^m u,$$

where m is a multi-index, M is a fixed positive integer and the c_m are real constants.

Our first type of problem consists of solving the partial differential equation

$$(15) \qquad\qquad\qquad\qquad L(u) = f$$

in $\mathbb{R}^n$, where $f : \mathbb{R}^n \to \mathbb{R}$ is a locally integrable function. We know that unless f satisfies some differentiability conditions the solution need not exist in the ordinary sense, so we shall look for a solution in the distribution sense. We are already aware of Green's work and of the concept of fundamental solution, introduced in §12.1, that we now define precisely.

Definition 12.13 *A* **fundamental solution** *of* (15) *is a distribution T such that* $L(T) = \delta$.

We have already shown, in Example 12.10 a fundamental solution of the one dimensional wave equation. As we already mentioned in §12.1, Leon Ehrenpreis in 1954 and independently Bernard Malgrange in 1955 proved that a fundamental solution of (15) always exists.[1] Such a solution is important because, as already suggested in §12.1, a solution of (15) in the distribution sense can always be found from it. We now formally prove this fact.

Theorem 12.13 *If T is a fundamental solution of* (15) *and either T or f has bounded support, then $T_f * T$ is a solution of* (15).[2]

Proof. It should be clear that, if f has bounded support, so does T_f and then, according to Theorem 12.12, $D^m (T_f * T) = T_f * D^m T$. Using now Theorem 12.11 (ii),

$$L(T_f * T) = \sum_{|m| \le M} c_m D^m (T_f * T) = \sum_{|m| \le M} c_m T_f * D^m T = T_f * \left(\sum_{|m| \le M} c_m D^m T \right)$$

$$= T_f * L(T) = T_f * \delta = T_f,$$

where the last equality was proved in Example 12.17, Q.E.D.

---

[1] For a proof of this fact see, for instance, W. Rudin, *Functional Analysis*, McGraw-Hill, New York, 1973, Theorem 8.5.

[2] The definition of a fundamental solution and Theorem 12.13 are extended in Exercise 7.5 to the case in which δ is replaced with δ^x.

Now let S be a sphere in $\mathbb{R}^3$ of radius a and centered at the origin and let D denote the interior of S. We pose the problem of finding a solution $u : D \cup S \to \mathbb{R}$ of

$$(16) \qquad\qquad u_{xx} + u_{yy} + u_{zz} = f \qquad \text{in } D$$

$$(17) \qquad\qquad u \equiv g \qquad\qquad \text{on } S,$$

where $f, g : S \to \mathbb{R}$ are continuous functions. It is enough to find a solution U of (16) with zero boundary values because, if v is a solution of the Laplace equation such that $v \equiv g$ on S (this solution was obtained in Theorem 11.7), then their sum $U + v$ is a solution of (16)–(17).

From now on we shall find it convenient to use the Laplacian operator Δ defined by

$$\Delta u = u_{xx} + u_{yy} + u_{zz}.$$

Postponing any consideration of boundary values for later, we shall now search for a solution of $\Delta u = f$, which according to Theorem 12.13 can be found from a fundamental solution. We would prefer this fundamental solution to be a regular distribution; that is, we would like to find a locally integrable function F such that $\Delta T_F = \delta$. Well, this seems to be rather a tall order, so we shall follow the usual advice in such cases: solve a simpler but related problem and see if this will provide any insight into the more difficult problem. A simpler problem would be to find a harmonic function in $\mathbb{R}^3$, and obvious symmetry reasons suggest that we should look for a function F that depends only on

$$r = \sqrt{x^2 + y^2 + z^2},$$

that is, a function of the form $F(s) = \Phi(r)$, where s shall denote the variable in $\mathbb{R}^n$ through the end of this section. If there is such a function,

$$F_x = \Phi'(r)\frac{\partial r}{\partial x} = \Phi'(r)\frac{x}{r}$$

and

$$F_{xx} = \Phi''(r)\frac{x^2}{r^2} + \Phi'(r)\left(\frac{1}{r} - \frac{x^2}{r^3}\right),$$

and we obtain similar equations for the partial derivatives with respect to y and z. Note that these equations are valid only outside the origin, a point at which these derivatives blow up. But this is good since it suggests the presence of a delta function at the origin, which is precisely what we are looking for in our search for a fundamental solution. So, we proceed. Adding the last partial derivative to similar derivatives with respect to y and z, we obtain

$$\Delta F = \Phi''(r) + \frac{2}{r}\Phi'(r)$$

and, since F is to be harmonic outside the origin,

$$r^2\Phi''(r) + 2r\Phi'(r) = 0.$$

This is a Cauchy-Euler equation whose general solution is

$$\Phi(r) = \frac{C}{r} + K,$$

where C and K are arbitrary constants. In particular, we can take $K = 0$ and then $F(s) = \Phi(r) = C/r$ is harmonic outside the origin for every constant C and blows up at the origin. On this basis we conjecture that there is a value of C such that the distribution $T_F : \mathcal{D}^3 \to \mathbb{R}$ defined by

$$T_F(\varphi) = \int_{\mathbb{R}^3} F\varphi$$

for any test function φ in $\mathcal{D}^3$ is a fundamental solution of $\Delta u = f$. Note that T_F is correctly defined as a distribution bacause F is locally integrable even though it blows up at the origin. This is clear away from the origin and, if B_ϵ is the open ball of radius ϵ centered at the origin, Gauss pointed out in his *Allgemeine Lehrsätze ...* of 1839 that the integral of F over B_ϵ can be considered proper (*eine wahre Integration*), since a change to polar coordinates yields

$$\int_{B_\epsilon} F = C \int_0^\epsilon \int_0^\pi \int_0^{2\pi} r \sin\phi \, d\theta \, d\phi \, dr = 2\pi\epsilon^2 C.$$

In contemporary terms the precise definition of T_F as an improper integral is

(18)
$$T_F(\varphi) = \lim_{\epsilon \to 0} \int_{\mathbb{R}^3 - B_\epsilon} F\varphi,$$

and to verify that it is a fundamental solution note that

(19)
$$\Delta T_F(\varphi) = (-1)^2 T_F(\Delta\varphi) = \lim_{\epsilon \to 0} \int_{\mathbb{R}^3 - B_\epsilon} F\Delta\varphi$$

by Definition 12.8. To evaluate the right-hand side recall that $\Delta F \equiv 0$ on $\mathbb{R}^3 - B_\epsilon$, and then, if S_ϵ is the sphere bounding B_ϵ and if $R > \epsilon$ is so large that $\varphi \equiv 0$ on and outside the sphere S_R of radius R centered at the origin, we can apply Green's theorem (*ii*) of Exercise 1.6 of Chapter 9 to the region V between S_ϵ and S_R to obtain

$$\int_{\mathbb{R}^3 - B_\epsilon} F\Delta\varphi = \int_V F\Delta\varphi = \int_V (F\Delta\varphi - \varphi\Delta F) = \int_{S_\epsilon} (FD_n\varphi - \varphi D_n F),$$

where D_n denotes derivative in the direction of the outward normal unit vector on S_ϵ.

The integral on the right-hand side is then split into the difference of integrals in the usual way, and the first of these is easy to deal with. If M_ϵ is the maximum value of $|D_n\varphi|$ on S_ϵ,

$$\left| \int_{S_\epsilon} FD_n\varphi \right| = \left| \int_{S_\epsilon} \frac{C}{\epsilon} D_n\varphi \right| \leq \frac{|C|M_\epsilon}{\epsilon} \int_{S_\epsilon} 1 = 4\pi|C|\epsilon M_\epsilon,$$

which approaches zero as $\epsilon \to 0$. To evaluate the second integral note that if s is any point of S_ϵ,

$$D_n F(s) = -\frac{\partial}{\partial r} \Phi(r)\bigg|_{r=\epsilon} = \frac{C}{\epsilon^2}.$$

Thus, if A_ϵ denotes the average value of φ on S_ϵ,

$$\int_{S_\epsilon} \varphi D_n F = \frac{C}{\epsilon^2} \int_{S_\epsilon} \varphi = \frac{C}{\epsilon^2} 4\pi \epsilon^2 A_\epsilon = 4\pi C A_\epsilon.$$

Since $A_\epsilon \to \varphi(0)$ as $\epsilon \to 0$, we conclude from (19) that

$$\Delta T_F(\varphi) = -4\pi C \varphi(0) = -4\pi C \delta(\varphi)$$

for all test functions φ, that is, $\Delta T_F = \delta$ if we choose $C = -1/4\pi$, which shows that

$$F(s) = -\frac{1}{4\pi r}$$

is a fundamental solution of $\Delta u = f$ in the distribution sense.

According now to Theorem 12.13 and given that f has support in $D \cup S$, $T_f * T_F$ is a solution of $\Delta u = f$. But for any test function φ,

$$T_f * T_F(\varphi) = T_f(\varphi_{T_F})$$

$$= \int_{\mathbb{R}^3} f(\sigma)\varphi_{T_F}(\sigma)\,d\sigma$$

$$= \int_D f(\sigma) T_F(\varphi^\sigma)\,d\sigma$$

$$= \int_D f(\sigma)\left(\int_{\mathbb{R}^3} F(\eta)\varphi(\eta + \sigma)\,d\eta\right) d\sigma$$

$$= \int_D f(\sigma)\left(\int_{\mathbb{R}^3} F(s - \sigma)\varphi(s)\,ds\right) d\sigma.$$

The order of integration can be reversed by Fubini's theorem, and we obtain

$$T_f * T_F(\varphi) = \int_{\mathbb{R}^3}\left(\int_D f(\sigma) F(s - \sigma)\,d\sigma\right)\varphi(s)\,ds,$$

showing that

$$\int_D f(\sigma) F(s - \sigma)\,d\sigma = -\frac{1}{4\pi}\int_D \frac{f(\sigma)}{\|s - \sigma\|}\,d\sigma$$

is a solution of $\Delta u = f$ in D in the distribution sense.

However, this solution does not vanish on S because f is arbitrary and $F(s - \sigma) \neq 0$ on S. The simplest way to overcome this difficulty is to replace $F(s - \sigma)$ in the preceding equation with a function of the form

$$G(s, \sigma) = F(s - \sigma) + w(s, \sigma),$$

where w is a harmonic function of s for each fixed σ in D and such that $G(s, \sigma) = 0$ for all s in S. Then, if we could find such a function w, differentiation under the integral sign, which is permissible because f is integrable and w has continuous partial derivatives, would show that

$$\Delta \int_D f(\sigma)w(s, \sigma)\, d\sigma = \int_D f(\sigma)\Delta w(s, \sigma)\, d\sigma = 0,$$

where Δ is the Laplacian operator with respect to s, and then

$$\Delta \int_D f(\sigma)G(s, \sigma)\, d\sigma = \Delta \int_D f(\sigma)F(s - \sigma)\, d\sigma = f(s)$$

in the distribution sense for σ in D. Since the integral on the left vanishes on S, it would provide a solution of our problem.

It is easy to think of one such possible w. If $\sigma \neq 0$ is fixed, if $\tilde\sigma$ is a point depending on σ but outside S, that is, such that $\|\tilde\sigma\| > a$, and if C is a constant, then the function

$$w(s, \sigma) = \frac{C}{4\pi \|s - \tilde\sigma\|},$$

which is of the same form as F, is easily shown to be harmonic in D by computing its partial derivatives. It remains to choose $\tilde\sigma$ and $C > 0$ such that

$$G(s, \sigma) = \frac{1}{4\pi}\left[\frac{C}{\|s - \tilde\sigma\|} - \frac{1}{\|s - \sigma\|}\right] = 0$$

for s in S. This is possible because the geometry of the sphere is very simple. In general, when the boundary of the domain of f is an arbitrary surface, it is not possible to find such a G.

To start with, choose $\tilde\sigma$ to be collinear with σ and the origin and let θ be the angle between s and σ as shown in Figure 12.2. Then $\tilde\sigma = k\sigma$ for some constant $k > a/\|\sigma\|$, and for $\|s\| = a$ we want

$$C\|s - \sigma\| = \|s - \tilde\sigma\| = \|s - k\sigma\|,$$

that is,

$$C^2\|s - \sigma\|^2 = \|s - k\sigma\|^2.$$

Using the law of cosines this is equivalent to

$$C^2(\|s\|^2 + \|\sigma\|^2 - 2\|s\|\|\sigma\|\cos\theta) = \|s\|^2 + k^2\|\sigma\|^2 - 2k\|s\|\|\sigma\|\cos\theta.$$

Putting $\|s\| = a$ and given that this equation must be satisfied for any angle θ, we must have

$$C^2(a^2 + \|\sigma\|^2) = a^2 + k^2\|\sigma\|^2 \qquad \text{and} \qquad C^2 = k.$$

The first of these equations then takes the form

$$k^2\|\sigma\|^2 - k(a^2 + \|\sigma\|^2) + a^2 = 0,$$

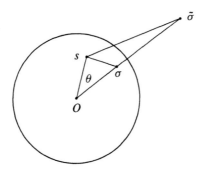

Figure 12.2

whose only solution larger than $a/\|\sigma\|$ is $k = a^2/\|\sigma\|^2$. Then $C = a/\|\sigma\|$ and

(20)
$$G(s, \sigma) = \frac{1}{4\pi} \left[\frac{a}{\|\sigma\| \|s - \tilde{\sigma}\|} - \frac{1}{\|s - \sigma\|} \right].$$

This is called the *Green function for the Poisson equation on the sphere*. The function U defined by

$$U(s) = \int_D f(\sigma) G(s, \sigma)\, d\sigma$$

is a solution of $\Delta u = f$ in D in the distribution sense that vanishes on S. When added to the solution of Theorem 11.7 with boundary values given by g, it provides a solution of the boundary value problem (16)–(17). However, the continuity of g is a requirement in Theorem 11.7. If g is only locally integrable, it is possible to use the Green function for the Poisson equation on the sphere to arrive at the same solution in the distribution sense (see Exercise 7.1).

§12.8 Distributions Depending on a Parameter

In applications to partial differential equations involving time, such as the heat or wave equations, we will have to deal with a family of distributions $\{T[t]\}$ depending on a real parameter $t \geq 0$.[1] In the rest of this section we shall implicitly assume this restriction on t.

Definition 12.14 *A family of distributions $\{T[t]\}$ is said to converge to a distribution T as $t \to t_0$ (it is allowed to replace t_0 with ∞) if and only if*

$$T[t_k](\varphi) \to T(\varphi)$$

---

[1] Such families of distributions are not considered in Schwartz' book. Our presentation in this section is partly based on that by I. M. Gel'fand and G. E. Shilov, *Generalized Functions*, Vol. 1, Academic Press, New York, 1964.

for any test function φ and any sequence $\{t_k\}$ such that $t_k \to t_0$ as $k \to \infty$. Then we write

$$\lim_{t \to t_0} T[t] = T$$

or $T[t] \to T$ as $t \to t_0$.

The proof of the following result is straightforward.

Theorem 12.14 *Let $\{S[t]\}$ and $\{T[t]\}$ be families of distributions depending on a parameter t. If $S[t] \to S$ and $T[t] \to T$ as $t \to t_0$, where S and T are distributions, and if c is a real constant, then*
  (i) *$S[t] + T[t] \to S + T$ and*
  (ii) *$cT[t] \to cT$*
as $t \to t_0$.

Lemma 12.6 *Let T be a distribution and let $\{T[t]\}$ be a family of distributions such that $T[t] \to T$ as $t \to t_0$. Then, for any test function φ and any sequence $\{t_k\}$ such that $t_k \to t_0$ as $k \to \infty$, we have*

$$D^m \varphi_{T[t_k]} \to D^m \varphi_T$$

as $k \to \infty$ in $C^\infty(\mathbb{R}^n)$ for any multi-index m. If, in addition, all $T[t]$ have support in the same bounded set, then

$$\varphi_{T[t_k]} \to \varphi_T$$

in $\mathcal{D}^n$.

Proof. For any multi-index m, Lemmas 12.3 and 12.2 impliy that

$$D^m \varphi_{T[t_k]}(x) = (D^m \varphi)_{T[t_k]}(x) = T[t_k](D^m \varphi^x)$$
$$\to T(D^m \varphi^x) = (D^m \varphi)_T(x) = D^m \varphi_T(x)$$

as $k \to \infty$ for all x in $\mathbb{R}^n$. If, in addition, all $T[t_k]$ and T have support in the same bounded set, then so do all the $\varphi_{T[t_k]}$ and φ. Then the convergence shown above is uniform on their common bounded support by the same argument used in the proof of Lemma 12.4, Q.E.D.

Theorem 12.15 *Let S and T be distributions, let $\{T[t]\}$ be a family of distributions such that $T[t] \to T$ as $t \to t_0$, and assume that either*
  (i) *S has bounded support or*
  (ii) *all the $T[t]$ have their supports in the same bounded set.*
Then

$$S * T[t] \to S * T$$

as $t \to t_0$.

Proof. Let $\{t_k\}$ be a sequence that converges to t_0 as $k \to \infty$. In case (i) $D^m \varphi_{T[t_k]} \to D^m \varphi_T$ in $C^\infty(\mathbb{R}^n)$ by Lemma 12.6 for any multi-index m and, since the extension of S to $C^\infty(\mathbb{R}^n)$ is continuous by Lemma 12.4, we conclude that

$$S * T[t_k](\varphi) = S(\varphi_{T[t_k]}) \to S(\varphi_T) = S * T(\varphi).$$

In case (ii), the same result follows from the fact that $\varphi_{T[t_k]} \to \varphi_T$ in $\mathcal{D}^n$, as shown by Lemma 12.6, and the continuity of S, Q.E.D.

Next we introduce the notion of derivative with respect to the parameter.

Definition 12.15 *If $\{T[t]\}$ is a family of distributions depending on a real parameter t, then we define its partial derivative with respect to t by*

$$D_t T[t] = \lim_{h \to 0} \frac{T[t+h] - T[t]}{h}$$

if the limit exists and is a distribution. Also, if k is any positive integer we define the kth partial derivative with respect to t by

$$D_t^k T[t] = D_t(D_t^{k-1} T[t])$$

if the stated partial derivatives exist as distributions.

It should be clear from the definition that, if $S[t]$ and $T[t]$ are families of distributions depending on a parameter t and if c is a constant, then

$$D_t(S[t] + c T[t]) = D_t S[t] + c D_t T[t].$$

Example 12.19 If H is the Heaviside function and c is a constant, then, in the notation introduced in Example 12.9, $H(x - ct) = H^{ct}(x)$. The regular distribution corresponding to this function is $T_{H^{ct}}$ and

$$D_t T_{H^{ct}} = \lim_{h \to 0} \frac{T_{H^{c(t+h)}} - T_{H^{ct}}}{h}.$$

But for any test function φ,

$$\frac{T_{H^{c(t+h)}} - T_{H^{ct}}}{h}(\varphi) = \frac{1}{h}\left[\int_{c(t+h)}^\infty \varphi - \int_{ct}^\infty \varphi \right] = -c\frac{1}{ch}\left[\int_0^{ct+ch} \varphi - \int_0^{ct} \varphi \right],$$

which approaches

$$-c\frac{d}{d(ct)}\int_0^{ct} \varphi = -c\varphi(ct) = -c\delta^{ct}(\varphi)$$

as $h \to 0$. A similar derivation can be carried out after replacing $H(x - ct)$ with $H(x + ct)$, showing that

$$D_t H(x \pm ct) = \pm c\delta^{\mp ct}$$

in the distribution sense. It can also be shown that

$$D_t \delta^{\pm ct} = \mp c \, (\delta^{\pm ct})'$$

(Exercise 8.2) and, therefore,

$$D_t^2 H(x \pm ct) = c^2 (\delta^{\mp ct})'$$

in the distribution sense.

Theorem 12.16 *Let S be a distribution, let $\{T[t]\}$ be a familiy of distributions, let k be a positive integer, and assume that $T[t]$ has partial derivatives with respect to t up to and including order k. Then, if either*

(i) *S has bounded support or*

(ii) *the $T[t+h]$ and their partial derivatives have their supports in the same bounded set for $|h|$ small enough,*

$$D_t^k (S * T[t]) = S * D_t^k T[t].$$

Proof. By Theorem 12.11(ii) and if $h \neq 0$,

$$\frac{S * T[t+h] - S * T[t]}{h} = \frac{S * (T[t+h] - T[t])}{h} = S * \frac{T[t+h] - T[t]}{h}.$$

Then, if $|h|$ is small enough, Theorem 12.15 implies that

$$D_t (S * T[t]) = \lim_{h \to 0} \left(S * \frac{T[t+h] - T[t]}{h} \right) = S * D_t T[t].$$

The general statement for arbitrary k follows by induction provided that $D_t^k T[t]$ exists, Q.E.D.

§12.9   The Cauchy Problem for Time-Dependent Equations

In §12.7 we dealt with an equation of the form $L(u) = f$, where L is a linear partial differential operator as described there. When u also depends on time, a more frequent type of problem consists of solving an equation of the form

$$(21) \qquad \qquad \frac{\partial^k u}{\partial t^k} + L(u) = 0$$

in $\mathbb{R}^n \times (0, \infty)$, where k is a positive integer, subject to initial conditions of the form

$$u(x, t) \to f_0(x) \qquad \qquad \text{as } t \to 0$$

$$\cdot \quad \cdot \quad \cdot \quad \cdot \quad \cdot \quad \cdot \quad \cdot \quad \cdot \quad \cdot \quad \cdot \quad \cdot$$

$$\frac{\partial^{k-2} u}{\partial t^{k-2}} u(x, t) \to f_{k-2}(x) \qquad \text{as } t \to 0$$

$$\frac{\partial^{k-1} u}{\partial t^{k-1}} u(x, t) \to f_{k-1}(x) \qquad \text{as } t \to 0,$$

where $f_0, \ldots, f_{k-1}$ are given locally integrable functions. This is usually called the *Cauchy problem* for (21) because in the 1840s Cauchy devoted several papers to the question of existence and uniqueness of solutions of a partial differential equation with given initial conditions. Similar but somewhat improved results were obtained later and independently by Sofia Vasilievna Kovalevskaia (1850–1891), a former student of Weierstrass.

Definition 12.16 *A* **fundamental solution** *of the Cauchy problem for* (21) *is a family of distributions* $T[t]$ *such that*

$$D_t^k T[t] + L(T[t]) = 0$$
$$T[t] \to 0 \qquad\qquad \text{as } t \to 0$$
$$. \quad . \quad . \quad . \quad . \quad . \quad . \quad . \quad . \quad .$$
$$D_t^{k-2} T[t] \to 0 \qquad\qquad \text{as } t \to 0$$
$$D_t^{k-1} T[t] \to \delta \qquad\qquad \text{as } t \to 0,$$

where 0 *denotes the distribution defined by* $0(\varphi) = 0$ *for any test function* φ.

If a fundamental solution is known we can immediately solve a simple case of the Cauchy problem for (21) as follows.

Lemma 12.7 *Let* $T[t]$ *be a fundamental solution of the Cauchy problem for* (21) *and let* f *be a locally integrable function. Then, if either*
  (i) f *has bounded support or*
  (ii) *the* $T[t + h]$ *and their partial derivatives with respect to* t *up to order* k *have their supports in the same bounded set for* $|h|$ *small enough,*

$$D_t^k (T_f * T[t]) + L(T_f * T[t]) = 0$$
$$T_f * T[t] \to 0 \qquad\qquad \text{as } t \to 0$$
$$. \quad . \quad . \quad . \quad . \quad . \quad . \quad . \quad . \quad .$$
$$D_t^{k-2}(T_f * T[t]) \to 0 \qquad\qquad \text{as } t \to 0$$
$$D_t^{k-1}(T_f * T[t]) \to T_f \qquad\qquad \text{as } t \to 0.$$

Proof. As in the proof of Theorem 12.13,

$$L(T_f * T[t]) = T_f * L(T[t]).$$

This and Theorem 12.16 yield

$$D_t^k(T_f * T[t]) + L(T_f * T[t]) = T_f * D_t^k T[t] + T_f * L(T[t]) = T_f * (D_t^k T[t] + L(T[t])).$$

Since $T[t]$ is a solution of (21), the right-hand side equals $T_f * 0 = 0$, and then $T_f * T[t]$ also satisfies (21). As for the initial conditions, Theorems 12.15 and 12.16 and the limits as $t \to 0$ in Definition 12.16 imply that

$$T_f * T[t] \to T_f * 0 = 0 \qquad \text{as } t \to 0$$

$$\cdots \cdots \cdots \cdots \cdots \cdots$$

$$D_t^{k-2}(T_f * T[t]) \to T_f * 0 = 0 \qquad \text{as } t \to 0$$

$$D_t^{k-1}(T_f * T[t]) \to T_f * \delta = T_f \qquad \text{as } t \to 0,$$

Q.E.D.

It is now an easy matter to solve the original Cauchy problem.

Theorem 12.17 *Let $T[t]$ be a fundamental solution of the Cauchy problem for* (21) *and let $f_0, \ldots, f_{k-1}$ be locally integrable functions. Then, if either*
(i) $f_0, \ldots, f_{k-1}$ have bounded support or
(ii) the $T[t + h]$ and their partial derivatives with respect to t up to order k have their supports in the same bounded set for $|h|$ small enough,

$$\sum_{j=1}^{k} D_t^{j-1}(T_{f_{k-j}} * T[t])$$

is a solution of the Cauchy problem for (21) *in the distribution sense.*

Proof. It is enough to show that the jth term in the stated sum, which we shall denote by $S_j[t]$, solves the problem if $f_{k-i} \equiv 0$ for $i \neq j$, and then the theorem follows by superposition of solutions. If we define

$$U_j[t] = T_{f_{k-j}} * T[t],$$

then, according to Lemma 12.7,

$$D_t^k U_j[t] + L(U_j[t]) = 0$$

$$U_j[t] \to 0 \qquad \text{as } t \to 0$$

$$\cdots \cdots \cdots \cdots \cdots \cdots \cdots$$

$$D_t^{k-2} U_j[t]) \to 0 \qquad \text{as } t \to 0$$

$$D_t^{k-1} U_j[t]) \to T_{f_{k-j}} \qquad \text{as } t \to 0.$$

Since $S_j[t] = D_t^{j-1} U_j[t]$ we have

$$D_t^k S_j[t] + L(S_j[t]) = D_t^k(D_t^{j-1} U_j[t]) + L(D_t^{j-1} U_j[t])$$
$$= D_t^{j-1}[D_t^k U_j[t] + L(U_j[t])]$$
$$= 0.$$

Also, as $t \to 0$,

$$D_t^{k-i} S_j[t] = D_t^{k-i+j-1} U_j[t] = D_t^{k-1-(i-j)} U_j[t] \to \begin{cases} 0 & \text{if } i > j \\ T_{f_{k-j}} & \text{if } i = j \end{cases}$$

and, if $i < j$,

$$D_t^{k-i} S_j[t] = D_t^{j-i-1}(D_t^k U_j[t]) = D_t^{j-i-1}[-L(U_j[t])] \to D_t^{j-i-1}(0) = 0$$

since $U_j[t] \to 0$ as $t \to 0$ and L is linear, Q.E.D.

In the rest of this section we shall apply these results to the problem of the vibrating string, assuming that the initial position and velocity are not differentiable at all points. As we well know, neither d'Alembert's nor Lagrange's solution can be the answer, and we shall look for a solution in the distribution sense. We are dealing with the Cauchy problem

$$\begin{aligned} u_{tt} - c^2 u_{xx} &= 0 & &\text{in } \mathbb{R} \times (0, \infty) \\ u(x, t) &\to f(x) & &\text{as } t \to 0 \\ u_t(x, t) &\to g(x) & &\text{as } t \to 0, \end{aligned}$$

where f and g are locally integrable.

To find a fundamental solution the initial conditions are replaced with $u(x, t) \to 0$ and $u_t(x, t) \to \delta$ in the distribution sense. Any constant function satisfies the wave equation and the first of these conditions, but not the second. We know that the singularity at $(0, 0)$ propagates along the characteristics $x = \pm ct$, which suggests that the fundamental solution that we seek is a piecewise constant function that is discontinuous at each point of these characteristics and only there. Such a function can be written in the form

$$C[H(x + ct) - H(x - ct)],$$

where C is an arbitrary constant and H is the Heaviside function. But, according to Example 12.19, its partial derivative with respect to t in the distribution sense is

$$Cc(\delta^{-ct} + \delta^{ct}).$$

Since this derivative must approach δ as $t \to 0$, $C = 1/2c$. In short, we conjecture that

$$F(x, t) = \frac{1}{2c}[H(x + ct) - H(x - ct)]$$

is a fundamental solution of the Cauchy problem for the wave equation in the distribution sense. That is, if we denote $T_{F(\cdot, t)}$ by $T_F[t]$, then $T_F[t]$ is a fundamental solution.

To verify our conjecture we first use the result of Example 12.19 to obtain

(22) $$D_t T_F[t] = \tfrac{1}{2}(\delta^{-ct} + \delta^{ct})$$

and

$$D_t^2 T_F[t] = \frac{c}{2}[(\delta^{-ct})' - (\delta^{ct})'].$$

Fixing now the value of the parameter t, $T_F[t]$ can be written in the form

(23) $$T_F[t] = \frac{1}{2c}(T_{H^{-ct}} - T_{H^{ct}}),$$

and, according to Example 12.9, its first and second derivatives are

$$(T_F[t])' = \frac{1}{2c}(\delta^{-ct} - \delta^{ct}) \qquad \text{and} \qquad (T_F[t])'' = \frac{1}{2c}[(\delta^{-ct})' - (\delta^{ct})'].$$

Therefore,

$$D_t^2 T_F[t] = c^2(T_F[t])'';$$

that is, $T_F[t]$ is a solution of the wave equation. Also, (23) and (22) show that

$$T_F[t] \to \frac{1}{2c}(T_H - T_H) = 0 \qquad \text{and} \qquad D_t T_F[t] = \tfrac{1}{2}(\delta + \delta) = \delta$$

as $t \to 0$. We conclude that $T_F[t]$ is a fundamental solution of the Cauchy problem for the wave equation. Since all $T_F[t+h]$ have support in the bounded interval $[-c(t+\tau), c(t+\tau)]$ if $0 < h < \tau$, it follows from Theorem 12.17 that

$$T_g * T_F[t] + T_f * D_t T_F[t]$$

is a solution of the Cauchy problem for the wave equation.

To show that $T_g * T_F[t]$ is a regular distribution and to find the function to which it corresponds, note that for any test function φ

$$T_g * T_F[t](\varphi) = T_g(\varphi_{T_F[t]})$$

$$= \int_{-\infty}^{\infty} g(s)\varphi_{T_F[t]}(s)\, ds$$

$$= \int_{-\infty}^{\infty} g(s)T_F[t](\varphi^s)\, ds$$

$$= \int_{-\infty}^{\infty} g(s)\left(\int_{-\infty}^{\infty} F(\sigma, t)\varphi(\sigma + s)\, d\sigma\right) ds$$

$$= \int_{-\infty}^{\infty} g(s)\left(\int_{-\infty}^{\infty} F(x - s, t)\varphi(x)\, dx\right) ds$$

$$= \int_{-\infty}^{\infty} \left(\int_{-\infty}^{\infty} g(s)F(x - s, t)\, ds\right)\varphi(x)\, dx,$$

where the change of order of integration is justified by Fubini's theorem. Thus, $T_g * T_F[t]$ is the regular distribution corresponding to the inner integral above, which equals

$$\frac{1}{2c}\int_{-\infty}^{\infty} g(s)[H(x - s + ct) - H(x - s - ct)]\, ds = \frac{1}{2c}\int_{x-ct}^{x+ct} g(s)\, ds.$$

Now, if φ is a test function, (22), Theorem 12.11 (ii), and (Example 12.17) yield

$$T_f * D_t T_F[t](\varphi) = \tfrac{1}{2}[T_f * (\delta^{-ct} + \delta^{ct})](\varphi)$$

$$= \tfrac{1}{2}[T_f * \delta^{-ct}(\varphi) + T_f * \delta^{ct}(\varphi)]$$

$$= \tfrac{1}{2}[T_f(\varphi^{-ct}) + T_f(\varphi^{ct})]$$

$$= \tfrac{1}{2}\left(\int_{-\infty}^{\infty} f(x)\varphi(x - ct)\,dx + \int_{-\infty}^{\infty} f(x)\varphi(x + ct)\,dx \right)$$

$$= \tfrac{1}{2}\left(\int_{-\infty}^{\infty} f(x + ct)\varphi(x)\,dx + \int_{-\infty}^{\infty} f(x - ct)\varphi(x)\,dx \right)$$

$$= \int_{-\infty}^{\infty} \tfrac{1}{2}[f(x + ct) + f(x - ct)]\varphi(x)\,dx.$$

This shows that $T_f * D_t T_F[t]$ is the regular distribution corresponding to

$$\tfrac{1}{2}[f(x + ct) + f(x - ct)].$$

Therefore $T_g * T_F[t] + T_f * D_t T_F[t]$ is the regular distribution corresponding to

$$u(x, t) = \tfrac{1}{2}[f(x + ct) + f(x - ct)] + \frac{1}{2c} \int_{x-ct}^{x+ct} g(s)\,ds,$$

which is a solution of the complete boundary value problem in the distribution sense.

The problem that we have solved refers to an ideal infinite string. The solution applies also to a finite string stretched between fixed point $x = 0$ and fixed point $x = a$ if f and g are the odd periodic extensions of its initial position and velocity of period $2a$ to the entire real line. In this way it generalizes the solution found in Chapter 5 to the case in which f and g are locally integrable.

§12.10 Conclusion

The foregoing presentation is only an introduction to the theory of distributions and its application to partial differential equations. A glaring omission, even at this level, is any discussion of integral transforms. It is not difficult to give at least an example if, for a few moments, we can proceed experimentally as taught by Fourier and Heaviside. If for each fixed value of ω in $\mathbb{R}^n$ we define a function ϕ_ω in $C^\infty(\mathbb{R}^n)$ by

$$\phi_\omega(x) = e^{-i\langle \omega, x \rangle},$$

then the Fourier transform of a nice enough function f is

$$\hat{f}(\omega) = \frac{1}{(2\pi)^n} \int_{\mathbb{R}^n} f(x)\phi_\omega(x)\,dx = \frac{1}{(2\pi)^n} T_f(\phi_\omega).$$

The expression on the right-hand side is well defined if f has compact support since we have seen that a distribution with compact support can be extended to $C^\infty(\mathbb{R}^n)$ (the fact that ϕ_ω is complex valued is no obstacle since we can write it as the sum of its real and imaginary parts and use linearity). Then, by analogy, we can define the Fourier transform of δ^s as

$$\frac{1}{(2\pi)^n}\,\delta^s(\phi_\omega) = \frac{1}{(2\pi)^n}\phi_\omega(s) = \frac{1}{(2\pi)^n}\,e^{-i\langle\omega,s\rangle}$$

for every s in $\mathbb{R}^n$. In particular, the Fourier transform of δ is

$$\frac{1}{(2\pi)^n}.$$

If we accept these facts, we can immediately find a fundamental solution of the Cauchy problem for the one-dimensional wave equation, that is, a function F such that

$$\begin{aligned}
F_{tt} &= c^2 F_{xx} & -\infty < x < \infty, \quad & t > 0 \\
F(x,0) &= 0 & -\infty < x < \infty & \\
F_t(x,0) &= \delta & -\infty < x < \infty &
\end{aligned}$$

in the distribution sense. Indeed, transforming this system and if $\widehat{F}_{tt} = \widehat{F_{tt}}$,

$$\begin{aligned}
\widehat{F}_{tt} &= -\omega^2 c^2 \widehat{F} & -\infty < \omega < \infty, \quad & t > 0 \\
\widehat{F}(\omega,0) &= 0 & -\infty < \omega < \infty & \\
\widehat{F}_t(\omega,0) &= \frac{1}{2\pi} & -\infty < \omega < \infty. &
\end{aligned}$$

Solving the given differential equation for $\widehat{F}$ subject to the given initial conditions,

$$\widehat{F}(\omega,t) = \frac{\sin\omega ct}{2\pi\omega c},$$

whose inverse Fourier transform is obtained from Example 7.1 to be

$$F(x,t) = \frac{1}{2c}[H(x+ct) - H(x-ct)],$$

as already shown in §12.9.

A more dramatic example is represented by the fundamental solution of the Cauchy problem for the three-dimensional wave equation. Proceeding as above we obtain

$$\widehat{F}(\omega,t) = \frac{\sin\|\omega\|ct}{(2\pi)^3\omega c},$$

whose inverse Fourier transform is

$$\int_{\mathbb{R}^n} \frac{1}{(2\pi)^3}\frac{\sin\|\omega\|ct}{\|\omega\|c}\,e^{i\langle\omega,x\rangle}\,d\omega.$$

But in Example 9.6 we showed that

$$\frac{\sin \|\omega\| ct}{\|\omega\| c} = \frac{1}{4\pi c^2 t} \int_{S_{ct}} e^{i\langle \omega, s\rangle} \, dS_s,$$

where S_{ct} is the sphere of radius ct centered at the origin, and then the fundamental solution that we seek is

$$
\begin{aligned}
T_F[t] &= \int_{\mathbb{R}^n} \frac{1}{(2\pi)^3} \left(\frac{1}{4\pi c^2 t} \int_{S_{ct}} e^{i\langle \omega, s\rangle} \, dS_s \right) e^{i\langle \omega, x\rangle} \, d\omega \\
&= \frac{1}{4\pi c^2 t} \int_{S_{ct}} \left(\int_{\mathbb{R}^n} \frac{1}{(2\pi)^3} e^{i\langle \omega, s\rangle} e^{i\langle \omega, x\rangle} \, d\omega \right) dS_s \\
&= \frac{1}{4\pi c^2 t} \int_{S_{ct}} \delta^{-s} \, dS_s
\end{aligned}
$$

since the inverse Fourier transform of $e^{i\langle \omega, s\rangle}/(2\pi)^3$ is δ^{-s}. This solution illustrates Huygens' principle rather sharply. A velocity impulse at the origin propagates as a spherical shell impulse with velocity c.

Even though we have been laboring under the disclaimer that we would proceed experimentally, we should not leave the right-hand side above without a word of explanation. The straightforward interpretation of this surface integral is as the family of distributions defined for each $t > 0$ by

$$\left(\int_{S_{ct}} \delta^{-s} \, dS_s \right)(\varphi) = \int_{S_{ct}} \delta^{-s}(\varphi) \, dS_s = \int_{S_{ct}} \varphi(-s) \, dS_s.$$

The result of Exercise 9.5 shows that with this interpretation $T_F[t]$ is, indeed, a fundamental solution of the Cauchy problem for the three-dimensional wave equation.

A theory of the Fourier and Laplace transforms of distributions, as well as all the statements made above about fundamental solutions, can be properly elaborated and justified but the boundedness of space, time, and energy on the part of the author forbid the inclusion of such topics at the moment. We refer the reader to Chapter VII in the second volume of Schwartz' original book.[1]

EXERCISES

2.1 Show that, if φ is a test function and if $f : \mathbb{R}^n \to \mathbb{R}$ is continuous and vanishes outside

---

[1] Or, for a more elementary presentation in English, see L. Schwartz, *Mathematics for the Physical Sciences*, Hermann, Paris, and Addison-Wesley, Reading, Massachusetts, 1966 (a translation of the 1955 French edition), Chapters V and VI.

a closed and bounded set, then their convolution,

$$f * \varphi(x) = \int_{\mathbb{R}^n} f(s)\varphi(x - s) \, ds,$$

is a test function (**Hint:** assume that a generalization of Theorem B.9 permits partial differentiation under the integral sign).

2.2 Let S be a spherical surface in $\mathbb{R}^n$. Show that the function T defined by

$$T(\varphi) = \int_S \varphi$$

for any test function φ is a distribution (a modification of this distribution will be used in Exercises 9.5 and 9.6).

2.3 If $f : \mathbb{R}^n \to \mathbb{R}$ is a continuous function, if φ is a test function whose integral over $\mathbb{R}^n$ has value 1, and if we define

$$\varphi_t(x) = \frac{1}{t^n} \varphi\left(\frac{1}{t} x\right)$$

for each $t > 0$, show that

$$\lim_{t \to 0} \int_{\mathbb{R}^n} f(s)\varphi_t(x - s) \, ds = f(x)$$

for all x in $\mathbb{R}^n$ (**Hint:** use the Jacobian of the function $y = x/t$ and the change of variable formula to show that the integral of φ_t over $\mathbb{R}^n$ is 1 for every $t > 0$; then use the commutativity of convolution for functions and show that

$$\int_{\mathbb{R}^n} [f(x - s) - f(x)]\varphi_t(s) \, ds$$

approaches zero as $t \to 0$ by splitting this integral into the sum of two: one over a small ball centered at the origin and one over the rest of the space).

2.4 Show that, if $n = 1$, the function T defined by

$$T(\varphi) = \mathrm{PV} \int_{-\infty}^{\infty} \frac{\varphi(x)}{x} \, dx \stackrel{\text{def}}{=} \lim_{\epsilon \to 0} \int_{|x| \geq \epsilon} \frac{\varphi(x)}{x} \, dx$$

for any test function φ, where PV stands for principal value, is a distribution. It is denoted by

$$\mathrm{PV} \frac{1}{x}$$

(**Hint:** if φ has support in $[-a, a]$,

$$\mathrm{PV} \int_{-\infty}^{\infty} \frac{\varphi(x)}{x} \, dx = \varphi(0) \, \mathrm{PV} \int_{-a}^{a} \frac{1}{x} \, dx + \mathrm{PV} \int_{-a}^{a} \frac{\varphi(x) - \varphi(0)}{x} \, dx$$

$$= \int_{-a}^{a} \frac{\varphi(x) - \varphi(0)}{x} \, dx,$$

and the absolute value of the quotient in the last integral is bounded by $\max |\varphi'|$ on $[-a, a]$ according to the mean value theorem).

JACQUES HADAMARD
David Eugene Smith Collection.
Rare Book and Manuscript Library.
Columbia University.
Reprinted with permission.

We have already mentioned two of Hadamard's crucial contributions to elementary or fundamental solution. A third, also included in his 1920 lectures at Yale,[1] is that of the *partie finie*
(finite part) of a divergent integral. For example, if $n = 1$, the Cauchy principal value

$$\text{PV} \int_0^\infty \frac{\varphi(x)}{x} \, dx = \lim_{\epsilon \to 0+} \int_\epsilon^\infty \frac{\varphi(x)}{x} \, dx$$

diverges. Indeed, if φ has support in $[-a, a]$, the right-hand side equals

$$\lim_{\epsilon \to 0+} \left[\varphi(0) \log a - \varphi(0) \log \epsilon + \int_\epsilon^a \frac{\varphi(x) - \varphi(0)}{x} \, dx \right]$$

and $\varphi(0) \log \epsilon$ diverges as $\epsilon \to 0+$. Hadamard's *partie finie* of this integral, indicated by the
symbol Pf, is obtained by discarding the divergent term in its principal value and is defined by

$$\text{Pf} \int_0^\infty \frac{\varphi(x)}{x} \, dx = \lim_{\epsilon \to 0+} \left[\int_\epsilon^\infty \frac{\varphi(x)}{x} \, dx + \varphi(0) \log \epsilon \right].$$

---

[1] Published in 1922 under the title *Lectures on Cauchy's Problem in Linear Partial Differential Equations*, and later reprinted by Dover, New York, in 1952.

Since any test function has domain in some bounded interval $[-a, a]$, the right-hand side above equals

$$\lim_{\epsilon \to 0+} \int_\epsilon^a \left[\frac{\varphi(x) - \varphi(0)}{x} \, dx + \varphi(0) \log a \right] = \int_0^a \frac{\varphi(x) - \varphi(0)}{x} \, dx + \varphi(0) \log a.$$

2.5 Show that the function defined by

$$\text{Pf} \int_0^\infty \frac{\varphi(x)}{x} \, dx,$$

as stated above, for any test function $\varphi : \mathbb{R} \to \mathbb{R}$ is a distribution. It is denoted by

$$\text{Pf} \, \frac{H(x)}{x},$$

where H is the Heaviside function.

2.6 As in Exercise 2.5, but integrating over $(-\infty, 0)$, define

$$\text{Pf} \, \frac{H(-x)}{x}$$

and show that it is a distribution. Then show that

$$\text{PV} \, \frac{1}{x} = \text{Pf} \, \frac{H(x)}{x} + \text{Pf} \, \frac{H(-x)}{x},$$

where the left-hand side is the distribution of Exercise 2.4.

2.7 If $\varphi : \mathbb{R} \to \mathbb{R}$ is a test function, then, by discarding the divergent term in

$$\text{PV} \int_0^\infty \frac{\varphi(x)}{x^{3/2}} \, dx,$$

show that

$$\text{Pf} \int_0^\infty \frac{\varphi(x)}{x^{3/2}} \, dx = \lim_{\epsilon \to 0+} \left[\int_\epsilon^\infty \frac{\varphi(x)}{x^{3/2}} \, dx - 2 \frac{\varphi(0)}{\sqrt{\epsilon}} \right]$$

and that this function of φ defines a distribution. Note that, if φ has a domain in some bounded interval $[-a, a]$, then the right-hand side above equals

$$\int_0^a \frac{\varphi(x) - \varphi(0)}{x^{3/2}} \, dx - 2 \frac{\varphi(0)}{\sqrt{a}}.$$

2.8 If $\varphi : \mathbb{R} \to \mathbb{R}$ is a test function, then, by discarding the divergent terms in

$$\text{PV} \int_0^\infty \frac{\varphi(x)}{x^2} \, dx,$$

show that

$$\text{Pf} \int_0^\infty \frac{\varphi(x)}{x^2} \, dx = \lim_{\epsilon \to 0+} \left[\int_\epsilon^\infty \frac{\varphi(x)}{x^2} \, dx - \frac{\varphi(0)}{\epsilon} + \varphi'(0) \log \epsilon \right]$$

and that this function of φ defines a distribution (**Hint:** according to Taylor's theorem with remainder,

$$\varphi(x) = \varphi(0) + x\varphi'(0) + \tfrac{1}{2} x^2 \varphi''(c),$$

where c is between zero and x).

2.9 If $\varphi : \mathbb{R} \to \mathbb{R}$ is a test function, then, by discarding any divergent terms in

$$\text{PV} \int_{-\infty}^{\infty} \frac{\varphi(x)}{x^2} \, dx,$$

find

$$\text{Pf} \int_{-\infty}^{\infty} \frac{\varphi(x)}{x^2} \, dx$$

and show that this function of φ defines a distribution.

2.10 If $\varphi : \mathbb{R} \to \mathbb{R}$ is a test function, then, by discarding the divergent term in

$$\text{PV} \int_{-\infty}^{0} \frac{\varphi(x)}{|x|} \, dx,$$

show that

$$\text{Pf} \int_{-\infty}^{0} \frac{\varphi(x)}{|x|} \, dx = \lim_{\epsilon \to 0+} \left[\int_{-\infty}^{-\epsilon} \frac{\varphi(x)}{|x|} \, dx + \varphi(0) \log \epsilon \right]$$

and that this function of φ defines a distribution. It is denoted by

$$\text{Pf} \frac{H(-x)}{|x|}.$$

Show that

$$\text{PV} \frac{1}{x} = \text{Pf} \frac{H(x)}{x} - \text{Pf} \frac{H(-x)}{|x|},$$

where the first two terms are the distributions of Exercises 2.4 and 2.5, respectively.

3.1 Show that, if f is a continuous function, then the support of the regular distribution T_f is the support of f.

3.2 Show that $x\delta$, that is, the product of f and δ, where $f(x) = x$, vanishes everywhere.

3.3 For $n = 1$, show that $(\sin)\delta$ vanishes everywhere and that $(\cos)\delta = \delta$.

3.4 Prove Theorem 12.3.

4.1 Prove that $D^m T$ is a distribution.

4.2 For $n = 1$, show that, if $f : \mathbb{R} \to \mathbb{R}$ is continuous and has a piecewise continuos derivative, then $(T_f)' = T_{f'}$.

4.3 Show that, if $f : \mathbb{R} \to \mathbb{R}$ is a locally integrable differentiable function and if H^x is as in Example 12.9, then

$$(T_{fH^x})' = f(x)\delta^x + T_{f'H^x}.$$

In particular, find the derivatives of $H(x) \sin x$ and $H(x) \cos x$ in the distribution sense.

4.4 By successive differentiation of the formula in Exercise 4.3, develop a formula for the kth derivative $(T_{fH^x})^{(k)}$, where k is a positive integer and f is k times differentiable, and use induction on k to prove it. In particular, find the second derivatives of $H(x) \sin x$ and $H(x) \cos x$ in the distribution sense.

4.5 Use the Leibniz rule to show that, if $f : \mathbb{R} \to \mathbb{R}$ is of class C^∞ and if k is a positive integer, then

$$f\delta^{(k)} = \sum_{l=0}^{k} (-1)^l \binom{k}{l} f^{(l)}(0) \, \delta^{(k-l)}.$$

Conclude that

$$(\sin)\delta' = -\delta \quad \text{and} \quad (\cos)\delta' = \delta'.$$

Show that $x\delta'$ does not vanish at the origin, and compare your result with the result of Exercise 3.2.

4.6 For each positive integer j define a function $p_j : \mathbb{R} \to \mathbb{R}$ by $p_j(x) = x^j$. Use the formula of Exercise 4.5 to show that

$$p_j \delta^{(k)} = \begin{cases} 0, & j > k \\ (-1)^j j! \delta, & j = k \\ (-1)^j \dfrac{k!}{(k-j)!} \delta^{(k-j)}, & j < k. \end{cases}$$

4.7 Fix a positive integer k, and for each integer i such that $-k \le i \le k$ define $f_i : \mathbb{R} \to \mathbb{R}$ by

$$f_i(x) = \begin{cases} -\dfrac{\pi + (x - 2i\pi)}{2}, & (2i-1)\pi < x < 2i\pi \\[2mm] \dfrac{\pi - (x - 2i\pi)}{2}, & 2i\pi < x < (2i+1)\pi \\[2mm] 0, & \text{otherwise} \end{cases}$$

and then

$$F_k = \sum_{i=-k}^{k} f_i.$$

Show that

$$(T_{F_k})' = -\tfrac{1}{2} T_{H-(2k+1)\pi} + \tfrac{1}{2} T_{H(2k+1)\pi} + \pi \sum_{i=-k}^{k} \delta^{2i\pi}.$$

4.8 Show that for $n = 1$ the derivative of $\log|x|$ in the distribution sense is the singular distribution

$$\text{PV} \frac{1}{x}$$

of Exercise 2.4 (**Hint:** $(\log \epsilon)[\varphi(\epsilon) - \varphi(0)] \to 0$ as $\epsilon \to 0+$ since $|\varphi(\epsilon) - \varphi(0)| \le \epsilon \max |\varphi'|$).

4.9 Show that the equality of second-order mixed partials always holds for distribution derivatives. Does this mean that higher order partial derivatives can be taken in any order?

4.10 Let k be a positive constant and let x and t be real variables. If we define a function $u : \mathbb{R}^2 \to \mathbb{R}$ to be identically zero for $t \le 0$ and

$$u(x,t) = \frac{1}{\sqrt{4\pi kt}} e^{-x^2/4kt}$$

for $t > 0$, evaluate $u_t - k u_{xx}$ in the distribution sense.

4.11 If H is the Heaviside function on $\mathbb{R}$, define a function $u : \mathbb{R}^2 \to \mathbb{R}$ by $u(x,y) = H(x)H(y)$. Show that u is a solution of the partial differential equation $u_{xy} = \delta$ in the distribution sense. Generalize to $\mathbb{R}^n$.

4.12 For $n = 2$, if $r = \sqrt{x^2 + y^2}$ and if we define the Laplace operator Δ by $\Delta u = u_{xx} + u_{yy}$, show that

$$r^2 \Delta \delta = 4\delta \qquad \text{and} \qquad \Delta \log \frac{1}{r} = -2\pi \delta$$

in the distribution sense. Which of these two formulas generalizes to arbitrary dimension and what is its form?

5.1 Let $\{f_k\}$ be a sequence of locally integrable functions on $\mathbb{R}^n$ that converges to a locally integrable function f, and assume that either

(i) $f_k \to f$ uniformly on every bounded subset of $\mathbb{R}^n$ or

(ii) $\int_S |f_k - f|^2 \to 0$ on every bounded subset S of $\mathbb{R}^n$.

Show that $T_{f_k} \to T_f$ as $k \to \infty$ (**Hint:** for (ii) use the n-dimensional version of the Schwarz inequality of Exercise 4.8 of Chapter 6).

5.2 For each positive integer k let $f_k(x) = \sin kx$ and $g_k(x) = \cos kx$. Find the limits of f_k and g_k in the distribution sense as $k \to \infty$.

5.3 For each positive integer k define $f_k : \mathbb{R} \to \mathbb{R}$ by

$$f_k(x) = \begin{cases} -k, & |x| < \dfrac{1}{2k} \\[2mm] 2k, & \dfrac{1}{2k} \le |x| \le \dfrac{1}{k} \\[2mm] 0, & |x| > \dfrac{1}{k}. \end{cases}$$

Show that $T_{f_k} \to \delta$ as $k \to \infty$ in the distribution sense. Note that $f_k(0) \to -\infty$ as $k \to \infty$, which shows that (8) cannot be justified in the distribution sense (or in any sense, because $\delta(0) = \infty$ would impliy that $2\delta(0) = \infty$ and then that $\delta = 2\delta$).

5.4 Find the derivative of $f(x) = |\sin x|$ in the distribution sense.

5.5 Show that the divergent series

$$\sum_{k=1}^{\infty} \cos kx$$

converges in the distribution sense to the sum

$$-\tfrac{1}{2} T_1 + \pi \sum_{k=-\infty}^{\infty} \delta^{2k\pi}$$

(**Hint:** prove first the convergence of the series

$$\sum_{k=1}^{\infty} \frac{\sin kx}{k},$$

which is the Fourier series on $(-\pi, \pi)$ of the function f of Example 12.12 in the distribution sense, and then use the result of Example 12.13).

5.6 Use the result of Exercise 5.5 to find the sum of the divergent series

$$\sum_{k=1}^{\infty} k \sin kx$$

in the distribution sense.

5.7 Show that the Fourier series of a piecewise continuous function $f : (-\pi, \pi) \to \mathbb{R}$ on $(-\pi, \pi)$ always converges in the distribution sense and find its sum.

5.8 Prove *Poisson's formula* for test functions:

$$\sum_{k=-\infty}^{\infty} \widehat{\varphi}(k) = \sum_{k=-\infty}^{\infty} \varphi(2k\pi),$$

where φ is an arbitrary test function on $\mathbb{R}$ and $\widehat{\varphi}$ is its Fourier transform (**Hint:** start with the result of Exercise 5.5, rewriting the cosine in terms of exponentials, and extend the definition of T_f to complex-valued functions in the obvious way).

6.1 Is the Poisson formula of Exercise 5.8 valid if φ is a just function of class C^∞ rather than a test function?

6.2 Show that, if $p(x)$ is a polynomial of degree k in the real variable x, then

$$\varphi_{T_p}(x) = \sum_{j=1}^{k} \frac{(-x)^j}{j!} T_{p^{(j)}}(\varphi);$$

that is, $\varphi_{T_p}(x)$ is also a polynomial of degree k.

6.3 Prove parts (i) to (iii) of Lemma 12.5.

6.4 If $n = 1$ and if T_1 is the regular distribution corresponding to the function $f \equiv 1$, find a distribution T such that

$$(T_1 * \delta') * T = 0 \qquad \text{and} \qquad T_1 * (\delta' * T) = T_1.$$

Explain why this does not contradict Theorem 12.11.

6.5 Show that, if $n = 1$, k is a positive integer, and the superscript (k) denotes a kth derivative, then

$$\delta^{(k)} * T = T * \delta^{(k)} = T^{(k)}$$

for any distribution T.

6.6 Show that
(i) $(\delta - \delta^1) * (\delta - \delta^1) = \delta - 2\delta^1 + \delta^2$ and
(ii) $(\delta - \delta^1) * (\delta - \delta^1) * (\delta - \delta^1) = \delta - 3\delta^1 + 3\delta^2 - \delta^3$.
Develop, on this basis, a formula for the convolution of $\delta - \delta^1$ with itself k times, and then prove it by induction on k.

7.1 Let S be a sphere in $\mathbb{R}^3$ of radius a and centered at the origin, let D denote the interior of S, let u be such that $\Delta u = 0$ in D and $u \equiv g$ on S, where g is a locally integrable function, and let G be the Green function for the Poisson equation on the sphere. Find u by applying Green's theorem (ii) of Exercise 1.6 of Chapter 9 to the functions u and G on the region inside S and outside a sufficiently small sphere in D centered at σ.

7.2 By analogy with the work done in the three-dimensional case, obtain the fundamental solution

$$\frac{1}{2\pi} \int_D f(\sigma) \log \|s - \sigma\| \, d\sigma$$

of the two-dimensional Poisson equation

$$u_{xx} + u_{yy} = f,$$

where f is a locally integrable function that vanishes outside a disc D centered at the origin.

7.3 By analogy with the work done in the three-dimensional case, show that, if $s = (x, y)$ is the variable in $\mathbb{R}^2$, then there is a $\tilde{\sigma}$ outside the disc D of radius a centered at the origin such that the function

$$G(s, \sigma) = \frac{1}{4\pi} \log \frac{a^2 \|s - \sigma\|}{\|\sigma\|^2 \|s - \tilde{\sigma}\|^2}$$

is a solution of the two-dimensional Poisson equation that vanishes on the boundary of D. It is called the *Green function for the Poisson equation on the disc*.

7.4 Let D be a circle in $\mathbb{R}^2$ of radius a, centered at the origin and let C be its bounding circumference. Use the results of Exercises 7.2 and 7.3 to solve the boundary value problem

$$\begin{aligned} u_{xx} + u_{yy} &= f & \text{in } D \\ u &\equiv g & \text{on } C \end{aligned}$$

(**Hint:** if g is continuous use the solution of Exercise 1.3 of Chapter 10, but, if g is only locally integrable, then, as in Exercise 7.1, use the Green function of Exercise 7.3 to obtain a function that is harmonic in D and coincides with g on C).

7.5 Define a *fundamental solution of* $L(u) = f$ *at* x to be a distribution T such that $L(T) = \delta^x$. Show that, if T is a fundamental solution of $L(u) = f$ at x and either T or f has bounded support, then $T_f^{-x} * T$ is a solution of $L(u) = f$ (refer to Example 12.17 for the definition of T_f^{-x}).

7.6 Let F be a fundamental solution of the Poisson equation in the distribution sense, let s denote the variable in either $\mathbb{R}^2$ or $\mathbb{R}^3$, and define a function F_σ by

(*i*) $F_\sigma(s) = -\dfrac{1}{4\pi \|s - \sigma\|}$ in $\mathbb{R}^3$ and

(*ii*) $F_\sigma(s) = \dfrac{1}{2\pi} \log \|s - \sigma\|$ in $\mathbb{R}^2$.

Show that T_{F_σ} is a fundamental solution of the Poisson equation at σ (**Hint:** in $\mathbb{R}^3$ replace the spheres S_ϵ and S_R in the text with spheres centered at σ to show that

$$\Delta T_{F_\sigma}(\varphi) = \varphi(\sigma) = \delta^\sigma(\varphi),$$

and in $\mathbb{R}^2$ carry out a similar modification of the argument in Exercise 7.2).

7.7 Let σ be a point in the open upper half-space in $\mathbb{R}^3$, and let $F_\sigma(s)$ be as in Exercise 7.6 (*i*). Find a solution of the Poisson equation in the upper half-space that vanishes on the xy-plane. By a modification of the argument in Exercise 7.1, derive the solution of Example 9.6, valid now without the continuity requirement on the boundary values (**Hint:** find a $\tilde{\sigma}$ that does for the half-space what $\tilde{\sigma} = a^2\sigma/\|\sigma\|^2$ did for the sphere, a general technique that is called *the method of images* and was introduced by William Thomson).

7.8 Let σ be a point in the open upper half-plane in $\mathbb{R}^2$, and let $F_\sigma(s)$ be as in Exercise 7.6 (*ii*). Find a solution of the Poisson equation in the upper half-plane that vanishes on the x-axis. By a modification of the argument in Exercise 7.7, derive the solution of Theorem 7.7.

8.1 Prove Theorem 12.14.

8.2 Show that

$$D_t \delta^{\pm ct} = \mp c(\delta^{\pm ct})'.$$

8.3   Let f be an integrable function on $\mathbb{R}^n$ such that

$$\int_{\mathbb{R}^n} f = 1,$$

if p is a fixed positive number define

$$f_t(x) = \frac{1}{t^{np}} f\left(\frac{1}{t^p} x\right)$$

for $t > 0$, and denote T_{f_t} by $T[t]$. Prove that $T[t] \to \delta$ as $t \to 0$.

8.4   Use the result of Exercise 8.3 to show that if

$$u(x, t) = \frac{1}{(2\pi)^n}\left(\frac{\pi}{kt}\right)^{n/2} e^{-\|x\|^2/4kt},$$

where x is in $\mathbb{R}^n$, $t > 0$, and k is a positive constant, and if $T[t]$ denotes $T_{u(\cdot, t)}$, then $T[t] \to \delta$ as $t \to 0$.

9.1   Show that for $n = 3$ the family of distributions $T[t]$ in Exercise 8.4 is a fundamental solution of the Cauchy problem for the three-dimensional heat equation. Use it to solve the boundary value problem

$$\begin{aligned} u_t &= k(u_{xx} + u_{yy} + u_{zz}) & x &\in \mathbb{R}^3, & t &> 0 \\ u(x, t) &\to f(x) & x &\in \mathbb{R}^3, & &\text{as } t \to 0, \end{aligned}$$

where f is a locally integrabl function with bounded support.

9.2   Show that for $n = 1$ the family of distributions $T[t]$ in Exercise 8.4 is a fundamental solution of the Cauchy problem for the one-dimensional heat equation. Use it to solve the boundary value problem

$$\begin{aligned} u_t &= ku_{xx} & x &> 0, & t &> 0 \\ u(x, t) &\to f(x) & x &> 0, & t &\to 0 \\ u(0, t) &= g(t) & & & t &> 0, \end{aligned}$$

where f is a locally integrable function with bounded support and g is a bounded function with a bounded continuous derivative on $(0, \infty)$, and such that $g(0) = 0$ (**Hint:** solve the problem for $g \equiv 0$ after extending f to $\mathbb{R}$ as an odd function, and then add this solution to that of Example 8.14).

9.3   If $a > 0$ is a constant, use the fundamental solution of Exercise 9.2 to show that the boundary value problem

$$\begin{aligned} u_t &= ku_{xx} & 0 &< x < a, & t &> 0 \\ u(0, t) &= 0 & & & t &> 0 \\ u(a, t) &= 0 & & & t &> 0 \\ u(x, t) &= f(x) & 0 &\le x \le a, & t &\to 0, \end{aligned}$$

where f is a piecewise continuous function with a piecewise continuous derivative, has the solution

$$u(x, t) = \frac{1}{\sqrt{4\pi kt}} \sum_{j=-\infty}^{\infty} \int_0^a f(s)\left[e^{-(x-s-2ja)^2/4kt} - e^{-[x+s-(2j+2)a]^2/4kt}\right] ds$$

(**Hint:** extend f to $\mathbb{R}$ as an odd periodic function of period $2a$.)

9.4   Compare the solution of Exercise 9.3 with that of Theorem 3.1 to establish the identity

$$\frac{2}{a} \sum_{n=1}^{\infty} e^{-n^2\pi^2 kt/a^2} \sin \frac{n\pi}{a} x \sin \frac{n\pi}{a} s$$

$$= \frac{1}{\sqrt{4\pi kt}} \sum_{j=-\infty}^{\infty} \left[e^{-(x-s-2ja)^2/4kt} - e^{-[x+s-(2j+2)a]^2/4kt} \right].$$

Note that the series on the left converges very rapidly for large t. Thus, for such values of t, the solution of the boundary value problem of Exercise 9.3 is adequetely represented by a few terms of the series in Theorem 3.1. By contrast, most of the terms of the series on the right are small even for t near zero. In fact, if x is not too close to either zero or a, this series is approximately equal to

$$e^{-(x-s)^2/4kt}.$$

Thus, for such values of x, the solution of the boundary value problem of Exercise 9.3 is adequetely represented by just one term in the series stated there.

9.5   Show that the family of distributions $T[t]$ defined on $\mathbb{R}^3$ by

$$T[t](\varphi) = \frac{1}{4\pi c^2 t} \int_{S_{ct}} \varphi(-s)\, dS_s,$$

where S_{ct} is the sphere of radius ct centered at the origin, is a fundamental solution of the Cauchy problem for the three-dimensional wave equation (**Hint:** make a change of variable so that the integral is over a sphere of radius 1 and then use Green's theorem (i) of Exercise 1.6 of Chapter 9).

9.6   Show that, if f and g are locally integrable functions and if $T[t]$ is the family of distributions of Exercise 9.5, then

$$T_g * T[t] + D_t(T_f * T[t]),$$

which, according to Theorem 12.17, is a solution of the boundary value problem

$$
\begin{array}{lll}
u_{tt} = c^2(u_{xx} + u_{yy} + u_{zz}) & (x, y, z) \in \mathbb{R}^3, & t > 0 \\
u(x, y, z, t) \rightarrow f(x, y, z) & (x, y, z) \in \mathbb{R}^3, & \text{as } t \rightarrow 0 \\
u_t(x, y, z, t) \rightarrow g(x, y, z) & (x, y, z) \in \mathbb{R}^3, & \text{as } t \rightarrow 0
\end{array}
$$

in the distribution sense, is the regular distribution corresponding to

$$\frac{1}{4\pi c^2 t} \int_{S_{ct}} g(x+s)\, dS_s + \frac{d}{dt}\left[\frac{1}{4\pi c^2 t} \int_{S_{ct}} f(x+s)\, dS_s \right].$$

Compare your result with the result of Example 9.6.

APPENDIX A

UNIFORM CONVERGENCE

We saw in Chapter 2 that Cauchy stated the first modern definition of the convergence of series in his *Cours d'analyse de l'École Royale Polytechnique* of 1821. The working version of this definition today was given by Heine in 1872, and in order to reproduce it in a form that will serve our purposes we shall accept the following notation. S will denote a set of points in the real line, in the plane, or in space. A point of S will be denoted by s except when we specifically mean a real number, in which case it will be denoted by x.

Definition A.1 *For each positive integer n, let $f_n : S \to \mathbb{R}$ be a real-valued function. The series*

$$\sum_{n=1}^{\infty} f_n$$

is said to be **convergent** *or to* **converge** *to a function $f : S \to \mathbb{R}$, called its* **sum**, *if and only if for every s in S and any $\epsilon > 0$ there is a positive number $N = N(\epsilon, s)$ such that*

$$\left| \sum_{k=1}^{n} f_k(s) - f(s) \right| < \epsilon$$

for every $n > N$.

The notation $N(\epsilon, s)$ is meant to emphasize the fact that this number generally depends on the choice of both ϵ and s. The same definition can be used for the convergence of a sequence $\{f_n\}$ rather than a series of functions. One simply replaces $\sum_{k=1}^{n} f_k$ with f_n.

After stating his definition and proving a number of convergence tests, Cauchy then stated an incorrect theorem: the sum of a convergent series of continuous functions is continuous. It was the Norwegian mathematician Niels Henrik Abel (1802–1829) who first called attention to this error by pointing out that the sum of the Fourier series of $f(x) = x$, whose terms are continuous functions, is discontinuous at each $x = (2n+1)\pi$, where n is any integer (see Example 2.9 of Chapter 2). Then, in 1829, Dirichlet's theorem on the convergence of Fourier series made this abundantly clear. This is not mentioned to show a blemish in Cauchy's work, but because of its connection with an important discovery. But before that, let us place this matter in a more general context.

We are interested in determining whether or not the properties of continuity, differentiability, or integrability of the f_n carry over to the sum of the series. Moreover, for functions with domain in the real line, we want to know if and under what conditions differentiation and integration can be carried out term by term; that is, whether or not the expressions

$$\left(\sum_{n=1}^{\infty} f_n\right)' = \sum_{n=1}^{\infty} f_n'$$

and

$$\int_a^b \sum_{n=1}^{\infty} f_n = \sum_{n=1}^{\infty} \int_a^b f_n$$

are valid. In general, the answer to each of our questions is no, as shown by the following very simple example.

Example A.1 The series

$$\sum_{n=1}^{\infty} x(1 - x^2)^n$$

converges in the interval $[-1, 1]$ to the function

$$f(x) = \begin{cases} 0 & \text{if } x = 0, \pm 1 \\ \dfrac{1}{x} - x & \text{if } x \neq 0, \pm 1. \end{cases}$$

This is clear for $x = 0, \pm 1$ because then every term of the series vanishes. Otherwise, put $r = 1 - x^2$ and observe that

$$(r + r^2 + \cdots + r^n)(1 - r) = r - r^{n+1} = r(1 - r^n).$$

Then

$$\left| \sum_{k=1}^{n} x(1 - x^2)^k - \frac{1 - x^2}{x} \right| = \left| x \frac{(1 - x^2)[1 - (1 - x^2)^n]}{1 - (1 - x^2)} - \frac{1 - x^2}{x} \right| = \frac{(1 - x^2)^{n+1}}{x},$$

and, since $0 < 1 - x^2 < 1$, the right-hand side can be made smaller than any given $\epsilon > 0$ by simply choosing n large enough. Therefore, the series converges, but, while every term is continuous, differentiable, and integrable on $[-1, 1]$, its sum is not.

Probably at Dirichlet's prompting, one of his students, Phillip Ludwig von Seidel (1821–1896), was led to investigate this situation. Consider, for instance, a series of continuous functions that converges to a function discontinuous at a point s_0. About 1848, Seidel and, independently, Sir George Gabriel Stokes (1819–1903) made the observation that, if $\epsilon > 0$ is given, no N can be chosen so that the inequality in Definition A.1

is satisfied for all s near s_0. More precisely, for each such s there is an N such that the inequality in Definition A.1 holds, but *the same N* does not work *for all s*, for, the closer s is to s_0, the larger N has to be. Seidel and Stokes failed to pursue the matter, but it seems clear that, if we include in the definition of convergence the stipulation that N be independent of s, then we have a new, more restrictive kind of convergence that, as we shall see, makes it impossible for the sum of the series to be discontinuous.

Definition A.2 *Let S and the f_n be as in Definition A.1. Then the series $\sum_{n=1}^{\infty} f_n$* **converges uniformly** *on S to a sum f if and only if for any $\epsilon > 0$ there is a positive number $N = N(\epsilon)$ such that*

$$\left| \sum_{k=1}^{n} f_k(s) - f(s) \right| < \epsilon$$

for every $n > N$ and all s in S.

The series in Example A.1 is not uniformly convergent. Indeed, if $0 < \epsilon \leq 1$ is given, if we assume that N can be chosen to satisfy Definition A.2, and if $x > 0$ is chosen to be so small that $0 < 2^{N+2} x < \epsilon$, then $x < 1/2$ so that $1 - x^2 > 1/2$. If we choose $n = N + 1$, we have

$$\frac{(1 - x^2)^{n+1}}{x} > \frac{1}{2^{N+2} x} > \epsilon,$$

a contradiction.

As it happens, the idea of a different kind of convergence was not entirely new. Already in 1838 Christof Gudermann (1798–1852) had referred to a kind of *convergence at the same rate* that is the precursor of uniform convergence. But the importance of the concept escaped him, as it would escape Seidel and Stokes later on. This realization was left to Gudermann's student Karl Theodor Wilhelm Weierstrass (1815–1897), one of the giants of modern mathematics. Not particularly inspired by the lectures at the University of Bonn, where he was a student, in 1839 he went to Münster to attend Gudermann's lectures, attracted by some notes from these that he had had the opportunity to see at Bonn. Gudermann was to influence Weierstrass' research on analytic functions, and it is quite likely that, while at Münster, they discussed the new concept of convergence. Weierstrass never finished his doctorate and became a *Gymnasium* teacher in 1841, a position that he held until 1854. During this time he produced an incredible amount of first-rate research in manuscript form but, unconcerned with questions of priority, it remained unpublished for the longest time. The fact that he referred to uniform convergence— *gleichmässige Convergenz*—in an 1841 manuscript supports the idea that he may have learned about it from Gudermann. Weierstrass' many research achievements eventually earned him a position at the University of Berlin in 1856, where he frequently discussed uniform convergence. He defined it formally, for functions of several variables, in a paper published in 1880. As in Cauchy's case regarding ordinary convergence, the importance of Weierstrass' contribution stems from the fact that he realized the usefulness of uniform

KARL WEIERSTRASS

Photograph by the author from Weierstrass'
Mathematische Werke of 1903.

convergence and incorporated it in theorems on the integrability and differentiability of series of functions term by term.

Before we show this, we must point out a shortcoming that Definitions A.1 and A.2 share. Application of these definitions requires an *a priori* knowledge of the sum f, but this may not be available. To obtain a result that does not require this knowledge we shall use the well-known *Cauchy's criterion* for the convergence of sequences of 1821. In spite of its name, this criterion had already been discovered in 1817 by the Bohemian priest Bernardus Placidus Johann Nepomuk Bolzano (1781–1848), a professor of the philosophy of religion at Prague.

Lemma A.1 (Cauchy's Criterion). *Let $\{x_n\}$ be a sequence of real numbers. If for every $\epsilon > 0$ there is a positive integer N such that*

$$|x_m - x_n| < \epsilon$$

for all m and $n > N$, then $\{x_n\}$ converges.

Proof. For a given $\epsilon > 0$ let N be as stated above. If all but a finite number of the x_n are equal, then they are equal to a certain number a. In particular, there is one

$n > N$ such that $x_n = a$. Then, if $m > N$, we have

$$|x_m - a| = |x_m - x_n| < \epsilon,$$

which shows that a is the limit of the sequence $\{x_n\}$.

If infinitely many of the x_n are distinct and if M is an integer larger than N, then the inequality $|x_M - x_n| < \epsilon$ for $n > N$ means that all the points x_n with $n > N$ are in the interval $[x_M - \epsilon, x_M + \epsilon]$. If we split this interval in half using its midpoint, then at least one of the resulting subintervals, call it $[a_1, b_1]$, contains infinitely many of the x_n. Splitting $[a_1, b_1]$ in half again, one of the resulting subintervals, call it $[a_2, b_2]$, contains infinitely many of the x_n. This splitting process can be continued indefinitely, obtaining for each positive integer k a subinterval $[a_k, b_k]$ that contains infinitely many of the x_n. We shall show that there is exactly one point contained in the intersection of these shrinking intervals and that this point is the limit of $\{x_n\}$.

One of the defining axioms of the real number system states that, if an infinite set S of real numbers is bounded above, then there is a smallest real number, called the *supremum* of S, that is not smaller than any of those in S.[1] Let a be the supremum of all the a_k, a set bounded above by b_1. Similarly, let b be the *infimum* of all the b_k, defined as the largest real number that is not larger than any of the b_k. It is clear that $a \leq b$ and that both a and b are contained in $[a_k, b_k]$ for each k. Since the length of $[a_k, b_k]$ approaches zero as $k \to \infty$, it follows that $a = b$ and that this is the only point contained in all the $[a_k, b_k]$. Now let k be so large that the length of $[a_k, b_k]$ is less than ϵ. Since $[a_k, b_k]$ contains a and infinitely many of the x_n, there is an $m > N$ such that $|x_m - a| < \epsilon$. Using the triangle inequality it follows that, if $n > N$,

$$|x_n - a| \leq |x_n - x_m| + |x_m - a| < \epsilon + \epsilon = 2\epsilon,$$

and this proves that $x_n \to a$ as $n \to \infty$, Q.E.D.

It was Bolzano's original and ingenious idea to repeatedly bisect an interval to close down on a single point that must be the limit of the sequence. We can now use this basic fact from the theory of the convergence of sequences to establish some results on the convergence of series.

Theorem A.1 (Cauchy's Criterion for Uniform Convergence). *A series of functions $\sum_{n=1}^{\infty} f_n$ converges uniformly on a set S if and only if for every $\epsilon > 0$ there is a positive number N such that*

$$\left| \sum_{k=m}^{n} f_k(s) \right| < \epsilon$$

for all $n \geq m > N$ and all s in S.

---

[1] See, for instance, W. Rudin, *Principles of Mathematical Analysis*, 3rd ed., McGraw-Hill, New York, 1976, page 8.

Proof. (if) Lemma A.1 applied to the sequence of partial sums of the given series shows that it converges for each s in S, and we can define

$$f(s) = \sum_{n=1}^{\infty} f_n(s).$$

Let ϵ and N be as above and, for brevity, define

$$F_n(s) = \sum_{k=1}^{n} f_k(s).$$

Since $F_n(s) \to f(s)$ for each s as $n \to \infty$, we can choose a positive integer $m = m(s) > N$ so large that

$$|F_m(s) - f(s)| < \epsilon.$$

If $n > N$ then, for any s in S and with m as above, we have

$$|F_n(s) - f(s)| \le |F_n(s) - F_m(s)| + |F_m(s) - f(s)| < \epsilon + \epsilon = 2\epsilon$$

for all s in S, showing that $F_n \to f$ uniformly on S.

(only if) If $\sum_{n=1}^{\infty} f_n = f$ uniformly on S and $\epsilon > 0$ is given, then

$$|F_n(s) - f(s)| < \frac{\epsilon}{2}$$

for n large enough and all s in S. Thus, if $n \ge m$,

$$\left| \sum_{k=m}^{n} f_n(s) \right| = |F_n(s) - F_{m-1}(s)| \le |F_n(s) - f(s)| + |f(s) - F_{m-1}(s)| < \epsilon$$

for m and n large enough and all s in S, Q.E.D.

The next theorem gives only a sufficient condition for uniform convergence, but it is quite useful in many applications.

Theorem A.2 (The Weierstrass M-test). *Let $\sum_{n=1}^{\infty} f_n$ be a series of functions defined on a set S, and suppose that for each n there is a number M_n such that*

$$|f_n(s)| \le M_n$$

for all s in S. If the series $\sum_{n=1}^{\infty} M_n$ converges, then $\sum_{n=1}^{\infty} f_n$ converges uniformly on S.

Proof. Given $\epsilon > 0$ and applying Lemma A.1 to the sequence of partial sums of the given series, there is an N such that

$$\left| \sum_{k=m}^{n} f_k(s) \right| \le \sum_{k=m}^{n} M_k < \epsilon$$

for $n \ge m > N$ and all s in S. The result follows from Theorem A.1, Q.E.D.

In a way, the Weierstrass M-test represents a case of overkill because, not only does it show that $\sum_{n=1}^{\infty} f_n$ converges uniformly, but also it shows that $\sum_{n=1}^{\infty} |f_n|$ converges uniformly. However, $\sum_{n=1}^{\infty} f_n$ may in some cases converge uniformly while $\sum_{n=1}^{\infty} |f_n|$ does not converge even at a single point (Exercise A.16). Even if $\sum_{n=1}^{\infty} f_n$ converges uniformly and $\sum_{n=1}^{\infty} |f_n|$ converges, this last series need not converge uniformly (Exercise A.15).

More refined convergence tests can be obtained, for series of a special form, using a formula due to Abel known as *summation by parts*. For each positive integer n let $f_n : S \to \mathbb{R}$ and $g_n : S \to \mathbb{R}$ be functions and define $F_n = \sum_{k=1}^{n} f_k$. If $n > m > 1$, we have

$$\sum_{k=m}^{n} f_k g_k = \sum_{k=m}^{n} (F_k - F_{k-1}) g_k$$

$$= \sum_{k=m}^{n} F_k g_k - \sum_{k=m-1}^{n-1} F_k g_{k+1}$$

$$= F_n g_n - F_{m-1} g_m - \sum_{k=m}^{n-1} F_k (g_{k+1} - g_k).$$

Using this formula—whose name refers to the fact that its general structure is similar to the formula for integration by parts—Abel gave a convergence test for numerical series, whose generalization for series of functions is still associated with his name.

Theorem A.3 (Abel's Test for Uniform Convergence). *For each positive integer n, let $f_n : S \to \mathbb{R}$ and $g_n : S \to \mathbb{R}$ be functions such that*
  (i) *$\sum_{n=1}^{\infty} f_n$ converges uniformly on S,*
  (ii) *either $g_n(s) \le g_{n+1}(s)$ or $g_{n+1}(s) \le g_n(s)$ for each n and all s in S, and*
  (iii) *there is a constant $M > 0$ such that $|g_n(s)| \le M$ for each n and all s in S.*
Then the series

$$\sum_{n=1}^{\infty} f_n g_n$$

converges uniformly on S.

Proof. Let f be the sum of $\sum_{n=1}^{\infty} f_n$. Since, for $n > m > 1$,

$$\sum_{k=m}^{n-1} (g_{k+1} - g_k) = g_n - g_m,$$

the summation by parts formula gives

$$\sum_{k=m}^{n} f_k g_k = F_n g_n - F_{m-1} g_m - \sum_{k=m}^{n-1} F_k (g_{k+1} - g_k) - f \left[g_n - g_m - \sum_{k=m}^{n-1} (g_{k+1} - g_k) \right]$$

$$= (F_n - f) g_n - (F_{m-1} - f) g_m - \sum_{k=m}^{n-1} (F_k - f)(g_{k+1} - g_k).$$

By Definition A.2, given $\epsilon > 0$ there is an N such that $|F_k(s) - f(s)| < \epsilon/4M$ for $k > N$ and all s in S. Then, if $n > m > N$, we have

$$\left| \sum_{k=m}^{n} f_k g_k \right| \le \frac{\epsilon}{4M}(|g_n| + |g_m| + |g_n - g_m|) \le \frac{\epsilon}{4M}2(|g_n| + |g_m|) \le \epsilon$$

on S. The result follows from Theorem A.1, Q.E.D.

The applicability of Abel's test is limited because we have to use another test first to determine the uniform convergence of $\sum_{n=1}^{\infty} f_n$. This cannot be the Weierstrass M-test, because, if this test applies to $\sum_{n=1}^{\infty} f_n$, then it also applies to $\sum_{n=1}^{\infty} f_n g_n$ in view of the condition $|g_n(t)| \le M$. Abel's test is very useful, however, when the f_n are constants rather than functions. We have the following corollary.

Corollary *If $\sum_{n=1}^{\infty} a_n$ is a convergent series of numbers and if $g_n : S \to \mathbb{R}$ are functions that satisfy conditions* (ii) *and* (iii) *of the theorem, then the series*

$$\sum_{n=1}^{\infty} a_n g_n$$

converges uniformly on S.

We are finally ready to state conditions under which continuity, integrability, and differentiability of the terms of a series of functions carry over to its sum. The first result is due to Abel who essentially proved it in 1826 but let the underlying concept of uniform convergence escape unnoticed.

Theorem A.4 *The sum of a uniformly covergent series of continuous functions on a set S is continuous on S.*

Proof. If $\sum_{n=1}^{\infty} f_n = f$ uniformly on S, if each f_n is continuous on S, if F_n denotes the nth partial sum of the series, and if s and s_0 are in S, then

$$|f(s) - f(s_0)| \le |f(s) - F_n(s)| + |F_n(s) - F_n(s_0)| + |F_n(s_0) - f(s_0)|.$$

By the uniform convergence of the series, the first and third terms on the right can be made arbitrarily small by choosing n large enough. The second term can be made arbitrarily small by choosing s close enough to s_0. Then the left-hand side can also be made arbitrarily small by choosing s close enough to s_0, Q.E.D.

The two remaining results are due to Weierstrass. The set S is now an interval in $\mathbb{R}$.

Theorem A.5 *If $\sum_{n=1}^{\infty} f_n$ is a uniformly convergent series of integrable functions defined on an interval $[a, b]$, then*

$$\int_a^b \sum_{n=1}^{\infty} f_n = \sum_{n=1}^{\infty} \int_a^b f_n;$$

that is, the sum of the series is integrable and may be integrated term by term.

Proof. If $F_n = \sum_{k=1}^{n} f_k$ and if f denotes the sum of the series, let s_n be the supremum of the set $\{|f(x) - F_n(x)| : a \le x \le b\}$ (this concept was introduced in the proof of Lemma A.1). If $a = x_0 < x_1 < \cdots < x_n = b$ is a partition of $[a, b]$, let M_i and m_i be the supremum and infimum of f on each (x_{i-1}, x_i), let I_m be the supremum of the sums $\sum_{i=1}^{n} m_i(x_i - x_{i-1})$ corresponding to f for all partitions of $[a, b]$, and let I_M be the infimum of the sums $\sum_{i=1}^{n} M_i(x_i - x_{i-1})$. We have

$$\int_a^b (F_n - s_n) \le I_m \le I_M \le \int_a^b (F_n + s_n),$$

and then $0 \le I_M - I_m \le 2s_n(b-a)$. Since $s_n \to 0$ as $n \to \infty$ because the convergence is uniform, we conclude that $I_m = I_M$. By Definition 2.2, this means that f is integrable. Finally,

$$\left| \int_a^b f - \int_a^b F_n \right| = \left| \int_a^b (f - F_n) \right| \le \int_a^b |f - F_n| \le s_n(b - a),$$

which approaches zero as $n \to \infty$, Q.E.D.

Theorem A.6 *If $\sum_{n=1}^{\infty} f_n$ converges on an interval $[a, b]$, if each f_n has a continuous derivative on $[a, b]$, and if the series $\sum_{n=1}^{\infty} f_n'$ converges uniformly on $[a, b]$, then $\sum_{n=1}^{\infty} f_n$ is differentiable on $[a, b]$ and*

$$\left(\sum_{n=1}^{\infty} f_n \right)' = \sum_{n=1}^{\infty} f_n';$$

that is, the series can be differentiated term by term.

Proof. Applying Theorem A.5 to the series of derivatives, which are integrable because they are continuous, and for each x in $[a, b]$ we obtain

$$\int_a^x \sum_{n=1}^{\infty} f_n' = \sum_{n=1}^{\infty} \int_a^x f_n' = \sum_{n=1}^{\infty} [f_n(x) - f_n(a)] = \sum_{n=1}^{\infty} f_n(x) - \sum_{n=1}^{\infty} f_n(a).$$

Then, differentiating with respect to the upper limit of integration,

$$\sum_{n=1}^{\infty} f_n' = \left(\sum_{n=1}^{\infty} f_n \right)'$$

on $[a, b]$, Q.E.D.

The hypotheses in Theorem A.6 are stronger than necessary. This was done for the sake of giving a simple proof, but the conclusion is valid if the f_n are assumed only to be differentiable on $[a, b]$ and if the series $\sum_{n=1}^{\infty} f_n$ converges at a single point.[1] Note also that, in contrast with Theorems A.4 and A.5, it is the uniform convergence of $\sum_{n=1}^{\infty} f_n'$ that is required rather than that of $\sum_{n=1}^{\infty} f_n$. The uniform convergence of $\sum_{n=1}^{\infty} f_n$ would not suffice, as illustrated by the following example.

---

[1] See, for instance, W. Rudin, *op. cit.*, page 152.

Example A.2 The series

$$\sum_{n=1}^{\infty} \frac{1}{n^2} \sin n^3 x$$

converges uniformly on any interval by the Weierstrass M-test because the series

$$\sum_{n=1}^{\infty} \frac{1}{n^2}$$

is known to converge. However, the series of derivatives

$$\sum_{n=1}^{\infty} n \cos n^3 x$$

diverges at every point $x = k\pi$, where k is an integer, because $|\cos n^3 k\pi| = 1$, and the nth term of the series does not approach zero as $n \to \infty$.

EXERCISES

A.1 Show that the series

$$\sum_{n=1}^{\infty} \frac{1}{2n^2} [\sin(n^3 + \pi)(x^2 + y^2) + x^3]$$

converges uniformly on $S = \{ (x, y) \in \mathbb{R}^2 : x^2 + y^2 < 1 \}$.

A.2 Prove that, if a series of real numbers $\sum_{n=1}^{\infty} a_n$ converges absolutely, then

$$\sum_{n=1}^{\infty} a_n \cos nx \qquad \text{and} \qquad \sum_{n=1}^{\infty} a_n \sin nx$$

converge uniformly on $\mathbb{R}$.

A.3 Prove that, if a series of real numbers $\sum_{n=1}^{\infty} a_n$ converges absolutely, then the series

$$\sum_{n=1}^{\infty} \frac{a_n}{n^x}$$

converges uniformly on $[0, \infty)$.

A.4 Show that the series

$$\sum_{n=1}^{\infty} \left(\frac{x}{10 \log x} \right)^n$$

converges uniformly on $[e, e^2]$.

A.5 Show that $\sum_{n=0}^{\infty}(x^n/n!)$ converges uniformly on $[-a, a]$ for any $a > 0$. Is the convergence uniform on $\mathbb{R}$?

A.6 Show that the series

$$\sum_{n=1}^{\infty}\frac{(-1)^n}{n}e^{-n^2(x^2+y^2)}$$

converges uniformly on $\mathbb{R}^2$.

A.7 Show that the series

$$\sum_{n=1}^{\infty}\frac{(-1)^n}{\sqrt{n}}\cos\frac{x}{n}$$

converges uniformly on $[-\pi/4, \pi/4]$.

A.8 Show that $\sum_{n=0}^{\infty}(x^n/n!)$ converges to a continuous function f on $\mathbb{R}$. Show that $f' = f$ on $\mathbb{R}$.

A.9 Show that

$$\int_0^{\pi}\sum_{n=1}^{\infty}\frac{1}{n^3}\sin nx = 2\sum_{n=1}^{\infty}\frac{1}{(2n-1)^4}.$$

A.10 Show that for $|x| < 1$

$$\log(1+x) = \sum_{n=1}^{\infty}\frac{(-1)^{n+1}}{n}x^n$$

(**Hint:** consider the geometric series $\sum_{n=0}^{\infty}(-x)^n$, which is uniformly convergent for $|x| \le r$ if $r < 1$).

A.11 Show that the series

$$\sum_{n=0}^{\infty}\frac{(-1)^n}{(2n+1)!}x^{2n+1} \quad\text{and}\quad \sum_{n=0}^{\infty}\frac{(-1)^n}{(2n)!}x^{2n}$$

can be differentiated term by term and find their derivatives. On what intervals are these derivatives valid?

A.12 Prove *Dirichlet's test for uniform convergence*: for every positive integer n, let $f_n: S \to \mathbb{R}$ and $g_n: S \to \mathbb{R}$ be functions such that
  (i) there is a constant M such that $\left|\sum_{k=1}^n f_k(s)\right| \le M$ for each n and all s in S,
  (ii) $g_{n+1}(s) \le g_n(s)$ for each n and all s in S, and
  (iii) $g_n \to 0$ uniformly on S.
Then the series $\sum_{n=1}^{\infty} f_n g_n$ converges uniformly on S.

A.13 Use Dirichlet's test to prove the uniform convergence of the series

$$\sum_{n=1}^{\infty}\frac{1}{n}\cos nx$$

on the interval $[\delta, \pi - \delta]$ for any $0 < \delta < \pi/2$.

A.14 Prove the following corollary of Dirichlet's test for uniform convergence. For each positive integer n, let $g_n: S \to \mathbb{R}$ be a function such that
  (i) $g_{n+1}(s) \le g_n(s)$ for each n and all s in S and
  (ii) $g_n \to 0$ uniformly on S.
Then $\sum_{n=1}^{\infty}(-1)^n g_n$ converges uniformly on S.

A.15 Define $f_n : [0, 1] \to \mathbb{R}$ by $f_n(x) = (-x)^n(1 - x)$. Show that $\sum_{n=1}^{\infty} f_n$ converges uniformly on $[0, 1]$. Does $\sum_{n=1}^{\infty} |f_n|$ converge on $[0, 1]$? Does it converge uniformly?

A.16 If $a > 0$, define $f_n : [0, a] \to \mathbb{R}$ by $f(x) = (-1)^n(x + n)/n^2$. Show that $\sum_{n=1}^{\infty} f_n$ converges uniformly on $[0, a]$. Does $\sum_{n=1}^{\infty} |f_n|$ converge at any $x \in \mathbb{R}$?

A.17 Show that the series

$$\sum_{n=1}^{\infty} (-x)^n(1 - x)e^{-n^2 y}$$

converges uniformly on $S = \{ (x, y) \in \mathbb{R}^2 : 0 < x < 1, \, y > 0 \}$ (**Hint:** use Exercise A.15).

A.18 Show that the series

$$\sum_{n=1}^{\infty} \left[(-1)^n(x + n) \sum_{k=1}^{n} \frac{1}{(k_n)^2} \right]$$

converges uniformly on $[0, 1]$ (**Hint:** use Exercise A.16).

APPENDIX B

IMPROPER INTEGRALS

We now consider the integration of real or complex valued functions whose domains may be unbounded intervals. We shall restrict our attention to functions that are *locally integrable*, that is, to functions that, if real valued, are bounded and integrable on every bounded subinterval of their domains in the sense of Definition 2.2, and that, if complex valued, are of the form $f = u + iv$, where u and v are real valued and locally integrable. In the second case, if $[a, b]$ is an interval in the domain of f, we define

$$\int_a^b f = \int_a^b u + i \int_a^b v.$$

This expression can be used to compute the integral of f if it is known how to compute those of u and v, but it is often possible to use the following version of the fundamental theorem of the integral calculus. If F is continuous on $[a, b]$ and such that $F' = f$ on (a, b), then

$$\int_a^b f = F(b) - F(a)$$

(see Exercise B.1).

Now, if the domain of f contains any of the intervals of integration considered below, we define

$$\int_a^\infty f = \lim_{b \to \infty} \int_a^b f,$$

$$\int_{-\infty}^b f = \lim_{a \to -\infty} \int_a^b f,$$

$$\int_{-\infty}^\infty f = \int_{-\infty}^0 f + \int_0^\infty f.$$

If, in each case, the right-hand side exists and is finite, the corresponding integral on the left is said to *converge*. If it does not converge, it is said to *diverge*. The integrals on the left are usually called *improper integrals*. Note that the last one cannot be defined as

(1)
$$\lim_{a \to \infty} \int_{-a}^{a} f,$$

as is shown by the case of $f(x) = x$. When, as in this example, the integral

(2)
$$\int_{-\infty}^{\infty} f$$

does not converge but the previous limit exists, this is the best we can have. Such a limit was frequently used by Cauchy, especially in his development of complex analysis, and for that reason it is now associated with his name.

Definition B.1 *If f is locally integrable on $\mathbb{R}$ and the limit (1) exists and is finite, then it is called the* **Cauchy principal value** *of (2) and is denoted by*

$$\text{PV} \int_{-\infty}^{\infty} f.$$

As is to be expected, if an improper integral on $(-\infty, \infty)$ converges, it coincides with its principal value.

Theorem B.1 *If (2) converges, then*

$$\int_{-\infty}^{\infty} f = \text{PV} \int_{-\infty}^{\infty} f.$$

The basic properties of improper integrals, whose proofs are easy from the definition, are as follows.

Theorem B.2 *Let f be locally integrable, let a and b be real numbers, and let c be a real or complex number.*

(i) *If $\int_{a}^{\infty} f$ and $\int_{a}^{\infty} g$ converge, then $\int_{a}^{\infty} (f \pm g)$ converges to $\int_{a}^{\infty} f \pm \int_{a}^{\infty} g$.*

(ii) *If $\int_{a}^{\infty} f$ converges, then $\int_{a}^{\infty} cf$ converges to $c \int_{a}^{\infty} f$.*

(iii) *If f and g are real valued, if $0 \le f \le g$, and if $\int_{a}^{\infty} g$ converges, then $\int_{a}^{\infty} f$ converges and*

$$\int_{a}^{\infty} f \le \int_{a}^{\infty} g.$$

(iv) *If $\int_{b}^{\infty} f$ converges for some b in $\mathbb{R}$, then $\int_{a}^{\infty} f$ converges for any a in $\mathbb{R}$ and*

$$\int_{a}^{\infty} f = \int_{a}^{b} f + \int_{b}^{\infty} f.$$

Similar results hold for the other two types of improper integrals, and we shall use them without further explanation.

Example B.1 $\int_1^\infty \left| \dfrac{\cos x}{x^2} \right| dx$ converges by Theorem B.2 (iii) because

$$0 \le \left| \frac{\cos x}{x^2} \right| \le \frac{1}{x^2},$$

and

$$\int_1^\infty \frac{1}{x^2}\, dx = \lim_{b \to \infty} \left[-\frac{1}{x} \right]_1^b = 1.$$

Note that we do not refer to a function f that has a convergent improper integral as integrable on the corresponding interval. That term is applied when $|f|$, rather than f, has a convergent improper integral.

Definition B.2 *A locally integrable function f is called* **integrable** *on* (a, ∞) *if and only if*

$$\int_a^\infty |f|$$

converges. This is also true for the other two types of intervals.

Theorem B.3 *If f is integrable on* (a, ∞), *then* $\int_a^\infty f$ *converges and*

$$\left| \int_a^\infty f \right| \le \int_a^\infty |f|.$$

Example B.2 $\int_1^\infty \dfrac{\cos x}{x^2}\, dx$ converges by Example B.1 and Theorem B.3.

Example B.3 For any positive constant a the integral

(3)
$$\int_0^\infty \frac{\sin ax}{x}\, dx$$

converges. It is enough to show it for $a = 1$ since the substitution $u = ax$ shows that the integral is independent of a. Note next that the integrand is piecewise continuous on any bounded subinterval and, therefore, locally integrable. Integrating first over $[1, \infty)$, we obtain

$$\int_1^\infty \frac{\sin x}{x}\, dx = \lim_{b \to \infty} \int_1^b \frac{\sin x}{x}\, dx$$

$$= \lim_{b \to \infty} \left[-\frac{\cos b}{b} + \cos 1 - \int_1^b \frac{\cos x}{x^2}\, dx \right]$$

$$= \cos 1 - \int_1^\infty \frac{\cos x}{x^2}\, dx,$$

where the right-hand side converges by Example B.2. Then (3) converges by Theorem B.2(iv). This is evaluated in Exercise B.3.

Next we consider improper integrals depending on a parameter y,

$$\int_a^\infty f(x, y)\, dx.$$

We assume that f is a locally integrable function of x for each y and ask whether or not this integral—if it converges—is a continuous, integrable, or differentiable function of y on a certain interval. From now on we shall restrict ourselves to integrals over $[a, \infty)$, but similar results are valid for the other two types. As in the case of series of functions, uniform convergence will play a role in the answer. We start with a precise definition.

Definition B.3 *Let I be an interval in $\mathbb{R}$ and assume that*

$$\tag{4} \int_a^\infty f(x, y)\, dx$$

converges for each y in I. Then (4) is said to **converge uniformly** *on I if and only if for any $\epsilon > 0$ there is a $\delta \geq a$ such that*

$$\left| \int_b^\infty f(x, y)\, dx \right| < \epsilon$$

for all $b \geq \delta$ and all y in I.

The proofs of some of the theorems in this appendix will be based on the following classical, but seldom stated, necessary and sufficient condition for uniform convergence.

Theorem B.4 *Let I be an interval in $\mathbb{R}$ and assume that (4) converges for each y in I. Then it converges uniformly on I if and only if for every sequence $\{x_n\}$ in $[a, \infty)$ such that*
   (i) $a = x_0 < x_1 < \cdots < x_n < \cdots$ *and*
   (ii) $x_n \to \infty$ *as $n \to \infty$*
the series

$$\tag{5} \sum_{n=0}^\infty \int_{x_n}^{x_{n+1}} f(x, y)\, dx$$

converges uniformly to (4) on I.

Proof. First note that, if $\{x_n\}$ is a sequence as stated above, then

$$\tag{6} \left| \int_a^\infty f(x, y)\, dx - \sum_{k=0}^{n-1} \int_{x_k}^{x_{k+1}} f(x, y)\, dx \right| = \left| \int_{x_n}^\infty f(x, y)\, dx \right|.$$

(*if*) If (5) converges uniformly to (4) on I, but (4) did not, then there would be an $\epsilon > 0$ and a sequence $x_1 < \cdots < x_n < \cdots$ with $x_n \to \infty$ as $n \to \infty$ such that the right-hand side of (6) is larger than or equal to ϵ for each x_n and some y in I. But this is impossible because the left-hand side of (6) must be less than ϵ for the given sequence if n is large enough. Therefore, (4) converges uniformly on I.

(*only if*) If $\{x_n\}$ is as stated and if the right-hand side of (6) is smaller than a given $\epsilon > 0$ for all x_n large enough and all y in I, then so is the left-hand side, proving that (5) converges uniformly to (4) on I if (4) converges uniformly on I, Q.E.D.

The following sufficient condition for the uniform convergence of (4), discovered by de la Vallée Poussin, is equivalent to the Weierstrass M-test for infinite series.

Theorem B.5 *Let I be an interval in $\mathbb{R}$ and assume that* (4) *converges for each y in I. If, in addition, there is a locally integrable function g such that*

(i) $|f(x, y)| \leq g(x)$ *for all y in I and*

(ii) $\displaystyle\int_a^\infty g$ *converges,*

then (4) *converges uniformly on I.*

Proof. By Theorem B.2(iii), condition (i) implies that

$$\int_a^\infty |f(x, y)|\, dx$$

converges for each y in I and does not exceed the integral of g on (a, ∞), which converges by (ii). Then, if $b > a$, Theorems B.3 and B.2(iii) imply that

$$\left| \int_b^\infty f(x, y)\, dx \right| \leq \int_b^\infty |f(x, y)|\, dx \leq \int_b^\infty g.$$

Since the right-hand side can be made less than any given ϵ by choosing b large enough, the same is true of the left-hand side, Q.E.D.

Using uniform convergence and Theorem B.5, the Riemann-Lebesgue theorem can be extended to unbounded intervals for integrable functions.

Theorem B.6 (**The Extended Riemann-Lebesgue Theorem**) *If f is integrable on (a, ∞), then*

$$\lim_{r \to \infty} \int_a^\infty f(x) \sin rx\, dx = \lim_{r \to \infty} \int_a^\infty f(x) \cos rx\, dx = 0.$$

Proof. For any $b > a$ we have

$$\int_a^\infty f(x) \sin rx\, dx = \int_a^b f(x) \sin rx\, dx + \int_b^\infty f(x) \sin rx\, dx.$$

The last integral is uniformly convergent with respect to r because $|f(x)\sin rx| \leq |f(x)|$ and f is integrable, and then its absolute value can be made arbitrarily small by choosing b large enough. For this b, and using the Riemann-Lebesgue theorem, the first integral on the right-hand side approaches zero as $r \to \infty$. This shows that the left-hand side above approaches zero as $r \to \infty$, Q.E.D.

For the applications that we have in mind it will suffice, from now on, to consider integrals of the form

$$\int_a^\infty f(x)g(x, y)\,dx,$$

where f is locally integrable, but g will be further restricted. While functions found in applications frequently have discontinuities, they seldom have wild kinds of discontinuities. For a function of two variables it is usually sufficient to allow jump discontinuities on a finite number of nice arcs of curve in each bounded rectangle. From now on we shall denote by $\bar{J}$ the closure of an interval J, that is, the union of J and its endpoints if any. In what follows, a function that has a range in $\mathbb{C}$ is allowed, in particular, to have only real values.

Definition B.4 *Let I and J be intervals. We say that a function $g: I \times J \to \mathbb{C}$ is* **patchwise continuous in** *y if and only if*

(i) the set $I \times J$ can be divided into disjoint subregions by a finite or countable collection of arcs of the form $h: J_s \to I$, where J_s is a closed subinterval of $\bar{J}$ and h is monotonic and continuous,

(ii) g is continuous in each of these subregions and can be extended to its boundary as a continuous function, and

(iii) each bounded rectangle in $I \times J$ intersects no more than a finite number of the arcs on which g is discontinuous.

Note that, if g is patchwise continuous in y on $I \times J$ and if $\bar{J}$ is the closure of J, then the extension of g to $I \times \bar{J}$ by subregions as described above is also patchwise continuous in y. Also, a function that is patchwise continuous in y is continuous on every horizontal segment except, possibly, at a finite number of points. However, no such restriction on the discontinuities of g is implied by the following weaker definition.

Definition B.5 *Let I and J be intervals. We say that a function $g: I \times J \to \mathbb{C}$ is* **piecewise continuous in** *y if and only if every bounded subinterval of J can be decomposed into the disjoint union of a finite number of subintervals J_s such that g is or can be extended to a patchwise continuous function of y on $I \times \bar{J}_s$.*

Example B.4 The function $g: [0, 1] \times [0, \infty) \to \mathbb{R}$ defined by

$$g(x, y) = \begin{cases} 0 & \text{if } x < 1 - y \text{ and } x < \frac{1}{2} \\ 1 & \text{if } x < y \text{ and } x > 1 - y \\ 2 & \text{if } x > y \text{ and } x > \frac{1}{2} \end{cases}$$

is patchwise continuous in y. Indeed, $[0, 1] \times [0, \infty)$ can be divided into three subregions by means of the segment $x = 1/2$ for $0 \leq y \leq 1/2$ and the arcs $x = 1 - y$ and $x = y$ for $1/2 \leq y \leq 1$. It is clear that g is continuous inside each of the subregions determined by these arcs and that it can be extended to its boundary as a continuous function.

Example B.5 The function $g : [0, \infty) \times [0, 1] \rightarrow \mathbb{R}$ defined by

$$
g(x, y) = \begin{cases} 0 & \text{if} \quad y < 1 - x \quad \text{and} \quad y < \frac{1}{2} \\ 1 & \text{if} \quad y < x \quad \text{and} \quad y > 1 - x \\ 2 & \text{if} \quad y > x \quad \text{and} \quad y > \frac{1}{2} \end{cases}
$$

is not patchwise continuous in y because the horizontal segment on the line $y = 1/2$ is not an arc of the form prescribed in Definition B.4, but it is piecewise continuous in y because $[0, \infty) \times [0, 1]$ can be divided into two horizontal strips by the line $y = 1/2$ and g is patchwise continuous on each strip.

We start by investigating the continuity properties of integrals with respect to a parameter, starting with the case of a bounded x-interval. We define a function of one variable to be *piecewise continuous* on an interval J if and only if it is piecewise continuous on every bounded subinterval of J.

Lemma B.1 *If $f : [a, b] \rightarrow \mathbb{C}$ is integrable, J is an interval, and $g : [a, b] \times J \rightarrow \mathbb{C}$ is piecewise continuous in y, then*

$$
(7) \qquad \int_a^b f(x) g(x, y) \, dx
$$

is a piecewise continuous function of y on J. If, in addition, g is patchwise continuous in y, then (7) is a continuous function of y on J.

Proof. Assume first that g is patchwise continuous in y. If y_0 is fixed in J and if $g(\cdot, y_0)$ denotes the function of x that results after fixing $y = y_0$ in $g(x, y)$, then $g(\cdot, y)$ can be discontinuous only at the points of intersection of the line $y = y_0$ and the arcs h of Definition B.4. That is, there is a partition $a = x_0 < \cdots < x_m = b$ of $[a, b]$ such that $g(\cdot, y_0)$ can be discontinuous only at x_i, $i = 0, \ldots, m$. From the conditions imposed on the arcs h it follows that, for any $\epsilon > 0$, there is a $\delta > 0$ so small that (i) the set

$$
S_{\epsilon\delta} = \{(x, y) \in [a, b] \times J : |y - y_0| \leq \delta, \ |x - x_i| \geq \epsilon \ \text{for} \ i = 0, \ldots, m\}
$$

is closed, (ii) g is continuous, and therefore uniformly continuous, on $S_{\epsilon\delta}$, and (iii)

$$
|g(x, y) - g(x, y_0)| < \epsilon
$$

for all (x, y) in $S_{\epsilon\delta}$. If I_ϵ denotes the set $\{x \in [a, b] : |x - x_i| \geq \epsilon \ \text{for} \ i = 0, \ldots, m\}$, if $l(I_\epsilon)$ denotes its length, if $|f| < M$, if $|y - y_0| \leq \delta$, and if we note that the length of $[a, b] - I_\epsilon$ is $2\epsilon(m - 1)$, then

$$\left| \int_a^b f(x)g(x, y) \, dx - \int_a^b f(x)g(x, y_0) \, dx \right|$$

$$\leq \int_{I_\epsilon} |f(x)||g(x, y) - g(x, y_0)| \, dx + \int_{[a,b]-I_\epsilon} |f(x)||g(x, y) - g(x, y_0)| \, dx$$

$$< \epsilon \int_{I_\epsilon} |f| + 2M \int_{[a,b]-I_\epsilon} |f|$$

$$< \epsilon \int_a^b |f| + 4\epsilon M^2(m - 1).$$

Since ϵ is arbitrary, this shows that the left-hand side approaches zero as $y \to y_0$, proving that (7) is continuous at $y = y_0$ and then on J since y_0 is arbitrary.

If g is only piecewise continuous in y, let (c, d) be a subinterval of J. Then there is a partition $c = y_0 < \cdots < y_n = d$ of $[c, d]$ such that, if J_j denotes the interval $[y_{j-1}, y_j]$ for $j = 1, \ldots, n$, g is patchwise continuous in y on each rectangle $[a, b] \times J_j$ and (7) is continuous on J_j. Since both j and (c, d) are arbitrary, (7) is piecewise continuous on J, Q.E.D.

Theorem B.7 *Let $f : [a, \infty) \to \mathbb{C}$ be locally integrable, let J be an interval, and let $g : [a, \infty) \times J \to \mathbb{C}$ be piecewise continuous in y. If*

(8)
$$\int_a^\infty f(x)g(x, y) \, dx$$

converges uniformly on J, then it is a piecewise continuous function of y on J. If, in addition, g is patchwise continuous in y, then (8) is a continuous function of y on J.

Proof. Let $\{x_n\}$ be a sequence in $[a, \infty)$ such that $a = x_0 < x_1 < \cdots < x_n < \cdots$ and $x_n \to \infty$ as $n \to \infty$, and assume first that g is patchwise continuous in y. By Lemma B.1,

$$\int_{x_n}^{x_{n+1}} f(x)g(x, y) \, dx$$

is continuous on J for each n. By Theorem B.4,

$$\int_a^\infty f(x)g(x, y) \, dx = \sum_{n=0}^\infty \int_{x_n}^{x_{n+1}} f(x)g(x, y) \, dx,$$

and the series on the right converges uniformly on J. By Theorem A.4, the left-hand side is continuous on J. The remaining assertion is proved as for Lemma B.1, Q.E.D.

Next we shall deal with the integrability of (7) and (8) with respect to the parameter y and with changing the order of integration in iterated integrals. Definitions B.4 and B.5 are not, and need not be, symmetric in x and y because of the asymmetric nature of the integrands in (7) and (8). However, we shall need some symmetry now since the change of order of integration should be a reversible process. First, we define patchwise continuous in x and piecewise continuous in x just as above but reversing to roles of x and y, and then the next move is predictable.

Definition B.6 *Let I and J be intervals. We say that a function $g : I \times J \to \mathbb{C}$ is* **patchwise (piecewise) continuous** *if and only if it is patchwise (piecewise) continuous in both x and y.*

Lemma B.2 *If $f : [a, b] \to \mathbb{C}$ is integrable and $g : [a, b] \times [c, d] \to \mathbb{C}$ is piecewise continuous, then*

$$\int_a^b f(x)g(x, y)\, dx \qquad and \qquad \int_c^d f(x)g(x, y)\, dy$$

are integrable functions on $[c, d]$ and $[a, b]$, respectively, and

$$\int_c^d \left(\int_a^b f(x)g(x, y)\, dx \right) dy = \int_a^b \left(\int_c^d f(x)g(x, y)\, dy \right) dx.$$

Proof. We shall assume that the functions are real valued or else apply the lemma to the real and imaginary parts of the integrand. Note that, by Lemma B.1, the first integral in this lemma is piecewise continuous, and therefore integrable, on $[c, d]$. Also

$$\int_c^d f(x)g(x, y)\, dy = f(x) \int_c^d g(x, y)\, dy$$

is integrable on $[a, b]$ as the product of two functions, the first of which is integrable and the second, again by Lemma B.1 but reversing the roles of x and y, is piecewise continuous and, therefore, also integrable.

It remains to be shown that the order of integration can be reversed. If f is piecewise continuous on $[a, b]$, then the result is well known and we shall accept it.[1] Otherwise, and according to Definition 2.2, for any $\epsilon > 0$ there is a piecewise constant function, $f_{pc} \leq f$, such that

$$\int_a^b (f - f_{pc}) < \epsilon,$$

and then

$$\left| \int_c^d \left(\int_a^b f(x)g(x, y)\, dx \right) dy - \int_a^b \left(\int_c^d f(x)g(x, y)\, dy \right) dx \right|$$

$$= \left| \int_c^d \left(\int_a^b [f(x) - f_{pc}(x)]g(x, y)\, dx \right) dy + \int_c^d \left(\int_a^b f_{pc}(x)g(x, y)\, dx \right) dy \right.$$

$$\left. - \int_a^b \left(\int_c^d f_{pc}(x)g(x, y)\, dy \right) dx - \int_a^b \left(\int_c^d [f(x) - f_{pc}(x)]g(x, y)\, dy \right) dx \right|.$$

---

[1] See, for instance, M. Spivak, *Calculus on Manifolds*, Benjamin, New York, 1965, page 58.

The two middle terms cancel each other out because f_{pc} is piecewise continuous and the order of integration can be reversed in that case. Then, if $g \le M$,

$$\left| \int_c^d \left(\int_a^b f(x)g(x,y)\, dx \right) dy - \int_a^b \left(\int_c^d f(x)g(x,y)\, dy \right) dx \right|$$

$$\le M \int_c^d \left(\int_a^b [f(x) - f_{pc}(x)]\, dx \right) dy + M \int_a^b \left(\int_c^d [f(x) - f_{pc}(x)]\, dy \right) dx$$

$$= 2M(d-c)\left(\int_a^b f - \int_a^b f_{pc} \right)$$

$$< 2M(d-c)\epsilon.$$

Since ϵ is arbitrary, the left-hand side must be zero, Q.E.D.

Theorem B.8 *If* $f : [a, \infty) \to \mathbb{C}$ *is locally integrable,* $g : [a, \infty) \times [c, d] \to \mathbb{C}$ *is piecewise continuous, and* (8) *converges uniformly on* $[c, d]$, *then it is integrable on* $[c, d]$ *and*

$$\int_c^d \left(\int_a^\infty f(x)g(x,y)\, dx \right) dy = \int_a^\infty \left(\int_c^d f(x)g(x,y)\, dy \right) dx.$$

Proof. By Theorem B.7, (8) is piecewise continuous, and therefore integrable, on $[c, d]$. If $\{x_n\}$ is a sequence as in Theorem B.4, then

$$\int_a^\infty f(x)g(x,y)\, dx = \sum_{n=0}^\infty \int_{x_n}^{x_{n+1}} f(x)g(x,y)\, dx,$$

and the series on the right converges uniformly on $[c, d]$. Since each of its terms is piecewise continuous on $[c, d]$ by Lemma B.1, its sum is integrable and the series can be integrated term by term from c to d by Theorem A.5. Since we may reverse the order of iterated integrals over bounded intervals by Lemma B.2,

$$\int_c^d \left(\int_a^\infty f(x)g(x,y)\, dx \right) dy = \sum_{n=0}^\infty \int_c^d \left(\int_{x_n}^{x_{n+1}} f(x)g(x,y)\, dx \right) dy$$

$$= \sum_{n=0}^\infty \int_{x_n}^{x_{n+1}} \left(\int_c^d f(x)g(x,y)\, dy \right) dx$$

$$= \int_a^\infty \left(\int_c^d f(x)g(x,y)\, dy \right) dx,$$

Q.E.D.

Theorem B.9 *Let $f : [a, \infty) \to \mathbb{C}$ be integrable, and let $g : [a, \infty) \times [c, \infty) \to \mathbb{C}$ be piecewise continuous. If (8) converges uniformly with respect to y on every bounded subinterval of $[c, \infty)$, if*

$$\int_c^\infty g(x, y) \, dy$$

converges uniformly with respect to x on $[a, \infty)$, and if the right-hand side below converges, then

$$\int_c^\infty \left(\int_a^\infty f(x) g(x, y) \, dx \right) dy = \int_a^\infty \left(\int_c^\infty f(x) g(x, y) \, dy \right) dx.$$

Proof. Given $\epsilon > 0$, let $d > c$ be so large that

$$\left| \int_d^\infty g(x, y) \, dy \right| < \epsilon$$

for all x in $[a, \infty)$. Then, by Theorem B.8,

$$\left| \int_a^\infty \left(\int_c^\infty f(x) g(x, y) \, dy \right) dx - \int_c^d \left(\int_a^\infty f(x) g(x, y) \, dx \right) dy \right|$$

$$= \left| \int_a^\infty \left(\int_c^\infty f(x) g(x, y) \, dy \right) dx - \int_a^\infty \left(\int_c^d f(x) g(x, y) \, dy \right) dx \right|$$

$$= \left| \int_a^\infty f(x) \left(\int_d^\infty g(x, y) \, dy \right) dx \right|$$

$$\le \int_a^\infty |f(x)| \left| \int_d^\infty g(x, y) \, dy \right| dx$$

$$< \epsilon \int_a^\infty |f|$$

because f is integrable. Since ϵ is arbitrary, this shows that

$$\lim_{d \to \infty} \int_c^d \left(\int_a^\infty f(x) g(x, y) \, dx \right) dy = \int_a^\infty \left(\int_c^\infty f(x) g(x, y) \, dy \right) dx,$$

Q.E.D.

Finally, we shall consider the differentiability of (7) and (8) with respect to the parameter y. We start with a preliminary result, a modified version of a well-known theorem.

Lemma B.3 (The Leibniz Rule) *Let h_1 and h_2 be two continuously differentiable curves defined on $[c, d]$ such that $h_1 \le h_2$, and let F be a continuous function defined on the region*

$$R = \{(x, y) \in \mathbb{R}^2 : h_1(y) \le x \le h_2(y), \ c \le y \le d\}$$

such that its partial derivative with respect to y, F_y, exists and can be extended to a rectangle containing R as a function that is patchwise continuous in y. Then

$$\frac{\partial}{\partial y} \int_{h_1(y)}^{h_2(y)} F(x, y)\, dx = \int_{h_1(y)}^{h_2(y)} F_y(x, y)\, dx + F(h_2(y), y)h_2'(y) - F(h_1(y), y)h_1'(y)$$

for all y in $[c, d]$.

Proof. Let $[a, b] \times [c, d]$ be a rectangle containing R and let $\widetilde{F}$ be the continuous extension of F to $[a, b] \times [c, d]$. If we define

$$I(x, y) = \int_a^x \widetilde{F}(s, y)\, ds,$$

then $I_x(x, y) = \widetilde{F}(x, y)$. Also, if $c \le y \le d$, reversing the order of integration in

$$\int_c^y \left(\int_a^x \widetilde{F}_y(s, t)\, ds \right) dt,$$

evaluating the resulting inner integral via the fundamental theorem of the integral calculus, and then differentiating both sides with respect to y, we obtain

$$I_y(x, y) = \int_a^x \widetilde{F}_y(s, y)\, ds$$

(see Exercise B.8). Finally, using the chain rule to compute

$$\frac{\partial}{\partial y} \int_{h_1(y)}^{h_2(y)} F(x, y)\, dx = \frac{\partial}{\partial y} [I(h_2(y), y) - I(h_1(y), y)]$$

and the partial derivatives of I computed above, the result follows (see Exercise B.9), Q.E.D.

To obtain the desired differentiability results we need to further restrict f.

Definition B.7 *Let I and J be intervals. We say that a function $g : I \times J \to \mathbb{C}$ is* **patchwise continuously differentiable in** y *if and only if*
    (i) *it is patchwise continuous in y,*
    (ii) *each arc h on which it has discontinuities is or can be extended to a continuously differentiable function on $\bar{J}$,*
    (iii) *these arcs do not intersect in the interior of $I \times J$, and*
    (iv) *the partial derivative g_y is patchwise continuous in y.*

Note that, if I is a bounded interval, then the arcs h on which g is discontinuous are necessarily finite in number. This is a consequence of Definition B.4(iii) because the domain of each h contains J and then its graph must intersect $I \times J_s$ for each bounded subinterval J_s of J.

Definition B.8 *Let I and J be intervals. We say that a function $g : I \times J \to \mathbb{C}$ is **piecewise continuously differentiable in** y if and only if every bounded subinterval of J can be decomposed into the disjoint union of a finite number of subintervals J_s such that g or its continuous extension is patchwise continuously differentiable on $I \times \bar{J}_s$.*

Lemma B.4 *Let $f : [a, b] \to \mathbb{C}$ be piecewise continuous, let J be an interval, and let $g : [a, b] \times J \to \mathbb{C}$ be piecewise continuously differentiable in y. Then*

$$(7) \qquad \int_a^b f(x)g(x, y)\, dx$$

has a piecewise continuous derivative with respect to y on J. If, in addition, g is patchwise continuously differentiable in y, then (7) is a continuously differentiable function of y on J, and if $h_1, \ldots, h_N : J \to [a, b]$ are the arcs on which g has discontinuities and if we define $F(x, y) = f(x)g(x, y)$, then

$$\frac{\partial}{\partial y} \int_a^b f(x)g(x, y)\, dx = \int_a^b f(x)g_y(x, y)\, dx + F(h_1(y)-, y)h_1'(y)$$

$$+ \sum_{i=1}^{N-1} [F(h_{i+1}(y)-, y)h_{i+1}'(y) - F(h_i(y)+, y)h_i'(y)] - F(h_N(y)+, y)h_N'(y).$$

Proof. Assume first that g is patchwise continuously differentiable in y, define $h_0 \equiv a$ and $h_{N+1} \equiv b$, and let (c, d) be a bounded subinterval of J. Then F may not be patchwise continuously differentiable in y on $[a, b] \times [c, d]$ because, if f has discontinuities, then F is discontinuous on some vertical segments that may intersect the h_i. However, there is a partition $c = y_0 < \cdots < y_n = d$ of $[c, d]$ such that, if J_j denotes the interval $[y_{j-1}, y_j]$ for $j = 1, \ldots, n$, F is patchwise continuously differentiable in y on each rectangle $[a, b] \times J_j$, with discontinuities on a finite number of arcs $\tilde{h}_1, \ldots, \tilde{h}_M$, $M \geq N$, that include all of the h_i and any additional vertical segments on which F is discontinuous. Then, if we define $\tilde{h}_0 \equiv h_0 \equiv a$ and $\tilde{h}_{M+1} \equiv h_{N+1} \equiv b$, we can use the additivity of the integral and Lemma B.3 to conclude that (7) is differentiable with respect to y on J_j and that

$$\frac{\partial}{\partial y} \int_a^b f(x)g(x, y)\, dx = \frac{\partial}{\partial y} \sum_{i=0}^M \int_{\tilde{h}_i(y)}^{\tilde{h}_{i+1}(y)} f(x)g(x, y)\, dx$$

$$= \sum_{i=0}^M \left[\int_{\tilde{h}_i(y)}^{\tilde{h}_{i+1}(y)} f(x)g_y(x, y)\, dx + F(\tilde{h}_{i+1}(y)-, y)\tilde{h}_{i+1}'(y) - F(\tilde{h}_i(y)+, y)\tilde{h}_i'(y) \right]$$

$$= \int_a^b f(x)g_y(x, y)\, dx + \sum_{i=0}^M [F(\tilde{h}_{i+1}(y)-, y)\tilde{h}_{i+1}'(y) - F(\tilde{h}_i(y)+, y)\tilde{h}_i'(y)]$$

$$= \int_a^b f(x)g_y(x, y)\, dx + \sum_{i=0}^N [F(h_{i+1}(y)-, y)h_{i+1}'(y) - F(h_i(y)+, y)h_i'(y)].$$

The possible discontinuities of f do not contribute any terms to the last sum because $h_i' \equiv 0$ for $i = 0, \ldots, M$. Therefore, the stated equation is independent of j and valid on $[c, dm]$. The right-hand side is continuous on $[c, d]$ by Lemma B.1 and the definition of the h_i. Since (c, d) is arbitrary and since $h_0' \equiv h_N' \equiv 0$, this proves the second assertion.

If g is only piecewise continuously differentiable in y, then $[a, b] \times [c, d]$ is the union of a finite number of horizontal strips on which g is patchwise continuously differentiable in y and, as above, (7) is a continuously differentiable function of y on each closed strip. Then it has a piecewise continuous derivative on J, Q.E.D.

It should be remarked that without the conditions imposed in Definition B.7 (ii) the conclusion of Lemma B.4 may not hold.

Example B.6 The function $g : [0, 1] \times [0, \infty) \to \mathbb{R}$ defined by

$$g(x, y) = \begin{cases} 1 & \text{if} \quad x \le \sqrt{y} \\ 0 & \text{if} \quad x > \sqrt{y} \end{cases}$$

is not piecewise continuously differentiable in y because the arc $x = \sqrt{y}$ is not differentiable at $y = 0$. With $f \equiv 1$ and this g, it is easy to see by direct computation that the integral in Lemma B.4 is not differentiable at $y = 0$.

The difficulty in extending Lemma B.4 to the case in which $[a, b]$ is replaced by an unbounded interval is that the finite sum in its statement may become an infinite series, whose convergence is then questionable. Of course, no such question arises if such a sum is finite, and this allows us to state the following result that is sufficient for our purposes.

Theorem B.10 *Let $f : [a, \infty) \to \mathbb{C}$ be integrable and piecewise continuous, let J be an interval, and let $g : [a, \infty) \times J \to \mathbb{C}$ be piecewise continuously differentiable in y with discontinuities restricted to a finite number of arcs in each horizontal strip on which it is patchwise continuously differentiable in y. If*

$$(8) \qquad \int_a^\infty f(x)g(x, y)\, dx$$

converges for all y in J and if

$$(9) \qquad \int_a^\infty f(x)g_y(x, y)\, dx$$

converges uniformly on J, then (8) has a piecewise continuous derivative with respect to y on J. If, in addition, g is patchwise continuously differentiable in y then (8) is a continuously differentiable function of y on J, and if $h_1, \ldots, h_N : J \to [a, \infty)$ are *the arcs on which g has discontinuities and if we define $F(x, y) = f(x)g(x, y)$, then*

$$\frac{\partial}{\partial y} \int_a^\infty f(x)g(x, y)\, dx = \int_a^\infty f(x)g_y(x, y)\, dx + F(h_1(y)-, y)h_1'(y)$$

$$+ \sum_{i=1}^{N-1} [F(h_{i+1}(y)-, y)h_{i+1}'(y) - F(h_i(y)+, y)h_i'(y)] - F(h_N(y)+, y)h_N'(y).$$

Proof. Assume first that g is patchwise continuously differentiable in y and let (c, d) be a bounded subinterval of J. Now let $\{x_n\}$ be a sequence in $[a, \infty)$ such that $a = x_0 < x_1 < \cdots < x_n < \cdots$ and $x_n \to \infty$ as $n \to \infty$, and take x_1 so large that g is continuous on $[x_1, \infty) \times [c, d]$. Such x_1 exists because each h_i has a maximum and a minimum value on $[c, d]$, an interval on which it must be defined by Definition B.7 (ii). By Lemma B.4,

$$(10) \quad \frac{\partial}{\partial y} \int_a^{x_1} f(x)g(x, y)\, dx = \int_a^{x_1} f(x)g_y(x, y) + F(h_1(y)-, y)h_1'(y)$$

$$+ \sum_{i=1}^{N-1} [F(h_{i+1}(y)-, y)h_{i+1}'(y) - F(h_i(y)+, y)h_i'(y)] - F(h_N(y)+, y)h_N'(y),$$

where $F(x, y) = f(x)g(x, y)$, while

$$(11) \qquad \frac{\partial}{\partial y} \int_{x_n}^{x_{n+1}} f(x)g(x, y)\, dx = \int_{x_n}^{x_{n+1}} f(x)g_y(x, y)\, dx$$

for $n \geq 1$, and the integrals on the right-hand sides are continuous on $[c, d]$ by Lemma B.1. Since

$$\int_a^\infty f(x)g(x, y)\, dx = \sum_{n=0}^\infty \int_{x_n}^{x_{n+1}} f(x)g(x, y)\, dx$$

and

$$\int_a^\infty f(x)g_y(x, y)\, dx = \sum_{n=0}^\infty \int_{x_n}^{x_{n+1}} f(x)g_y(x, y)\, dx$$

and since the last series on the right converges uniformly on $[c, d]$ by Theorem B.4, it follows from Theorem A.6 that (8) is differentiable and, using (10) and (11), that

$$\frac{\partial}{\partial y} \int_a^\infty f(x)g(x, y)\, dx = \sum_{n=0}^\infty \frac{\partial}{\partial y} \int_{x_n}^{x_{n+1}} f(x)g(x, y)\, dx$$

$$= \int_a^{x_1} f(x)g_y(x, y)\, dx + F(h_1(y)-, y)h_1'(y)$$

$$+ \sum_{i=1}^{N-1} [F(h_{i+1}(y)-, y)h_{i+1}'(y) - F(h_i(y)+, y)h_i'(y)] - F(h_N(y)+, y)h_N'(y)$$

$$+ \sum_{n=1}^{\infty} \int_{x_n}^{x_{n+1}} f(x)g_y(x, y)\, dx$$

$$= \int_a^{\infty} f(x)g_y(x, y)\, dx + F(h_1(y)-, y)h_1'(y)$$

$$+ \sum_{i=1}^{N-1} [F(h_{i+1}(y)-, y)h_{i+1}'(y) - F(h_i(y)+, y)h_i'(y)] - F(h_N(y)+, y)h_N'(y)$$

on $[c, d]$, where the right-hand side is a continuous function of y on $[c, d]$ by Theorem B.7 and the definition of the h_i. Since (c, d) is arbitrary this proves the second assertion, and the first is proved as in Lemma B.4, Q.E.D.

EXERCISES

B.1   Show that, if $f = u + iv$, where u and v are real valued and integrable on $[a, b]$, and if there is a complex valued function F that is continuous on $[a, b]$ and such that $F' = f$ on (a, b), then

$$\int_a^b f = F(b) - F(a).$$

B.2   Give an example of a function that is integrable on $(-\infty, \infty)$ but whose square is not.

B.3   Show that for any positive constant a

$$\int_0^{\infty} \frac{\sin ax}{x}\, dx = \frac{\pi}{2}$$

(**Hint:** make the substitution $x = \left(N + \frac{1}{2}\right)t$ and rewrite the new integrand in terms of the Dirichlet kernel).

B.4   Give an example of a function $g : [0, 1] \times [0, \infty) \to \mathbb{R}$ that is piecewise continuous in y but is not piecewise continuous in x.

B.5   Prove or disprove that

$$\int_0^{\infty} \left(\int_0^1 \cos xy\, dx \right) dy = \int_0^1 \left(\int_0^{\infty} \cos xy\, dy \right) dx.$$

B.6 Show that

$$\int_1^\infty \left(\int_1^\infty \frac{x^2 - y^2}{(x^2 + y^2)^2} \, dx \right) dy \neq \int_1^\infty \left(\int_1^\infty \frac{x^2 - y^2}{(x^2 + y^2)^2} \, dy \right) dx.$$

Explain why this does not contradict Theorem B.8.

B.7 Prove or disprove that

$$\int_1^\infty \left(\int_1^\infty \frac{x - y}{(x + y)^3} \, dx \right) dy = \int_1^\infty \left(\int_1^\infty \frac{x - y}{(x + y)^3} \, dy \right) dx.$$

B.8 Complete the proof of the equation

$$I_y(x, y) = \int_a^x \widetilde{F}_y(s, y) \, ds$$

in the proof of Lemma B.3.

B.9 Complete the proof of Lemma B.3.

B.10 Show by direct computation that, if g is either the function of Example B.4 or that of Example B.6, then

$$\int_0^1 g(x, y) \, dx$$

is not a differentiable function of y on $[0, \infty)$. Find its derivative with respect to y in each horizontal strip on which g is patchwise continuous.

B.11 Give an example of a continuous function F on $[0, 1] \times [0, \infty)$ such that F_y is not patchwise continuous in y and for which the equation in Lemma B.3 does not hold (**Hint:** refer to the function g of Example B.6 but use two parabolas through the origin instead of one).

B.12 If $f : [0, \infty) \to \mathbb{R}$ is integrable and piecewise continuous and if

$$g(x, y) = e^{-|x - y|},$$

apply Theorem B.10 to evaluate the first and second derivatives of

$$\int_0^\infty f(x) g(x, y) \, dx$$

with respect to y.

APPENDIX C

TABLES OF FOURIER AND LAPLACE TRANSFORMS

Table of Fourier Transforms

	$f(x) \qquad (a, \delta > 0)$	$\hat{f}(\omega)$
1	$\begin{cases} 1 & \text{if} \quad \lvert x \rvert < \delta \\ 0 & \text{if} \quad \lvert x \rvert \geq \delta \end{cases}$	$\dfrac{\sin \omega\delta}{\pi\omega}$
2	$\begin{cases} -1 & \text{if} \quad -\delta < x < 0 \\ 1 & \text{if} \quad 0 < x < \delta \\ 0 & \text{if} \quad \lvert x \rvert \geq \delta \end{cases}$	$\dfrac{1 - \cos \omega\delta}{\pi i\omega}$
3	$\begin{cases} 1 - \frac{1}{2}\lvert x \rvert & \text{if} \quad \lvert x \rvert < 2 \\ 0 & \text{if} \quad \lvert x \rvert \geq 2 \end{cases}$	$\dfrac{\sin^2 \omega}{\pi\omega^2}$
4	$\dfrac{2a}{x^2 + a^2}$	$e^{-a\lvert\omega\rvert}$
5	$\begin{cases} 0 & \text{if} \quad x < 0 \\ e^{-ax} & \text{if} \quad x \geq 0 \end{cases}$	$\dfrac{1}{2\pi(a + i\omega)}$
6	$\begin{cases} 0 & \text{if} \quad x < 0 \\ xe^{-ax} & \text{if} \quad x \geq 0 \end{cases}$	$\dfrac{1}{2\pi(a + i\omega)^2}$
7	$e^{-a\lvert x \rvert}$	$\dfrac{a}{\pi(\omega^2 + a^2)}$
8	$xe^{-a\lvert x \rvert}$	$-\dfrac{2ia\omega}{\pi(a^2 + \omega^2)^2}$

Table of Fourier Transforms (Continued)

	$f(x)$ $(a, \delta > 0)$	$\hat{f}(\omega)$
9	e^{-ax^2}	$\dfrac{1}{2\sqrt{a\pi}} e^{-\omega^2/4a}$
10	xe^{-ax^2}	$-\dfrac{i\omega}{4a\sqrt{a\pi}} e^{-\omega^2/4a}$
11	$x^2 e^{-ax^2}$	$-\dfrac{\omega^2}{8a^2\sqrt{a\pi}} e^{-\omega^2/4a}$
12	$\begin{cases} 0 & \text{if } x < 0 \\ x^n e^{-ax} & \text{if } x \geq 0 \end{cases}$	$\dfrac{n!}{2\pi(a+i\omega)^{n+1}}$
13	$\begin{cases} \cos ax & \text{if } |x| < \delta \\ 0 & \text{if } |x| \geq \delta \end{cases}$	$\dfrac{1}{2\pi}\left[\dfrac{\sin(\omega-a)\delta}{(\omega-a)} + \dfrac{\sin(\omega+a)\delta}{(\omega+a)}\right]$
14	$\begin{cases} \cos^2 ax & \text{if } |x| < \pi/2a \\ 0 & \text{if } |x| \geq \pi/2a \end{cases}$	$\dfrac{2a^2}{\pi\omega(4a^2-\omega^2)} \sin\dfrac{\omega\pi}{2a}$
15	$e^{-|x|}\sin x$	$-\dfrac{2\omega i}{\pi(\omega^4+4)}$

Table of Laplace Transforms

	$f(t)$	$F(s)$
1	1	$\dfrac{1}{s}$
2	$\begin{cases} 0 & \text{if } 0 \leq t < a \\ 1 & \text{if } a \leq t < \infty \end{cases}$	$\dfrac{e^{-as}}{s}$
3	$\begin{cases} 1 & \text{if } 0 \leq t < a \\ 0 & \text{if } a \leq t < \infty \end{cases}$	$\dfrac{1-e^{-as}}{s}$
4	$\begin{cases} t & \text{if } 0 \leq t < 1 \\ 1 & \text{if } 1 \leq t < \infty \end{cases}$	$\dfrac{1-e^{-s}}{s^2}$

Table of Laplace Transforms (Continued)

	$f(t)$	$F(s)$
5	t^n	$\dfrac{n!}{s^{n+1}}$
6	$\sqrt{t}$	$\dfrac{1}{2s}\sqrt{\dfrac{\pi}{s}}$
7	$\dfrac{1}{\sqrt{t}}$	$\sqrt{\dfrac{\pi}{s}}$
8	e^{at}	$\dfrac{1}{s-a}$
9	$\begin{cases} e^t & \text{if } 0 \le t < 1 \\ e & \text{if } 1 \le t < \infty \end{cases}$	$\dfrac{e^{1-s}-s}{s(1-s)}$
10	$t^n e^{at}$	$\dfrac{n!}{(s-a)^{n+1}}$
11	$\dfrac{a}{\sqrt{\pi t}}e^{-a^2/t}$	$\dfrac{a}{\sqrt{s}}e^{-2a\sqrt{s}}$
12	$\dfrac{a}{t\sqrt{\pi t}}e^{-a^2/t}$	$e^{-2a\sqrt{s}}$
13	$\sin at$	$\dfrac{a}{s^2+a^2}$
14	$\cos at$	$\dfrac{s}{s^2+a^2}$
15	$t\sin at$	$\dfrac{2as}{(s^2+a^2)^2}$
16	$t\cos at$	$\dfrac{s^2-a^2}{(s^2+a^2)^2}$
17	$\sin at + at\cos at$	$\dfrac{2as^2}{(s^2+a^2)^2}$
18	$\sin at - at\cos at$	$\dfrac{2a^3}{(s^2+a^2)^2}$
19	$\dfrac{1}{t}\sin at$	$\arctan\dfrac{a}{s}$

Table of Laplace Transforms (Continued)

	$f(t)$	$F(s)$
20	$|\sin at|$	$\dfrac{a}{s^2 + a^2} \coth \dfrac{\pi s}{2a}$
21	$\sin at \sin bt$	$\dfrac{2abs}{[s^2 + (a-b)^2][s^2 + (a+b)^2]}$
22	$\dfrac{1}{a} \sin at - \dfrac{1}{b} \sin bt$	$\dfrac{b^2 - a^2}{(s^2 + a^2)(s^2 + b^2)}$
23	$\cos at - \cos bt$	$\dfrac{(b^2 - a^2)s}{(s^2 + a^2)(s^2 + b^2)}$
24	$e^{at} \sin bt$	$\dfrac{b}{(s-a)^2 + b^2}$
25	$e^{at} \cos bt$	$\dfrac{s-a}{(s-a)^2 + b^2}$
26	$te^{at} \sin bt$	$\dfrac{2b(s-a)}{[(s-a)^2 + b^2]^2}$
27	$te^{at} \cos bt$	$\dfrac{(s-a)^2 - b^2}{[(s-a)^2 + b^2]^2}$
28	$\sinh at$	$\dfrac{a}{s^2 - a^2}$
29	$\cosh at$	$\dfrac{s}{s^2 - a^2}$
30	$\sinh at - \sin at$	$\dfrac{2a^3}{s^4 - a^4}$
31	$\cosh at - \cos at$	$\dfrac{2a^2 s}{s^4 - a^4}$
32	$\sin at \sinh at$	$\dfrac{2a^2 s}{s^4 + 4a^4}$

APPENDIX D

HISTORICAL BIBLIOGRAPHY

APOSTOL, TOM M., *Mathematical Analysis*, 2nd ed., Addison-Wesley, Reading, Massachusetts, 1974.

BELHOSTE, BRUNO, *Augustin-Louis Cauchy: A Biography*, Springer-Verlag, New York, 1991.

BELL, ERIC T., *The Development of Mathematics*, 2nd ed., McGraw-Hill, New York, 1945; reprinted by Dover Publications, New York, 1992.

BIRKHOFF, GARRETT, *A Source Book in Classical Analysis*, Harvard University Press, Cambridge, Massachusetts, 1973.

BOTTAZZINI, UMBERTO, *The Higher Calculus: A History of Real and Complex Analysis from Euler to Weierstrass*, Springer-Verlag, New York, 1986.

BOYER, CARL B., *A History of Mathematics*, John Wiley and Sons, 1968; reprinted by Princeton University Press, Princeton, New Jersey, 1985.

CAJORI, FLORIAN, *History of Mathematics*, 3rd ed., The Macmillan Company, New York, 1919; reprinted by Chelsea Publishing Company, New York, 1980.

CAJORI, FLORIAN, *A History of Mathematical Notations*, 2 vols., Open Court, Chicago, 1928 and 1929.

CANNELL, DORIS M., *George Green: Mathematician and Physicist, 1793–1841*, The Athlone Press, London, England, & Atlantic Highlands, New Jersey, 1993.

CARSLAW, HORATIO S., *An Introduction to the Theory of Fourier's Series and Integrals*, 3rd ed., Macmillan and Company, New York, 1931; reprinted by Dover Publications, New York, 1950.

CARSLAW, HORATIO S. AND JOHN C. JAEGER, *Operational Methods in Applied Mathematics*, 2nd ed., Oxford University Press, 1948; reprinted by Dover Publications, New York, 1963.

CLEATOR, P. E. *Lost Languages*, The John Day Company, New York, 1959.

DAUBECHIES, INGRID, "Orthonormal Bases of Compactly Supported Wavelets," *Comm. Pure Appl. Math.*, **41** (1988) 909–996.

DIEUDONNÉ, JEAN A., *History of Functional Analysis*, North-Holland, Amsterdam, The Netherlands, 1981.

EVES, HOWARD W., *An Introduction to the History of Mathematics*, 5th ed., Saunders College Publishing, Philadelphia, 1983.

GRABINER, JUDITH V., *The Origins of Cauchy's Rigorous Calculus*, The MIT Press, Cambridge, Massachusetts, 1981.

GRATTAN-GUINNESS, IVOR, *The Development of the Foundations of Mathematical Analysis from Euler to Riemann*, Cambridge, Massachusetts, 1970.

GRATTAN-GUINNESS, IVOR, *Joseph Fourier 1768–1830*, in collaboration with JEROME R. RAVETZ, The MIT Press, Cambridge, Massachusetts, 1972.

GRATTAN-GUINNESS, IVOR, *Convolutions in French Mathematics, 1800–1840*, 3 vols., Birkhäuser Verlag, Basel, 1990.

GRATTAN-GUINNESS, IVOR, ED., *From the Calculus to Set Theory*, Duckworth, London, England, 1980.

GRAY, JEREMY J., "Rodrigues Transformations Groups," *Arch. Hist. Exact Sci.*, **21** (1979) 375–385.

GRAY, ANDREW, *Lord Kelvin: An Account of His Scientific Life and Work*, J. M. Dent & Company, London, England, and E. P. Dutton & Company, New York, 1908; reprinted by Chelsea Publishing Company, New York, 1973.

HANKINS, THOMAS L., *Jean d'Alembert: Science and the Enlightenment*, The Clarendon Press, Oxford, England, 1970.

HAWKINS, THOMAS, *Lebesgue's Theory of Integration*, 2nd ed., The Regents of the University of Wisconsin, 1975; reprinted by Chelsea Publishing Company, New York, 1979.

HEIMS, STEVE J., *John von Neumann and Norbert Wiener*, The MIT Press, Cambridge, Massachusetts, 1980.

HERIVEL, JOHN, *Joseph Fourier: The Man and the Physicist*, Oxford University Press, Oxford, 1975.

HEWITT, EDWIN AND ROBERT E. HEWITT, "The Gibbs-Wilbraham Phenomenon: An Episode in Fourier Analysis," *Arch. Hist. Exact Sci.*, **21** (1979) 129–160.

HILLE, EINAR, *Analytic Function Theory*, Vol. 2, Ginn and Company, New York, 1962; reprinted by Chelsea Publishing Company, New York, 1977.

HILLE, EINAR, *Lectures on Ordinary Differential Equations*, Addison-Wesley, Reading, Massachusetts, 1969.

HOBSON, ERNEST W., *The Theory of Functions of a Real Variable and the Theory of Fourier's Series*, 2 vols., Cambridge University Press, London, England, 1927; reprinted by Dover Publications, New York, 1957.

HOBSON, ERNEST W., *The Theory of Spherical and Ellipsoidal Harmonics*, Cambridge University Press, London, England, 1931; reprinted by Chelsea Publishing Company, New York, 1955.

JOURDAIN, PHILIP E. B., "The Development of the Theory of Transfinite Numbers," *Arch. Math. Phys.*, **10** (1906) 254–281 and **14** (1908–1909) 287–311.

KLINE, MORRIS, *Mathematical Thought from Ancient to Modern Times*, Oxford University Press, New York, 1972.

KÖRNER, THOMAS W., *Fourier Analysis*, Cambridge University Press, Cambridge, 1989.

LANGER, WILLIAM L., ED., *An Encyclopedia of World History*, 4th ed., Houghton Mifflin, Boston, 1968.

LIEBERSTEIN, H. MELVIN, *Theory of Partial Differential Equations*, Academic Press, New York, 1972.

LORIA, GINO, "Le mathématicien J. Liouville et ses œuvres," *Archeion*, **18** (1936); English translation: "J. Liouville and His Work," *Scripta Mathematica*, **4** (1936) 147–154, 257–262, and 301–306.

LÜTZEN, JESPER, "Heaviside's Operational Calculus," *Arch. Hist. Exact Sci.*, **21** (1979) 161–200.

LÜTZEN, JESPER, *Joseph Liouville 1809–1882: Master of Pure and Applied Mathematics*, Springer-Verlag, New York, 1990.

LÜTZEN, JESPER, *The Prehistory of the Theory of Distributions*, Springer-Verlag, New York, 1982.

MACDONALD, DAVID K. C., *Faraday, Maxwell, and Kelvin*, Anchor Books, Garden City, New York, 1964.

MAY, KENNETH O., *Bibliographical and Research Manual of the History of Mathematics*, University of Toronto Press, Toronto, 1973.

MEYER, HERBERT W., *A History of Electricity and Magnetism*, Burndy Library, Norwalk, Connecticut, 1972.

MEYER, YVES, "Book Reviews," *Bull. Am. Math. Soc.*, **28** (1993) 350–360.

MONASTYRSKY, MICHAEL, *Riemann, Topology and Physics*, Birkhäuser Verlag, Boston, 1987.

MONNA, A. F., *Dirichlet's Principle: A Mathematical Comedy of Errors and Its Influence on the Development of Analysis*, Oosthoek, Scheltema & Holkema, Utrech, The Netherlands, 1975.

MONNA, A. F., *Functional Analysis in Historical Perspective*, John Wiley & Sons, New York, 1973.

NACHBIN, LEOPOLDO, ED., *Mathematical Analysis and Applications. Essays Dedicated to Laurent Schwartz on the Occasion of His 65th Birthday. Part A*, Academic Press, New York, 1981.

Ore, Øystein, *Niels Henrik Abel*, University of Minnesota, 1974; reprinted by Chelsea Publishing Company, New York, 1974

Reid, Constance, *Hilbert-Courant*, Springer-Verlag, New York, 1986.

Rey Pastor, Julio, *Elementos de la Teoría de Funciones*, 3ª ed., Editorial Ibero-Americana, Madrid, 1953.

Riesz, Frigyes and Béla Szökefalvi-Nagy, *Functional Analysis*, Ungar Publishing Company, New York, 1955; reprinted by Dover Publications, New York, 1990.

Rowe, David E. and John McCleary, eds., *The History of Modern Mathematics*, 2 vols., Academic Press, San Diego, 1989.

Rudin, Walter, *Functional Analysis*, McGraw-Hill, New York, 1973.

Sagan, Hans, *Boundary and Eigenvalue Problems of Mathematical Physis*, John Wiley & Sons, New York, 1961; reprinted by Dover Publications, New York, 1989.

Sansone, Giovanni, *Orthogonal Functions*, Interscience, New York, 1959; reprinted by Dover Publications, New York, 1991.

Smith, David E., *A Source Book in Mathematics*, McGraw-Hill, New York, 1929; reprinted by Dover Publications, New York, 1959.

Smith, David E., *History of Mathematics*, Ginn and Company, Boston, 1923; reprinted by Dover Publications, New York, 1958.

Spivak, Michael, *Calculus on Manifolds*, W. A. Benjamin, New York, 1965.

Szegö, Gabor, *Orthogonal Polynomials*, 3rd ed., American Mathematical Society, Providence, Rhode Island, 1967.

Struik, Dirk J., *A Concise History of Mathematics*, 4th ed., Dover Publications, New York, 1987.

Struik, Dirk J., ed., *A Source Book in Mathematics, 1200–1800*, Harvard University Press, Cambridge, Massachusetts, 1969.

Thompson, Silvanus P., *The Life of William Thomson, Baron Kelvin of Largs*, 2 vols., Macmillan and Company, London, England, 1910.

Tietze, Heinrich, *Famous Problems of Mathematics*, Greylock Press, New York, 1965.

Tolstov, Giorgi P., *Fourier Series*, Prentice-Hall, Englewood Cliffs, New Jersey, 1962; reprinted by Dover Publications, New York, 1976.

Tucciarone, John, "The Development of the Theory of Summable Divergent Series from 1880 to 1925," *Arch. Hist. Exact Sci.*, **10** (1973) 1–40.

Whittaker, Edmund T., "Oliver Heaviside," *Bull. Calcutta Math. Soc.*, **20** (1928–1929) 199–220.

Watson, George N., *A Treatise on the Theory of Bessel Functions*, 2nd ed., Cambridge University Press, London, England, 1958.

INDEX

CPSIA information can be obtained at www.ICGtesting.com
Printed in the USA
LVOW090737191211

260090LV00004B/64/A